# Recent Titles in This Series

162 **Ciro Ciliberto, E. Laura Livorni, and Andrew J. Sommese, Editors,** Classification of algebraic varieties, 1994

161 **Paul A. Schweitzer, S. J., Steven Hurder, Nathan Moreira dos Santos, and José Luis Arraut, Editors,** Differential topology, foliations, and group actions, 1994

160 **Niky Kamran and Peter J. Olver, Editors,** Lie algebras, cohomology, and new applications to quantum mechanics, 1994

159 **William J. Heinzer, Craig L. Huneke, and Judith D. Sally, Editors,** Commutative algebra: Syzygies, multiplicities, and birational algebra, 1994

158 **Eric M. Friedlander and Mark E. Mahowald, Editors,** Topology and representation theory, 1994

157 **Alfio Quarteroni, Jacques Periaux, Yuri A. Kuznetsov, and Olof B. Widlund, Editors,** Domain decomposition methods in science and engineering, 1994

156 **Steven R. Givant,** The structure of relation algebras generated by relativizations, 1994

155 **William B. Jacob, Tsit-Yuen Lam, and Robert O. Robson, Editors,** Recent advances in real algebraic geometry and quadratic forms, 1994

154 **Michael Eastwood, Joseph Wolf, and Roger Zierau, Editors,** The Penrose transform and analytic cohomology in representation theory, 1993

153 **Richard S. Elman, Murray M. Schacher, and V. S. Varadarajan, Editors,** Linear algebraic groups and their representations, 1993

152 **Christopher K. McCord, Editor,** Nielsen theory and dynamical systems, 1993

151 **Matatyahu Rubin,** The reconstruction of trees from their automorphism groups, 1993

150 **Carl-Friedrich Bödigheimer and Richard M. Hain, Editors,** Mapping class groups and moduli spaces of Riemann surfaces, 1993

149 **Harry Cohn, Editor,** Doeblin and modern probability, 1993

148 **Jeffrey Fox and Peter Haskell, Editors,** Index theory and operator algebras, 1993

147 **Neil Robertson and Paul Seymour, Editors,** Graph structure theory, 1993

146 **Martin C. Tangora, Editor,** Algebraic topology, 1993

145 **Jeffrey Adams, Rebecca Herb, Stephen Kudla, Jian-Shu Li, Ron Lipsman, and Jonathan Rosenberg, Editors,** Representation theory of groups and algebras, 1993

144 **Bor-Luh Lin and William B. Johnson, Editors,** Banach spaces, 1993

143 **Marvin Knopp and Mark Sheingorn, Editors,** A tribute to Emil Grosswald: Number theory and related analysis, 1993

142 **Chung-Chun Yang and Sheng Gong, Editors,** Several complex variables in China, 1993

141 **A. Y. Cheer and C. P. van Dam, Editors,** Fluid dynamics in biology, 1993

140 **Eric L. Grinberg, Editor,** Geometric analysis, 1992

139 **Vinay Deodhar, Editor,** Kazhdan-Lusztig theory and related topics, 1992

138 **Donald St. P. Richards, Editor,** Hypergeometric functions on domains of positivity, Jack polynomials, and applications, 1992

137 **Alexander Nagel and Edgar Lee Stout, Editors,** The Madison symposium on complex analysis, 1992

136 **Ron Donagi, Editor,** Curves, Jacobians, and Abelian varieties, 1992

135 **Peter Walters, Editor,** Symbolic dynamics and its applications, 1992

134 **Murray Gerstenhaber and Jim Stasheff, Editors,** Deformation theory and quantum groups with applications to mathematical physics, 1992

133 **Alan Adolphson, Steven Sperber, and Marvin Tretkoff, Editors,** $p$-Adic methods in number theory and algebraic geometry, 1992

132 **Mark Gotay, Jerrold Marsden, and Vincent Moncrief, Editors,** Mathematical aspects of classical field theory, 1992

(*Continued in the back of this publication*)

# Classification of Algebraic Varieties

CONTEMPORARY
MATHEMATICS

162

# Classification of Algebraic Varieties

Algebraic Geometry Conference on
Classification of Algebraic Varieties
May 22–30, 1992
University of L'Aquila
L'Aquila, Italy

Ciro Ciliberto
E. Laura Livorni
Andrew J. Sommese
Editors

American Mathematical Society
Providence, Rhode Island

The Algebraic Geometry Conference on Classification of Algebraic Varieties was held May 22–30, 1992, at the University of L'Aquila, L'Aquila, Italy, with support provided by the following institutions: Ambasciata di Spagna; Azienda Autonoma del Soggiorno e Turismo di L'Aquila; Comune di L'Aquila; C.N.R. (Consiglio Nazionale delle Ricerche); Consiglio Regionale di L'Aquila; Dipartimento di Matematica dell'Università degli Studi di L'Aquila; Dipartimento di Matematica dell'Università degli Studi di Pisa; Dipartimento di Matematica dell'Università degli Studi di Roma II; Fondazione Cassa di Risparmio della Provincia di L'Aquila; G.N.S.A.G.A (Gruppo Nazionale per le Strutture Algebriche e Geometriche e loro Applicazioni); Università degli Studi di L'Aquila.

1991 *Mathematics Subject Classification*. Primary 14–06.

---

**Library of Congress Cataloging-in-Publication Data**

Algebraic Geometry Conference on Classification of Algebraic Varieties (1992: University of L'Aquila)

Classification of algebraic varieties: proceedings of the Algebraic Geometry Conference on Classification of Algebraic Varieties, May 22–30, 1992, University of L'Aquila, L'Aquila, Italy/Ciro Ciliberto, E. Laura Livorni, Andrew J. Sommese, editors.

p. cm. — (Contemporary mathematics; v. 162)

ISBN 0-8218-5179-9 (acid-free)

1. Algebraic varieties–Classification–Congresses. I. Ciliberto, D. (Ciro), 1950– . II. Livorni, E. L. (Elvira Laura), 1951– . III. Sommese, Andrew John. IV. Title. V. Series: Contemporary mathematics (American Mathematical Society); .

QA564.A37 1992 — 93–44549

516.3′5—dc20 — CIP

---



The paper used in this book is acid-free and falls within the guidelines
established to ensure permanence and durability.
Printed on recycled paper.
All articles in this volume were printed from copy prepared by the authors.
Some articles were typeset using AMS-TeX or AMS-LaTeX,
the American Mathematical Society's TeX macro systems.

10 9 8 7 6 5 4 3 2 1 — 99 98 97 96 95 94

# Contents

## Introduction

The vitality of algebraic geometry belies its long history and great depth. This is because of the ongoing wide ranging, interaction that algebraic geometry has with the rest of mathematics.

This activity, was clear at the L'Aquila Conference, which was attended by numerous mathematicians from all the word (see enclosed list of participants) and is illustrated by the breadth of the problems treated in the seminars (see also the enclosed list of seminars held during the Conference) and studied in the proceedings. Just to mention a few of the topics in the proceedings, there are articles on:

- the topology of algebraic varieties and on braid groups,
- Riemann surfaces and their jacobians,
- projective classification theory,
- toric varieties,
- enumerative formulae,
- generalizations of Kähler differential geometry,
- the theory of algebraic surfaces,
- the Calabi-Yau threefolds (of much current interest to physicists),
- the spectacular mathematics growing out of the work of Mori.

The conference in L'Aquila was not only a place for fruitful mathematical interaction, but was an occasion where old friendships were renewed and many new friendships were made. The atmosphere was very stimulating and enjoyable.

The conference and this proceedings owe their great success to all who attended for their active discussions and their endurance.

The editors

## List of participants


**A. Alzati**, Dipartimento di Matematica - Università degli Studi di Milano - Via C. Saldini 50 - 20133 MILANO - Italy. E-mail address: ALZATI @ UMIMAT.MAT.UNIMI.IT

**M. Andreatta**, Dipartimento di Matematica - Università degli Studi di Trento - 38050 POVO - Trento - Italy. E-mail address: ANDREATTA @ ITNCISCA.BITNET

**E. Arbarello**, Dipartimento di Matematica - Università degli Studi di Roma I - Piazzale Aldo Moro 5 - 00185 ROMA - Italy.

**E. Arrondo**, Departamento de Algebra Facultad de Matematicas - Universidad Complutense - E 28040 MADRID - Spain. E-mail address: W 258 @ EMDUCM11.BITNET

**L. Bădescu**, Institute of Mathematics of the Romania Academy - P.O. BOX 1-764, Ro-70700 - BUCHAREST- Romania

**E. Ballico**, Dipartimento di Matematica - Università degli Studi di Trento - 38050 POVO - Trento - Italy. Fax number 0039-461881624. E-mail address: BALLICO @ ITNCISCA. BITNET

**F. Bardelli**, Dipartimento di Matematica - Università degli Studi di Pisa - Via Buonarroti 2 - 56127 PISA - Italy. E-mail address: BARDELLI @ DM.UNIPI.IT

**M. A. Barja**, Departamento Matematica Aplicada 1 - Etseib Universitat Politecnia Catalunya - Diagonal 647 - BARCELONA - Spain.

**I. Bauer**, Dipartimento di Matematica - Università degli Studi di Pisa - Via Buonarroti 2 - 56127 PISA - Italy. E-mail address: BAUER @ BIANCHI.DM.UNIPI.IT

**S. Bauer**, Mathematisches Institut der Universität - Universität Göttingen - Bunsenstrasse 3 - 5 - 3400 GÖTTINGEN - Germany. E-mail address: BAUER @ CFGAUSS. UNIMATH.GWDG.DE

**M. C. Beltrametti**, Dipartimento di Matematica - Università degli Studi di Genova - Via L. B. Alberti 4 - 16132 GENOVA - Italy. E-mail address: MBELTRA @ IGECUNIV. CISI.UNIGE.IT

**L. Berardi**, Dipartimento di Energetica - Università degli Studi di L'Aquila - 67100 L'AQUILA - Italy

**M. Bertolini**, Dipartimento di Matematica - Università degli Studi di Milano - Via C. Saldini 50 - 20133 MILANO - Italy. E-mail address: BERTOLINI @ UMIMAT.MAT.UNIMI.IT

**G. M. Besana**, Dipartimento di Matematica - Università degli Studi di L'Aquila - Via Vetoio, Loc. Coppito - 67100 L'AQUILA - Italy. E-mail address: BESANA @ VXSCAQ.CINECA.IT

**A. Biancofiore**, Dipartimento di Matematica - Università degli Studi di L'Aquila - Via Vetoio, Loc. Coppito - 67100 L'AQUILA - Italy. E-mail address: BIANCOFIORE @ VXSCAQ.CINECA.IT

**G. Bolondi**, Dipartimento di Matematica - Università degli Studi di Trento - 38050 POVO - Trento - Italy. E-mail address: BOLONDI @ ITNCISCA.BITNET

**G. Brown**, Mathematics Institute - University of Warwick- COVENTRY CV4 7AL - England E-mail address: GDB @ MATHS.WARWICK.AC.UK

**A. Bruno**, Dipartimento di Matematica - Università degli Studi di Roma I- Piazzale Aldo Moro 5 - 00185 ROMA - Italy

**F. Campana**, Départment de Mathématiques - Université Nancy 1 - BP 239, 54506 VANDOEIVRE CEDEX - France.E-mail address: CAMPANA @ IECH.FR

**G. Casnati**, Dipartimento di Matematica Pura ed Applicata- Università di Padova- Via Belzoni 7 - 35131 PADOVA - Italy. E-mail address: CASNATI @ PDMAT1.UNIPD.IT

**F. Catanese**, Dipartimento di Matematica - Università degli Studi di Pisa - Via Buonarroti 2 - 56127 PISA - Italy. E-mail address: CATANESE @ ICNUVEVM.BITNET

**C. Ciliberto**, Dipartimento di Matematica-Università degli Studi di Roma II- Via della Ricerca Scientifica - 00133 ROMA - Italy. E-mail address: CILIBERTO @ MAT.UTOVRM.IT

**E. Colombo**, Dipartimento di Matematica - Università degli Studi di Milano - Via C. Saldini 50 - 20133 MILANO - Italy. E-mail address: COLOMBO @ UMIMAT.MAT.UNIMI.IT

**M. Cornalba**, Dipartimento di Matematica - Università degli Studi di Pavia - Strada Nuova 65 - 27100 PAVIA - Italy. E-mail address: CORNALBA @ IPV85.UNIPV.IT

**A. Corti**, Scuola Normale Superiore di Pisa - Piazza dei Cavalieri 7 - 56127 PISA - Italy. E-mail address: CORTI @ IPISNSVA.BITNET

**M. de Cataldo**, Department of Mathematics - University of Notre Dame - Notre Dame, Indiana 46556 - U.S.A. E-mail address: MARKAN @ CARTAN.MATH.ND.EDU

**S. Di Rocco**, Department of Mathematics - University of Notre Dame - Notre Dame, Indiana 46556 - U.S.A. E-mail address: SANDRADR @ CARTAN.MATH. ND.EDU

**I. Dolgachev**, Department of Mathematics - University of Michigan - ANN ARBOR, MI 48109 - U.S.A. E-mail address: IGOR_DOLGACHEV @ UM.UMICH.EDU

**L. Ein**, Department of Mathematics - University of Illinois at Chicago - CHICAGO, IL 60680 - U.S.A.

**G. Ellingsrud**, Matematisk Institutt - Universitet i Bergen - Allegaten 55 - 5007 BERGEN - Norway. E-mail address: GEIR @ MI.UIB.NO

**F. Eugeni**, Dipartimento di Energetica- Università degli Studi di L'Aquila - 67100 L'AQUILA - Italy

**M. L. Fania**, Dipartimento di Matematica - Università degli Studi di L'Aquila - Via Vetoio, Loc. Coppito- 67100 L'AQUILA - Italy. E-mail address: FANIA @ VXSCAQ. CINECA.IT

**B. Fantechi**, Dipartimento di Matematica - Università degli Studi di Trento - 38050 POVO - Trento - Italy. E-mail address: FANTECHI @ ITNCISCA.BITNET

**H. Flenner**, Mathematisches Institut der Universität - Universität Göttingen - Bunsenstrasse 3- 5 - 3400 GÖTTINGEN - Germany. E-mail address: HFLENNER @ CFGAUSS. UNI-MATH.GUDG.DE

**P. Francia,** Dipartimento di Matematica - Università degli Studi di Genova - Via L.B. Alberti 4 - GENOVA - Italy. E-mail address: FRANCIA @ IGECUNIV.BITNET

**E. Frigerio,** Dipartimento di Matematica - Università degli Studi di Milano - Via C. Saldini 50 - 20133 MILANO - Italy

**L. Guerra,** Dipartimento di Matematica - Università di Perugia - Via Vanvitelli 1 - 06100 PERUGIA - Italy

**L. Göttsche,** Max-Planck-Institut für Mathematik - Gottfried Claren Strasse 26 - 5300 BONN 3 - Germany. E-mail address: LOTHAR @ MPIM-BONN.MPG.DBP.DE

**K. Hulek,** Institut für Mathematik - Universität Hannover - Welfengarten 1 W 3000 HANNOVER 1 - Germany. E-mail address: KLAUS.HULEK@CDC2.C. MATHEMATIK.UNI-HANNOVER.DBP.DE

**P. Ionescu,** Department of Mathematics - University of Bucharest - St.ACADEMIEI, NR. 14 - 70109 BUCHAREST - Romania

**V. Iskovskih,** Steklov Mathematician Institute- Vivilov Str. 42 - MOSCOW - GSP I.II 7 966 - RUSSIA

**K. Jaczewski,** Institute of Mathematics - Warsaw University - Banache 2 - 02-097 WARSZAWA - Poland. E-mail address: MAS 1 @ PLEARN.BITNET

**T. Kato,** Katholieke Universiteit Leuven-Department of Mathematics-fac. of Sci. - Yamaguchi University - YAMAGUCHI 753 - JAPAN

**Y. Kawamata,** Department of Mathematics - University of Tokyo - Hongo TOKYO 113 - Japan. E-mail address: KAWAMATA @ UTS5.S.U-TOKYO.AC.JP

**S. L. Kleiman,** Department of Mathematics - 2-278 MIT - CAMBRIDGE - Massachusetts 02139 - U.S.A. E-mail address: KLEIMAN @ MATH. MIT. EDU

**J. Kollár,** Department of Mathematics - University of Utah - SALT LAKE CITY - UT 84112 - U.S.A. E-mail address: Kollar @ MATH.UTAH.EDU

**K. Konno**, College of General Education, Kyushu University, Ropponmatsu, Chuo-ku, FUKUOKA 810, Japan. E-mail address: KONNO @ EMMY.DM.UNIP.IT

**H. H. Lange**, Mathematische Institut - University of Erlangen - Bismarckstr. 1 1/2 - 8520 ERLANGEN - Germany. E-mail address: LANGE @ CNVE.RRZE.UNI-ERLANGEN. DBP.DE

**A. Lanteri**, Dipartimento di Matematica - Università degli Studi di Milano - Via C. Saldini 50 - 20133 MILANO - Italy. E-mail address: LANTERI @ UMIMAT.MAT.UNIMI.IT

**E. L. Livorni**, Dipartimento di Matematica - Università degli Studi di L'Aquila - Via Vetoio, Loc. Coppito-67100 L'AQUILA-Italy. E-mail address:LIVORNI @ VXSCAQ. CINECA.IT

**A. Lopez**, Dipartimento di Matematica - Università degli Studi di Pavia - Strada Nuova 65 - 27100 PAVIA - Italy. E-mail address: LOPEZ @ IPVIAN.BITNET

**M. Manetti**, Scuola Normale Superiore di Pisa - Piazza dei Cavalieri - 56126 PISA - Italy. E-mail address: MANETTI @ IPISNSVA.BITNET

**N. Manolache**, Institute of Mathematics of the Romania Academy - P.O. BOX 1-764, Ro-70700 - BUCHAREST- Romania

**G. Marini**, Dipartimento di Matematica - Dottorato di Ricerca - Università degli Studi di Roma I - Piazzale Aldo Moro 5 - 00185 ROMA - Italy. E-mail address: MARINI @ VAXRMA. INFN.IT

**G. Martens**, Mathematisches Institut - University of Erlangen- Nürnberg - Bismarckstr. 1 1/2 - 8520 ERLANGEN - Germany. E-mail address: MARTENS @ CNVE.RRZE. UNI-ERLANGEN.DBP.DE

**G. Megyesi**, Trinity College Cambridge - CB2 ITQ - England. E-mail address: 9M 115 @ PHX.CAM.AC.UK.

**K. Meeyoung**, Department of Mathematics - University of Notre Dame - Notre Dame, Indiana 46556 - U.S.A. E-mail address: MYING @ CARTAN.MATH.ND.EDU

**R. Miranda**, Department of Mathematics - Colorado State University - FORT COLLINS, CO 80523 - U.S.A. E-mail address: MIRANDA @ RIEMANN.MATH.COLOSTATE. EDU

**Y. Miyaoka**, Department of Mathematics - Rikkyo University - Nishi-Ikeburo, Toshimaku - 171 TOKYO - Japan

**B. Moishezon**, Department of Mathematics - Columbia University - NEW YORK, N.Y. 10027 - U.S.A.

**J. Murre**, Department of Mathematics - University of Leiden - P.O. BOX 9512 - 2300 RA LEIDEN - The Netherlands. E-mail address: MURRE @ RULCRI.LEIDENUNIV.NL

**K. O'Grady**, Institute for Advanced Study - Princeton - New Jersey 08540 - U.S.A.

**P. Oliverio**, Dipartimento di Matematica - Università della Calabria - 87036 ARCAVACATA DI RENDE (Cosenza) - Italy

**C. Okonek**, Mathematisches Institut - Universität Zürich - Rämistrasse 74 - CH-8001 Zürich - Switzerland

**G. Ottaviani**, Dipartimento di Matematica Applicata - Università degli Studi di Firenze - Via S. Marta 3 - 50139 Firenze - Italy.

**M. Palleschi**, Dipartimento di Matematica - Università degli Studi di Milano - Via C. Saldini 50 - 20133 MILANO - Italy. E-mail address: APELLES @ UMIMAT.MAT.UNIMI.IT

**C. Pedrini**,Dipartimento di Matematica - Università degli Studi di Genova - Via L.B. Alberti 4 - GENOVA - Italy. E-mail address: PEDRINI @ IGECUNIV.BITNET

**T. Peternell**, Mathematisches Institut - Universität Bayreuth - Universitätsstr. 30, Postf. 10 12 51 - 8580 BAYREUTH - Germany. E-mail address: THOMAS. PETERNELL @ UNI-BAYREUTH.DBP.DE

**R. Piene**, Matematisk Institutt - University of Oslo - P.B. 1053 Blindern - 0136 OSLO - Norway. E-mail address: RAGNIP @ MATH.uìo.no

**G. Pirola**, Dipartimento di Matematica - Università degli Studi di Pavia - Strada Nuova 65 - 27100 PAVIA - Italy. E-mail address: PIRO23 @ IPVIAN.BITNET

**M. Reid**, Mathematics Institute - University of Warwick - COVENTRY CV4 7AL - England. E-mail address: MILES @ MATHS.WARWICK.AC.UK

**I. Reider**, Department of Mathematics - University of Oklahoma-NORMAN, OK 73019 - U.S.A.

**J. Roberts**, Department of Mathematics - University of Minnesota - 206 Church St. S.E. - MINNEAPOLIS - MN 55455 - U.S.A. E-mail address: ROBERTS @ S6. MATH.UMN.EDU

**E. Rogora**, Dipartimento di Matematica - Università degli Studi di Roma I - Piazzale Aldo Moro 5 - 00185 ROMA - Italy. E-mail address: DOTTORAT @ ITCASPUR

**M. Rossi**, Dipartimento di Matematica - Università degli Studi di Firenze - Via Buonarroti 2 - Italy. E-mail address: UDINI @ IFIIDG.BITNET

**F. Russo**, Dipartimento di Matematica - Università degli Studi di Milano - Via C. Saldini 50 20133 MILANO - Italy

**F. Sakai**, Department of Mathematics - Faculty of Science - Saitama University - Shimo-Okubo 255, URAWA 338 - Japan. E-mail address: FSAKAI @ RIMATH. SAITAMA-U.AC.JP

**G. K. Sankaran**, DPMMS 16 - Mill Lane - CAMBRIDGE CB2 1SB - England. E-mail address: GKS1 @ PHX.CAM.AC.UK

**M. Schneider**, Mathematisches Institut - Universität Bayreuth - Universitätsstr. 30, Postf. 10 12 51 - 8580 BAYREUTH - Germany. E-mail address: MICHAEL.SCHNEIDER @ UNI-BAYREUTH.DE

**E. Sernesi**, Dipartimento di Matematica - Università degli Studi di Roma III - Via Corrado Segre 2 - 00146 ROMA - Italy. E-mail address: SERNESI @ ITCASPUR.BITNET

**F. Serrano**, Departament Algebra I Geometria - Universitat DE BARCELONA - Gran Via 585 - 08007 BARCELONA - Spain. E-mail address: UBAAGX06 @ EBCESCA1

**N. Shepherd-Barron**, Department of Mathematics - University of Cambridge - 16 Mill Lane - CAMBRIDGE CB 2 1S B - England N.I.SHEPHERD-BARRON @ PMMS.CAM. AC.UK

**H. Sterk**, Faculteit der Wiskunde en Informatica - Universiteit van Amsterdam - Plantage Muidergracht 24 - 1018 TV AMSTERDAM - The Netherlands. E-mail address: STERK @ FWI.UVA.NL

**P. Supino**, Dipartimento di Matematica - Dottorato di Ricerca - Università degli Studi di Roma I - Piazzale Aldo Moro 5 - 00185 ROMA - Italy. E-mail address: SUPINO @ VAXRMA.INFN.IT

**Sir P. Swinnerton-Dyer**, The Dower House - Thriplow - Royston - HERTS SG8 7RJ - England

**G. van der Geer**, Matematisch Instituut - Universiteit van Amsterdam - Plantage Muidergracht 24 - NL 1018 TV AMSTERDAM - The Netherlands. E-mail address: GEER @ FWI.UVA.NL

**A. Verra**, Dipartimento di Matematica - Università degli Studi di Roma I - Piazzale Aldo Moro 5 - 00185 ROMA - Italy.

**H. Verrill**, Department Mathematics of Cambridge 16 - Mill Lane - CAMBRIDGE CB2 ISB - England

**M. Teicher**, Department of Mathematics - BAR-ILAN University - Bamat - Gan - Israel. E-mail address: TEICHER @ BIMACS

**J. Wahl**, Department of Mathematics - University of North Carolina - Chapel Hill - NC 27599-3250 - U.S.A. E-mail address: JW @ MATH.UNC.EDU

**P.M.H. Wilson**, Trinity College Cambridge - CB2 ITQ - England. E-mail address: pmhW @ pmms.cam.ac.uk

**J.A. Wisniewski**, Institute of Mathematics - Warsaw University - Banacha 2 - 02-097 WARSZAWA - Poland. E-mail address: JAREKW @ PLEARN.BITNET

**F.L. Zak**, CEMI RAN - Krasikov st. 32, Moscow 117418, RUSSIA. E-mail address: Zak @ cemi.msk.su C10201 @ sucemi

**R. Zollo**, Dipartimento di Matematica - Università degli Studi di L'Aquila - Via Vetoio Loc. Coppito - 67100 L'AQUILA - Italy

**F. Zucconi**, Dipartimento di Matematica - Università degli Studi di Pisa - Via Buonarroti 2 - 56127 PISA - Italy

## List of seminars held during the conference

**Marco Andreatta** : Elementary contractions whose supporting divisors are $K_X \otimes L^{\otimes r}$

**Enrique Arrondo** : Obstructions for embedding surfaces in codimension two

**Lucian Bădescu** : Formal functions along zero loci of sections of ample vector bundles and applications

**Fabio Bardelli** : Algebraic cycles on certain Calabi-Yau threefolds

**Ingrid Bauer** : Generic lemniscates of algebraic functions

**Stefan Bauer** : Moduli of bundles on elliptic surfaces

**Gian Mario Besana** : On the geometry of conic bundles arising in adjunction theory

**Giorgio Bolondi** : Hilbert schemes of surfaces in $\mathbf{P}^4$- Liaison classes of surfaces in $\mathbf{P}^4$

**Frédéric Campana** : The class $\mathcal{C}$ is not stable by small deformations

**Alessio Corti** : Birational models for threefolds fibered in del Pezzo surfaces

**Igor Dolgachev** : Ogg-Shafarevich theory of elliptic threefolds

**Lawrence Ein** : Some Reider-type theorem for smooth threefolds

**Geir Ellingsrud** : Twisted cubics on a general quintic threefold

**Barbara Fantechi** : Deformations of Hilbert schemes of points of a surface

**Hubert Flenner** : Rational surfaces in 3-folds with ample normal bundle

**Lothar Göttsche** : Hilbert schemes of points and formulas for contacts with linear subspaces and trisecants

**Klaus Hulek** : New irregular surfaces in $\mathbf{P}^4$

**Paltin Ionescu** : Boundedness for some special families of embedded manifolds

**Vassilii Iskovskih** : Relations among generators in 2-dimensional Cremona group over a non-closed field

**Krzysztof Jaczewski** : Generalized Euler Sequence and Toric Varieties

**Yujiro Kawamata** : On the log flips and applications

**Steven L. Kleiman** : A generalized Teissier-Plücker formula

**János Kollár** : Effective base point freeness

**Kazushiro Konno** : Even canonical surfaces with small $K^2$

**Angelo Lopez** : Noether-Lefschetz, space curves and mathematical instantons

**Marco Manetti** : Deformation equivalence of surfaces

**Nicolae Manolache** : Multiple structures on smooth support

**Rick Miranda** : Encoding number for torsion sections of elliptic surfaces

**Yoichi Miyaoka** : Rational curves of small degree on a Fano manifold

**Boris Moishezon**: Cuspidal Curves and Algebraic Surfaces

**Jacob Murre** : Motives and Chow groups

**Kieran O'Grady** : On the smooth classification of projective surfaces

**Christian Okonek** : On the topology of non-Kählerian surfaces

**Giorgio Ottaviani** : On the relatives of the Horrocks bundle on $\mathbf{P}^5$

**Marino Palleschi** : Double covers of $\mathbf{P}^n$ as very ample divisors

**Claudio Pedrini** : 0-cycles on singular projective surfaces

**Thomas Peternell** : Manifolds with semi-positive curvature

**Ragni Piene** : Projective surfaces with small osculating spaces

**Giampietro Pirola** : Projections on abelian variety and applications

**Miles Reid** : Inductive Mori theory and applications

**Igor Reider** : On the Chern numbers and number of moduli of surfaces of general type

**Fumio Sakai** : Remarks on the irregularity of cyclic multiple planes

**Michael Schneider** : Bounds for stable bundles and Weierstrass schemes

**Edoardo Sernesi** : Torelli type theorems

**Nicholas Shepherd-Barron** : Some foliations on tori

**Sir Peter Swinnerton-Dyer** : Diophantine problems on rational varieties

**Gerard van der Geer** : Reed-Muller Codes and Supersingular Curves

**Alessandro Verra** : Prym varieties of plane sextics

**Jonathan Wahl** : The Miyaoka-Yau inequality for normal surfaces and local analogues

**Peltan M.H. Wilson** : Calabi-Yau threefolds

**Jarek Wisniewski** : On Fano manifolds of dimension $\geq 4$

**Fyodor Zak** : Quadratic normality and higher secant varieties

Contemporary Mathematics
Volume 162, 1994

# ON CURVES IN $C^{(2)}$ GENERATING PROPER ABELIAN SUBVARIETIES OF J(C)

A.ALZATI - G.P.PIROLA

ABSTRACT. Let $C$ be an algebraic smooth curve on complex numbers, in this paper we assume that there exists a curve in the symmetric product of $C$ such that its image in $J(C)$ generates a proper abelian subvariety. By studying this situation we obtain many results concerning the existence of low genus curves in the product of two curves and the $S(2,h,g)$ problem.

## 1. Introduction

Let C be an algebraic projective smooth curve defined over $\mathbb{C}$; let g be its genus and assume that $g \geq 3$. Let $C^{(2)}$ be the symmetric product of C with itself. Let $Pic^r(C)$ be the Picard variety of the degree r line bundles on C; $Pic^0(C) = J(C)$ will be the Jacobian variety of C. Let A ba <u>proper</u> Abelian subvariety of J(C).

In a previous paper, (see [2]), we have proved the following fact: if we fix $\alpha \in Pic^2(C)$ and we consider the map $\Phi_\alpha : C^{(2)} \to J(C)$ such that $\Phi_\alpha(P,Q) = P + Q - \alpha$, then $\dim[\Phi_\alpha^{-1}(A)] = 1$ if and only if there exists a map between C and another non rational curve K admitting an involution. For instance we are in this situation when $C^{(2)}$ contains a curve whose geometric genus is strictly less than g.

In this paper by applying the results of [2] and some elementary monodromy arguments, we can prove the

---

1991 Mathematics Subject classification. Primary 14H35,14H45

Supported by the MURST of the Italian Government (40% project "Geometria Algebrica").

This paper is in final form and no version of it will be submitted for publication elsewhere.

following:

THEOREM (1,1). *Let $C$ be a smooth projective curve over $\mathbb{C}$ of genus $g \geq 3$; let $H$ be an irreducible curve in $C^{(2)}$; if $g(H) \leq (g-1)/2$ then $C$ is a double covering of $H$.*

COROLLARY (1,2). *In the previous assumptions, if $g(H) = 1$, $g(C) \geq 2$, then $C$ is hyperelliptic or bielliptic, and if $g(C) = 3$, $C$ is both.*

Corollary (1,2) was proved by Silverman and Abramovich-Harris (see [1]) by different methods. By theorem (1,1) we can prove some other results on this problem, (see [1]): let C be a smooth curve, we say that C is of type $(2,h)$ if it admits a map of degree 2 or less to a smooth curve of genus $h$ or less. Let $\psi: C' \to D$ be a morphism of smooth curves, assume that $\deg(\psi) = 2$, assume also that $g(D) \leq h$, we say that the property $S(2,h,g)$ holds if, for any non constant map $\varphi: C' \to C$, with $g = g(C)$, C is of type $(2,h)$. In this paper we will show that $S(2,2,g)$ holds except one case at most in which $g(C) = 3$, (see [1]), and that $S(2,h,g)$ holds if $h \leq (g-1)/2$, (this last inequality is also contained in a preprint of Abramovich).

We were suggested to consider [2] from this new point of view by the paper [1] which contains some interesting results concerning the arithmetic of algebraic curves. We believe that, when it is possible to apply the techniques of [1], in the case of 2-symmetric product of curves, our approach is more precise and we believe that it could give better bounds.

Moreover, by using the previous results and two powerful finiteness theorems, (see [3] and [6]), we can show the following:

THEOREM (1,3). *Let $C$ be a non hyperelliptic, non bielliptic, smooth curve, $g(C) \geq 3$ and let $D$ be a smooth curve of genus $g' \leq g$. Then there exists only a finite number of families of irreducible curves $\Gamma$ in $D \times C$ of type $(m,2)$, (i.e. $\Gamma(p \times C) = 2$, $\Gamma(D \times q) = m$, $p \in D$, $q \in C$) for any positive integer $m$.*

In §3 we consider the case in which $\Phi_\alpha^{-1}(A)$ is a reduced curve $\Gamma$ which can be reducible and/or singular.

In §4 we study the case in which the map between C and K has degree 2. In §5 we analize the cases in which $g(C) = 3, 4$. In §6, by using the previous analysis, we prove theorem (1,1) and corollary (1,2) and in §7 the results related to the $S(2,h,g)$ property. In §8 we prove theorem (1,3).

In §9 we give a further result: for any curve C let $C \times C$ be the Cartesian product of C with itself, let $\Delta$ be the diagonal of $C \times C$ and let $\pi$ be the natural double covering $\pi : C \times C \to C^{(2)}$. Let B be the divisor of $C^{(2)}$ such that $\pi^* B = \Delta$, let us call C also the divisor of $C^{(2)}$ such that $\pi^* C = H_1 + H_2$ where $H_1$, $H_2$ are the divisors $\mathbf{p} \times C$ and $C \times \mathbf{q}$, $\mathbf{p} \in C, \mathbf{q} \in C$, respectively. It is well known that, for generic C, the Neron-Severi group of $C^{(2)}$ is generated from C and B over $\mathbb{Z}$.

Now let C be a <u>generic</u> curve of genus 3, recall that there is an involution i on $C^{(2)}$ such that $i(P,Q) = (R,S)$ if and only if the images of P,Q,R,S under any fixed canonical embedding of C (as a smooth plane quartic) are on the same line. We have:

THEOREM (1,4). *Let C be a generic curve of genus 3, let $f : C \to C^{(2)}$ be an immersion, then $\Gamma = f(C)$ can belong only to 4 families of (irreducible) curves of genus 3 in $C^{(2)}$, i.e. in $NS(C^{(2)})$ we can have only: $\Gamma = C$, $\Gamma = 2B$, $\Gamma = i(C)$, $\Gamma = i(2B)$.*

## 2. Notation and conventions

| | |
|---|---|
| curve | :algebraic projective reduced curve defined over $\mathbb{C}$ |
| $g(X)$ | :(geometric) genus of the curve X |
| $\Gamma'$ | :$\pi^{-1}(\Gamma)$ for any curve $\Gamma$ in $C^{(2)}$ |
| $\gamma'$ | :$g(\Gamma')$, when $\Gamma'$ is irreducible |
| $C/\sim$ | :normalization of the quotient curve of C by its equivalence relation "$\sim$" |
| $j_*$ | :if $j : C \to K$ is a map between curves, $j_*$ is the induced map from $J(C)$ to $J(K)$ |
| $j^*$ | :induced map from $J(K)$ to $J(C)$ |
| $j^{\vee}$ | :induced map from $C^{(2)}$ to $K^{(2)}$ |

| | |
|---|---|
| $B(j)$ | :abelian subvariety of $J(C)$ corresponding to the connected component of $\mathrm{kern}(j^*)$ containing 0 |
| $b_j$ | :total branching order of $j$ |
| $g^r{}_t$ | :linear series whose dimension is $r$ whose degree is $t$ |
| $Z_q$ | :cyclic group of order $q$. |

## 3. The general case

First of all we recall what we have proved in [2]: let $C$ be a non rational curve, let $A$ be a proper abelian subvariety of $J(C)$ suppose that $\Phi_\alpha{}^{-1}(A)$ contains a curve $\Gamma$ for some $\alpha \in \mathrm{Pic}^2(C)$, then there exist two smooth curves $K$ and $K'$, $g(K) > 0$, and two non constant maps $f:C \to K$ and $h:K \to K'$ such that:

i) $\deg(h) = 2$

ii) $f^\smile(\Gamma) = \{(P,Q) \in K^{(2)} \mid h(P) = h(Q)\}$

iii) $A \supseteq (h\circ f)^*[J(K')] + B(f)$;

The assertion ii) means that $f^\smile$ sends $\Gamma$ onto the locus $\Lambda$ of $K^{(2)}$ which is the image of the graph of the involution $i:K \to K$, with quotient curve $K'$, under the natural map $K \times K \to K^{(2)}$. It may occur that $K = C$, $f = I_C$, in this case $\Gamma = \Lambda$. It may occur that $A = 0$, in this case $K = C$, $f = I_C$, $g(K') = 0$, $C$ is hyperelliptic and $\Gamma = \Lambda$ is the graph of the hyperelliptic involution.

In this paper we assume that $g(C) \geq 3$ and that there exists a curve $\Gamma$ in $C^{(2)}$ such that, for some $\alpha \in \mathrm{Pic}^2(C)$, $\Phi_\alpha(\Gamma)$ is a curve which generates a proper abelian subvariety $A$ of $J(C)$ with $1 \leq \dim(A) \leq g-1$. If $\Gamma$ is irreducible we can assume $\gamma = g(\Gamma) \geq 1$, if $\Gamma$ is reducible we can assume that it has a non rational component at least. Sometimes we will also assume $g(\Gamma_i) \geq 2$ for any irreducible component $\Gamma_i$ of $\Gamma$.

Now our point of view is slightly different from that one in [2]: firstly we consider $\Gamma$ and then we consider $A$: the minimal proper abelian subvariety of $J(C)$, if it exists, generated by $\Phi_\alpha(\Gamma)$; so that we can always suppose that $A = B(f) + (h\circ f)^*[J(K')]$, hence: $\dim(A) = g - g(K) + g(K')$. Note that if $\dim(A) = 1$ then $\Gamma$ is an irreducible elliptic curve.

In this paragraph we want to describe the curves $\Gamma$ such that $\Phi_\alpha(\Gamma)$ generates a proper abelian subvariety A of J(C) for a suitable $\alpha$.

The curve $\Gamma$ in $C^{(2)}$ defines a symmetric correspondence T over C such that for any point $P \in C$, $T(P) = \{Q_1, Q_2, \ldots, Q_d\}$ if and only if $(P, Q_i) \in \Gamma$ $\forall i = 1, 2, \ldots d$. We recall, see [2], that on C there exists a non trivial equivalence relation "$\sim$" such that: for any $P_1, P_2 \in C$, $P_1 \sim P_2$ if and only if there exists $R \in C$ such that $R \in T(P_1) \cap T(P_2)$. Moreover $K = C/\sim$ so that $\deg(f) = d$. The general fibres of $h \circ f$ can be considered as bipartite graphs (in the following picture we describe the case d=4)

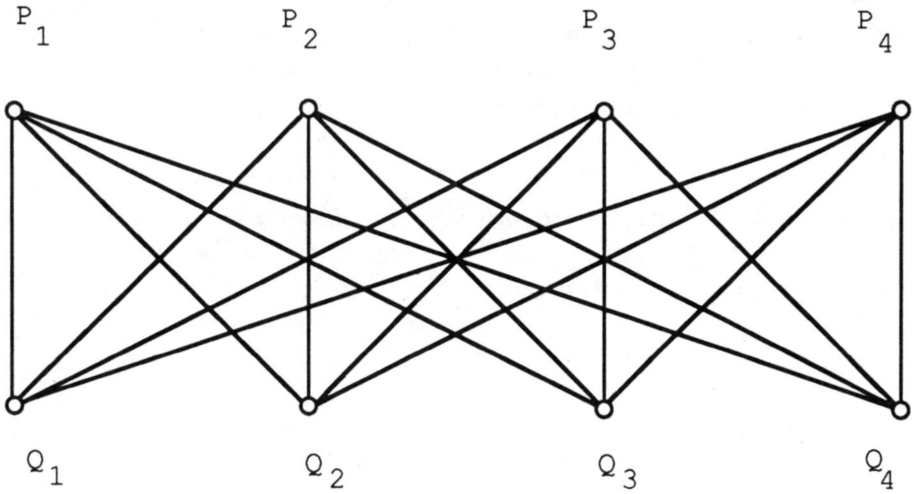


in which the two distinguished sets of vertices, $\{P_i\}$ and $\{Q_i\}$, represent fibres of f. Two vertices $\mathfrak{a}, \mathfrak{b}$ are adjacent if and only if $(\mathfrak{a}, \mathfrak{b}) \in \Gamma$; for any vertex $P_i$, $T(P_i) = \{Q_1, Q_2, \ldots, Q_d\}$; for any vertex $Q_i$, $T(Q_i) = \{P_1, P_2, \ldots, P_d\}$

If we consider $\Gamma'$in $C \times C$, ($\Gamma'$ is the double covering of $\Gamma$ in $C^{(2)}$), we can define the map $\varphi: \Gamma' \to C$ as the natural projection onto the first factor. For any point $P \in C$, $\varphi^{-1}(P) = T(P)$ so that if we look at the bipartite graph corresponding to a generic fibre of $h \circ f$, we see that each one of the distinguished set of vertices is the fibre, with respect to $\varphi$, of any vertex not belonging to this set.

The above description is very useful to study the links between the branching points of $f, h, h\circ f$ and the singular points of $\Gamma'$.

By Riemann-Hurwitz formula we have:

$$(3,1)\qquad \begin{aligned} b_h &= 2[g(K)-1] - 4[g(K')-1] \geq 0 \\ b_f &= 2(g-1) - 2d[g(K)-1] \geq 0 \\ b_{h\circ f} &= 2(g-1) - 4d[g(K')-1] = b_f + db_h \geq 0. \end{aligned}$$

For any branching point $\mathbf{p}$ of h, the fibre $(h\circ f)^{-1}(\mathbf{p})$ degenerates in such a way that every $P_i$ collapses to the corresponding $Q_i$ and the d points $(P_i, P_i)$ belong to $\Gamma'$ and $\Delta$, so that, if we put $\delta = \Gamma'\Delta$, we get $\delta = db_h$; but they are not singular points for $\Gamma'$. On the contrary, if $h^{-1}(\mathbf{p})$ is a branching point for f too, then we get singular points for $\Gamma'$.

For any branching point $\mathbf{q}$ of f the fibre of $h\circ f$ containing $f^{-1}(\mathbf{q})$, in general, degenerates in such a way that some of the $P_i$ collapse among them and we do not get singular points for $\Gamma'$; but if in addition, in the same fibre of $h\circ f$ we have another collapse among the $Q_i$ (corresponding to another branching point $\mathbf{r}$ of f such that $h(\mathbf{q}) = h(\mathbf{r})$), then we have couples of singular points for $\Gamma'$. In any case the branching points of f, alone, do not cause collapse of any $P_i$ with any $Q_i$.

It is easy to see that the singular points of $\Gamma'$ occur only in the previously described cases. This fact and (3,1) allow us to show:

$$(3,2)\qquad \Gamma'^2 + \Gamma' K_{C\times C} = 2d(g-1) + db_f.$$

Now we can make the following remarks which will be very useful in the sequel.

REMARK (3,3). If $d = 1$ then $f = I_C$, $K = C$, $B(f) = 0$, $A = h^*[J(K')]$, $\Gamma'$ is the graph of the involution i in $C\times C$; hence the projection $\varphi$ has degree 1 (h has degree 2), and therefore $\Gamma'$ is irreducible, smooth, isomorphic to C and $\gamma' = g$. By the Riemann-Hurwitz formula applied to $\pi$ we get $b_h = 2(g-1)-4(\gamma-1)$ and by using (3,1) we also get $(g+1)/2 \geq \gamma = g(K')$.

REMARK (3,4). If $\dim(A) = 1$ we have $1 \geq g-g(K)+g(K')$. If $g-g(K)=0$, then $g(K') = 0,1$ and we are in the previous case: $d = 1$. If $g-g(K)=1$, then $g(K') = 0$ and from (3,1) we easily get that $g \leq 2$, or $g = 3$, $g(K) = 2$, $g(K') = 0$, $d = 2$ and $b_f = 0$, but we are assuming $g \geq 3$, so $g \leq 2$ is impossible. We will return to the other case more later.

From now on we will suppose $d \geq 2$ and $\dim(A) \geq 2$.

We fix some notations: if $\Gamma'$ is reducible we call $\Gamma'_i$ its irreducible components and $\gamma'_i$ their geometrical genus, analogously we will have $\Gamma_i$ and $\gamma_i$. Pay attention: in this section we assume that there is only one irreducible component $\Gamma'_i$ for any irreducible component $\Gamma_i$. We also have morphisms $\varphi_i:\Gamma'_i \to C$, defined on the normalization of $\Gamma'_i$, so that $b_{\varphi_i}$ does not depend on the singular points of each $\Gamma'_i$.

Let $d_i = \Gamma'_i H_1 = \Gamma'_i H_2$, $\Sigma d_i = d$; $\Gamma'_i K_{C\times C} = 4d_i(g-1)$.

We have the following 4 cases:

A) $\Gamma'$ is irreducible and smooth

B) $\Gamma'$ is irreducible and singular

C) $\Gamma'$ is reducible and its components are smooth

D) $\Gamma'$ is reducible and some of its components are singular.

For each case we state what we get from (3,1) and (3,2) by recalling also that $\delta = db_h$ and by using the Riemann-Hurwitz formula applied to $\pi$ or $\pi_i$.

Case A):

$db_f = b_\varphi$

$2(\gamma-1) + d(d+1)[g(K)-1] = 2d(g-1) + 2d[g(K')-1]$.

Case B): let $\nu$ be the double of the difference between the aritmetic and geometric genus of $\Gamma'$, then:

$db_f = b_\varphi + \nu$

$2(\gamma-1) + d(d+1)[g(K)-1] \leq 2d(g-1) + 2d[g(K')-1]$.

Case C):

$db_f = 2\Sigma(\Gamma'_i\Gamma'_j) + \Sigma b_{\varphi_i}$

$4\Sigma(\gamma_i-1) + \delta = 2d(g-1) + db_f - 2\Sigma(\Gamma'_i\Gamma'_j)$

$2\Sigma(\gamma_i-1) + d(d+1)[g(K)-1] \leq 2d(g-1) + 2d[g(K')-1]$.

Case D): let $\nu_i$ be the double of the difference between the aritmetic and geometric genus of $\Gamma'_i$, then:

$db_f = 2\Sigma(\Gamma'_i\Gamma'_j) + \Sigma b_{\varphi_i} + \Sigma\nu_i$

$4\Sigma(\gamma_i-1) + \delta = 2d(g-1) + db_f - 2\Sigma(\Gamma'_i\Gamma'_j) - \Sigma\nu_i$

$2\Sigma(\gamma_i-1) + d(d+1)[g(K)-1] \leq 2d(g-1) + 2d[g(K')-1]$.

REMARK (3,5). If for any proper abelian subvariety A of J(C), dim(A) $\geq 2$, we look for the curves $\Gamma$ in $C^{(2)}$, smooth and irreducible, with the smaller possible genus, such that $\Phi_\alpha(\Gamma)$ generates A for a suitable $\alpha$, then we can assume that $\gamma = \dim(A) = g - g(K) + g(K')$. In this case, obviously, $\Gamma'$ can not be reducible, otherwise $\gamma = g$. If, moreover, we assume that $\Gamma'$ is smooth, by considering the case A) above, it is not difficult to see that $\gamma = g/2$ or $\gamma = (g+1)/2$ and $d \leq 4$, (recall that $g \geq 3$ and $d \geq 2$).

## 4. The case d = 2.

In this paragraph we study the case $d = 2$ in great detail because it is necessary for our future proofs.

If $d = 2$ and $\Gamma'$ is singular it can have only double points because $\Gamma'H_1 = \Gamma'H_2 = 2$. As $\Gamma'$ is symmetric for any double point (P,Q) with P≠Q there is another double point (Q,P) so that the number of double points of $\Gamma'$ is $2\mathcal{B}+\mathcal{A}$ where $\mathcal{A}$ is the number of double points on the diagonal; a local computation shows that at any double point (P,P) of $\Gamma'$ $\Gamma$ is smooth so that the number of double points of $\Gamma$ is $\mathcal{B}$. In fact, take $z_1,z_2$ as local coordinates in $C\times C$ around (P,P) = (0,0), take $w_1,w_2$ as local coordinates in $K\times K$ around (f(P),f(P)) = (0,0) and $\zeta_1,\zeta_2$ in $K'\times K'$ around (h[f(P)],h[f(P)]) = (0,0), then the local equations of $f\times f$ are: $w_i = z_i{}^2$ and the local equations of $h\times h$ are: $\zeta_i = w_i{}^2$.

Hence: $\zeta_1-\zeta_2 = (z_1)^4 - (z_2)^4=(z_1-z_2)(z_1+z_2)[(z_1)^2+(z_2)^2]$ and the local equation of $\Gamma'$ is $[(z_1)^2+(z_2)^2] = 0$, because $z_1-z_2 = 0$ and $z_1+z_2 = 0$ are, respectively, the local equations of $\Delta$ and of the locus $\{\mathbf{p},\mathbf{q} \in C \mid f(\mathbf{p}) = f(\mathbf{q}),\ \mathbf{p} \neq \mathbf{q}\}$. So that the local equation of $\Gamma$ is $\rho^2-\sigma = 0$, where $\rho = z_1+z_2$ and $\sigma = z_1z_2$, hence $\Gamma$ is smooth at $(0,0)$.

As $d = 2$, any branching point of $f$ has branch order 1, so that we can write $b_f = 2\mathcal{B} + \mathcal{E} + \mathcal{A}$ where $\mathcal{E}$ is the number of branching points of $f$ not coupled in a fibre of $h \circ f$, (see §3) and $b_h = \mathcal{D} + \mathcal{A}$ where $\mathcal{D}$ is the number of branching points of $h$ alone.

If $\Gamma'$ is reducible the only possibility is $\Gamma' = \Gamma'_1 + \Gamma'_2$ and each $\Gamma'_i$ is isomorphic to $C$ by the projections $\varphi_i$, moreover $b_{\varphi_i} = 0$, $d_i = 1$, $i = 1,2$; so that $\Gamma'_1$ and $\Gamma'_2$ are smooth, genus $g$ curves and case D) coincides with C).

Now we consider the three cases A),B),C) when $d = 2$.

Case A): in general there is nothing new to say. We have only: $(\gamma-1) + 3[g(K)-1] = 2(g-1) + 2[g(K')-1]$.

Case B): a priori $\Gamma'$ has $\mathcal{A} + 2\mathcal{B}$ singular double points and $\Gamma$ has $\mathcal{B}$ singular double points, but it is easy to show that $\mathcal{A} = 0$, so that: $2(\gamma-1) = \Gamma^2 + \Gamma K_C(2) - 2\mathcal{B}$.

Case C): every $\Gamma'_i$ determines an automorphism of $C$ and the points of $\Gamma'_i$ are the ordered couples $(P,\kappa_i(P))$, $i = 1,2$ where the $\kappa_i$ are the automorphisms. If any $\Gamma'_i$ is symmetric then, for any $P \in C$ we have $\kappa_1{}^2(P) = \kappa_2{}^2(P) = P$ so that we have an order 4 group of involutions acting on $C$, $K = C/(\kappa_1\kappa_2)$, and the generic fibre of $h \circ f$ is described in the following picture, on the left:

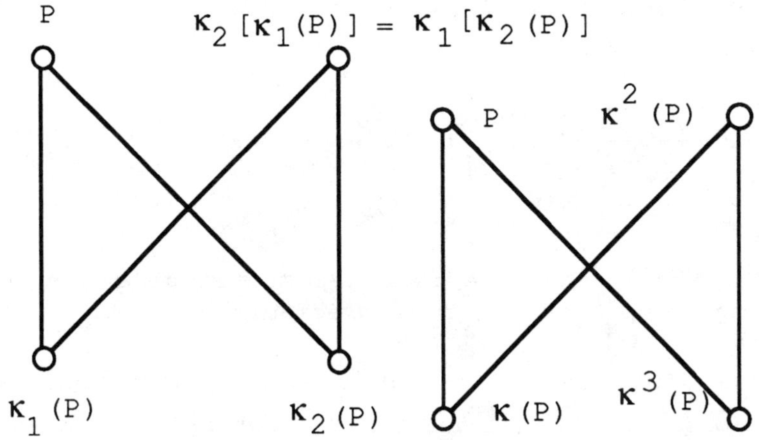

In this subcase it is easy to see that $\mathcal{A} = \mathcal{E} = 0$.

On the contrary, if $\Gamma'_2$ is symmetric to $\Gamma'_1$, then for any $P \in C$ we have $\kappa_2[\kappa_1(P)] = P$, i.e. $\kappa_2 = \kappa_1^{-1}$; if we call $\kappa_1 = \kappa$ then we have $\kappa \neq \kappa^{-1}$ but $\kappa^2 = \kappa^{-2}$ so that $\kappa$ is an automorphism of C whose period is 4; in the cyclic group generated by $\kappa$ we get the involution $\kappa^2$, $K = C/(\kappa^2)$ and the generic fibre of h∘f is described in the above picture, on the right. In this subcase $\mathcal{D} = \mathcal{E} = 0$, then it is easy to see that it must be: $b_f = b_h = 0$, moreover $\Gamma$ is irreducible, $g(K') \neq 0$ and $\gamma = \gamma'_1 = \gamma'_2 = g$.

We can also discuss the monodromy group of h∘f when d = 2. In the following picture we have marked the generical fibre of h∘f

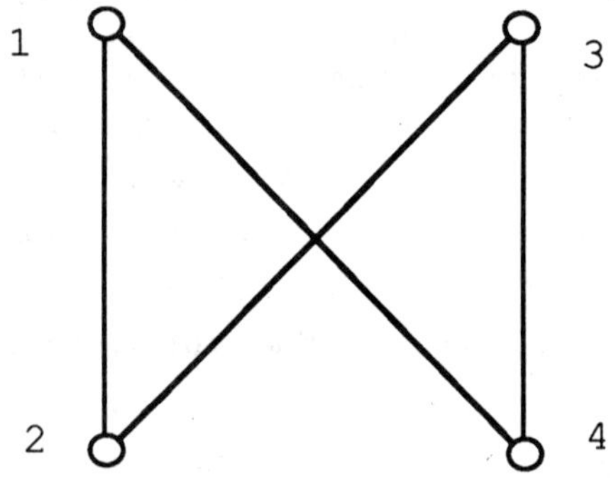


so that we can describe in the following list the elements of the group $S_4$ which correspond to the given branching points:

$\mathcal{A}$ ⟶ (1234)
$\mathcal{B}$ ⟶ (13)(24)
$\mathcal{D}$ ⟶ (12)(34)
$\mathcal{E}$ ⟶ (13) or (24).

If $K' = \mathbb{P}^1$ the previous elements generate the monodromy group G of h∘f, more precisely we have:

PROPOSITION (4,1). *In the previous notations and assumptions, if $K' = \mathbb{P}^1$ and $b_f = 0$ then only the case C) holds.*

PROOF. In case A): $G \simeq Z_2$ or $G \simeq D_8$ (one of the Sylow subgroups of $S_4$) according to $\mathcal{E} = 0$ or not; but $G \simeq Z_2$ is not possible, so we always have $\mathcal{E} \neq 0$ and $G \simeq D_8$; it is not possible that $b_f = 0$.

In case B) recall that $\mathcal{A} = 0$ and $\mathcal{B} \neq 0$, hence it is not possible that $b_f = 0$. □

REMARK (4,2). When $K' = \mathbf{P}^1$ we can describe G in all cases: in case A) $G \simeq D_8$. In case B) we have: $G \simeq D_8$ unless $\mathcal{E} = 0$; if $\mathcal{E} = 0$ and $\mathcal{D} = 0$ then $G \simeq Z_2$: impossible; if $\mathcal{E} = 0$ and $\mathcal{D} \neq 0$ then $G \simeq Z_2 \oplus Z_2$, but in this case it is easy to see that $\Gamma$ has two components and so we are in case C). Hence in case B) we always have $\mathcal{E} \neq 0$ and $G \simeq D_8$.

In case C): $G \simeq Z_2 \oplus Z_2$ or $G \simeq Z_4$ according to the existence of two involutions on $\Gamma'$ or not.

## 5. The cases g = 3 and g = 4.

In this paragraph we study our problem when g(C) is small. First of all we consider, for any g, the case in which $g(K) = 1$ and therefore $g(K') = 0$ (recall that $\dim(A) = g - g(K) + g(K') < g$).

If K is an elliptic curve $f^{\vee}(\Gamma) = \Lambda = \{(P,Q) \in K^{(2)} \mid h(P) = h(Q)\}$ is the graph of the natural involution on K, and $A = B(f)$ which is a (g-1)-dimensional abelian subvariety of J(C). For curves C admitting a map towards an elliptic curve the existence of a curve such $\Gamma$ is always assured: we can take $\Gamma = (f^{\vee})^{-1}(\Lambda)$ and $A = B(f)$ to get that $\Phi_\alpha(\Gamma)$ is contained in B(f) for a suitable $\alpha$.

Therefore in §5 we will suppose $g(K) \geq 2$ (and $d \geq 2$); we can also suppose that $\dim(A) \geq 2$ as the cases $\dim(A) = 0,1$ were previously studied in general, moreover, if $\Gamma$ is reducible, we will assume that it has not rational or elliptic components, but we will consider also the case in which some irreducible component $\Gamma_i$ of $\Gamma$ has $\Gamma'_i$ reducible. The possible case are the following:

1) $g = 3$, $\dim(A) = 2$, $g(K) = 2$, $g(K') = 1$, $d = 2$, $b_f = 0$, $b_h = 2$, $\delta = 4$, $\gamma = 2$ and $\Gamma$ is smooth.

2) $g = 4$, $\dim(A) = 2$, $g(K) = 2$, $g(K') = 0$, $d = 2$, $b_f = 2$, $b_h = 6$, $\delta = 12$, $\gamma = 2$ and $\Gamma$ is smooth.

3) $g = 4$, $\dim(A) = 3$, $g(K) = 2$, $g(K') = 1$, $b_h = 2$, $b_{h\circ f} = 6$ and:

3a) $d = 2$, $b_f = 2$, $\delta = 4$, and $\Gamma$ is smooth of genus 4 or it is singular, with one double point, of genus 3, or it is reducible in two smooth, genus 2 components, intersecting in one point;

3b) $d = 3$, $b_f = 0$, $\delta = 6$, and $\Gamma$ is smooth of genus 4 or it is reducible in three smooth, genus 2, components without intersections, or it is reducible in two smooth components, one of genus 2 and the other one of genus 3, without intersections.

## 6. Proof of theorems (1,1) and (1,2)

PROOF OF THEOREM (1,1). For a suitable $\alpha$ we have that $\Phi_\alpha(H)$ generates a proper abelian subvariety A of J(C) such that $\dim(A) \leq (g-1)/2$. We can use our results of [2], (see §1), and we get that: $g - g(K) + g(K') \leq (g-1)/2$.

If $g - g(K) < (g-1)/2$ then $d = 1$, so that, by remark (3,3), C is isomorphic to $\Gamma'$ which is a double covering of $\Gamma = H$.

If $g - g(K) = (g-1)/2$ then $d = 2$, $g(K') = 0$, $b_f = 0$; we check the cases examined in §4, by proposition (4,1) we conclude that case C) holds, so there is a reducible curve $\Gamma$ in $C^{(2)}$, $\Phi_\alpha(\Gamma) = \Phi_\alpha(H)$, $\Gamma = \Gamma_1 \cup \Gamma_2$ with $\Gamma_i$ irreducible, $H = \Gamma_1$; hence C, which is isomorphic to $\Gamma_1'$, is a double covering of $\Gamma_1 = H$. □

REMARK (6,1). Actually, when $g-g(K) = (g-1)/2$, we can also prove our statement by considering that, in this case, C is an unramified double covering of a hyperelliptic curve and by using the related results of Mumford (see [4]).

PROOF OF COROLLARY (1,2). If $g(C) = 2$ we have nothing to prove. If $g(C) \geq 3$ we apply our results of [2] to $A = \Phi_\alpha(H)$; in this case $\dim(A) = 1$ hence, by remark (3,4), $C = K$, $g(K') \leq 1$, or $g = 3$, $g(K) = 2$, $g(K') = 0$, $d = 2$, $b_f = 0$. In this last case in $C^{(2)}$ there exists a curve $\Gamma$ reducible into two smooth, non intersecting components and it is easy to see (§4, case C)), that one component is H and the other one is rational, hence C is bielliptic and hyperelliptic. □

## 7. The S(d, h, g) property

Now we show the following (essentially contained in [1]):

PROPOSITION (7,1). *The property S(2,2,g) holds except one case at most in which g(C) = 3.*

PROOF. We can assume $g \geq 3$ otherwise C is obviously of type (2,2).

Let A be $\varphi_*\psi^*[J(D)]$, A is an abelian subvariety of J(C), $\dim(A) \leq 2$.

Let $\Lambda = \{(P,Q) \in C'^{(2)} \mid \psi(P) = \psi(Q)\}$, the points of $\Lambda$ are the fibres of $\psi$. Let $\Gamma$ be the curve $\varphi^{\vee}(\Lambda)$, it is easy to see that, for a suitable $\alpha$, $\Phi_\alpha(\Gamma)$ is contained in A. If $\Phi_\alpha(\Gamma)$ is a point, for instance if $\dim(A) = 0$, then C admits a $g^1 2$ so that C is obviously of type (2,2); otherwise we can apply our results, see §3, and we get that there exist two smooth curves K and K', $g(K) > 0$, and two non constant maps $f: C \to K$ and $h: C \to K'$; moreover $d \leq \deg(\varphi)$ as $\deg(\psi) = 2$.

If $\dim(A) = 0$ $\Phi_\alpha(\Gamma)$ is a point and C admits a $g^1 2$.

If $\dim(A) = 1$, then $g-g(K)+g(K') = 1$. As $g \geq 3$ we have: $g = g(K)$, $g(K') = 1$ and C is of type (2,2); or $g = 3$, $g(K) = 2$, $g(K') = 0$, $d = 2$, $b_f = 0$: this is the case considered in the proof of (1,2), as we saw there, C is hyperelliptic.

If $\dim(A) = 2$, then $g-g(K)+g(K') = 2$. If $g \geq 4$ we have: $g = g(K)$, $g(K') = 2$ and C is of type (2,2), or $g-g(K) = 2$, $g(K') = 0$; then we have: $g = 4$, $d = 2$, $b_f = 2$ and C is of type (2,2), or: $g = 4$, $d = 3$, $b_f = 0$, or: $g = 5$, $d = 2$, $b_f = 0$. In the first case we can argue as in §5 (by forgetting the condition $\gamma_i \geq 2$) and conclude that $\Gamma$ contains a rational or elliptic component, then C is (2,2), (recall corollary (1,2)). In the second case we can argue as in §4 and we use proposition (4,1): necessarily $\Gamma = \Gamma_1 \cup \Gamma_2$ with $\Gamma_1, \Gamma_2$ smooth, $\Gamma_1\Gamma_2 = 0$, $\gamma_1 = 2$, $\gamma_2 = 0$, so that C is hyperelliptic, (note that $g(\Gamma_i) \neq 1$, otherwise we could apply corollary (1,2), whose proof shows that $g = 5$, $d = 2$ is impossible).

If $g = 3$, we have: $g(K) = 2$, $g(K') = 0$, $d = 2$ and $C$ is $(2,2)$, or: $g(K) = 1$, $g(K') = 0$ and in this case we can say nothing about the property $(2,2)$ for $C$. □

REMARK (7,2). As above, you can prove the same result of [1] concerning the case $g(D) = 1$.

By our previous results it is very easy to show that:

THEOREM (7,3). *$S(2,h,g)$ holds if $h \leq (g-1)/2$ (or, obviously $g = 0$).*

PROOF. If $h = 0,1$ see [1]; let us assume $h \geq 2$, hence $g \geq 3$. By arguing as above we get maps $f:C \to K$, $h:K \to K'$ with $K,K'$ smooth, $g(K) > 0$, $\deg(h) = 2$ and $h \geq g - g(K) + g(K')$. If $h < (g-1)/2$ we have: $2g(K)-1 > g$, hence $g(K) \geq 3$ and, by theorem (3,1), we get $\deg(f) = 1$ so that the map $h:C \to K'$ shows that $C$ is of type $(2,h)$.

If $h = (g-1)/2$ we get: $g-g(K) = (g-1)/2$, $g(K) = (g+1)/2$, $g(K') = 0$, $d = 2$, $b_f = 0$, $\Gamma$ reducible into two components $\Gamma_1$ and $\Gamma_2$, corresponding to two involutions of $C$, with $\gamma_1+\gamma_2 = (g-1)/2$. Hence $C$ admits a degree 2 map to a curve of genus $h = (g-1)/2$ or less. □

## 8. Proof of theorem (1,3)

LEMMA (8,1). *Let $S$ be a surface of general type with irregularity $q > 2$; let $C$ be a fixed, smooth, genus $g$ curve such that $g \leq q$; let $f:C \to S$ be a morphism then $C' = f(C)$ belong to a finite number of families of curves on $S$.*

PROOF. Let $\varphi$ be the induced map $\varphi:J(C) = \mathrm{Alb}(C) \to \mathrm{Alb}(S)$. If $\mathrm{Im}(\varphi) \neq \mathrm{Alb}(S)$ we consider $\psi:S \to \mathrm{Alb}(S) \to \mathrm{Alb}(S)/\mathrm{Im}(j)$, as $\psi(C')$ is a point we have $(C')^2 \leq 0$ so we get our thesis by Bogomolov's theorem (see [3], cor.3.4.6.). If $\mathrm{Im}(\varphi) = \mathrm{Alb}(S)$ then $\varphi$ is an isogeny; we can consider the dual isogeny $\varphi'$ such that $\varphi'\circ\varphi = \eta$ is the multiplication for a fixed integer $n$. Let $j:C \to J(C)$ be a fixed embedding, $\eta[j(C)] = C''$ is a curve in $J(C)$ with at most a finite number of ordinary singularities and this number is bounded independently from $\varphi$ and therefore from $f$. In fact, as $j$ is an embedding and $\eta$ is etale, $C''$ has a singular point, necessarily an ordinary singular point, if and only if there exist a couple of points P,Q on C such that n(P-Q) is linearly equivalent to 0. But Tors[J(C)] intersects C-C in a finite number of points only, (see [6]), hence we have that C" have only a finite number of (ordinary) singularities and this number is bounded independently from n, hence $\varphi$, hence f.

Then the aritmethic genus of C' is bounded independently from f, so that $K_S C' + (C')^2$ is bounded too; as $K_S C' \geq 0$ we get that $(C')^2$ is bounded and then we can conclude by Bogomolov's theorem. □

REMARK (8,2). In the previous assumptions, if $S = C^{(2)}$, $(g(C) \geq 2)$, $g' < g$, $(C')^2 < 0$ we also have that there are only a finite number of curves C' because in this case they are rigid.

Now we can prove theorem (1,3):

PROOF. To any such $\Gamma$ we can associate a morphism $f_\Gamma : D \to C^{(2)}$: for any $\mathbf{p} \in D$, $f_\Gamma(\mathbf{p}) = (\mathbf{q}_1, \mathbf{q}_2)$ where $\{(\mathbf{p},\mathbf{q}_1),(\mathbf{p},\mathbf{q}_2)\} = \Gamma(\mathbf{p} \times C)$. By lemma (8,1) we have only a finite number of families of curves $f_\Gamma(D)$ in $C^{(2)}$ and it is not possible that infinitely many families of curves $\Gamma$ give rise to the same family of $f_\Gamma(D)$ in $C^{(2)}$ otherwise we would have infinitely many different morphisms between D and one of its image in $C^{(2)}$, but, by corollary (1,2), $C^{(2)}$ does not contain rational or elliptic curves, so that, by the De Franchis theorem (see [4]), we can have only a finite number of such morfisms. □

REMARK (8,3). Note that the use of the De Franchis theorem in the previous proof, is crucial: if $C^{(2)}$ contains a rational or an elliptic curve E, we can construct infinitely many morphisms between D and E, each one of these corresponding to different curves $\Gamma$ in $D \times C$ of type (m,2).

QUESTION (8,4). Theorem (1,3) is true for curves of type (m,k) with $g > k \geq 3$ ?

## 9. Proof of theorem (1,4)

We begin by showing that $\Gamma$ has to be smooth. Let $\varphi$ be the induced map $\varphi: J(C) = Alb(C) \to Alb[C^{(2)}] = J(C)$. As C is generic, J(C) has not proper abelian subvarities, hence $\varphi$ is an isogeny; we can consider the dual isogeny $\varphi': J(C) \to J(C)$ such that $\varphi' \circ \varphi = \eta$ is the multiplication for a fixed integer n. Let $j: C \to J(C)$ a fixed embedding, if $\Gamma$ is singular $\eta[j(C)] = C''$ is singular too; but, as in the proof of lemma (8,1), it implies that there exist a couple of points P,Q on C such that n(P-Q) is linearly equivalent to 0. In this case there exists a degree n branched covering between C and $\mathbb{P}^1$ with at least two points whose branching index is n-1; by counting moduli we get that C can not be generic; hence $\Gamma$ is smooth.

To get our assertion we have only to show that there are only finitely many classes aC+bB in $NS(C^{(2)})$ containing an effective, irreducible, smooth, curve $\Gamma$ such that $g(\Gamma) = 3$.

We recall that $K_C(2) = 4C-B$, $C^2 = CB = 1$, $B^2 = -2$, $2\gamma-2 = \Gamma^2+\Gamma K_C(2)$ as $\Gamma$ is smooth. Moreover we recall that the involution i on $C^{(2)}$ is such that i(P,Q) = (R,S) if and only if the images of P,Q,R,S under any fixed canonical embedding of C (as a smooth plane quartic) are on the same line. As a smooth plane quartic has class 12 and 24 two-tangent lines it is easy to see that i(C) = = 3C-B and i(B) = 8C-3B.

If we choose $\Gamma$ = C, $\Gamma$ = 2B, $\Gamma$ = i(C) = 3C-B, $\Gamma$ = i(2B) = 16C-6B, we get 4 families of curves each one containing a smooth, irreducible, genus 3 curve as we want, (recall that B and i(B) are not effective). We have only to show that there are no other possibilities. In fact we must have: $2\gamma-2 = 4$, $\Gamma C \geq 0$, $\Gamma K_C(2) \geq 0$, $i(C)\Gamma \geq 0$, $\Gamma(2B) \geq 0$ (except when $\Gamma$ is a multiple of B) and $i(2B)\Gamma \geq 0$ (except when $\Gamma$ is a multiple i(B)).

Therefore we have: $a^2+2ab-2b^2+3a+6b = 4$, $a+b \geq 0$, $3a+6b \geq 0$, $2a+ +5b \geq 0$, $a-2b \geq 0$, $5a+14b \geq 0$; the only solutions are: C,B,2B,i(C),i(B),i(2B), as B and i(B) are not effective we get our assertion. □

## REFERENCES

1.D.Abramovich-J.Harris: "Abelian varieties and curves in $W_d(C)$" Comp. Math. **78** (1991) 227-238.

2.A.Alzati-G.P.Pirola: "On Abelian subvarieties generated by symmetric correspondences" Math. Z. **205** (1990) 333-342

3.M.Deschamps: "Courbes de genre géométrique borné sur une surface de type général (d'apres F.A.Bogomolov)" Sem. Bourbaki 30e année 1977/78 **519** 233-247.

4.A.Howard-A.J.Sommese: "On the theorem of de Franchis" Ann. Sc. Norm.Sup.Pisa Cl.Sc.IV ser.**10** (1983) 429-436.

5.D.Mumford: "Prym varieties I" Contribution to Analysis, New york, Academic press 1974.

6.M.Raynaud: "Sous variétés d'une varieté abélienne et points de torsion" on "Arithmetic and Geometry" ed. by J.Coates and S.Helgason Progr. in Math. **35** Birkhauser 1983.

A.ALZATI Dip. di Mat. Università di Milano via C. Saldini 50 20133 MILANO.
E-mail address: ALZATI@VMIMAT.MAT.UNIMI.IT
G.P.PIROLA Dip. di Mat. Università di Pavia via Strada Nuova 65 27100 Pavia.
E-mail address: PIRO23@IPVIAN.BITNET

Contemporary Mathematics
Volume **162**, 1994

# LEFSCHETZ TYPE RESULTS FOR PROPER INTERSECTIONS

LUCIAN BĂDESCU

ABSTRACT. We prove two theorems of Lefschetz type for the pair $(X, f^{-1}(\Delta))$, where $f : X \to \mathbb{P}^n \times \mathbb{P}^n$ is a finite morphism from a smooth projective variety $X$ such that $\dim(X) \geq n+2$ and $f(X)$ and the diagonal $\Delta$ have a proper intersection in $\mathbb{P}^n \times \mathbb{P}^n$. These results involve algebraic invariants of the varieties in question such as the Picard or Abhyankar-Grothendieck fundamental groups.

## INTRODUCTION AND STATEMENTS OF RESULTS

Let us begin by recalling two important generalizations of the classical theorems of Lefschetz on hyperplane sections. In 1970 Barth proved the following result (see [2]).

**Theorem A.** *Let $A$ and $B$ be two smooth projective connected subvarieties of the complex projective space $\mathbb{P}^n_{\mathbb{C}}$ such that $a+b \geq n+r$ and $2a \geq n+s$, where $a = \dim(A)$ and $b = \dim(B)$. Assume that $A$ and $B$ have a proper intersection, i.e. $\dim(A \cap B) = a+b-n$. Then the natural map*

$$H^q(B, \mathbb{Q}) \longrightarrow H^q(A \cap B, \mathbb{Q})$$

*on singular cohomology groups with rational coefficients is an isomorphism for every $q$ such that $q \leq \min\{r-1, s\}$.*

**Note.** Taking $B = \mathbb{P}^n_{\mathbb{C}}$ one gets Barth's famous result asserting that the maps $H^q(\mathbb{P}^n, \mathbb{Q}) \to H^q(A, \mathbb{Q})$ are isomorphisms for every $q \leq s$, provided that $2a \geq n+s$. Later on, Barth and Larsen extended this statement to analogous assertions concerning the homotopy groups or singular homology (or cohomology) groups with coefficients in $\mathbb{Z}$. In particular, if $2a \geq n+2$ one gets (via the exponential sequence) the important fact that the natural map $\mathrm{Pic}(\mathbb{P}^n) \to \mathrm{Pic}(A)$ is an

1991 *Mathematics Subject Classification.* Primary 14C22; Secondary 14E20, 14K30.
This paper is in final form and no version of it will be submitted for publication elsewhere.

isomorphism. Algebraic proofs of Barth's results (including the assertion about the Picard group) have been subsequently given by Ogus (see [18] and [19]).

In 1981 Fulton and Lazarsfeld (see [7]) proved the following theorem by using a deep result conjectured by Deligne and proved by Goresky and MacPherson (whose proof involves the stratified Morse theory).

**Theorem B.** *Let $X$ be a complex projective local complete intersection variety of pure dimension $m$, and let $f : X \to \mathbb{P}^n_{\mathbb{C}} \times \mathbb{P}^n_{\mathbb{C}}$ be a finite morphism. Denote by $\Delta$ the diagonal of $\mathbb{P}^n_{\mathbb{C}} \times \mathbb{P}^n_{\mathbb{C}}$, and assume that $m \geq n+2$. Then $\pi_i(X, f^{-1}(\Delta)) = 0$ for $i = 1$ and for $3 \leq i \leq m-n$, and there is a canonical exact sequence*

$$\pi_2(f^{-1}(\Delta)) \to \pi_2(X) \to \mathbb{Z} \to \pi_1(f^{-1}(\Delta)) \to \pi_1(X) \to 1,$$

*where $\pi_i(X,Y)$ (resp. $\pi_i(X)$) denotes the i-th homotopy group of a pair $(X,Y)$ (resp. of $X$).*

As a consequence of Theorem B, Fulton and Lazarsfeld find:

**Corollary.** *Let $X$ be a complex projective local complete intersection of pure dimension $m$, let $f : X \to \mathbb{P}^n_{\mathbb{C}}$ be a finite morphism, and let $Y \subset \mathbb{P}^n_{\mathbb{C}}$ be a closed local complete intersection of pure codimension $d$. Then the induced map $f_* : \pi_i(X, f^{-1}(Y)) \to \pi_i(\mathbb{P}^n_{\mathbb{C}}, Y)$ is bijective if $i \leq m-d$ and surjective if $i = m-d+1$.*

Theorem B and its corollary generalize many previous results of Barth, Larsen and Ogus. Related results were also proved by Milnor, Lazarsfeld, Newstead, Sommese, and others. We refer the reader to [7] and its references, [21], as well as to the recent paper by Lyubeznik [17]. Now, let k be an arbitrary algebraically closed ground field. Then we have the following connectivity result of Fulton and Hansen [6] (see also [3] for a special case).

**Theorem C.** *Let $f : X \to \mathbb{P}^n \times \mathbb{P}^n$ be a finite morphism from an irreducible variety $X$ over $k$ such that $\dim(X) \geq n+1$. Then $f^{-1}(\Delta)$ is connected (and of positive dimension), where $\Delta$ is the diagonal of $\mathbb{P}^n \times \mathbb{P}^n$.*

If in theorem C we assume in addition that $f(X)$ and $\Delta$ have a proper intersection (i.e. $\dim(f(X) \cap \Delta) = \dim f(X) - n = \dim(X) - n$) then we have the following stronger conclusion involving the $\mathbf{G}_3$ condition of Hironaka and Matsumura (see [14]):

**Theorem D.** *In the hypotheses of theorem C assume in addition that $f(X)$ and $\Delta$ have a proper intersection. Then $f^{-1}(\Delta) \setminus W$ is $\mathbf{G}_3$ in $X \setminus W$ for every closed subscheme $W$ of $f^{-1}(\Delta)$ such that $\dim(W) \leq \dim(f^{-1}(\Delta)) - 2$, i.e. the canonical map $\mathbf{K}(X \setminus W) \to \mathbf{K}(\widehat{X \setminus W}_{/f^{-1}(\Delta)\setminus W})$ from the field of rational functions of*

*$X \setminus W$ to the ring of formal rational functions of $X \setminus W$ along $f^{-1}(\Delta) \setminus W$, is an isomorphism. Moreover, $f^{-1}(\Delta)$ meets every closed subvariety of $X$ of dimension $\geqslant n$.*

**Note.** Theorem D is proved in [1] and implies a result of Faltings [5] which asserts that $A \cap B \setminus W$ is $\mathbf{G}_3$ in $A \setminus W$ and in $B \setminus W$ for every projective irreducible subvarieties $A$ and $B$ of $\mathbb{P}^n$ such that $\dim(A) + \dim(B) \geq n + 1$ and $A$ and $B$ have a proper intersection, and for every closed subscheme $W$ of $A \cap B$ such that $\dim(W) \leq \dim(A \cap B) - 2$. The proof of theorem D given in [1] uses global methods and relies on some results of Hironaka and Matsumura [14], while Faltings' proof of his result requires local methods. Note also that the complex-analytic analogue of Faltings' result (in case when $W$ is empty) was previously known (see [4]).

The aim of this paper is, using algebraic methods, to get results of Lefschetz type for pairs of the form $(X, f^{-1}(\Delta))$, where $f : X \to \mathbb{P}^n \times \mathbb{P}^n$ is a finite morphism from a smooth irreducible variety $X$ such that $\dim(X) \geq n + 2$, $\mathrm{char}(k) = 0$ and $f(X)$ and $\Delta$ have a proper intersection. These results will involve the Picard group and the Abhyankar-Grothendieck fundamental group $\pi_1^{alg}$ (see [10]). They could be considered as complementing theorem D (in the case when $\dim(X) \geq n + 2$), on the one hand, and in some sense as partial algebraic analogues of theorem B, on the other hand.

**Theorem 1.** *Let $f : X \to \mathbb{P}^n \times \mathbb{P}^n$ be a finite morphism from a smooth irreducible variety $X$ over $k$. Assume $\mathrm{char}(k) = 0$, $\dim(X) \geq n + 2$ and $Y := f^{-1}(\Delta)$ normal (where the inverse image of the diagonal is taken in the scheme-theoretical sense). Assume also that $f(X)$ and $\Delta$ have a proper intersection (i.e. $\dim(Y) = \dim(X) - n$). Then the natural maps*

$$\underline{Pic}^0(X) \to \underline{Pic}^0(Y) \text{ and } Alb(Y) \to Alb(X).$$

*of Picard and Albanese varieties are isomorphisms. In particular, $q(X) = q(Y)$, where $q(V) = h^1(V, O_V)$ is the irregularity of a projective variety $V$. If in addition $\dim(X) \geq n + 3$ and $Y$ is smooth in codimension two, then there is a canonical exact sequence*

$$0 \longrightarrow \mathbb{Z} \longrightarrow Pic(X) \longrightarrow Pic(Y) \longrightarrow 0.$$

Combining theorem 1 with some results of Barth and Larsen we get:

**Corollary.** *Let $g : A \to \mathbb{P}^n$ be a finite morphism from a smooth irreducible variety $A$ of dimension $a$ over $k$, and let $B$ be a smooth irreducible closed subvariety of dimension $b$ of $\mathbb{P}^n$. Assume $\mathrm{char}(k) = 0$, $a + b \geq n + 2$ and $2b \geq n + 2$. Assume also that $g(A)$ and $B$ have a proper intersection in $\mathbb{P}^n$, and that $C := g^{-1}(B)$ (in the scheme-theoretical sense) is normal. Then the natural maps*

$$\underline{Pic}^0(A) \to \underline{Pic}^0(C) \text{ and } Alb(C) \to Alb(A).$$

*are isomorphisms. If in addition $a + b \geq n + 3$ and $C$ is smooth in codimension two, then the natural map $Pic(A) \to Pic(C)$ is an isomorphism.*

*Proof of the corollary.* Since $2b \geq n+2$ the results of Barth and Larsen mentioned above (which have also been proved by algebraic methods by Ogus) imply that

$q(B) = 0$ and $\mathrm{Pic}(B) \cong \mathbb{Z}[\mathcal{O}_B(1)]$. Set $X = A \times B$ and $f := g \times i : X \to \mathbb{P}^n \times \mathbb{P}^n$, with i the inclusion of $B$ in $\mathbb{P}^n$. Then $C \cong f^{-1}(\Delta)$, and since $q(B) = 0$, $\mathrm{Pic}(X) = \mathrm{Pic}(A \times B) \cong \mathrm{Pic}(A) \times \mathrm{Pic}(B) \cong \mathrm{Pic}(A) \times \mathbb{Z}$ (by [8], III, (4.6.5) and the base change theorems, or also by [13], p. 292, Ex. 12.6). Then the statements about $\underline{\mathrm{Pic}}^0$ and Pic follow from theorem 1. Since the Albanese variety is the dual of the Picard variety, the statement about Alb also follows.

**Theorem 2.** *Let $f : X \to \mathbb{P}^n \times \mathbb{P}^n$ be a finite morphism from a smooth irreducible variety $X$ over $k$. Assume $char(k) = 0$, $\dim(X) \geq n + 2$, $Y := f^{-1}(\Delta)$ contained in the smooth locus $X_0 = Reg(X)$ of $X$, and that $f(X)$ and $\Delta$ have a proper intersection. Then there is a canonical exact sequence of profinite groups*

$$\hat{\mathbb{Z}} \longrightarrow \pi_1^{alg}(Y) \longrightarrow \pi_1^{alg}(X_0) \longrightarrow 1,$$

*where $\pi_1^{alg}(V)$ is the algebraic fundamental group of an algebraic variety $V$ (see [10]), the second map is induced by the inclusion $Y \subset X_0$, and $\hat{\mathbb{Z}}$ is the profinite completion of $\mathbb{Z}$.*

**Corollary.** *Let $g : A \to \mathbb{P}^n$ be a finite morphism from a smooth irreducible variety $A$ of dimension $a$ over $k$, and let $B$ be a smooth irreducible subvariety of dimension $b$ of $\mathbb{P}^n$. Assume $char(k) = 0$, $a + b \geq n + 2$ and $2b \geq n$. Assume also that $g(A)$ and $B$ have a proper intersection in $\mathbb{P}^n$. Then there is a canonical exact sequence of profinite groups*

$$\hat{\mathbb{Z}} \longrightarrow \pi_1^{alg}(g^{-1}(B)) \longrightarrow \pi_1^{alg}(A) \longrightarrow 1.$$

*Proof of the corollary.* We proceed similarly as in the proof of the corollary of theorem 1. Since $2b \geq n$, $B$ is algebraically simply connected (this fact is already a consequence of theorem C, see [7]). Then use the isomorphism $\pi_1^{alg}(A \times B) \cong \pi_1^{alg}(A) \times \pi_1^{alg}(B)$ (see [10]), and the right side is isomorphic to $\pi_1^{alg}(A)$, because B is algebraically simply connected.

**Acknowledgement.** A preliminary version of this paper was written during my stay at the University Paris VII in the Fall 1991. I want to thank this Institution, and especially the members of "Equipe de Physique Mathématique et Géométrie" for hospitality and stimulating atmosphere, and the CNRS for financial support. I also want to thank the organizers of the L'Aquila Conference for their kind invitation and support.

## Proofs of theorems 1 and 2

The proofs of theorems 1 and 2 rely on the Grothendieck-Lefschetz theory as developed in [11] (see also [12]), and on an idea of Deligne's (see [7]) consisting in using the following elementary and remarkable geometric construction (which was also used in other classical or modern problems, see e. g. [16]). In the projective space $\mathbb{P}^{2n+1}$ of coordinates $(x_0, ..., x_{2n+1})$ consider the linear subspaces

$$L_1 = \{x = (x_0, ..., x_{2n+1}) \in \mathbb{P}^{2n+1} / x_0 = ... = x_n = 0\} \text{ and}$$

$$L_2 = \{x = (x_0, ..., x_{2n+1}) \in \mathbb{P}^{2n+1} / x_{n+1} = ... = x_{2n+1} = 0\},$$

and set $U = \mathbb{P}^{2n+1} \setminus (L_1 \cup L_2)$. Consider the morphism $g : U \to \mathbb{P}^n \times \mathbb{P}^n$ defined by

$$g(x_0, ..., x_{2n+1}) = ((x_0, ..., x_n), (x_{n+1}, ..., x_{2n+1})).$$

Then $g$ is the projection of a locally trivial $\mathbb{G}_m$-bundle (in the Zariski topology), where $\mathbb{G}_m$ is the multiplicative group of $k$. Let $H$ be the linear subspace of $\mathbb{P}^{2n+1}$ defined by the equations

$$x_0 = x_{n+1}, x_1 = x_{n+2}, ..., x_n = x_{2n+1}.$$

Then $H \subset U$ and $g/H$ defines an isomorphism between $H$ and $\Delta$. In particular, if we set $X' := f(X)$, we get the commutative diagram

$$\begin{array}{ccccc} Y' := H \cap U_{X'} & \longrightarrow & U_{X'} := g^{-1}(X') & \longrightarrow & U \\ \downarrow & & \downarrow g_{X'} & & \downarrow g \\ X' \cap \Delta & \longrightarrow & X' & \longrightarrow & \mathbb{P}^n \times \mathbb{P}^n \end{array}$$

in which $g_{X'}$ is the projection of a locally trivial $\mathbb{G}_m$-bundle, the horizontal arrows are closed immersions and the left vertical arrow is an isomorphism. Set $U_X := U_{X'} \times_{X'} X$, $Y := Y' \times_{X' \cap \Delta} f^{-1}(\Delta)$, and denote by $f' : U_X \to U_{X'}$ the morphism induced by $f : X \to X'$ (or else, the first projection of $U_X$). Then we get the commutative square

$$\begin{array}{ccc} Y & \longrightarrow & U_X \\ \downarrow & & \downarrow g_X \\ f^{-1}(\Delta) & \longrightarrow & X \end{array}$$

in which $g_X$ is again the projection of a locally trivial $\mathbb{G}_m$-bundle, the horizontal arrows are closed immersions and the left vertical arrow is an isomorphism. Let $Z'$ be the closure of $U_{X'}$ in $\mathbb{P}^{2n+1}$. Then $Y'$ is the zero locus of the section $s' = (s'_0, ..., s'_n) \in H^0(Z', \mathcal{O}_{Z'}(1)^{\oplus(n+1)})$, with $s'_i = (x_i - x_{n+i+1})/Z'$, $i = 0, 1, ..., n$. Since by the hypotheses of theorems 1 and 2, $X'$ and $\Delta$ have a proper intersection and since $Y' \cong X' \cap \Delta$, we get $\operatorname{codim}_{Z'}(Y') = \operatorname{codim}_{U_{X'}}(Y') = \operatorname{codim}_{X'}(X' \cap \Delta) + 1 = n + 1$. In other words, $Y'$ is a complete intersection of $Z'$ with the $n + 1$ hyperplanes of $\mathbb{P}^{2n+1}$ given by $x_i = x_{n+i+1}$, $i = 0, 1, ..., n$.

**Lemma 1.** *In the above notations, the weak Grothendieck-Lefschetz condition $Lef(U_X, Y)$ holds for the pair $(U_X, Y)$, i.e. for every open subset $V$ of $U_X$ such that $Y \subset V$ and for every locally free sheaf $F$ on $V$ of finite rank, there is an open subset $V'$ with $Y \subset V' \subseteq V$ such that the natural map $H^0(V', F) \to H^0(\hat{U}_{X/Y}, \hat{F})$ is an isomorphism, where $\hat{U}_{X/Y}$ is the formal completion of $U_X$ along $Y$.*

*Proof of lemma 1.* We shall give two proofs.
**First proof.** Since $f : X \to X'$ is finite and surjective, the morphism $f' : U_X \to$

$U_{X'}$ is also finite and surjective. In particular, it makes sense to speak about the finite extension $\mathbf{K}(Z') = \mathbf{K}(U_{X'}) \to \mathbf{K}(U_X)$ of fields of rational functions. Let then $Z$ be the normalization of $Z'$ in $\mathbf{K}(U_X)$, and set $\mathcal{O}_Z(1) := u^*(\mathcal{O}_{Z'}(1))$, where $u : Z \to Z'$ is the canonical morphism. Since $u$ is finite $\mathcal{O}_{Z'}(1)$ ample, $\mathcal{O}_Z(1)$ is also ample. On the other hand, in both theorems 1 and 2, $f^{-1}(\Delta)$ is contained in $X_0 = \text{Reg}(X)$ (the regular locus of X). Because the morphism $g_X$ is the projection of a $\mathbb{G}_m$-bundle and $Y \cong f^{-1}(\Delta)$ (via $g_X$), it follows that $Y \subset \text{Reg}(U_X)$. Therefore $\text{Reg}(U_X) \subset Z$. In other words, the embeddings $Y \hookrightarrow U_X$ and $Y \hookrightarrow Z$ are equivalent in the sense that both $U_X$ and $Z$ contain a common open neighbourhood of $Y$ (namely, $\text{Reg}(U_X)$). Set $s_i = u^*(s_i')$ and $s = (s_0, \ldots, s_n)$. Then $s$ is a global section of the ample vector bundle $\mathcal{O}_Z(1)^{\oplus(n+1)}$ and the zero locus of $s$ coincides with $Y$ (by the first commutative diagram). Then $Z \setminus Y$ is covered by the $n+1$ affine open subsets $D_+(s_0), \ldots, D_+(s_n)$. In particular, computing the cohomology of $Z \setminus Y$ by Čech, we get that $\text{cd}(Z \setminus Y) \leq n$, where $\text{cd}(Z \setminus Y)$ is Hartshorne's cohomological dimension defined by $\text{cd}(Z \setminus Y) = \max\{q \geq 0 / \exists G \in \text{Coh}(Z \setminus Y) \text{ such that } H^q(Z \setminus Y, G) \neq 0\}$ (see e.g. [12]). Recalling that $\dim(Z) = \dim(X) + 1 \geq n + 3$, we infer that $\text{cd}(Z \setminus Y) \leq \dim(Z) - 2$. If $Z$ were smooth then by a result of Hartshorne (see [12], p. 96, (3.4)) we would get that $Lef(Z,Y)$ holds. But in general $Z$ is not smooth. However, $Z$ is normal, $Y \subset \text{Reg}(Z)$, and $Y$ is a local complete intersection in $Z$ (since $Y$ is even a global complete intersection in $Z$). Then by an appropriate change of Hartshorne's arguments (see [1], theorem (1.3)), the inequality $\text{cd}(Z \setminus Y) \leq \dim(Z) - 2$ still implies $Lef(Z,Y)$ in our situation.

**Second proof.** Since $Y'$ is the intersection of $Z'$ with a linear subspace of $\mathbb{P}^{2n+1}$ of codimension $n+1$ in $\mathbb{P}^{2n+1}$ and $\text{codim}_{Z'}(Y') = n+1$, by [14], lemma 4.3, $Y'$ is $\mathbf{G}_3$ in $Z'$, or else, $Y'$ is $\mathbf{G}_3$ in $U_{X'}$. Because $f' : U_X \to U_{X'}$ is proper and surjective, by [14], theorem 2.7, we infer that $Y = f'^{-1}(Y')$ is also $\mathbf{G}_3$ in $U_X$, and hence $Y$ is $\mathbf{G}_3$ in $Z$ because $Y \subset \text{Reg}(U_X) \subseteq Z$.

Since $Y$ is the zero locus of a section of an ample vector bundle of rank $n+1 < \dim(Z)$, it follows that $Y$ meets every hypersurface of $Z$. Then essentially applying a result of Speiser (see [22], theorem 1, or also [12], p. 202, cor. 2.2) we deduce that $Lef(Z,Y)$ holds. In loc. cit. the variety $Z$ was assumed to be smooth. However, Speiser's proof still works, provided $Z$ is smooth along $Y$.

**Lemma 2.** *In the above situation and notations (including those of the proof of lemma 1), for every locally free sheaf $G$ of finite rank on $\hat{Z} := \hat{Z}_Y$, the sheaf $G(m) := G \otimes \widehat{\mathcal{O}_Z(m)}$ is generated by its global sections.*

*Proof of lemma 2.* Lemma 2 is a direct consequence of [12], p.168, prop. 1.3 (which is the key point in proving the effectivity part of $Leff(Z,Y)$). In fact prop. 1.3 of [12] assumes that $Z$ is smooth. Since in our situation $Y \subset \text{Reg}(Z)$ and $\text{char}(k) = 0$, one can replace $Z$ by a resolution of singularities $\tilde{Z}$ of $Z$ such that $\text{Reg}(Z)$ remains unchanged, and then we can apply prop.1.3 of [12] to $(\tilde{Z}, Y)$. However, we do not really need to use the resolution of singularities because, if we look carefully at the proof of prop.1.3 of [12], we see that the hypothesis that $Y \subset \text{Reg}(Z)$ is actually sufficient to apply this result directly to $(Z, Y)$.

**Corollary (of lemmas 1 and 2).** *In the situation of lemma 1 there is a commutative diagram*

$$(*)\qquad \begin{array}{ccc} Y & \longrightarrow & U_X \\ \downarrow & & \downarrow{\scriptstyle g_X} \\ f^{-1}(\Delta) & \longrightarrow & X \end{array}$$

*such that the first vertical map is an isomorphism, $g_X$ is the projection of a locally trivial $\mathbb{G}_m$-bundle (in the Zariski topology), $Y \subset Reg(U_X)$, and the effective Grothendieck-Lefschetz condition $Leff(U_X, Y)$ holds, i.e. $Lef(U_X, Y)$ holds and for every locally free sheaf $G$ of finite rank on $\hat{U}_{X/Y}$ there is an open neighbourhood $V$ of $Y$ in $U_X$ and a locally free sheaf $F$ of finite rank on $V$ with the property that $\hat{F}_{/Y} \cong G$.*

*Proof of the corollary.* The existence of diagram $(*)$ was proved above. Moreover, by lemma 1 we know that $Lef(U_X, Y)$ holds. It remains to prove the effective part of $Leff(U_X, Y)$. We proceed exactly as in the proof theorem 1.5 of [12], pp.172-173. For the convenience of the reader we repeat the argument here. Let $G$ be a locally free sheaf on $\hat{Z} = \hat{Z}_Y = \hat{U}_Y$ of finite rank. Using lemma 2 twice one gets an exact sequence of the form

$$\widehat{\mathcal{O}_Z(-p)}^{\oplus b} \xrightarrow{\ \phi\ } \widehat{\mathcal{O}_Z(-m)}^{\oplus a} \longrightarrow G \longrightarrow 0,$$

for some positive integers $m$, $p$, $a$ and $b$. Considering the locally free sheaf

$$E = \underline{\mathrm{Hom}}_Z(\mathcal{O}_Z(-p)^{\oplus b}, \mathcal{O}_Z(-m)^{\oplus a}),$$

from $Lef(Z, Y)$ we infer that there is an open neighbourhood $V'$ of $Y$ in $U_X$ such that the natural map $H^0(V', E) \to H^0(\hat{Z}, \hat{E})$ is an isomorphism. This implies that the homomorphism $\phi \in H^0(\hat{Z}, \hat{E})$ comes from a homomorphism

$$\mu \in \mathrm{Hom}_{V'}(\mathcal{O}_{V'}(-p)^{\oplus b}, \mathcal{O}_{V'}(-m)^{\oplus a}) = H^0(V', E),$$

i.e. $\hat{\mu} = \phi$. Then it is clear that $F' := \mathrm{Coker}(\mu)$ is a coherent sheaf on $V'$ such that $\hat{F'} \cong G$. Since $G$ is locally free, $F'$ is also locally free along $Y$. Therefore, shrinking $V'$ if necessary, we get the desired neighbourhood $V$ of $Y$ in $U_X$ such that $F := F'/V$ is locally free and satisfies $\hat{F} \cong G$.

**Note.** Using theorem D and arguments similar to those of the second proof of lemma 1, one can prove that, under the hypotheses of theorems 1 and 2, the weak Grothendieck-Lefschetz condition $Lef(X, f^{-1}(\Delta))$ also holds.

**Lemma 3.** *In the hypotheses of theorem 1 and in the above notations, let $P_X := \mathbf{P}(\mathcal{O}_X(1,0) \oplus \mathcal{O}_X(0,1))$ be the projective bundle associated to the pull-back of the vector bundle $\mathcal{O}(1,0) \oplus \mathcal{O}(0,1)$ on $\mathbb{P}^n \times \mathbb{P}^n$ via the morphism f,and denote by $p_X : P_X \to X$ the canonical projection of $P_X$. Then the variety $U_X$ can be canonically embedded in $P_X$ as an open dense subset such that $p_X$ extends the morphism $g_X$, the complement of $U_X$ in $P_X$ is the union of two* *irreducible effective divisors* $E_1$ *and* $E_2$ *with the property that* $p_X/E_i$ *defines an isomorphism between* $E_i$ *and* $X$, $i = 1, 2$, *and there is a canonical morphism* $h_X : P_X \to Z$ *which is an isomorphism on* $U_X$ *(recall that since* $X$ *is smooth,* $U_X \subset Z$*), and which blows down* $E_1$ *and* $E_2$ *to subvarieties of* $Z$ *of codimension* $\geq 2$.

*Proof.* Consider the blowing up $h : P \to \mathbb{P}^{2n+1}$ of $\mathbb{P}^{2n+1}$ along $L_1 \cup L_2$ (where $L_1$ and $L_2$ are the linear subspaces considered above). Then it is well known (and easy to prove, see e.g. [16]) that $P$ is isomorphic to the projective bundle $\mathbf{P}(\mathcal{O}(1,0) \oplus \mathcal{O}(0,1))$, $g \circ h$ is a morphism defined everywhere on $P$ which coincides to the canonical projection $p$ of $\mathbf{P}(\mathcal{O}(1,0) \oplus \mathcal{O}(0,1))$, and $p/L_i'$ defines an isomorphism between $L_i' := h^{-1}(L_i)$ and $\mathbb{P}^n \times \mathbb{P}^n$, $i = 1, 2$. Consider the cartesian diagram

$$\begin{array}{ccccc} P_X & \longrightarrow & P_{X'} & \longrightarrow & P \\ {\scriptstyle p_X}\downarrow & & \downarrow{\scriptstyle p_{X'}} & & \downarrow{\scriptstyle p} \\ X & \longrightarrow & X' & \longrightarrow & \mathbb{P}^n \times \mathbb{P}^n \end{array}$$

The existence of the open immersion $U_X \hookrightarrow P_X$ comes from the fact that $U_X$ is the pull-back of $U = \mathbb{P}^{2n+1} \setminus (L_1 \cup L_2)$, while the existence of $U_X \hookrightarrow Z$ from the fact that $Z$ is the normalization of $Z'$ in $\mathbf{K}(U_X)$ and from the smoothness of $U_X$ (recall that the smoothness of $X$ implies the smoothness of $U_X$ because $g_X : U_X \to X$ is a $\mathbb{G}_m$-bundle). Set $E_i := L_i' \times_{\mathbb{P}^n \times \mathbb{P}^n} X$, $i = 1, 2$. Then it is clear that $E_1 \cup E_2$ is the complement of $U_X$ in $P_X$. Since $p/L_i'$ defines an isomorphism between $L_i'$ and $\mathbb{P}^n \times \mathbb{P}^n$, it follows that $p_X/E_i$ defines an isomorphism between $E_i$ and $X$, $i = 1, 2$. In particular, $E_1$ and $E_2$ are irreducible effective divisors whose intersection is empty. Since $P_{X'}$ and $Z'$ are both irreducible and contain $U_{X'} = U \cap Z'$ as an open subset, we infer that $h(P_{X'}) = Z'$. Recalling that $P_X$ is smooth (as a $\mathbb{P}^1$-bundle over the smooth variety $X$) and the definition of $Z$, the equality $h(P_{X'}) = Z'$ implies that $P_X$ dominates $Z$ (and this yields the morphism $h_X$). Finally, it remains to prove that $h_X(E_i)$ are subvarieties of Z of codimension $\geq 2$, $i = 1, 2$. This follows from $h_X(E_i) \cap Y = \emptyset$ together with the observation that $Y$ is the zero locus of a section of an ample vector bundle on $Z$ whose rank is $< \dim(Z)$.

*Proof of theorem 1.* Because $X$ is smooth and $p_X : P_X \to X$ is a $\mathbb{P}^1$-bundle, $P_X$ (and hence also $U_X$) is smooth. Since $Y = f^{-1}(\Delta)$ is the zero locus of a section $s = (s_0, \ldots, s_n) \in H^0(Z, \mathcal{O}_Z(1)^{\oplus(n+1)})$ and $\text{codim}_Z(Y) = n + 1$, the normal bundle $N_{Y,Z}$ of $Y$ in $Z$ is isomorphic to $\mathcal{O}_Y(1)^{\oplus(n+1)} := \mathcal{O}_Z(1)^{\oplus(n+1)}/Y$, and hence $I_Y/I_Y^2 \cong \mathcal{O}_Y(-1)^{\oplus(n+1)}$, where $I_Y$ denotes the sheaf of ideals of $Y$ in $Z$. Therefore for every $m \geq 1$ we get:

$$(1) \qquad I_Y^m/I_Y^{m+1} \cong \mathbf{S}^m(I_Y/I_Y^2) \cong \mathbf{S}^m(\mathcal{O}_Y(-1)^{\oplus(n+1)}),$$

where $\mathbf{S}^m$ denotes the $m$-th symmetric power. In particular, for every $m \geq 1$, $I_Y^m/I_Y^{m+1}$ is a direct sum of line bundles of the form $\mathcal{O}_Y(-j)$ with $j > 0$. Recalling that $Y = f^{-1}(\Delta)$ is normal of dimension $\geq 2$ and $\text{char}(k) = 0$ by hypotheses, a well known theorem of Kodaira-Mumford implies that for all $m \geq 1$ one has

$$(2) \qquad H^1(Y, I_Y^m/I_Y^{m+1}) = 0.$$

According to lemma 3 we get that the map $a : \mathrm{Pic}(P_X) \to \mathrm{Pic}(\hat{U}_{X/Y}) = \mathrm{Pic}(\hat{Z}_{/Y})$ factors as

$$\mathrm{Pic}(P_X) \xrightarrow{\ b\ } \mathrm{Pic}(U_X) \xrightarrow{\ c\ } \mathrm{Pic}(\hat{U}_{X/Y}),$$

with all the arrows the natural ones. By the corollary of lemmas 1 and 2 one gets that the effective Grothendieck-Lefschetz condition $Leff(U_X, Y)$ holds. Taking into account that $U_X$ is smooth and Y meets every divisor of $U_X$ (in fact, $Y$ meets every hypersurface of $Z$), then $Leff(U_X, Y)$ implies that the map $c$ is an isomorphism. Using lemma 3 one gets an exact sequence of the form

$$0 \longrightarrow \mathbb{Z}E_1 \oplus \mathbb{Z}E_2 \longrightarrow \mathrm{Pic}(P_X) \xrightarrow{\ b\ } \mathrm{Pic}(U_X) \longrightarrow 0.$$

The injectivity of the second arrow comes from the projectivity of $P_X$ and the fact that the morphism $h_X$ blows $E_1$ and $E_2$ down to subvarieties of Z of codimension $\geq 2$. Recalling that the map $c$ is an isomorphism, the above exact sequence translates into

$$(3) \quad 0 \longrightarrow \mathbb{Z}E_1 \oplus \mathbb{Z}E_2 \longrightarrow \mathrm{Pic}(P_X) \xrightarrow{\ a\ } \mathrm{Pic}(\hat{U}_{X/Y}) \longrightarrow 0.$$

The exact sequence (3) yields in particular the isomorphism

$$(4) \qquad \underline{\mathrm{Pic}}^0(P_X) \cong \underline{\mathrm{Pic}}^0(\hat{U}_{X/Y}).$$

For all $m \geq 1$ consider the infinitesimal neighbourhood $Y_m = (Y, \mathcal{O}_{U_X}/I_Y^m)$ of order $m$ of Y in $U_X$ and the standard exact sequence

$$H^1(Y, I_Y^m/I_Y^{m+1}) \to \mathrm{Pic}(Y_{m+1}) \to \mathrm{Pic}(Y_m) \to H^2(Y, I_Y^m/I_Y^{m+1}).$$

By (1) the first space is zero. Since $\mathrm{char}(k) = 0$, by Lefschetz's principle we may assume that $k$ is the field $\mathbb{C}$ of complex numbers. Then $NS(Y_m) := \mathrm{Pic}(Y_m)/\underline{\mathrm{Pic}}^0(Y_m) \subseteq H^2(Y, \mathbb{Z})$ for every $m \geq 1$. (Alternatively, instead of using Lefschetz's principle we could use analogous facts in terms of étale cohomology.) This implies that for every $m \geq 1$ the maps $NS(Y_{m+1}) \to NS(Y_m)$ are injective. Therefore one has the exact sequence of algebraic groups

$$(5) \quad 0 \longrightarrow \underline{\mathrm{Pic}}^0(Y_{m+1}) \xrightarrow{\ a_m\ } \underline{\mathrm{Pic}}^0(Y_m) \xrightarrow{\ b_m\ } H_m = H^2(Y, I_Y^m/I_Y^{m+1}),$$

with $H_m$ a linear algebraic group. By a result of Grothendieck [9], the normality of $Y = Y_1$ implies that $\underline{\mathrm{Pic}}^0(Y_1)$ is an abelian variety. Since all the maps $a_m$ are injective we infer that $\underline{\mathrm{Pic}}^0(Y_m)$ is an abelian variety for all $m \geq 1$. Therefore the maps $b_m$ are all zero because $H_m$ is a linear algebraic group. This implies that the map $a_m$ is an isomorphism for every $m \geq 1$, whence

$$(6) \qquad \underline{\mathrm{Pic}}^0(\hat{U}_{X/Y}) \cong \operatorname*{proj\,lim}_m \underline{\mathrm{Pic}}^0(Y_m) \cong \underline{\mathrm{Pic}}^0(Y).$$

From (4) and (6) we get that the natural map $\underline{\mathrm{Pic}}^0(P_X) \to \underline{\mathrm{Pic}}^0(Y)$ is an isomorphism. Finally, since $P_X$ is a $\mathbb{P}^1$-bundle over $X$ we get the first statement of theorem 1. The second statement follows from the first one because the Albanese variety is the dual of the Picard variety (in the sense of the theory of abelian varieties). The equality $q(X) = q(Y)$ is a consequence of the first assertion and the fact that in characteristic zero the irregularity is the dimension of the Picard variety (see [9]).

Assume now that $\dim(X) \geq n+3$ and $\dim(\mathrm{Sing}(Y)) \leq \dim(Y) - 3$, i.e. $\dim(Y) \geq 3$ and $Y$ is smooth in codimension two. Notice that $Y$ is a Cohen-Macaulay variety because Y is a complete intersection in $U_X$ and $U_X$ is smooth. Since $Y$ is a Cohen-Macaulay projective variety which is smooth in codimension two, and $I_Y^m/I_Y^{m+1}$ is a direct sum of line bundles of the form $\mathcal{O}_Y(-j)$ with $j > 0$, by a generalization of the Kodaira vanishing theorem (see [21], 7.80) for every $m \geq 1$ we get

$$H^2(Y, I_Y^m/I_Y^{m+1}) = 0.$$

Having these vanishings, the above arguments (and especially (5)) show that the natural maps $\mathrm{Pic}(U_X) \to \mathrm{Pic}(\hat{U}_X/Y) \to \mathrm{Pic}(Y)$ are both isomorphisms. Now look at the commutative diagram

$$\begin{array}{ccccccccc}
 & & 0 & & 0 & & & & \\
 & & \downarrow & & \downarrow & & & & \\
 & & \mathbb{Z}E_1 & \xrightarrow{id} & \mathbb{Z}E_1 & & & & \\
 & & \downarrow & & \downarrow & & & & \\
0 & \longrightarrow & \mathbb{Z}E_1 \oplus \mathbb{Z}E_2 & \longrightarrow & \mathrm{Pic}(P_X) & \longrightarrow & \mathrm{Pic}(Y) & \longrightarrow & 0 \\
 & & \downarrow & & \downarrow & & \downarrow{\scriptstyle id} & & \\
0 & \longrightarrow & \mathbb{Z}E_2 & \longrightarrow & \mathrm{Pic}(P_X \setminus E_1) & \longrightarrow & \mathrm{Pic}(Y) & \longrightarrow & 0 \\
 & & \downarrow & & \downarrow & & & & \\
 & & 0 & & 0 & & & &
\end{array}$$

By (3) and the isomorphism $\mathrm{Pic}(\hat{U}_{X,Y}) \cong \mathrm{Pic}(Y)$, the middle row is exact. Since the first two columns are obviously exact, we infer that the third row is also exact. Finally, since $E_1$ is a section of $p_X$ we have an isomorphism $\mathrm{Pic}(P_X) \cong \mathrm{Pic}(X)$. In other words, we get the exact sequence

$$0 \longrightarrow \mathbb{Z}E_2 \longrightarrow \mathrm{Pic}(X) \longrightarrow \mathrm{Pic}(Y) \longrightarrow 0,$$

which completes the proof of theorem 1.

*Proof of theorem 2.* In the diagram $(*)$ the morphism $g_X : U_X \to X$ is a locally trivial $\mathbb{G}_m$-bundle. Restricting it to $X_0$ we get the locally trivial $\mathbb{G}_m$-bundle $g_0 : U_0 := U_{X_0} \to X_0$. Then consider the homotopy sequence of $g_0$ in the algebraic version given by Grothendieck (see [10], XIII, Exemple 4.4 and Prop. 4.1)

$$\text{(7)} \qquad \pi_1^{alg}(\mathbb{G}_m) \longrightarrow \pi_1^{alg}(U_0) \longrightarrow \pi_1^{alg}(X_0) \longrightarrow 1,$$

in which the first map is induced by the inclusion $\mathbb{G}_m \hookrightarrow U_0$ (with $\mathbb{G}_m$ regarded as a fibre of $g_0$), and the second map by $g_0$. Since $X_0$ is smooth, $U_0$ is also smooth. By the corollary of lemmas 1 and 2 we get that the effective Grothendieck-Lefschetz condition $Leff(U_0, Y)$ holds. Then proceeding exactly as in the proof of theorem 2.1, p. 175 of [12], we infer that $\pi_1^{alg}(U_0) \cong \pi_1^{alg}(Y)$. Therefore the exact sequence (7) becomes

$$\pi_1^{alg}(\mathbb{G}_m) \longrightarrow \pi_1^{alg}(Y) \longrightarrow \pi_1^{alg}(X_0) \longrightarrow 1.$$

Finally, since $\pi_1^{alg}(\mathbb{G}_m) \cong \hat{\mathbb{Z}}$ we get the result.

## REFERENCES


1. L. Bădescu and E. Ballico, *Formal rational functions along zero loci of sections of ample vector bundles*, Rev. Roumaine Math. Pures Appl. **38** (1993) (to appear).
2. W. Barth, *Transplanting cohomology classes in complex projective space*, Amer. J. Math. **92** (1970), 951-967.
3. ———, *Verallgemeinerung des Bertinischen Theorems in Abelschen Mannigfaltigkeiten*, Annali Sc. Norm. Sup. Pisa **23**, 317-330.
4. W. -L. Chow, *On meromorphic maps of algebraic varieties*, Annals of Math. **89** (1969), 391-403.
5. G. Faltings, *Some theorems about formal functions*, Publ. Res. Inst. Math. Sci. Kyoto Univ. **16** (1980), 721-732.
6. W. Fulton and J. Hansen, *A connectedness theorem for projective varieties, with applications to intersections and singularities of mappings*, Annals of Math. **110** (1979), 159-166.
7. W. Fulton and R. Lazarsfeld, *Connectivity and its applications in algebraic geometry*, Lecture Notes in Math., vol. 862, Springer Verlag, Berlin-Heidelberg-New York, 1981, pp. 26-92.
8. A. Grothendieck and J. Dieudonné, *Eléments de Géométrie Algébrique III*, vol. 8, Paris, 1961.
9. A. Grothendieck, *Fondements de la Géométrie Algébrique (éxposés no. 232 and 236)*, Extraits du Séminaire Bourbaki 1957-1962, Paris, 1961/62.
10. ———, *Revêtements étales et groupe fondamental*, Lecture Notes Math., vol. 224, Springer Verlag, Berlin-Heidelberg-New York, 1971.
11. ———, *Cohomologie locale des faisceaux cohérents et théorèmes de Lefschetz locaux et globaux*, North Holland, Amsterdam, 1968.
12. R. Hartshorne, *Ample subvarieties of algebraic varieties*, Lecture Notes in Math., vol. 156, Springer Verlag, Berlin-Heidelberg-New York, 1970.
13. ———, *Algebraic Geometry*, Springer Verlag, Berlin-Heidelberg-New York, 1977.
14. H. Hironaka and H. Matsumura, *Formal functions and formal embeddings*, J. Math. Soc. Japan **20** (1968), 52-82.
15. M. Larsen, *On the topology of complex projective manifolds*, Invent. Math. **19** (1973), 251-260.
16. A. T. Lascu and D. B. Scott, *An algebraic correspondence with applications to blowing up Chern classes*, Annali Mat. Pura Appl. **102** (1975), 1-36.
17. G. Lyubeznik, *Etale cohomological dimension and the topology of algebraic varieties*, Annals of Math. (to appear).
18. A. Ogus, *Local cohomological dimension of algebraic varieties*, Annals of Math. **98** (1973), 327-365.
19. ———, *On the formal neighbourhood of a subvariety of a projective space*, Amer. J. Math. **97** (1975), 1085-1107.
20. B. Shiffman and A. J. Sommese, *Vanishing theorems on complex manifolds*, Progress in Math. No. 56, Birkhauser, Boston Basel Stuttgart, 1985.
21. A. J. Sommese, *Submanifolds of abelian varieties*, Math. Ann. **233** (1978), 229-256.

22. R. Speiser, *Cohomological dimension of abelian varieties*, Amer. J. Math. **95** (1973), 1-34.

INSTITUTE OF MATHEMATICS OF THE ROMANIAN ACADEMY, P.O. BOX 1-764, RO-70700 BUCHAREST, ROMANIA

*E-mail address*: lbadescu@ roearn.bitnet

Contemporary Mathematics
Volume **162**, 1994

# SOME EFFECTS OF THE SPECTRAL VALUES ON REDUCTIONS

Mauro C. Beltrametti and Andrew J. Sommese

**Introduction.** Let $X$ be a connected, $n$-dimensional algebraic submanifold of a complex projective space. Let $L$ denote the restriction of the hyperplane section bundle to $X$.

In [S1] the second author studied the restriction map $\Gamma(K_X \otimes L) \to \Gamma((K_X \otimes L)_\ell)$ where $\ell \subset X$ is a line and $X$ is a surface. Using the fact that $L-\ell$ is spanned it was shown that if $\ell^2 < 0$ then $K_X \otimes L$ has global sections spanning $(K_X \otimes L)_\ell$. At the time the need to study lines seemed a technical problem special to a certain approach to the adjunction mapping and not intrinsically connected with the study of $K_X \otimes L$.

In [BSS1] it was shown that except for degree $2^n - 1$, no $n$-fold, $X \subset \mathbb{P}^N$ with $2n \geq N+1$ contains any linear $\mathbb{P}^{n-1}$, $\mathcal{P}$, with normal bundle, $\mathcal{N}_{\mathcal{P}}^X = \mathcal{O}_{\mathbb{P}^{n-1}}(-1)$. This simple fact made a very helpful simplification when studying projective 3-folds in $\mathbb{P}^5$: if for such an $X$, with $n = 3$, the degree is not 7 and $K_X + (n-1)L$ nef and big, then $K_X + (n-1)L$ is very ample.

In [BFS1], Theorem (0.3.1), it was shown that if $X$ possessed a linear $\mathbb{P}^{n-1}$, $\mathcal{P}$, then $h^0(K_X + tL) > 0$ for appropriate $t$ depending on $n = \dim X$ and the integer $m$ such that the normal bundle of $\mathcal{P}$ is $\mathcal{O}_{\mathbb{P}^{n-1}}(m)$. This result is used to conclude in certain important cases from certain global information that if $K_X + (n-1)L$ is nef and big, then it is very ample. Moreover if $K_X + (n-2)L$ is nef and big, then the result is used to conclude in certain important cases from certain global information that the image of $X$ under the sections of a high power of $K_X + (n-2)L$ is factorial.

This article grew out of a realization that a version of these results for linear and quadric subvarieties (not necessarily of codimension 1) would have very useful consequences in adjunction theory, and give a means of approaching the spectral value conjectures the authors pose in the appendix to [BSW].

The scheme of this paper is as follows.

In §0, we give some background material connected with the *unnormalized spectral value*. In the case of a very ample line bundle $L$ on a smooth projective manifold $X$, this is defined to be

$$u(X, L) := \sup\{t \in \mathbb{Q} | \kappa(K_X + tL) = -\infty\},$$

---

1991 *Mathematics Classification*. Primary 14C20; Secondary 14J99, 14J40.

The second author's research was partially supported by NSF Grant DMS 89-21702.

This paper is in final form and no version of it will be submitted for publication elsewhere.

where for a $\mathbb{Q}$-Cartier divisor, $D, \kappa(D)$ is the Kodaira dimension of $D$. In particular we show how the spectral value of certain natural $\mathbb{Q}$-Cartier divisors on a projective manifold and the first and the second reduction from adjunction theory are behaved.

In §1, we show in Proposition (1.3) that given a codimension $r$ subvariety, $Y$, of $X$ then $h^0(K_X + qL) > 0$ if $r$, the conormal bundle of $Y$, and $K_{X|Y}$ satisfy appropriate conditions. The important point is that the conclusion is global, but the hypotheses are local to $Y$. This is especially so since the condition of the conormal bundle is by nature a weak positivity condition. Proposition (1.4) is a version of this involving the nef value of the pair. Corollary (1.5) is a version of the result when $Y$ is a linear $\mathbb{P}^{n-r}$.

In §2 somewhat stronger results, Proposition (2.1) and (2.2), are proved in the case when $Y$ is a divisor.

In §3 an application, Proposition (3.1), of these results is made to the first and second reductions of a pair. Some examples and open questions (see (3.3)) are included.

We would both like to thank the University of Notre Dame for making this research possible. We would also like to thank the Max-Planck-Institut für Mathematik in Bonn for its support. This paper was conceived and partially worked out during the author's stay at the Max-Planck-Institut in July, 1991. Finally, we would like to thank Ms. Cinzia Matri for her nice typing.

## §0. Background material.

**(0.1) Notation.** We work over the complex field $\mathbb{C}$. By *variety* we mean an irreducible and reduced projective scheme, $V$. We denote its structure sheaf by $\mathcal{O}_V$.

Basically we use the standard notation from algebraic geometry. We almost always follow the notation of [BS]. We refer to it and to [KMM] for definitions of the following: $\mathbb{Q}$-divisors, $\mathbb{Q}$-Cartier divisors, $\mathbb{Q}$-factorial, numerically effective (nef, for short), big, numerical equivalence (denoted by $\sim$), linear equivalence (denoted by $\approx$) of $\mathbb{Q}$-divisors, intersection of cycles, canonical divisor, terminal and canonical singularities.

We say that a normal variety, $V$, is *e-Gorenstein* if $eK_V$ is a line bundle, where $e$ is a positive integer and $K_V$ is the canonical divisor. The smallest positive integer, $e$, such that $eK_V$ is a line bundle is called the *index* of $V$.

Notice that if $V$ has terminal singularities, then $V$ is *e-Gorenstein*.

Recall also that for a positive integer $m, mK_V$ stands for

$$mK_V := \mathcal{O}_V((K)^{\otimes m})^{**} = \mathcal{O}_V(mK) := K_V^{[m]}$$

where $K$ is the canonical (Weil) divisor of $V$ defined by $\mathcal{O}_{\text{reg}(V)}(K) = \wedge^n \Omega^1_{\text{reg}(V)}$ and reg$(V)$ denotes the nonsingular locus of $V$.

Let $\mathcal{F}$ be a sheaf of $\mathcal{O}_V$-modules on a variety $V$ of dimension $n$. Then
$\chi(\mathcal{F}) := \sum_{i=0}^n (-1)^i h^i(\mathcal{F})$, the Euler characteristic of $\mathcal{F}$, where $h^i(\mathcal{F})$ denotes the complex dimension of $H^i(V, \mathcal{F}), i \geq 0$.

We also say that $\mathcal{F}$ is *spanned* if it is spanned at each point by the space, $\Gamma(\mathcal{F})$, of its global sections.

Let $L$ be a line bundle on a variety $V$ and let $Y$ be a subvariety of $V$. Then
$L_Y :=$ the restriction of $L$ to $Y$;

$\mathcal{J}_Y$ := the ideal sheaf of $\mathcal{O}_V$ defining $Y$ in $V$.
For a reflexive sheaf $\mathcal{L}$ of rank 1 we denote by $|\mathcal{L}|$ the complete linear system associated to $\mathcal{L}$.

For a $\mathbb{Q}$-Cartier divisor $D$ (i.e. some integer multiple of $D$ is an integral Cartier divisor),

$\kappa(D)$ := the Kodaira dimension of $D$ (note that if $D$ is not a Cartier divisor but $ND$ is Cartier for $N > 0$, then $\kappa(D) := \kappa(ND)$).

For a real number, $a$, we denote by $\lceil a \rceil$ the smallest integer $\geq a$.

**(0.2) Nef values.** We state the following result in the case of terminal singularities which occur in the adjunction theory (see e.g. [BFS1], [BS]) even though it holds true in the more general case of log-terminal singularities (see [KMM], 4.1).

**(0.2.1) Kawamata Rationality Theorem.** *Let $V$ be a normal variety of dimension $n$ with terminal singularities and let $e$ be the index of $V$. Let $\psi : V \to S$ be a projective morphism onto a variety $S$. Let $L$ be a $\psi$-ample line bundle on $V$. If $K_V$ is not $\psi$-nef, then*

$$\tau := \min\{t \in \mathbb{R}, K_V + tL \text{ is } \psi\text{-nef}\}$$

*is a positive rational number. Moreover, expressing $e\tau = u/v$ with $u$, $v$ positive coprime integers, we have $u \leq e(\max_{s \in S}\{\dim \psi^{-1}(s)\} + 1)$.*

With the notation and assumptions as in (0.2.1) we say that the rational number $\tau$ is the *$\psi$-nef value* of $(V, L)$. If $S$ is a point, $\tau$ is called the *nef value* of $(V, L)$. Note also that, if $S$ is a point, then $K_V + \tau L$ is nef and hence by the Kawamata-Shokurov base point free theorem ([KMM], §3) we know that for $m \gg 0$, with $m/e$, $m\tau$ integral, $|m(K_V + \tau L)|$ is base point free and defines a morphism $\phi : V \to W$, with $W$ normal and $\phi$ having connected fibers. This morphism, $\phi$, is called the *nef value morphism* of $(V, L)$.

**(0.2.2) Lemma** ([BS], (0.8.3)). *Let $V$ be a normal variety with terminal singularities. Let $L$ be an ample line bundle on $V$. Assume that $K_V$ is not nef. A rational number, $\tau$, is the nef value of $(V, L)$ if and only if $K_V + \tau L$ is nef but not ample.*

**(0.2.3) Lemma.** *Let $V$ be a normal variety with terminal singularities and let $L$ be an ample line bundle on $V$. Assume that $K_V$ is not nef, let $\tau$ be the nef value of $(V, L)$ and let $\phi : V \to W$ be the nef value morphism of $(V, L)$. Let $D := (\phi_* L)^{**}$. Assume that $\phi$ is birational, $W$ is $\mathbb{Q}$-Gorenstein and $D$ is $\mathbb{Q}$-Cartier. Then $K_V + \tau L \approx \phi^*(K_W + \tau D)$.*

**Proof.** There is an algebraic subset $Z \subset W$ of codimension $\geq 2$ such that the restriction $\phi : V - \phi^{-1}(Z) \to W - Z$ is an isomorphism. Denote $U = W - Z$. Let $\tau = p/q$ for positive integers $p, q$ with $e|q$ where $e$ is the index of $V$. We also have $qK_V + pL \approx \phi^* H$ for some ample line bundle $H$ on $W$. The restriction $H_U$ of $H$ to $U$ satisfies the condition $H_U \cong (qK_W + pD)_U \cong (qK_V + pL)_{\phi^{-1}(U)}$. Since $qK_W + pD$ is a rank 1 reflexive sheaf it thus follows that $H \approx qK_W + pD$, and hence $K_V + \tau L \approx \phi^*(K_W + \tau D)$. Q.E.D.

**(0.3) Reductions** ([S3], (0.5), [FS], §0 and [BFS1], §0). Let $\mathcal{M}$ be a smooth connected variety of dimension $n \geq 3$ and let $\mathcal{L}$ be a very ample line bundle on

$\mathcal{M}$. We say that a pair $(M, L)$ with $M$ smooth is a *(first) reduction* of $(\mathcal{M}, \mathcal{L})$ if $L$ is ample and

*i)* there exists a morphism $\pi : \mathcal{M} \to M$ expressing $\mathcal{M}$ as $M$ with a finite set $B$ blown up, $L := (\pi_* \mathcal{L})^{**}$;

*ii)* $\mathcal{L} \approx \pi^* L - [\pi^{-1}(B)]$ (equivalently, $K_{\mathcal{M}} + (n-1)\mathcal{L} \approx \pi^*(K_M + (n-1)L)$).

Except for an explicit list of well understood pairs $(\mathcal{M}, \mathcal{L})$ we can assume (see [S3], [SV] and [FS]):

*a)* $K_{\mathcal{M}} + (n-1)\mathcal{L}$ is spanned and big and $K_M + (n-1)L$ is ample. Note that in this case such a reduction $(M, L)$ is unique up to isomorphism. We will refer to this reduction, $(M, L)$, as *the reduction* of $(\mathcal{M}, \mathcal{L})$.

*b)* $K_M + (n-2)L$ is nef and big.

Then from the Kawamata-Shokurov base point free theorem ([KMM], §3) we know that $|m(K_M + (n-2)L)|$ gives rise to a morphism for some $m > 0$, say $\varphi : M \to X$, and for $m$ large enough we can assume that $\varphi$ has connected fibers and normal image. Thus there is an ample line bundle $\mathcal{K}$ on $X$ such that $\varphi^* \mathcal{K} \approx K_M + (n-2)L$. The pair $(X, \mathcal{K})$ is known as the *second reduction* of $(\mathcal{M}, \mathcal{L})$. Let $\mathcal{D} := (\varphi_* L)^{**}$. Then $\mathcal{D}$ is a 2-Cartier divisor and $\mathcal{K} \approx K_X + (n-2)\mathcal{D}$ (see [BS], (4.4)). The morphism $\varphi$ is very well behaved (in [S2], [Fa], [FS], [BFS1], [BS] and [Fu2]). Let us recall the following results we need.

**(0.3.1) Theorem** ([BFS1], (0.2.2) and [BS], (4.1.5)). *Let $(\mathcal{M}, \mathcal{L})$, $(M, L)$ be as in* (0.3) *with $n \geq 4$ and let $(X, \mathcal{K}), \varphi : M \to X$, be the $2^{nd}$ reduction of $(\mathcal{M}, \mathcal{L})$. Then $\varphi$ is an isomorphism outside of an algebraic subset $Z \subset X$ such that* $\dim Z \leq 1$. *We have:*

a) *Any 1-dimensional component of $Z$ is a smooth curve $C$ such that $C \subset$* reg$(X)$ *and in a neighborhood $\mathcal{U}$ of $C$ the restriction of $\varphi$ to $\varphi^{-1}(\mathcal{U})$ is simply the blow up along $C$. The divisor $D = \varphi^{-1}(C)$ is a $\mathbb{P}^{n-2}$ bundle over $C$ with $L_{\mathbb{P}^{n-2}} \cong \mathcal{O}_{\mathbb{P}^{n-2}}(1)$ and $\mathcal{N}^M_{D|\mathbb{P}^{n-2}} \cong \mathcal{O}_{\mathbb{P}^{n-2}}(-1)$ for each fiber $\mathbb{P}^{n-2}$.*

b) *If $x$ is a 0-dimensional component of $Z$ then $\varphi^{-1}(x)$ is a reduced divisor, $D$, and either*

b1) *$(D, L_D) \cong (\mathbb{P}^{n-1}, \mathcal{O}_{\mathbb{P}^{n-1}}(1))$ and $\mathcal{N}^M_D \cong \mathcal{O}_{\mathbb{P}^{n-2}}(-2)$, or*

b2) *$(D, L_D) \cong (Q, \mathcal{O}_Q(1))$ where $Q$ is an irreducible hyperquadric in $\mathbb{P}^n$ and $\mathcal{N}^M_D \cong \mathcal{O}_Q(1)$.*

**(0.3.2) Proposition** ([BFS1], (0.2.4)). *Let $(\mathcal{M}, \mathcal{L}), (M, L)$ and $(X, \mathcal{K})$ be as in* (0.3) *with $n \geq 3$. Then $X$ has isolated rational terminal singularities. If $n = 3, X$ is 2-Gorenstein, while, for $n \geq 4, X$ is 2 -factorial and it is Gorenstein in even dimensions.*

**(0.3.3) Lemma.** *Let $(\mathcal{M}, \mathcal{L}), (M, L), (X, \mathcal{K}), \mathcal{K} \approx K_X + (n-2)\mathcal{D}$ be as in* (0.3).*Let $B$ be the finite subset of $M$ where the first reduction map, $\pi : \mathcal{M} \to M$, is not an isomorphism and let $Z \subset X$ be the algebraic subset with $\dim Z \leq 1$ where the second reduction map, $\varphi : M \to X$, is not an isomorphism. Then we have*

(0.3.3.1) *If $\pi$ is not isomorphism, no rational curves, $\ell$, on $M$ with $L \cdot \ell = 1$ can meet $B$;*

(0.3.3.2) *If $\varphi$ is not an isomorphism, no rational curves, $\gamma \not\subset Z$, on $X$ with $\mathcal{D} \cdot \gamma = 1$ can meet $Z$.*

**Proof.** See [BSW], (A.1.1), (A.1.3).

**(0.4) (Unnormalized) spectral value.** Let $V$ be a normal projective variety of dimension $n$ and let $D$ be a $\mathbb{Q}$-Cartier divisor on $V$. Let us point out the following general fact.

**(0.4.1)** $\kappa(D) = -\infty$ if and only if $\kappa(mD) = -\infty$ for some positive integer $m$.

To see this we have of course only to show the "if" part. Indeed, $\kappa(mD) = -\infty$ means that $h^0(mND) = 0$ for all integers $N > 0$ such that $mND$ is an integral Cartier divisor. Assume $\kappa(D) \geq 0$. Then $h^0(N'D) > 0$ for some $N' > 0$ such that $N'D$ is an integral Cartier divisor. Hence clearly $h^0(mN'D) > 0$, which contradicts the assumption $\kappa(mD) = -\infty$.

Now let $D$ be a $\mathbb{Q}$-Cartier divisor on $V$ such that $\kappa(D) = n$. Let $e$ be the index of $V$. Let us define $u(V, D) := \sup\{t \in \mathbb{Q}, \kappa(K_V + tD) = -\infty\}$, or

$$u(V, D) := \sup\{t \in \mathbb{Q}, h^0(N(K_V + tD)) = 0 \text{ for all } N > 0 \text{ such that } N(K_V + tD) \text{ is an integral Cartier divisor}\}.$$

We are taking the sup as a real number. It is an open question whether it is rational (see [S3], §1). Note that $u(V, D) < +\infty$ since $\kappa(D) = n$. This real number, $u(V, D)$, is called the *(unnormalized) spectral value* of $(V, D)$. Note also that in the case of a semi-ample and big line bundle $L$, the *(normalized) spectral value*, $\sigma(V, L)$, defined in [S3], §1, is $\sigma(V, L) = n + 1 - u(V, L)$.

**(0.4.2) Remark.** Let $L$ be an ample line bundle on a variety $V$ as in (0.4). According to [Fu2], the invariant $\kappa\varepsilon(V, L) := -\inf\{t \in \mathbb{Q}, \kappa(K_V + tL) \geq 0\}$ is called *Kodaira energy* of $(V, L)$. Note that $u(V, L) = -\kappa\varepsilon(V, L) = n+1-\sigma(V, L)$.

**(0.4.3) Lemma.** *Let $L$ be an ample line bundle on a normal variety $V$ with terminal singularities. Assume that $K_V$ is not nef and let $\tau, u = u(V, L)$ be the nef value and the unnormalized spectral value of $(V, L)$ respectively. Then $u \leq \tau$ with equality if and only if the nef value morphism, $\phi$, of $(V, L)$ is not birational.*

**Proof.** The inequality $u \leq \tau$ is obvious since $h^0(N(K_V + \tau L)) > 0$ for some $N > 0$ with $N\tau$ integral. Assume $u = \tau$. If $\phi$ is birational then we have a decomposition, for $N >> 0$, with $N\tau$ integer, $N(K_V + \tau L) = A + E$ for some Cartier divisors $A, E$ such that $A$ is ample and $E$ is effective. Therefore for any rational number $\varepsilon, 0 < \varepsilon << 1$, one has $N(K_V + (\tau - \varepsilon)L) = A - \varepsilon NL + E$ where $A - \varepsilon NL$ is still ample. Thus $h^0(m(K_V + (\tau - \varepsilon)L) > 0$ for $m >> 0$ and hence $u < \tau - \varepsilon < \tau$, which is a contradiction.

To show the converse, assume $\phi$ is not birational. Note that a general fiber, $F$, of $\phi$ is a local complete intersection and has terminal singularities. Then $K_F + \tau L_F \sim \mathcal{O}_F$. If $\tau > u$, there exist positive coprime integers $p, q$ such that $\tau > p/q > u$ and $-(qK_F + pL_F)$ is ample. Therefore $h^0(N(K_F + (p/q)L_F) = 0$ for all $N > 0$ with $N/q$ integer. This implies that $p/q \leq u$, which contradicts $\tau > p/q > u$. Q.E.D.

**(0.4.4) Lemma.** *Let $L$ be an ample line bundle on a normal variety $V$ with terminal singularities. Assume that $K_V$ is not nef and let $\tau, \phi : V \to W$ be the nef value and the nef value morphism of $(V, L)$ respectively. If $\phi$ is birational, $W$ is $\mathbb{Q}$-Gorenstein and $D := (\phi_* L)^{**}$ is $\mathbb{Q}$-Cartier, then $u(W, D) \leq u(V, L)$.*

*If there exist effective divisors $\{E_i\}_i$ in $V$ such that either*

**) $K_V \approx \phi^* K_W + \Sigma a_i E_i$ for rational $a_i \geq 0$, or*

***) $L \approx \phi^* D - \Sigma_i b_i E_i$ for rational $b_i \geq 0$,* *then* $u(W, D) = u(V, L)$.

**Proof.** There is an algebraic subset $Z \subset W$ of codimension $\geq 2$ such that the restriction $\phi : V - \phi^{-1}(Z) \to W - Z$ is an isomorphism. Fix a (positive) rational number $\lambda > u(V, L)$. Choose an integer $N >> 0$ such that

*i)* $N(K_V + \lambda L)$ is Cartier and has non-trivial sections, $NK_V, N\lambda L$ are Cartier;

*ii)* $NK_W, N\lambda D$ are Cartier divisors.

Let $s$ be a non-trivial section of $N(K_V + \lambda L)$ as in *i*). Then, since $V - \phi^{-1}(Z) \cong W - Z$, the restriction of $s$ to $V - \phi^{-1}(Z)$ gives rise to a non-trivial $s' \in H^0(W - Z, N(K_W + \lambda D))$. Since $W$ is normal, by Levi's Extension Theorem $s'$ extends to a global section of $N(K_W + \lambda D)$. Therefore we have $h^0(N(K_V + \lambda L)) \leq h^0(N(K_W + \lambda D))$ and hence $u(V, L) \geq u(W, D)$.

If the equality doesn't hold, we have $u(V, L) > u(W, D)$ and there exists a rational number $\lambda$ such that $u(V, L) > \lambda > u(W, D)$. Therefore $h^0(m(K_V + \lambda L)) = 0$ for all $m > 0$ such that $m(K_V + \lambda L)$ is an integral Cartier divisor and $h^0(m'(K_W + \lambda D)) > 0$ for some $m'$ with $m'(K_W + \lambda D)$ an integral Cartier divisor.

From Lemma (0.2.3) we have $K_V + \tau L \approx \phi^*(K_W + \tau D)$. Thus the assumptions *) and **) are equivalent. Note that

$$\begin{aligned} K_V + \lambda L &= K_V + \tau L + (\lambda - \tau)L \approx \phi^*(K_W + \tau D) + (\lambda - \tau)L \\ &\approx \phi^*(K_W + \tau D) + (\lambda - \tau)(\phi^* D - \Sigma_i b_i E_i) \\ &\approx \phi^*(K_W + \lambda D) + (\tau - \lambda)\Sigma_i b_i E_i, \end{aligned}$$

where the last linear equivalence follows from **). Note also that $\tau - \lambda > 0$ since $\lambda < u(V, L) \leq \tau$ by Lemma (0.4.3). Therefore $h^0(m'(K_W + \lambda D)) > 0$ implies that $h^0(m(K_V + \lambda L)) > 0$ for any multiple $m$ of $m'$ such that $m(\tau - \lambda)b_i$ is integer for all $i$. This contradicts the assertion $h^0(m(K_V + \lambda L)) = 0$ in the paragraph above. Q.E.D.

Let us recall the following standard fact.

**(0.4.5) Lemma.** *Let* $\psi : V \to W$ *be a birational morphism of normal varieties,* $V$, $W$, *with canonical singularities. Let* $e_V, e_W$ *be the indices of* $V, W$ *respectively and let* $m$ *be a positive integer such that* $e_V, e_W$ *divide* $m$. *Then* $\psi_*(mK_V) \cong mK_W$.

**Proof.** Let $\sigma : V^\wedge \to V$ be a desingularization of $V$. Since $V$ has canonical singularities we have (see e.g. [R]) that

$$\begin{array}{ccc} \sigma_*(K_{V^\wedge}^{\otimes e_V N}) & \subset & K_V^{[e_V N]} \\ \beta \uparrow & & \uparrow \alpha \\ \sigma_*(K_{V^\wedge}^{\otimes e_V})^{\otimes N} & = & (K_V^{[e_V]})^{\otimes N} \end{array}$$

Note that $(K_V^{[e_V]})^{\otimes N}$ and $K_V^{[e_V N]}$ are invertible and coincide on $\mathrm{reg}(V)$. Since $V - \mathrm{reg}(V)$ has codimension $\geq 2$ in $V$, we conclude that $\alpha$ is an isomorphism. Then all the morphisms in the diagram above are isomorphisms and hence in particular

$$\sigma_*(e_V N K_{V^\wedge}) = \sigma_*(K_{V^\wedge}^{\otimes e_V N}) \cong K_V^{[e_V N]} = e_V N K_V.$$

Let now $\rho := \psi \circ \sigma : V^\wedge \to W$. Then $\rho$ is a desingularization of $W$. By assumption we have $m = e_V a = e_W b$ for some positive integers $a$, $b$. Thus by the above we have

$$mK_W \cong \rho_*(e_W bK_{V^\wedge}) = \psi_*(\sigma_*(e_V aK_{V^\wedge})) \cong \psi_*(e_V aK_V) = \psi_*(mK_V).$$

Q.E.D.

**(0.4.6) Theorem.** *Let $L$ be an ample line bundle on a projective variety, $V$. Assume that $V$ has terminal singularities with $K_V$ not nef and let $\tau, \phi : V \to W$ be the nef value and the nef value morphism respectively of $(V, L)$. Assume that $\phi$ is birational, $W$ has terminal singularities, $D := (\phi_* L)^{**}$ is $\mathbb{Q}$-Cartier. Then $\mathcal{H} := K_W + \tau D$ is $\mathbb{Q}$-Cartier and*

$$u(W, \mathcal{H}) = \frac{u(V, L)}{\tau - u(V, L)}, \quad or \quad u(V, L) = \frac{u(W, \mathcal{H})\tau}{1 + u(W, \mathcal{H})}.$$

**Proof.** Let $\tau = p/q$ for positive integers $p, q$ with $e_V | q$. Note that $qK_V + pL \approx \phi^* H$ for some ample line bundle $H$ on $W$ by the Kawamata-Shokurov Base Point Free Theorem. Note also that $H \approx qK_W + pD := q\mathcal{H}$ by the proof of (0.2.3). Thus $\mathcal{H}$ is $\mathbb{Q}$-Cartier.

Choose integers $a, b$ such that $b/a < u(W, \mathcal{H})$. For any positive integer $N$ chosen so that $NaK_W + Nb\mathcal{H}$ is Cartier, $h^0(N(aK_W + b\mathcal{H})) = 0$ since $b/a < u(W, \mathcal{H})$. Thus one has $h^0(N(a + b)qK_W + NpbD) = 0$. By Levi's Extension Theorem (compare with the proof of (0.4.4)),

$$h^0(N(a + b)qK_V + NpbL) \le h^0(N(a + b)qK_W + NpbD) = 0.$$

Then we conclude that $pb/q(a + b) \le u(V, L)$, i.e. $\frac{pb/qa}{1+b/a} \le u(V, L)$, or $\frac{\tau b/a}{1+b/a} \le u(V, L)$. In the limit as $b/a \to u(W, \mathcal{H})$ we get $\frac{\tau u(W,\mathcal{H})}{1+u(W,\mathcal{H})} \le u(V, L)$. We can assume without loss of generality that we don't have equality. Chose integers $a, b$ such that

$$\frac{\tau u(W, \mathcal{H})}{1 + u(W, \mathcal{H})} < \frac{b}{a} < u(V, L).$$

Now consider any integer $N > 0$ such that $NK_V, NK_W$ are Cartier, i.e. $e_V, e_W$ divide $N$. By the projection formula and Lemma (0.4.5) we have

$$\begin{aligned}\phi_*(N(pa - bq)K_V + Nb(qK_V + pL)) &\cong \phi_*(N(pa - bq)K_V + Nb\phi^* H) \\ &\cong N(pa - bq)K_W + NbH.\end{aligned}$$

Thus, recalling that $H \approx q\mathcal{H}$,

$$\begin{aligned}0 = &h^0(Np(aK_V + bL)) = h^0(N(pa - bq)K_V + Nb(qK_V + pL)) = \\ &h^0(\phi_*(N(pa - bq)K_V + Nb(qK_V + pL))) = h^0(N(pa - bq)K_W + Nbq\mathcal{H}).\end{aligned}$$

Therefore $\frac{b/a}{\tau - b/a} = \frac{bq}{pa - bq} \le u(W, \mathcal{H})$. Hence in the limit as $b/a \to u(V, L)$ we get

$$\frac{u(V, L)}{\tau - u(V, L)} \le u(W, \mathcal{H}).$$

Note that, since $\phi$ is birational, $u(V, L) < \tau$ by Lemma (0.4.3). Thus we can rewrite the inequality above as

$$u(V,L) \leq \frac{\tau u(W,\mathcal{H})}{1+u(W,\mathcal{H})}.$$

This proves the Theorem. Q.E.D.

The following is an easy consequence of (0.4.4), (0.4.6) (compare with [S4], (0.3.1), [BFS1], (0.2.7) and [F2], (1.11), (2.7)).

**(0.4.7) Corollary.** *Let $\mathcal{M}$ be a smooth projective variety of dimension $n \geq 3$ and let $\mathcal{L}$ be a very ample line bundle on $\mathcal{M}$. If a first reduction, $(M, L)$, of $(\mathcal{M}, \mathcal{L})$ exists ,then*

i) $u(\mathcal{M},\mathcal{L}) = u(M,L)$.

*If the second reduction, $(X,\mathcal{K}), \mathcal{K} \approx K_X + (n-2)\mathcal{D}$, of $(\mathcal{M},\mathcal{L})$ exists, then*

ii) $u(\mathcal{M},\mathcal{L}) = u(M,L) = u(X,\mathcal{D})$.

*Furthermore*

iii) $u(X,\mathcal{D}) = \frac{u(X,\mathcal{K})(n-2)}{1+u(X,\mathcal{K})}$.

**Proof.** From (0.4.4) we have i), ii). From (0.4.5) we have $u(M,L) = \frac{u(X,\mathcal{K})(n-2)}{1+u(X,\mathcal{K})}$. Thus iii) follows from ii). Q.E.D.

## §1. The general case.

First, let us fix the following

**(1.0) Assumptions.** Let $X$ be a normal projective $n$-dimensional variety, $n \geq 2$, with isolated irrational singularities and let $L$ be a very ample line bundle on $X$. Let $Y$ be a smooth subvariety of $X$ of dimension $n-r$ and such that $Y \subset \mathrm{reg}(X)$. Let $d = L^{n-r} \cdot Y = L_Y^{n-r}$ be the *degree* of $Y$.

We need two preparatory Lemmas. The first one is a very general fact (the assumptions in (1.0) are not needed in (1.1)).

**(1.1) Lemma (Mumford [Mum]).** *Let $L$ be a very ample line bundle on an irreducible projective variety, $X$. Let $Y \subset X$ be an irreducible subvariety of degree $d$ relative to $L$. If either*

(1.1.1) *$Y$ is a smooth, or*

(1.1.2) *$Y \subset \mathrm{reg}(X)$ and $\mathrm{cod}_X Y = 1$*

*then $L^d \otimes \mathcal{J}_Y$ is spanned by global sections.*

**Proof.** First assume that $Y$ is smooth. We have $X$ embedded in $\mathbb{P}^N$ by $\Gamma(L)$ with $\dim\Gamma(L) = N+1$. Let $\mathcal{I}_Y$ denote the ideal sheaf of $Y$ in $\mathbb{P}^N$, where we identify $Y$ with its image in $\mathbb{P}^N$ by $\Gamma(L)_Y$. Note then $\mathcal{O}_{\mathbb{P}^N}(d) \otimes \mathcal{I}_Y$ is spanned by global sections implies that $L^d \otimes \mathcal{J}_Y$ is spanned by global sections. This follows since degree $d$ homogeneous polynomials cutting out $Y$ scheme theoretically on $\mathbb{P}^N$ restrict to sections of $L^d$ cutting out $Y$ scheme theoretically on $X$.

We proceed by induction on $c := N - \dim Y$, the codimension of $Y$ in $\mathbb{P}^N$. If $c = 1$, i.e. if $\dim Y = N-1$, then there is nothing to prove. Therefore we can assume that $\dim Y \leq N-2$. Fix an arbitrary point $y \in Y$. From $\dim Y \leq N-2$ it follows that the closure of the union of the set of secant lines containing $y$ plus the tangent cone of $Y$ through $y$ is a proper closed algebraic subset, $Z$, of $\mathbb{P}^N$. Let $U = \mathbb{P}^N - Z$. Let $\pi_p : Y \to \mathbb{P}^{N-1}$ denote the restriction to $Y$ of the projection $\pi : \mathbb{P}^N \to \mathbb{P}^{N-1}$ from $p \in U$. Note that $\pi_p(Y)$ is a degree $d$ subvariety of $\mathbb{P}^{N-1}$, $\pi_p^{-1}(\pi_p(y)) = y$, and $\pi_p$ is an isomorphism on a Zariski neighborhood of $y$ in $Y$. Let $\tau_p$ denote the tangent vector at $y$ to the line in $\mathbb{P}^N$ connecting $p$

and $y$. Let $(\mathcal{N}_Y^{\mathbb{P}^N})_y^*$ denote the conormal space of $Y$ in $\mathbb{P}^N$ at $y$ and let $V_p$ denote the codimension 1 subspace of $(\mathcal{N}_Y^{\mathbb{P}^N})_y^*$ that annihilates $\tau_p$.

Noting that $\dim \mathbb{P}^{N-1} - \dim \pi_p(Y) < N - \dim Y$, we can assume by the induction hypothesis that degree $d$ polynomials on $\mathbb{P}^{N-1}$ cut out $\pi_p(Y)$. The pullbacks of these polynomials to $\mathbb{P}^N$ are degree $d$ and vanish on $Y$. The differentials of the polynomials in local coordinates span $V_p$. Since $U$ is nonempty and Zariski open, a choice of $p' \in U$ not on the line connecting $p$ and $y$ will lead to a subspace $V_{p'} \neq V_p$. Thus we can find a set of degree $d$ homogeneous polynomials on $\mathbb{P}^N$, vanishing on $Y$ and whose differentials span $(\mathcal{N}_Y^{\mathbb{P}^N})_y^*$. Thus global sections of $\mathcal{I}_Y \otimes \mathcal{O}_{\mathbb{P}^N}(d)$ span $\mathcal{I}_Y \otimes \mathcal{O}_{\mathbb{P}^N}(d)$ at an arbitrary point $y \in Y$.

To finish we must show that if $q \in \mathbb{P}^N - Y$ then there is a degree $d$ polynomial that vanishes on $Y$, but not on $q$. Since $\dim Y < \dim \mathbb{P}^{N-1}$ it follows that the closure, $W$, of the union of lines from $q$ to points of $Y$ is a proper subset of $\mathbb{P}^N$. If $p \in (\mathbb{P}^N - W) \cap U$ then the image of $q$ under the projection, $\pi_p$, to $\mathbb{P}^{N-1}$ from $p$ is disjoint from the image, $\pi_p(Y)$, of $Y$. Thus the pullbacks of the degree $d$ polynomials on $\mathbb{P}^{N-1}$ vanishing on $\pi_p(Y)$ don't all vanish on $q$.

Now let us turn to case (1.1.2) of the lemma. Let $n := \dim X$. If $n = N$, there is nothing to prove. Therefore we can assume that $n < N$. Fix an arbitrary $y \in Y$ and an arbitrary $x \in X - Y$. We can choose a projection $\pi : \mathbb{P}^N - \mathbb{P}^{N-n-1} \to \mathbb{P}^n$ from a linear $\mathbb{P}^{N-n-1}$ such that $\mathbb{P}^{N-n-1} \subset \mathbb{P}^N - X$ and such that the restriction, $\pi_X$, of $\pi$ to $X$ satisfies the conditions:

1) $\pi_X^{-1}(\pi_X(y)) \cap Y = y$ and $x \notin \pi_X^{-1}(\pi_X(Y))$;

2) $\pi_X$ is an immersion in some neighborhood of $y$ in $X$.

Since $\pi(Y)$ is a degree $d$ hypersurface of $\mathbb{P}^n$, it is defined scheme theoretically by a single degree $d$ homogeneous polynomial. Let $f$ denote the pullback of this polynomial to $\mathbb{P}^N$. The restriction of $f$ to $X$ spans $L^d \otimes \mathcal{J}_Y$ in a neighborhood of $y$ by 2). Moreover $f(x) \neq 0$ by property 1). Since $y$ is an arbitrary point of $Y$ and $x$ is an arbitrary point of $X - Y$ this proves the lemma. Q.E.D.

It is natural to ask the following

**(1.1.3) Question.** Let $\mathcal{I}_Y$ denote the ideal sheaf of $Y$, an irreducible degree $d$ subvariety of $\mathbb{P}^N$. When does it follow that $\mathcal{I}_Y \otimes \mathcal{O}_{\mathbb{P}^N}(d)$ is spanned by global sections?

**(1.2) Lemma.** *Let $X, L, Y, d, r$ be as in* (1.0). *Assume that $h^0((\mathcal{N}_Y^X)^* \otimes L^{-m}) > 0$ for some $m \geq 0$. If $t \geq r$ and $h^0(K_{X|Y} + (t(d+m) + 1 - rm)L_Y) > 0$ then $h^0(K_X + qL) > 0$ for $q \geq td + 1$.*

**Proof.** Denote $\mathcal{N}_Y^X$ by $\mathcal{N}_Y$. It suffices to set $q = td + 1$. Let $p : V \to X$ be the blowing up of $X$ along $Y$. Let $E = p^{-1}(Y)$. Since $Y$ is smooth and $Y \subset \mathrm{reg}(X)$ we have the general facts

$$(1.2.1) \qquad E \cong \mathbb{P}(\mathcal{N}_Y^*) \quad ; \quad \mathcal{N}_E^V \cong \mathcal{O}_E(-1).$$

Let $\overline{L} = p^*L$. Since $L^d \otimes \mathcal{J}_Y$ is spanned by Lemma (1.1), it thus follows that $d\overline{L} - E$ is spanned on $V$. For any integer $t$, let $\mathcal{F}_t := K_V + t(d\overline{L} - E) + \overline{L}$. By tensoring the exact sequence,

$$0 \to K_V \to K_V \otimes \mathcal{O}_V(E) \to K_E \to 0,$$

with $t(d\overline{L}-E)+\overline{L}$ we get the exact sequence

$$(1.2.2)\qquad 0\to\mathcal{F}_t\to\mathcal{F}_t\otimes\mathcal{O}_V(E)\to K_E\otimes\overline{L}^{dt}\otimes\mathcal{O}_E(-tE)\otimes\overline{L}_E\to 0.$$

We have, for $t\geq 1$,

$$(1.2.3)\qquad \begin{aligned} h^0(K_X+(td+1)L)&=h^0(K_V+(td+1)\overline{L})\\ &\geq h^0(\mathcal{F}_t+E)\geq h^0(K_E+t(d\,\overline{L}-E)_E+\overline{L}_E).\end{aligned}$$

Indeed, the equality in (1.2.3) follows by noting that $p_*K_V\cong K_X$, since $Y\subset\operatorname{reg}(X)$, and

$$h^0(K_V+(td+1)\overline{L})=h^0(p_*(K_V+(td+1)\overline{L}))=h^0(K_X+(td+1)L).$$

The first inequality is obvious and the second inequality follows from the sequence (1.2.2) since $t(d\overline{L}-E)+\overline{L}$ is nef and big and hence $h^1(\mathcal{F}_t)=0$ by a variant of the Kawamata-Viehweg vanishing theorem (see [S3],(0.2)).

To prove the lemma it is therefore enough to show that

$$(1.2.4)\qquad h^0(K_E+t(d\overline{L}-E)_E+\overline{L}_E)>0.$$

From the canonical bundle formula (for the $\mathbb{P}^{r-1}$ bundle $p_E:E\to Y$) we have $K_E\approx p^*(K_Y+\det\mathcal{N}_Y^*)-r\xi$ where $\xi$ is the tautological line bundle $\xi:=\mathcal{O}_E(1)$. Hence we find

$$\begin{aligned}(*)\quad &(K_E+t(d\overline{L}-E)_E+\overline{L}_E\\ &\approx p^*K_Y+p^*(\det\mathcal{N}_Y^*)-r\xi+t(d\overline{L}-E)_E+\overline{L}_E\\ &=p^*K_Y+p^*(\det\mathcal{N}_Y^*)+(1+td)\overline{L}_E+(t-r)\xi=\\ &=p^*K_Y+p^*(\det\mathcal{N}_Y^*)+(1+td)\overline{L}_E+(t-r)(\xi-m\overline{L}_E)+m(t-r)\overline{L}_E.\end{aligned}$$

Since $p_*\xi=p_*((\mathcal{N}_E^V)^*)$ by (1.2.1) and $h^0(\mathcal{N}_Y^*-mL)>0$ by assumption, we have

$$h^0(\xi-m\overline{L}_E)=h^0(p_*(\xi-m\overline{L}_E))=h^0(\mathcal{N}_Y^*-mL)>0.$$

Thus we can forget the term $(t-r)(\xi-m\overline{L}_E)$ in (*) and to show (1.2.4) we are reduced to proving that

$$h^0(p^*K_Y+p^*(\det\mathcal{N}_Y^*)+(1+td+mt-mr)\overline{L}_E)>0.$$

Since $K_Y+\det\mathcal{N}_Y^*\cong K_{X|Y}$ and $p_*\overline{L}_E\cong L_Y$ this is equivalent to

$$h^0(K_{X|Y}+(1+t(m+d)-mr)L_Y>0.$$

This gives the result. Q.E.D.

**(1.2.5) Remark.** Let $\mathcal{E}$ be a rank $r$ vector bundle on an $n$-dimensional, connected projective manifold, $X$. Assume that $\mathcal{E}$ is very ample in the sense that the tautological bundle, $\xi_{\mathcal{E}}$, on $\mathbb{P}(\mathcal{E})$ is very ample, e.g. assume that $\mathcal{E}\otimes\mathcal{L}^{-1}$ is spanned by global sections for some very ample line bundle, $\mathcal{L}$, on $X$. Then the above proof applied to the fibers of $p:\mathbb{P}(\mathcal{E})\to X$ with $m=0$ and $d=1$ shows that $K_{\mathbb{P}(\mathcal{E})}\otimes\xi_{\mathcal{E}}^q$ is spanned by global sections if $q\geq\max\{\operatorname{rank}\mathcal{E},\dim X+1\}$. In particular if $\operatorname{rank}\mathcal{E}\geq\dim X+1$ then $K_X\otimes\det\mathcal{E}$ is spanned by global sections since $K_{\mathbb{P}(\mathcal{E})}\otimes\xi_{\mathcal{E}}^{\operatorname{rank}\mathcal{E}}\approx p^*(K_X\otimes\det\mathcal{E})$. As $(X,\mathcal{E})=(\mathbb{P}^n,\oplus_1^{n+1}\mathcal{O}_{\mathbb{P}^n}(1))$ shows this is sharp. (See [BSS2] for a related result for ample and spanned vector bundles on 3-folds.) The same sort of reasoning shows that appropriate adjoint bundles on quadric fibrations and varieties covered with linear $\mathbb{P}^k$'s are spanned. For example if $L$ is a very ample line bundle of defect $k \geq 1$ on a connected, projective manifold, $X$, (see [BFS2] and [BSW]), then $K_X + ((n+k)/2+1)L$ is spanned by global sections and not ample if $k \geq n/3$. We will return to a careful study of such spannedness questions in a sequel to this article.

As a consequence of Lemma (1.2) we can show that $K_X + qL$ is effective for suitable integers $q$. The following are the main results of this section.

**(1.3) Proposition.** *Let $X$, $L$, $Y$, $d$, $r$ be as in* (1.0). *Assume that for some $m \geq 0$, $h^0((\mathcal{N}_Y^X)^* \otimes L^{-m}) > 0$. Let*

$$\alpha := \min\{s \in \mathbb{Z}\,;\, h^0(K_Y + sL_Y) > 0,\ s \geq 1\}.$$

*Assume that $q \geq \lceil(\alpha-1)/(m+d)\rceil d + rd + 1$. Then $h^0(K_X + qL) > 0$.*
**Proof.** Let $\mathcal{N}_Y$ be the normal bundle of $Y$ in $X$. From Lemma (1.2) we know that $h^0(K_X + (td+1)L) > 0$, with $t \geq r$, as soon as

$$h^0(K_Y + \det\mathcal{N}_Y^* + (1+td+tm-mr)L_Y) > 0. \tag{1.3.1}$$

Since $\det(\mathcal{N}_Y^* \otimes L^d) \cong \det\mathcal{N}_Y^* \otimes L^{rd}$, (1.3.1) is equivalent to

$$h^0(K_Y + \det(\mathcal{N}_Y^* + dL) + (1+td+tm-mr-rd)L_Y) > 0. \tag{1.3.2}$$

Look at the exact sequence

$$0 \to \mathcal{J}_Y^2 \otimes L^d \to \mathcal{J}_Y \otimes L^d \to \mathcal{N}_Y^* \otimes L^d \to 0.$$

Since $L^d \otimes \mathcal{J}_Y$ is spanned by Lemma (1.1), we conclude that $\mathcal{N}_Y^* \otimes L^d$ is spanned by the image of $\Gamma(\mathcal{J}_Y \otimes L^d)$ and hence $\det(\mathcal{N}_Y^* \otimes L^d)$ is spanned too. Therefore, to show that $h^0(K_X + (td+1)L) > 0$, we can forget the term $\det(\mathcal{N}_Y^* \otimes L^d)$ in (1.3.2). Thus (1.3.2) holds true as soon as

$$1 + t(d+m) - r(m+d) \geq \alpha \quad \text{or} \quad t \geq \lceil(\alpha-1)/(d+m)\rceil + r.$$

For $q \geq td+1$ we get the result. Q.E.D.

**(1.4) Proposition.** *Let $X$, $L$, $Y$, $d$, $r$ be as in* (1.0), *with $h^0((\mathcal{N}_Y^X)^* \otimes L^{-m}) > 0$ for some $m \geq 0$. Assume that $K_{X|Y} \cong aL_Y$ for some integer $a$. Further assume that $K_X$ is not nef and let $\tau$ be the nef value of $(X, L)$. If*

$$q \geq \max\{d\lceil(rm+\tau-1)/(d+m)\rceil + 1, rd+1\}$$

*then $h^0(K_X + qL) > 0$.*
**Proof.** Let $\mathcal{L} := K_X + (t(d+m)+1-rm)L$, with $t \geq r$, and let $\mathcal{L}_Y$ be the restriction of $\mathcal{L}$ to $Y$. Assume that $t(d+m)+1-rm \geq \tau$. Then $\mathcal{L}$ is nef and hence $\mathcal{L}_Y$ is nef too. By the assumption on $K_{X|Y}$ we have $\mathcal{L}_Y \approx L_Y^b$ for some integer $b$. Note also that $b \geq 0$ since $\mathcal{L}_Y$ is nef. Therefore $h^0(\mathcal{L}_Y) > 0$ since $L_Y$ is very ample. Thus by Lemma (1.2) we conclude that $h^0(K_X + (td+1)L) > 0$ under the present assumption $t(d+m)+1-rm \geq \tau$, or $t \geq \lceil(\tau+mr-1)/(m+d)\rceil$. For $q \geq td+1$ we have the result. Q.E.D.

**(1.4.1) Note.** The assumption $K_{X|Y} \cong aL_Y$ of (1.4) is satisfied e.g. when $\mathrm{Pic}(Y) \cong \mathbb{Z}[L_Y]$.

**(1.5) Corollary** (linear $\mathbb{P}^{n-r}$ case). *Let $X$, $L$, $Y$, $d$, $r$ be as assumed in* (1.0). *Assume that $h^0((\mathcal{N}_Y^X)^* \otimes L^{-m}) > 0$ for some $m \geq 0$. Further assume that $Y \cong \mathbb{P}^{n-r}$ (with $d = 1$), i.e. $L_Y \cong \mathcal{O}_{\mathbb{P}^{n-r}}(1)$. If $q \geq \lceil (\tau + m(r+1))/(m+1) \rceil$, then $h^0(K_X + qL) > 0$.*

**Proof.** Since $\mathrm{Pic}(Y) \cong \mathbb{Z}[L_Y]$, we can apply Proposition (1.4) to $Y$ with $d = 1$. Then $h^0(K_X + qL) > 0$ if $q \geq \lceil (rm + \tau - 1)/(m+1) \rceil + 1$. Note that

$$\lceil (rm + \tau - 1)/(m+1) \rceil + 1 = \lceil (rm + \tau - 1)/(m+1) + 1 \rceil =$$

$$\lceil (rm + \tau - 1 + m + 1)/(m+1) \rceil = \lceil (\tau + (r+1)m)/(m+1) \rceil.$$

Q.E.D.

## §2. The divisorial case.

First, let us fix the following

**(2.0) Assumptions.** Let $X$ be a normal projective $n$-dimensional variety, $n \geq 2$, with isolated irrational singularities and let $L$ be a very ample line bundle on $X$. Let $D$ be a reduced irreducible divisor on $X$, contained in the set of smooth points, $\mathrm{reg}(X)$. Let $d = L^{n-1} \cdot D = L_D^{n-1}$ be the *degree* of $D$.

We need the following numerical fact.

**(2.1) Lemma.** *Let $X$, $L$, $D$, $d$ be as in* (2.0). *Assume that $\mathcal{N}_D^X \cong -mL_D$ for some positive integer $m$. Then $(dL - D)^n = 0$ if and only if $L^n = ((m+d)^n - d^n)/md^{n-1}$. Note that $((m+d)^n - d^n)/md^{n-1}$ is an integer if and only if $d$ divides $m$. Hence in particular $dL - D$ is nef and big if $d$ does not divide $m$.*

**Proof.** We have

$$(dL - D)^n = \sum_{k=0}^{n} \binom{n}{k} (-D)^k (dL)^{n-k}. \tag{2.1.1}$$

Note that, for $k \geq 1$,

$$D^k \cdot (dL)^{n-k} = (D_D)^{k-1} \cdot (dL_D)^{n-k} = (-mL_D)^{k-1} \cdot (dL_D)^{n-k} =$$
$$= (-1)^{k-1} m^{k-1} d^{n-k} L_D^{n-1} = (-1)^{k-1} m^{k-1} d^{n-k+1}$$

and hence, for $k \geq 1$,

$$\binom{n}{k} (-D)^k \cdot (dL)^{n-k} = -\binom{n}{k} m^{k-1} d^{n-k+1}. \tag{2.1.2}$$

Combining (2.1.1) and (2.1.2) we get $(dL - D)^n = d^n L^n - \sum_{k=1}^{n} \binom{n}{k} m^{k-1} d^{n-k+1}$. Therefore $(dL - D)^n = 0$ if and only if

$$m(dL - D)^n/d = md^{n-1}L^n - \sum_{k=1}^{n} \binom{n}{k} m^k d^{n-k}$$
$$= md^{n-1}L^n - ((m+d)^n - d^n) = 0.$$

Thus we conclude that $(dL-D)^n = 0$ if and only if $L^n = ((m+d)^n - d^n)/md^{n-1}$. Recall that $dL - D$ is spanned by Lemma (1.1). Q.E.D.

We can show now that $K_X + qL$ is effective, for suitable integers $q$.

**(2.2) Proposition** (the normal case). *Let $X, L, D, d$ be as in* (2.0). *Assume that $h^0((\mathcal{N}_D^X)^* \otimes L^{-m}) > 0$ for some integer $m \geq 0$.*
(2.2.1) (compare with [BFS1], (0.3.1)). *Let $(D, L_D) \cong (\mathbb{P}^{n-1}, \mathcal{O}_{\mathbb{P}^{n-1}}(1))$. Then $h^0(K_X + qL) > 0$ if $q \geq \lceil (n+m)/(m+1) \rceil$. Furthermore, if $\mathcal{N}_D^X \cong -mL_D$ with $m > 0$ and $L^n \neq ((m+1)^n - 1)/m$, then $h^0(K_X + qL) > 0$ for $q \geq \lceil n/(m+1) \rceil$.*
(2.2.2) *Assume that $D$ is normal and let $(D, L_D) \ncong (\mathbb{P}^{n-1}, \mathcal{O}_{\mathbb{P}^{n-1}}(1))$. Then $h^0(K_X + qL) > 0$ if $q \geq d\lceil (n-2)/(d+m) \rceil + 1$. Furthermore if $\mathcal{N}_D^X \cong -mL_D$ with $m > 0$ and $L^n \neq ((m+d)^n - d^n)/md^{n-1}$, then $h^0(K_X + qL) > 0$ for $q \geq d\lceil (n-1)/(m+d) \rceil$.*
**Proof.** For an integer $t$, let $\mathcal{F}_t := K_X + t((dL - D) + L$. By tensoring the exact sequence

$$0 \to K_X \to K_X \otimes D \to K_D \to 0,$$

with $t(dL - D) + L$ we get the exact sequence

$$(2.2.3) \qquad 0 \to \mathcal{F}_t \to \mathcal{F}_t \otimes \mathcal{O}_X(D) \to K_D \otimes L^{dt} \otimes \mathcal{O}_X(-tD) \otimes L_D \to 0.$$

We have, for $t \geq 1$,

$$(2.2.4) \qquad h^0(K_X + (td+1)L) \geq h^0(\mathcal{F}_t + D) \geq h^0(K_D + t(dL-D)_D + L_D).$$

Indeed, the first inequality in (2.2.4) is obvious. Furthermore note that $t(dL - D) + L$ is nef and big since $dL - D$ is spanned by Lemma (1.1). Hence $h^1(\mathcal{F}_t) = 0$ by the Kawamata-Viehweg vanishing theorem so that the sequence (2.2.3) gives the second inequality.

Thus, to show that $K_X + (td+1)L$ is effective, it is enough to prove that

$$h^0(K_D + t((dL-D)_D + L_D) > 0.$$

Write $K_D + t((dL - D)_D + L_D = K_D + (td+1)L_D + t(-D - mL)_D + tmL_D$ and note that $h^0((-D - mL)_D) = h^0((\mathcal{N}_D^X)^* \otimes L^{-m}) > 0$ by assumption. Therefore we are reduced to showing

$$(2.2.5) \qquad h^0(K_D + (td + tm + 1)L_D) > 0.$$

First, assume $(D, L_D) \cong (\mathbb{P}^{n-1}, \mathcal{O}_{\mathbb{P}^{n-1}}(1))$, so in particular $d = 1$. Then by combining (2.2.4) and (2.2.5) we see that $h^0(K_X + (t+1)L) > 0$ if $-n + t(1 + m) + 1 \geq 0$, or $t \geq \lceil (n-1)/(m+1) \rceil$. Writing $q = t + 1$ and noting that $\lceil (n-1)/(m+1) \rceil + 1 = \lceil (m+n)/(m+1) \rceil$ we have the first statement of (2.2.1).

Furthermore if $\mathcal{N}_D^X \cong -mL_D$ and $L^n \neq ((m+1)^n - 1)/m$, $m > 0$, Lemma (2.1) applies to say that $L - D$ is nef and big. Thus exactly the same argument above shows that $h^0(K_X + tL) > 0$ if $-n + t(1+m) > 0$, or $t \geq \lceil n/(m+1) \rceil$. This gives the second statement of (2.2.1).

From now on, we can assume $(D, L_D) \ncong (\mathbb{P}^{n-1}, \mathcal{O}_{\mathbb{P}^{n-1}}(1))$. Suppose that $h^0(K_D + (n-1)L_D) = 0$. Then, since $L_D$ is spanned, we have $h^0(K_D + sL_D) = 0$ for $s = 1, \ldots, n-1$. Let $\varphi : M \to D$ be a desingularization of $D$. Hence $h^0(K_M + s\varphi^* L_D) = 0$ for $s = 1, \ldots, n-1$. Thus, since $D$ is normal, Theorem (2.2) of [Fu1] applies to say that there exists a birational morphism $f : D \to \mathbb{P}^{n-1}$

such that $L_D \cong f^*\mathcal{O}_{\mathbb{P}^{n-1}}(1)$. Since $L_D$ is ample, in fact very ample, $f$ is an isomorphism in our case. This contradicts the assumption $D \not\cong \mathbb{P}^{n-1}$. Therefore we conclude that $h^0(K_D + (n-1)L_D) > 0$ and hence (2.2.5) holds true as soon as $t(d+m)+1 \geq n-1$, or $t \geq \lceil (n-2)/(d+m) \rceil$. Writing $q = td+1$ we have the first statement of (2.2.2). Furthermore, if $\mathcal{N}_D^X \cong -mL_D$ and $L^n \neq ((m+d)^n - d^n)/md^{n-1}$, $m > 0$, we have from Lemma (2.1) that $dL - D$ is nef and big. Therefore exactly the same argument as in the beginning of the proof, by looking at the exact sequence

$$0 \to \mathcal{F}_t \otimes L^{-1} \to \mathcal{F}_t \otimes \mathcal{O}_X(D) \otimes L^{-1} \to K_D \otimes L^{dt} \otimes \mathcal{O}_X(-tD) \to 0$$

instead of (2.2.3), shows that $h^0(K_X + tdL) > 0$ as soon as $h^0(K_D + t(d+m)L_D) > 0$. Then exactly the same argument, as above, by using Fujita's theorem, shows that this is true if $t(d+m) \geq n-1$, or $t \geq \lceil (n-1)/(m+d) \rceil$. Writing $q = td$ we have the second part of (2.2.2). Q.E.D.

**(2.3) Proposition** (the non-normal case). *Let $X$, $L$, $D$, $d$ be as in (2.0). Assume that $D$ is not normal. Further assume that $h^0((\mathcal{N}_D^X)^* \otimes L^{-m}) > 0$ for some integer $m \geq 0$. Then $h^0(K_X + qL) > 0$ if $q \geq d\lceil (d+n-1)/(d+m) \rceil + 1$.*

**Proof.** Recall that $K_D \approx K_{X|D} + \mathcal{N}_D^X$ and consider the exact sequence

$$0 \to K_X \otimes L^s \otimes \mathcal{O}_X(-D) \to K_X \otimes L^s \to K_D \otimes L_D^s \otimes (\mathcal{N}_D^X)^* \to 0.$$

Note that $sL - D$ is nef and big for $s \geq d+1$ since $dL - D$ is spanned by Lemma (1.1). Therefore by the Kawamata-Viehweg vanishing theorem $h^i(K_X + sL - D) = 0$ , for $s \geq d+1$, $i \geq 1$, as well as $h^i(K_X + sL - D) = 0$ , for $s \geq 1$. Thus $h^i(K_D + sL_D - [D]_D) = 0$ for $s \geq d+1$, $i \geq 1$ and hence, for $s \geq d+1$,

$$p(s) := \chi(K_D + sL_D - [D]_D) = h^0(K_D + sL_D - [D]_D).$$

Clearly, since $L_D$ is effective, if $p(s) = 0$ for $s \geq d+n$, then $p(s) = 0$ for $d+1 \leq s \leq d+n$, that is $p(s)$ has $n$ roots. This is not possible since $p(s)$ is a polynomial of degree $\leq n-1$. Therefore we conclude that $h^0(K_D + sL_D - [D]_D) > 0$ for $s \geq d+n$ and hence

$$h^0(K_D + sL_D) > 0 \quad \text{for } s \geq d+n. \tag{2.3.1}$$

Exactly the same argument as in the first part of the proof of Proposition (2.2), by combining (2.3.1) above with (2.2.5) in the proof of (2.2), shows that $h^0(K_X + (td+1)L) > 0$ if $t(d+m)+1 \geq d+n$, or $t \geq \lceil (d+n-1)/(m+1) \rceil$. Writing $q = td+1$, we have the result. Q.E.D.

## §3. Applications to reductions and further remarks.

Let $\mathcal{M}$ be a smooth $n$-dimensional projective variety and let $\mathcal{L}$ be a very ample line bundle on $\mathcal{M}$. The results of the previous sections apply to show that if the (unnormalized) spectral value of $(\mathcal{M}, \mathcal{L})$ satisfies suitable numerical conditions, then the structure of the first and the second reduction is very special. Through the proof of the following theorem we use over and over, without explicitly referring to it, the structure theorem (0.3.1) for the second reduction.

**(3.1) Theorem.** *Let $\mathcal{M}$ be a smooth $n$-dimensional projective variety, $n \geq 3$, $\mathcal{L}$ a very ample line bundle on $\mathcal{M}$. Let $(M, L), \pi : \mathcal{M} \to M$ be the first reduction of*

*$(\mathcal{M},\mathcal{L})$ and let $(X,\mathcal{K}), \mathcal{K} \approx K_X + (n-2)\mathcal{D}$, $\varphi : M \to X$ be the second reduction of $(\mathcal{M},\mathcal{L})$. Let $u$ be the (unnormalized) spectral value of $(\mathcal{M},\mathcal{L})$. Then we have (note that for* (3.1.1) *below the existence of the second reduction is not needed):*
(3.1.1) *If $u > \lceil (n+1)/2 \rceil$ ( or $u > \lceil n/2 \rceil$ if $\mathcal{L}^n \neq 2^n - 1$) then $(\mathcal{M},\mathcal{L}) \cong (M,L)$;*
(3.1.2) *if $u > \lceil (n+2)/3 \rceil$ ( or $u > \lceil n/3 \rceil$ if $\mathcal{L}^n \neq (3^n-1)/2$) then $X$ is factorial;*
(3.1.3) *if $u > \lceil (n+1)/2 \rceil$ then $\mathcal{D}$ is ample and the only positive dimensional fibers of $\varphi$ are $(n-1)$-dimensional irreducible quadrics, $Q$, with $\mathcal{N}_Q^M \cong \mathcal{O}_Q(-1)$;*
(3.1.4) *if $u > 2\lceil (n-1)/3 \rceil$ then $(\mathcal{M},\mathcal{L}) \cong (M,L) \cong (X,\mathcal{D})$.*

**Proof.** First note that the assumptions in each case (3.1.1)–(3.1.4) imply that $n \geq 4$. Indeed, for $n = 3$, $\lceil (n+1)/2 \rceil = \lceil (n+2)/3 \rceil = 2\lceil (n-1)/3 \rceil = 2$. Hence in particular $h^0(N(K_{\mathcal{M}} + 2\mathcal{L})) = 0$ for each $N > 0$. This is clearly impossible since, by assumption, there exists the first reduction, $M$, of $\mathcal{M}$.

To prove (3.1.1) note that if $\mathcal{M}$ is not isomorphic to $M$, there exists $D \cong \mathbb{P}^{n-1}$ in $\mathcal{M}$ with $\mathcal{N}_D^{\mathcal{M}} \cong \mathcal{O}_{\mathbb{P}^{n-1}}(-1)$. Then (2.2.1) with $m = 1$ applies to $\mathcal{M},\mathcal{L},D$ to say that $h^0(K_{\mathcal{M}} + q\mathcal{L}) > 0$ if $q \geq \lceil (n+1)/2 \rceil$. This contradicts the assumption that $u > \lceil (n+1)/2 \rceil$.

Assume $\mathcal{L}^n \neq 2^n - 1$. Then again by (2.2.1) with $m = 1$, we know that $h^0(K_{\mathcal{M}} + q\mathcal{L}) > 0$ for $q \geq \lceil n/2 \rceil$. This contradicts the present assumption $u > \lceil n/2 \rceil$.

To prove (3.1.2) note that if $X$ is not factorial there exists in $M$ a divisor $D \cong \mathbb{P}^{n-1}$ which contracts under $\varphi$ to a 2-factorial point with $\mathcal{N}_D^M \cong \mathcal{O}_{\mathbb{P}^{n-1}}(-2)$. By (0.3.3) such a divisor lies in $\mathcal{M}$ with $\mathcal{N}_D^{\mathcal{M}} \cong \mathcal{O}_{\mathbb{P}^{n-1}}(-2)$. Then (2.2.1) with $m = 2$ applies to $\mathcal{M},\mathcal{L},D$ to say that $h^0(K_{\mathcal{M}} + q\mathcal{L}) > 0$ if $q \geq \lceil (n+2)/3 \rceil$. This contradicts the assumption that $u > \lceil (n+2)/3 \rceil$.

Assume $\mathcal{L}^n \neq (3^n-1)/2$. Then, again by (2.2.1) with $m = 2$, we know that $h^0(K_{\mathcal{M}} + q\mathcal{L}) > 0$ for $q \geq \lceil n/3 \rceil$. This contradicts the present assumption $u > \lceil n/3 \rceil$.

To prove (3.1.3) note that $(\mathcal{M},\mathcal{L}) \cong (M,L)$ by (3.1.1). Note also that, by (3.1.2), there are no divisors $D \cong \mathbb{P}^{n-1}$ in $\mathcal{M}$ with $\mathcal{N}_D^M \cong \mathcal{O}_{\mathbb{P}^{n-1}}(-2)$ contracting to points under $\varphi : \mathcal{M} \to X$. Therefore to prove (3.1.3) it is enough to show that there are no divisors $D$ in $\mathcal{M}$ which contract to smooth curves $C \subset \mathrm{reg}(X)$. Any such a divisor $D$ is a $\mathbb{P}^{n-2}$ bundle over $C$, $L_{\mathbb{P}^{n-2}} \cong \mathcal{O}_{\mathbb{P}^{n-2}}(1)$ and $\mathcal{N}_{D|\mathbb{P}^{n-2}}^{\mathcal{M}} \cong \mathcal{O}_{\mathbb{P}^{n-2}}(-1)$ for each fiber $\mathbb{P}^{n-2}$. Thus, let us assume that a divisor $D$ as above exists. Note that the exact sequence

$$0 \to \mathcal{N}_{\mathbb{P}^{n-2}}^D \cong \mathcal{O}_{\mathbb{P}^{n-2}} \to \mathcal{N}_{\mathbb{P}^{n-2}}^{\mathcal{M}} \to \mathcal{N}_{D|\mathbb{P}^{n-2}}^{\mathcal{M}} \to 0$$

splits, i.e. $\mathcal{N}_{\mathbb{P}^{n-2}}^{\mathcal{M}} \cong \mathcal{O}_{\mathbb{P}^{n-2}} \oplus \mathcal{O}_{\mathbb{P}^{n-2}}(-1)$ and hence $h^0((\mathcal{N}_{\mathbb{P}^{n-2}}^{\mathcal{M}})^* \otimes \mathcal{L}^{-1})) > 0$. Note also that the nef value of $(\mathcal{M},\mathcal{L}) \cong (M,L)$ is $n-2$ since $K_{\mathcal{M}} + (n-2)\mathcal{L}$ is nef and big but not ample (see (0.2.2)). Therefore (1.5) with $r = 2, m = 1, \tau = n-2$ applies to say that $h^0(K_{\mathcal{M}} + q\mathcal{L}) > 0$ if $q \geq \lceil (n+1)/2 \rceil$. This contradicts the assumption $u > \lceil (n+1)/2 \rceil$.

(3.1.4) First note that we can assume $n \geq 5$. Indeed, for $n = 4$, the assumption on $u$ gives $u > 2$. Hence in particular $h^0(N(K_{\mathcal{M}} + 2\mathcal{L})) = 0$ for each $N > 0$. From [S3], (0.3.1) (see also [BFS1], (0.2.7)) we know that, for any $N > 0$, $h^0(N(K_{\mathcal{M}} + 2\mathcal{L})) = h^0(N(K_M + 2L))$. Since, by assumption, there exists the $2^{nd}$ reduction $(X,\mathcal{K})$, we have $h^0(N(K_M + 2L)) > 0$ for some $N >> 0$. This leads to a contradiction.

Thus we can assume $n \geq 5$. Hence in particular $n > 2\lceil (n-1)/3 \rceil$ implies $u > \lceil (n+1)/2 \rceil$, so that the isomorphism $\mathcal{M} \cong M$ follows from (3.1.1) and the only positive dimensional fibers of the $2^{nd}$ reduction map $\varphi : \mathcal{M} \to X$ are $(n-1)$-dimensional quadrics $Q$ with $\mathcal{N}_Q^{\mathcal{M}} \cong \mathcal{O}_Q(-1)$ by (3.1.3). Assume that there exists such a quadric, $Q$. Then (2.2.2) with $d = 2, m = 1$ applies to $\mathcal{M}, \mathcal{L}, Q$. Note that $\mathcal{L}^n \neq (3^n - 2^n)/2^{n-1}$ since $(3^n - 2^n)/2^{n-1} \notin \mathbb{Z}$ for $n \geq 2$. Thus we conclude that $h^0(K_{\mathcal{M}} + q\mathcal{L}) > 0$ for $q \geq 2\lceil (n-1)/3 \rceil$. Once again, this contradicts our assumption $u > \lceil (n-1)/3 \rceil$. Q.E.D.

**(3.2) Examples.** The examples (3.2.1) and (3.2.2) below show that the bounds in (3.1.1) are sharp. The example (3.2.3) show sharpness for the bound in (2.2.1).
(3.2.1) Let $(M, L) := (\mathbb{P}^n, \mathcal{O}_{\mathbb{P}^n}(2))$, with $n \geq 4$. Let $\pi : \mathcal{M} \to M$ be the blowing up at a single point, $x$. Let $E = \pi^{-1}(x)$. Then $\mathcal{L} := \pi^* L \otimes \mathcal{O}_{\mathcal{M}}(-E)$ is very ample and $K_{\mathcal{M}} + (n-1)\mathcal{L} \approx \pi^*(K_M + (n-1)L) = \pi^* \mathcal{O}_{\mathbb{P}^n}(n-3)$, so that $(M, L)$ is the first reduction of $(\mathcal{M}, \mathcal{L})$ if $n \geq 4$. Here $\mathcal{L}^n = L^n - 1 = 2^n - 1$. We have

$$\begin{aligned} u(\mathcal{M}, \mathcal{L}) = u(M, L) &= \sup\{t \in \mathbb{Q}, h^0(N(K_M + tL)) > 0 \text{ for all } N > 0\} \\ &= \sup\{t \in \mathbb{Q}, h^0(\mathcal{O}_{\mathbb{P}^n}(N(2t - n - 1)) > 0 \text{ for all } N > 0\} \\ &= \sup\{t \in \mathbb{Q}, t < (n+1)/2\} = (n+1)/2. \end{aligned}$$

This shows that the bound $u(\mathcal{M}, \mathcal{L}) > \lceil (n+1)/2 \rceil$ in the case $\mathcal{L}^n = 2^n - 1$ of (3.1.1) is sharp.
(3.2.2) Let $(M, L) := (Q, \mathcal{O}_Q(2))$ where $Q$ is a smooth quadric in $\mathbb{P}^{n+1}$. Let $\pi : \mathcal{M} \to M$ be the blowing up at a single point, $x$. Let $E = \pi^{-1}(x)$. Then $\mathcal{L} := \pi^* L \otimes \mathcal{O}_{\mathcal{M}}(-E)$ is very ample and $K_{\mathcal{M}} + (n-1)\mathcal{L} \approx \pi^*(K_M + (n-1)L) = \pi^* \mathcal{O}_Q(n-1)$ so that $(M, L)$ is the first reduction of $(\mathcal{M}, \mathcal{L})$. Here $\mathcal{L}^n = L^n - 1 = 2^{n+1} - 1 \neq 2^n - 1$. We have

$$\begin{aligned} u = u(\mathcal{M}, \mathcal{L}) = u(M, L) &= \sup\{t \in \mathbb{Q}, h^0(N(K_M + tL)) > 0 \text{ for all } N > 0\} \\ &= \sup\{t \in \mathbb{Q}, h^0(\mathcal{O}_Q(N(2t - n)) > 0 \text{ for all } N > 0\} \\ &= \sup\{t \in \mathbb{Q}, t < n/2\} = n/2. \end{aligned}$$

This shows that the bound $u > \lceil n/2 \rceil$ in the case $\mathcal{L}^n \neq 2^n - 1$ of (3.1.1) is sharp.
(3.2.3) Let $X = \mathbb{P}(\mathcal{O}_{\mathbb{P}^{n-1}} \oplus \mathcal{O}_{\mathbb{P}^{n-1}}(m))$. Let $p : X \to \mathbb{P}^{n-1}$ the $\mathbb{P}^1$ bundle projection. Let $\xi := \mathcal{O}_X(1), H := p^* \mathcal{O}_{\mathbb{P}^{n-1}}(1)$ and let $L := \xi + H$. Then $L$ is very ample. Since $\xi^2 - mH \cdot \xi \sim 0$ we have $\xi^a \sim m^{a-1} H^{a-1} \cdot \xi$ for any positive integer $a$. Compute

$$\begin{aligned} L^n &= \sum_{j=0}^{n} \binom{n}{j} \xi^{n-j} \cdot H^j = \sum_{j=0}^{n-1} \binom{n}{j} m^{n-j-1} H^{n-1} \cdot \xi = \\ &= \sum_{j=0}^{n-1} \binom{n}{j} m^{n-j-1} = (\sum_{j=0}^{n-1} \binom{n}{j} m^{n-j})/m = ((m+1)^n - 1)/m. \end{aligned}$$

Then from (2.2.1) we have $h^0(K_X + qL) > 0$ for $q \geq \lceil (n+m)/(m+1) \rceil$. Note that $\lceil (n+m)/(m+1) \rceil \geq 2$.

From the canonical bundle formula for the $\mathbb{P}^1$ bundle $p : X \to \mathbb{P}^{n-1}$ we have

$$K_X + 2\xi \approx p^*(K_{\mathbb{P}^{n-1}} \otimes \det(\mathcal{O}_{\mathbb{P}^{n-1}} \oplus \mathcal{O}_{\mathbb{P}^{n-1}}(m))) \approx (m-n)H.$$

Since $p_*(q-2)\xi \cong \mathcal{O}_{\mathbb{P}^{n-1}} \oplus \mathcal{O}_{\mathbb{P}^{n-1}}(m) \oplus \cdots \oplus \mathcal{O}_{\mathbb{P}^{n-1}}((q-2)m)$ and $H = p_* \mathcal{O}_{\mathbb{P}^{n-1}}(1)$, we get by the projection formula, for $q \geq 2$,

$$h^0(K_X + qL) = \oplus_{i=0}^{q-2} h^0(\mathcal{O}_{\mathbb{P}^{n-1}}(im + q + m - n)).$$

Thus $h^0(K_X + qL) = 0$ if and only if $h^0(\mathcal{O}_{\mathbb{P}^{n-1}}((q-2)m + q + m - n)) = 0$, or, if and only if $q \geq \lceil (n+m)/(m+1) \rceil$. This proves the sharpness in (2.2.1).

Note also that $\lceil (m+n)/(m+1) \rceil \geq \lceil n/(m+1) \rceil$ with equality if and only if $n \equiv 1 \bmod (m+1)$. To see this write

$$\lceil (m+n)/(m+1) \rceil = \lceil (m+1+n-1)/(m+1) \rceil = 1 + \lceil (n-1)/(m+1) \rceil.$$

Let us conclude the section with a few remarks.

**(3.3) Remark.** The following variant of the conjecture in the appendix to [BSW] (compare also with [Fu2]) is a motivating conjecture for this paper.
*Conjecture.* Let $X$ be a smooth projective variety of dimension $n$ and $\mathcal{L}$ an ample line bundle on $X$. If $u := u(X, \mathcal{L}) > (n+1)/2$, then $u \in \mathbb{Q}$, and $K_X + u\mathcal{L}$ is nef.

Note that if $n \leq 7$ or $n \geq 8$ with $u \geq n-3$ this conjecture follows from the results of [BFS1] and the appendix of [BSW].

An interesting problem is to classify the pairs $(X, L)$ where $L$ is very ample, $X$ is smooth and $Y$ is a submanifold of $X$, with $\pi : V \to X$ the blowing up of $X$ along $Y$ and such that $(\pi^*L - \pi^{-1}(Y))^n = 0$, $n = \dim X$. If $n = 2$ and $Y$ is a line this is done in [S1]. The case when $Y$ is a divisor $\cong \mathbb{P}^{n-1}, \mathcal{N}_Y^X \cong \mathcal{O}_Y(-1)$ and the restriction $L_Y$ of $L$ to $Y$ is $\mathcal{O}_Y(1)$ (and $\pi$ is an isomorphism) is of particular interest. In (3.2.1) we give a particular example which is the only possible example if $n = 2$. Is this the only possibility for $n \geq 3$?

## REFERENCES


[BFS1] M.C. Beltrametti, M.L. Fania, A.J. Sommese, *On the adjunction theoretic classification of projective varieties,* Math. Ann., 290 (1991), 31–62.

[BFS2] M.C. Beltrametti, M.L. Fania, A.J. Sommese, *On the discriminant variety of a projective manifold,* Forum Math., 4 (1992), 529–547.

[BS] M.C. Beltrametti, A.J. Sommese, *On the adjunction theoretic classification of polarized varieties,* Jour. für die reine und angew. Math., 427 (1992), 157–192.

[BSS1] M.C. Beltrametti, M. Schneider, A.J. Sommese, *Threefolds of degree* 11 *in* $\mathbb{P}^5$, Complex Projective Geometry, London Math. Soc. Lecture Note Series, 179 (1992), 59–80.

[BSS2] M.C. Beltrametti, M. Schneider, A.J. Sommese, *Applications of the Ein-Lazarsfeld criterion for spannedness of adjoint bundles,* to appear Math. Z.

[BSW] M.C. Beltrametti, A.J. Sommese, J.A. Wiśniewski, *Results on varieties with many lines and their application to adjunction theory,* with an appendix by M.C. Beltrametti and A.J. Sommese, Proceedings of the 1990 Bayreuth Conference on "Complex Algebraic Varieties," Springer Lecture Notes in Math., 1507 (1992), 16–38.

[Fa] M.L. Fania, *Configurations of −2 rational curves on sectional surfaces of n-folds,* Math. Ann., 275 (1986), 317–325.

[FS] M.L. Fania, A.J. Sommese, *On the projective classification of smooth n-folds with n even,* Arkiv för Matematik, 27 (1989), 245–256.

[Fu1] T. Fujita, *Remarks on quasi-polarized varieties,* Nagoya Math. J., 115 (1989), 105–123.

[Fu2] T. Fujita, *On Kodaira energy and adjoint reduction of polarized manifolds,* Manuscripta Math., 76 (1992), 59–84.

[KMM] Y. Kawamata, K. Matsuda, K. Matsuki, *Introduction to the minimal model problem,* Algebraic Geometry, Sendai, 1985, Advanced Studies in Pure Math., 10 (1987), 283–360.

[Mum] D. Mumford, *Varieties defined by quadratic equations*, in: Questions on algebraic varieties, C.I.M.E., Cremonese, Rome (1970), 30–100.

[R] M. Reid, *Canonical 3-folds,* Algebraic Geometry, Angers 1979 (ed. A. Beauville), Sijthoff and Noordhoff, (1980), 273–310.

[S1] A.J. Sommese, *Hyperplane sections of projective surfaces, I. The adjunction mapping,* Duke Mathematical Journal, 46 (1979), 377–401.

[S2] A.J. Sommese, *Configurations of −2 rational curves on hyperplane sections of projective threefolds,* Classification of Algebraic and Analytic Manifolds, (ed. K. Ueno), Progress in Mathematics, 39 (1983), Birkhäuser.

[S3] A.J. Sommese, *On the adjunction theoretic structure of projective varieties,* Complex Analysis and Algebraic Geometry, Proceedings Göttingen, 1985 (ed. H. Grauert), Springer Lecture Notes in Math., 1194 (1986), 175–213.

[S4] A.J. Sommese, *On the nonemptiness of the adjoint linear system of a hyperplane section of a threefold,* Jour. für die reine und angew. Math., 402 (1989), 211–220; *Erratum,* ibid., 411 (1990), 122–123.

[SV] A.J. Sommese, A. Van de Ven, *On the adjunction mapping,* Math. Ann., 278 (1987), 593–603.

M.C. Beltrametti
Dipartimento di Matematica
Universitá degli Studi di Genova
Via L.B. Alberti, 4, I-16132 GENOVA (ITALY)
fax: Italy + 10–3538769

A.J. Sommese
Department of Mathematics
University of Notre Dame
NOTRE DAME, Indiana 46556, U.S.A.
sommese@nd.edu
fax: U.S.A. + 219–631–6579

Contemporary Mathematics
Volume **162**, 1994

# Surfaces in $\mathbf{P}^4$ and deficiency modules

GIORGIO BOLONDI

ABSTRACT. We describe how liaison techniques can help in the study of surfaces in $\mathbf{P}^4$, giving an explicit description of the cohomological liaison invariants "à la Horrocks" of an even liaison class, and explaining how to derive from them a minimal surface in the class.

## 0. Introduction and preliminaries

Let $\mathbf{k}$ be an algebraically closed field of characteristic zero, and let $S=\mathbf{k}[x_0,x_1,x_2,x_3,x_4]$. With "surface" we will always mean a locally Cohen-Macaulay equidimensional two-codimensional subscheme of $\mathbf{P}^4$.

A common and useful way for constructing surfaces in $\mathbf{P}^4$ with given properties can be sketched as follows: 1) translate these properties into properties of the graded modules $M_i(X) = \bigoplus_{n\in\mathbf{Z}} H^i(\mathbf{P}^4,\mathcal{J}_X(n))$, i=1,2; 2) show that such modules exist; 3) then derive from these modules one or more suitable vector bundles (via syzygies); 4) and finally get X from suitable morphisms of these vector bundles. This has been done, for instance, in [6] and [10], and it is in fact a typical liaison procedure. We will show here a general approach to it in the framework of liaison theory.

In paragraphs 1 and 2 we will see how these modules are involved into liaison classification. In particular, in paragraph 1 we make explicit for surfaces in $\mathbf{P}^4$ the Horrocks-Rao classification, getting as liaison invariant of an even liaison class a triple $(M_1,M_2,b)$, with $M_1$ and $M_2$ graded module of finite length and $b\in \mathrm{Ext}^2(M_2,M_1)$.

In paragraph 2 we describe an analogue, for surfaces in $\mathbf{P}^4$, of the Rao construction for curves in $\mathbf{P}^3$. In that situation, the liaison invariant of a curve C is exactly the deficiency module $M(C)= \bigoplus_{n\in\mathbf{Z}} H^1(\mathbf{P}^3,\mathcal{J}_C(n))$ , and Rao construction shows that every finite length graded module is actually the invariant of an even

---

1991 Mathematics Subject Classification. Primary 14J10, 14M06.
This paper is in final form and no version will be submitted for publication elsewhere

liaison class; i.e. every finite length graded module is actually - up to shift- the deficiency module of a (even smooth) curve in $\mathbf{P}^3$. Here we show how to derive from a triple $(M_1,M_2,b)$ as above an even liaison class having that triple as liaison invariant.

In paragraph 3, following the approach of [9], we construct explicitly a minimal surface (in the sense of the LR-property, see def. 3.2) in a fixed even liaison class.

A useful output of these descriptions is in the study of the stratification via cohomology of the Hilbert scheme of surfaces in $\mathbf{P}^4$, and in particular in the determination of non-void strata. Note that two surfaces having the same cohomology (see below §3) and in the same even liaison class are contained in the same irreducible component of the Hilbert scheme.

The author thanks I.O. Kleppe and Kr. Ranestad since this work started from discussions with them, and the University of Oslo and the Oslo College of Engineering for their hospitality.

## 1. From surfaces to cohomology

In this paragraph we describe explicitly the liaison invariants of a surface $X \subset \mathbf{P}^4$. The general frame of this approach is contained in [11] and [7].

Let $X \subset \mathbf{P}^4$ be a surface, $I_X$ its homogeneous ideal and take the sheafification of the beginning of a minimal free resolution of $I_X$:

(*) $$0 \to \mathcal{F}_X \to \mathcal{P} \to \mathcal{J}_X \to 0.$$

It is easy to check that $\mathcal{P}$ is a direct sum of line bundles and $\mathcal{F}$ is a locally free sheaf ([5], p.3) whose graded cohomology modules are

$$H^1_*(\mathbf{P}^4,\mathcal{F})=0$$

$$H^2_*(\mathbf{P}^4,\mathcal{F}) = M_1(X) = \bigoplus_{n\in\mathbf{Z}} H^1(\mathbf{P}^4,\mathcal{J}_X(n))$$

$$H^3_*(\mathbf{P}^4,\mathcal{F}) = M_2(X) = \bigoplus_{n\in\mathbf{Z}} H^2(\mathbf{P}^4,\mathcal{J}_X(n)).$$

It is known by [11] that two surfaces X and Y are evenly linked by liaison (i.e. one goes from X to Y through en *even* number of linkages) if and only if the corresponding sheaves $\mathcal{F}_X$ and $\mathcal{F}_Y$ are *stably equivalent*, i.e. there exist $a_i$'s, $b_j$'s, and c integers such that

$$\mathcal{F}_X \oplus [\bigoplus_i \mathcal{O}_\mathbf{P}(-a_i)] \cong \mathcal{F}_Y(c) \oplus [\bigoplus_j \mathcal{O}_\mathbf{P}(-b_j)].$$

But stable equivalence classes of vector bundles on $\mathbf{P}^n$ are classified by [7] via equivalence of complexes, getting the following invariants if n=4 and $H^1_*(\mathbf{P}^4,\mathcal{F})=0$.

Let $F=H^0_*(\mathbf{P}^4,\mathcal{F})$, and take the dual of a projective resolution of $F^*$ (note that the sheafification of F is $\mathcal{F}$)

$$0 \to F \to V^0 \overset{\lambda^0}{\to} V^1 \overset{\lambda^1}{\to} V^2 \overset{\lambda^2}{\to} V^3 \overset{\lambda^3}{\to} 0$$

$$V^2 \searrow \operatorname{Im}\lambda^2 \nearrow V^3, \quad 0 \nearrow \operatorname{Im}\lambda^2, \quad V^3 \searrow H^3 \searrow 0$$

where $F=\mathrm{Ker}\lambda^0$, and $H^3$ is the third cohomology module of the complex and it is isomorphic to $M_2$.

This resolution gives a monad of the form

$$\begin{array}{ccccccc}
 & 0 & & 0 & & 0 & \\
 & \downarrow & & \downarrow & & \downarrow & \\
0\to & \mathrm{Im}\lambda^1 & \to & \mathrm{Ker}\lambda^2 & \to & H^2 & \to 0\\
 & \downarrow & & \downarrow & & \downarrow & \\
0\to & \mathrm{Im}\lambda^1 & \to & V^2 & \to & T^2 & \to 0\\
 & \downarrow & & \downarrow & & \downarrow & \\
 & 0 & \to & \mathrm{Im}\lambda^2 & \to & \mathrm{Im}\lambda^2 & \to 0\\
 & & & \downarrow & & \downarrow & \\
 & & & 0 & & 0 &
\end{array}$$

where $H^2$ is the second cohomology module of the complex, and hence is isomorphic to $M_1$, and the third column represents an extension $b\in \mathrm{Ext}^1(\mathrm{Im}\lambda^2,H^2)$.

The other monad of the complex is

$$\begin{array}{ccccccc}
 & 0 & & 0 & & 0 & \\
 & \downarrow & & \downarrow & & \downarrow & \\
0\to & \mathrm{Im}\lambda^0 & \to & \mathrm{Ker}\lambda^1 & \to & H^1 & \to 0\\
 & \downarrow & & \downarrow & & \downarrow & \\
0\to & \mathrm{Im}\lambda^0 & \to & V^1 & \to & T^1 & \to 0\\
 & \downarrow & & \downarrow & & \downarrow & \\
 & 0 & \to & \mathrm{Im}\lambda^1 & \to & \mathrm{Im}\lambda^1 & \to 0\\
 & & & \downarrow & & \downarrow & \\
 & & & 0 & & 0 &
\end{array}$$

but this monad is trivial, since $H^1$ (the first cohomology module of the complex) vanishes.

As a consequence of Horrocks'classification, we get that in our situation the stable equivalence class of $\mathcal{F}_X$ is individuated by $M_1$, $M_2$ and b.

On the other hand, we have an exact sequence

$$0\to \mathrm{Im}\lambda^2 \to V^3 \to H^3 \to 0$$

(since $H^3=\mathrm{Ker}\lambda^3/\mathrm{Im}\lambda^2=V^3/\mathrm{Im}\lambda^2$), and hence we get a long exact sequence

$$\mathrm{Ext}^1(V^3,H^2)\to \mathrm{Ext}^1(\mathrm{Im}\lambda^2,H^2)\to \mathrm{Ext}^2(H^3,H^2)\to \mathrm{Ext}^2(V^3,H^2);$$

since $\mathrm{Ext}^1(V^3,H^2)=\mathrm{Ext}^2(V^3,H^2)=0$ ($V^3$ is free), we get canonical isomorphisms

$$(**)\qquad \mathrm{Ext}^1(\mathrm{Im}\lambda^2,H^2)\cong \mathrm{Ext}^2(H^3,H^2)=\mathrm{Ext}^2(M_2,M_1).$$

We can summarize the above discussion in the fact that the even liaison invariant of a surface in $\mathbf{P}^4$ is the triple $(M_1,M_2,b)$, i.e.:

THEOREM 1.1. X *is evenly linked to* Y <=> $\mathcal{F}_X$ *is stably equivalent to* $\mathcal{F}_Y$ <=> $M_1(X)\cong M_1(Y)$, $M_2(X)\cong M_2(Y)$, $b(X)=b(Y)$ [*up to (the same) shift of degrees*].

Note that in this triple is included the information about the *relative* shift of the deficiency modules $M_1(X)$ and $M_2(X)$ of the surface (which is of course invariant by liaison). Note moreover that in order to obtain the triple we can use *any* locally free resolution of $\mathcal{J}_X$ having the form (*), i.e. where the middle term is a direct sum of line bundles and the left term has trivial first cohomology module.

## 2. From cohomology to surfaces

Now we reverse the construction, showing that for every triple $(M_1,M_2,b)$, where $M_1$ and $M_2$ are graded S-module of finite length and $b \in \mathrm{Ext}^2(M_2,M_1)$, there exists a liaison class whose invariants are exactly $(M_1,M_2,b)$, i.e. there exists a surface (and even a smooth one) whose invariants are $(M_1,M_2,b)$, up to the same shift of degrees. This is the analogue of the Rao construction for curves in $\mathbf{P}^3$. As in that situation, of course, we have to deal with sheafification of syzygies.

We fix some notation. Let us take a minimal free resolution of $M_1$:

$$\begin{array}{ccccccccccccccccccc}
0 \to & P_5 & \overset{\sigma_5}{\to} & P_4 & \overset{\sigma_4}{\to} & & P_3 & & \overset{\sigma_3}{\to} & P_2 & \overset{\sigma_2}{\to} & & P_1 & & \overset{\sigma_1}{\to} & P_0 & \to M_1 \to 0 \\
& & & & \searrow & \nearrow & & \searrow & \nearrow & & \searrow & \nearrow & & \searrow & \nearrow & & \\
& & & & & K_3 & & & K_2 & & & K_1 & & & K_0 & & \\
& & & & \nearrow & & \searrow & \nearrow & & \searrow & \nearrow & & \searrow & \nearrow & & \searrow & \\
& & & & 0 & & & 0 & & & 0 & & & 0 & & & 0
\end{array}$$

As usual, if F is a graded S-module, we denote with $\widetilde{F}$ the sheafification of F. From the above resolution, we get an exact sequence

$$0 \to \widetilde{K}_1 \to \widetilde{P}_1 \to \widetilde{P}_0 \to 0$$

and the following cohomology modules

$$H^2_*(\mathbf{P}^4, \widetilde{K}_1) = H^3_*(\mathbf{P}^4, \widetilde{K}_1) = 0$$

$$H^1_*(\mathbf{P}^4, \widetilde{K}_2) = H^3_*(\mathbf{P}^4, \widetilde{K}_2) = 0$$

$$H^1_*(\mathbf{P}^4, \widetilde{K}_3) = H^2_*(\mathbf{P}^4, \widetilde{K}_3) = 0$$

$$H^1_*(\mathbf{P}^4, \widetilde{K}_1) = H^2_*(\mathbf{P}^4, \widetilde{K}_2) = H^3_*(\mathbf{P}^4, \widetilde{K}_3) = M_1.$$

Let us take a minimal free resolution of $M_2$:

$$\begin{array}{ccccccccccccccccccc}
0 \to & Q_5 & \overset{\tau_5}{\to} & Q_4 & \overset{\tau_4}{\to} & & Q_3 & & \overset{\tau_3}{\to} & Q_2 & \overset{\tau_2}{\to} & & Q_1 & & \overset{\tau_1}{\to} & Q_0 & \to M_2 \to 0 \\
& & & & \searrow & \nearrow & & \searrow & \nearrow & & \searrow & \nearrow & & \searrow & \nearrow & & \\
& & & & & L_3 & & & L_2 & & & L_1 & & & L_0 & & \\
& & & & \nearrow & & \searrow & \nearrow & & \searrow & \nearrow & & \searrow & \nearrow & & \searrow & \\
& & & & 0 & & & 0 & & & 0 & & & 0 & & & 0
\end{array}$$

From the above resolution,we get the following cohomology modules

$$H^2_*(\mathbf{P}^4, \tilde{L_1}) = H^3_*(\mathbf{P}^4, \tilde{L_1}) = 0$$
$$H^1_*(\mathbf{P}^4, \tilde{L_2}) = H^3_*(\mathbf{P}^4, \tilde{L_2}) = 0$$
$$H^1_*(\mathbf{P}^4, \tilde{L_3}) = H^2_*(\mathbf{P}^4, \tilde{L_3}) = 0$$
$$H^1_*(\mathbf{P}^4, \tilde{L_1}) = H^2_*(\mathbf{P}^4, \tilde{L_2}) = H^3_*(\mathbf{P}^4, \tilde{L_3}) = M_2$$

Now we interpret, via standard homological arguments, the element $b \in \mathrm{Ext}^2(M_2,M_1)$ as an element of other groups. This may be heavy, but these canonical isomorphisms will be useful later on. Recall that if M is an S-module of finite length, then $\mathrm{Ext}^i(M,S)=0$ for $i \neq 5$, and that with our notation $P_i$ and $Q_i$ are free for every i.

We have:

$\mathrm{Ext}^2(M_2,M_1) \cong \mathrm{Ext}^3(M_2,K_0)$ (from the exact sequence $0 \to K_0 \to P_0 \to M_1 \to 0$, since $\mathrm{Ext}^2(M_2,P_0)=\mathrm{Ext}^3(M_2,P_0)=0$, being $M_2$ of finite length)

$\cong \mathrm{Ext}^4(M_2,K_1)$ (from the exact sequence $0 \to K_1 \to P_1 \to K_0 \to 0$, since $\mathrm{Ext}^3(M_2,P_1)=\mathrm{Ext}^4(M_2,P_1)=0$ )

$\cong \mathrm{Ext}^3(L_0,K_1)$ (from the exact sequence $0 \to L_0 \to Q_0 \to M_2 \to 0$, since $\mathrm{Ext}^3(Q_0,K_1)=\mathrm{Ext}^4(Q_0,K_1)=0$)

$\cong \mathrm{Ext}^2(L_1,K_1)$ (from the exact sequence $0 \to L_1 \to Q_1 \to L_0 \to 0$, since $\mathrm{Ext}^2(Q_1,K_1)=\mathrm{Ext}^3(Q_1,K_1)=0$)

$\cong \mathrm{Ext}^1(L_2,K_1)$ (from the exact sequence $0 \to L_2 \to Q_2 \to L_1 \to 0$, since $\mathrm{Ext}^1(Q_2,K_1)=\mathrm{Ext}^2(Q_2,Q_1)=0$).

Now we have an (uniquely well-determined in a canonical way) element $b_1$ which lives in $\mathrm{Ext}^1(L_2,K_1)$.

Note that $\mathrm{Ext}^1(L_2,P_2)$ is zero, since it is isomorphic to $\mathrm{Ext}^1_{\mathcal{O}_\mathbf{P}}(\tilde{L_2}, \tilde{P_2})$ (see, for instance, the exact sequence of comparison, [9], p. 22: it gives

$$0 \to \mathrm{Ext}^1(L_2,P_2) \to \mathrm{Ext}^1_{\mathcal{O}_\mathbf{P}}(\tilde{L_2}, \tilde{P_2}) \to \mathrm{Hom}_S(L_2, H^1_*(\mathbf{P}^4, \tilde{P_2})),$$

and this last group is zero since $H^1_*(\mathbf{P}^4, \tilde{P_2}) =0$ ).

Now consider the exact sequence

$$0 \to K_2 \to P_2 \to K_1 \to 0.$$

We have an injection

$$0 \to \mathrm{Ext}^1(L_2,K_1) \to \mathrm{Ext}^2(L_2,K_2),$$

and this last group is isomorphic to

$$\mathrm{Ext}^1(L_3,K_2)$$

(from the exact sequence $0 \to L_3 \to Q_3 \to L_2 \to 0$, since $\mathrm{Ext}^1(Q_3,K_2)=\mathrm{Ext}^2(Q_3,K_2)=0$).

Hence $b_1$ determines uniquely an element $b_2 \in \mathrm{Ext}^1(L_3,K_2)$.

On the other hand, from the exact sequence

$$0 \to L_3 \to Q_3 \to L_2 \to 0$$

we get a long exact sequence

$$\mathrm{Hom}(Q_3,K_1) \xrightarrow{g} \mathrm{Hom}(L_3,K_1) \xrightarrow{f} \mathrm{Ext}^1(L_2,K_1) \to \mathrm{Ext}^1(Q_3,K_1)=0$$

and there exists $b_3 \in \mathrm{Hom}(L_3,K_1)$, such that $f(b_3)=b_1$. Note that another element $c \in \mathrm{Hom}(L_3,K_1)$ is such that $f(c)=b_1$ if and only if $b_3-c \in \mathrm{Im}(g)$, i.e. if and only if $b_3-c$ factorizes through $\alpha: L_3 \to Q_3$, where $\alpha$ is the morphism appearing in the minimal free resolution of $M_2$.

Now we look to the associated sheaves, i.e. to $\tilde{K}_1$ and $\tilde{L}_3$, and we consider the morphism induced by $b_3$, let us denote it by $\beta$. Up to adding direct sums of line bundles $\mathcal{A}$ and $\mathcal{B}$, one can find a morphism of sheaves $\beta'$: $\tilde{L}_3 \oplus \mathcal{A} \to \tilde{K}_1 \oplus \mathcal{B}$ (containing as a minor $\beta$) which is injective and drops rank in codimension two, hence along a surface X (see, f.i., [1]). That is to say, there exists a surface X and an exact sequence

$$0 \to \tilde{L}_3 \oplus \mathcal{A} \xrightarrow{\beta} \tilde{K}_1 \oplus \mathcal{B} \xrightarrow{d} \mathcal{I}_X(t) \to 0$$

with $M_1(X) = H^1_*(\mathbf{P}^4, \tilde{K}_1) = M_1$, $M_2(X) = H^3_*(\mathbf{P}^4, \tilde{L}_3) = M_2$,

from which one can easily derive in a standard way a resolution of $\mathcal{I}_X(t)$ as in (*).

Take the beginning of a free resolution of $\tilde{K}_1 \oplus \mathcal{B}$, which will have the form

$$\begin{array}{llllll} \ldots\ldots \to & \tilde{P}_2 \oplus \mathcal{B} & \xrightarrow{\psi'} & \tilde{K}_1 \oplus \mathcal{B} & \to 0, & \quad \psi'=(\psi,1_B) \\ \nearrow & & & & & \\ \tilde{K}_2 & & & & & \end{array}$$

and look to the diagram

$$\begin{array}{ccccccccc} & & 0 & & 0 & & 0 & & \\ & & \downarrow & & \downarrow & & \downarrow & & \\ 0 & \to & \tilde{K}_2 & \xrightarrow{id} & \tilde{K}_2 & \to & 0 & \to & 0 \\ & & \downarrow & & \downarrow \chi & & \downarrow & & \\ 0 & \to & \mathcal{F} & \to & \tilde{P}_2 \oplus \mathcal{B} & \to & \mathcal{I}_X(t) & \to & 0 \\ & & \downarrow & & \downarrow \psi' & & \downarrow id & & \\ 0 & \to & \tilde{L}_3 \oplus \mathcal{A} & \xrightarrow{\beta} & \tilde{K}_1 \oplus \mathcal{B} & \xrightarrow{d} & \mathcal{I}_X(t) & \to & 0 \\ & & & & \downarrow & & \downarrow & & \\ & & & & 0 & & 0 & & \end{array}$$

(the first column is the exact sequence of the kernels of the second and the third columns).
Hence

$$0 \to \mathcal{F} \to \tilde{P}_2 \oplus \mathcal{B} \to \mathcal{I}_X(t) \to 0$$

is a resolution of $\mathcal{I}_X(t)$ having the form (*). One can see the first column as an element of $\mathrm{Ext}^1_{\mathcal{O}_{\mathbf{P}}}(\tilde{L}_3 \oplus \mathcal{A}, \tilde{K}_2)$, and since the exact sequence

$$0 \to \mathcal{A} \to \tilde{L}_3 \oplus \mathcal{A} \to \tilde{L}_3 \to 0$$

induces an exact sequence

$$\ldots \to \mathrm{Ext}^1_{\mathcal{O}_{\mathbf{P}}}(\tilde{L}_3 , \tilde{K}_2) \to \mathrm{Ext}^1_{\mathcal{O}_{\mathbf{P}}}(\tilde{L}_3 \oplus \mathcal{A}, \tilde{K}_2) \to \mathrm{Ext}^1_{\mathcal{O}_{\mathbf{P}}}(\mathcal{A}, \tilde{K}_2)$$

and this last group is zero, it determines an element in $\mathrm{Ext}^1_{\mathcal{O}_{\mathbf{P}}}(\tilde{L}_3, \tilde{K}_2)$ ; this group is isomorphic to $\mathrm{Ext}^1(L_3,K_2)$ (again thanks to the exact sequence of comparison), and this element by construction (everything is functorial) is exactly $b_2$. It follows moreover that the element of $\mathrm{Ext}^2(M_2,M_1)$ individuated by $\mathcal{F}$ is exactly b.
In fact, it arises as an element $b_4 \in \mathrm{Ext}^1(\mathrm{Im}\lambda^2,M_1)$, but in our notation now $\mathrm{Im}\lambda^2=L_0$, and the isomorphism

$$\mathrm{Ext}^1(\mathrm{Im}\lambda^2, H^2) \cong \mathrm{Ext}^2(M_2,M_1)$$

of (**) is exactly the isomorphism

$$\mathrm{Ext}^1(L_0,M_1) \cong \mathrm{Ext}^2(M_2,M_1)$$

induced by the exact sequence

$$0 \to L_0 \to Q_0 \to M_2 \to 0.$$

This completes the proof of the following

PROPOSITION 2.1. *Every triple* $(M_1,M_2,b)$ *is actually the triple of liaison invariants of a liaison class of surfaces in* $\mathbf{P}^4$.

REMARK 2.2. This means that up to a same shift of degrees there exists a surface S whose cohomological liaison invariants are exactly $(M_1,M_2,b)$. The fact that therefore there exists a *smooth* surface with these invariants is standard , but we will see it directly in the following paragraph.

## 3. About the minimal element

We showed in paragraph 2 that, given a triple $(M_1,M_2,b)$, there exists a surface $S \subset \mathbf{P}^4$ having a locally free resolution

$$0 \to \mathcal{F} \to \mathcal{P} \to \mathcal{I}_S \to 0,$$

where $\mathcal{P}$ is a direct sum of line bundles and $\mathcal{F}$ is a locally free sheaf such that

$$H^1_*(\mathbf{P}^4,\mathcal{F})=0$$
$$H^2_*(\mathbf{P}^4,\mathcal{F}) = M_1(S) = M_1(t)$$
$$H^3_*(\mathbf{P}^4,\mathcal{F}) = M_2(S) = M_2(t)$$
$$b(\mathcal{F}) = b \text{ (up to the same shift of degrees).}$$

Note that, in principle, t may be large, in particular if we want a smooth surface. So it is important to look at the better shift, and this is related to the so-called LR-property. We recall here something about this property.

An even liaison class $\mathcal{L}$ of two-codimensional subschemes of $\mathbf{P}^n$ (different from the class of arithmetically Cohen-Macaulay subschemes) can be "graduated" as follows. The collection of graded modules $M_i(X)$, $1\leq i\leq n-2$ is a set of even liaison invariants up to shift of degrees (as we saw here, even for n=4 they are not sufficient but only necessary); if $\mathcal{L}$ is not the liaison class of arithmetically Cohen-Macaulay subschemes (which is well-known), then there is a well-defined notion of *leftmost shift* of $\{M_i(X)\}_{1\leq i\leq n-2}$ that can actually occur in $\mathcal{L}$ (see [3], prop.1.4). For a subscheme $X\in \mathcal{L}$, we say that $X\in \mathcal{L}^h$ ($h\geq 0$) if the collection $\{M_i(X)\}_{1\leq i\leq n-2}$ is shifted h places to the right of the leftmost possible one (note that $\mathcal{L}^h\neq\emptyset$ $\forall h\geq 0$).

DEFINITION 3.1. Let X be a two-codimensional subscheme of $\mathbf{P}^n$ and let $F\in I_X$. Choose a general form A of degree d, and consider the subscheme Y whose homogenrous ideal is given by $I_Y= A\cdot I_X + (F)$. Y is then called a *basic double link* of X via the hypersurfaces zero sets of F and A.

In a naive way, Y is simply the union of X and the complete intersection $V(F)\cap V(A)$ of the hypersurfaces given by F and A. Note that Y is then evenly linked to X, and that the cohomology of Y depends only on the cohomology of X, d=degA and f=degF.

In particular, if $\mathcal{I}_X$ has a locally free resolution

$$0 \to \mathcal{P} \to \mathcal{F} \to \mathcal{I}_X \to 0$$

where $H^1_*(\mathbf{P}^4,\mathcal{F})=0$ , then $\mathcal{I}_Y$ has a locally free resolution of the form

$$0 \to \mathcal{P}(-d)\oplus \mathcal{O}_{\mathbf{P}}(-d-f) \to \mathcal{F}(-d)\oplus \mathcal{O}_{\mathbf{P}}(-f) \to \mathcal{I}_Y \to 0$$

(see [3]). Note moreover that if $X\in \mathcal{L}^h$, then $Y\in \mathcal{L}^{h+d}$.

DEFINITION 3.2. Let $\mathcal{L}$ be an even liaison class of two-codimensional subschemes of $\mathbf{P}^n$. We say that $\mathcal{L}$ has the *LR-property* if the following conditions hold:

a) if $M,N\in \mathcal{L}^0$, then there is a deformation from one to the other through subschemes all in $\mathcal{L}^0$;

b) given $V_0 \in \mathcal{L}^0$ and $V \in \mathcal{L}^h$ ($h \geq 1$), then there exists a sequence of subschemes $V_0$, $V_1,\ldots,V_r$ such that for all i, $1 \leq i \leq r$, $V_i$ is a basic double link of $V_{i-1}$ and V is a deformation of $V_r$ through subschemes all in $\mathcal{L}^h$.

This property gives a description of the structure of an even liaison class in terms of one element in the minimal shift; note that the elements of the minimal shift form an irreducible family, hence they have the same cohomology; in particular, for surfaces in $\mathbf{P}^4$ they have the same d, $p_a$ and $\pi$. For other details about this property, the reader is referred to [3]; we quote simply here some of the main consequences (others can be found in [3] and [4]):

1) The basic double links in b) can be performed via complete intersections where d=1, i.e. the LR-property says that every surface in the even liaison class can be deformed to the union of a surface in the minimal shift with h degenerate surfaces

2) The elements of $\mathcal{L}$ are distributed into irreducible, disjoint, non void subsets $\mathcal{L}(b;g_2\ldots g_p)$, where $b \geq 0$ and $g_p > g_{p-1} > \ldots > g_2 > \alpha(V_0)$ (the minimal degree of an hypersurface containing $V_0$; it does not depends on $V_0$), such that two elements M,N of $\mathcal{L}$ lie in the same subset if and only if they have the same cohomology, i.e. $h^i(\mathbf{P}^n,\mathcal{J}_M(t)) = h^i(\mathbf{P}^n,\mathcal{J}_N(t))$ for every t and for every i, $0 \leq i \leq n-1$.

3) In every subset $\mathcal{L}(b;g_2\ldots g_p)$ there is an element which is obtained by adding "via liaison addition" (see [3] for definition and references) to $V_0$ an arithmetically Cohen-Macaulay subscheme.

4) The cohomology of $M \in \mathcal{L}(b;g_2\ldots g_p)$ is a function of the cohomology of $V_0$ and of $b,g_i$, and it can be computed easily via the mapping cone procedure.

In [2] it was proved that

THEOREM 3.3. *Every even liaison class of two-codimensional subschemes of* $\mathbf{P}^n$ *has the LR-property.*

REMARK 3.4. More generally, this property holds for every even liaison class of two-codimensional subschemes of an arithmetically Gorenstein variety, see [5].

Since the LR-property holds for surfaces in $\mathbf{P}^4$, if one knows the cohomology of an element in the minimal shift, one can describe in a purely algorithmical way all possible cohomologies of surfaces in $\mathcal{L}$. In order to determine explicitly a minimal element, we follow the line of attack of [9], there developed for curves in $\mathbf{P}^3$; see also [2], remark at page 127. In some sense, this paragraph makes the previous one superfluous; nevertheless, we believe that both constructions may be useful, depending on the situation; and the construction of §2 is surely often easier to handle.

It is better to work here with resolutions of the ideal sheaf of a surface where the free part is at the beginning. In fact, any surface $X \subset \mathbf{P}^4$ has a locally free resolution of the following form:

$$0 \to \mathcal{S} \to \mathcal{N}_X \to \mathcal{J}_X \to 0,$$

where $\mathcal{S}$ is a direct sum of line bundles and $\mathcal{N}_X$ is a locally free sheaf whose graded cohomology modules are

$$H^1_*(\mathbf{P}^4,\mathcal{N}_X)=M_1(X)$$
$$H^2_*(\mathbf{P}^4,\mathcal{N}_X) = M_2(X)$$
$$H^3_*(\mathbf{P}^4,\mathcal{N}_X) = 0$$

(take S directly linked to X, consider a resolution of $\mathcal{J}_S$ of the form (*) of paragraph 1, and apply the mapping cone procedure).

If now Y is evenly linked to X, then $\mathcal{J}_Y$ has the following resolution

$$0 \to \mathcal{S}\oplus\mathcal{A} \to \mathcal{N}_X\oplus\mathcal{B} \to \mathcal{J}_Y(h) \to 0$$

where $\mathcal{A}$ and $\mathcal{B}$ are direct sums of line bundles. Our procedure will obtain $\mathcal{N}_X$ as the sheafification (up to stable equivalence again) of the graded module individuated by $b_1\in \mathrm{Ext}^1(L_2,K_1)$.

We need the following results, which are proved in [9],§IV, for an $\mathcal{O}_{\mathbf{P}^3}$-module; we now have $S=\mathbf{k}[x_0,x_1,x_2,x_3,x_4]$ and $\mathbf{P}=\mathbf{P}^4$, but there is no substantial difference in the proofs; also the basic definitions are taken from [9],IV, just adapted to our situation. Hence what follows is mostly a re-writing of [9], IV.

DEFINITION 3.5. Let $\mathcal{A}$ and $\mathcal{B}$ be coherent $\mathcal{O}_\mathbf{P}$-modules, $\mathcal{B}\subset\mathcal{A}$. $\mathcal{B}$ is said to be *maximal* if for every $\mathcal{B}'$, $\mathcal{O}_\mathbf{P}$-module, such that $\mathcal{B}\subset\mathcal{B}'\subset\mathcal{A}$, $\mathrm{rk}\mathcal{B} = \mathrm{rk}\mathcal{B}'$, it is $\mathcal{B}= \mathcal{B}'$.

Let $\mathcal{M}$ be an $\mathcal{O}_\mathbf{P}$-module with no-torsion (in our application this will be always true), and let $M=H^0_*(\mathbf{P}^4,\mathcal{M})$. If a is an integer, we denote by $M_{\leq a}$ the submodule of M generated by elements of degree $\leq$a, and by $\mathcal{M}_{\leq a}$ the $\mathcal{O}_\mathbf{P}$-submodule of $\mathcal{M}$, image of the natural morphism

$$\bigoplus_{n\leq a} \mathcal{O}_\mathbf{P}(-n)^{t(n)} \to \mathcal{M}, \quad \text{where } t(n)=h^0(\mathbf{P}^4,\mathcal{M}(n)).$$

A minimal free resolution of M begins with

$$\sigma\colon \bigoplus_{n\in\mathbf{Z}} S(-n)^{l_M(n)} \to M$$

where $l_M:\mathbf{Z} \to \mathbf{N}$ is a function which does not depend on the resolution chosen. We denote with $\sigma_a$ the restriction of $\sigma$ to $\bigoplus_{n\leq a} S(-n)^{l_M(n)}$ and with

$\tilde{\sigma}_a : \bigoplus_{n\leq a} \mathcal{O}_\mathbf{P}(-n)^{l_M(n)} \to \mathcal{M}$ the restriction of $\tilde{\sigma}$ (the sheafification of $\sigma$).

DEFINITION-NOTATION 3.6.

$\alpha_a = \mathrm{rank}\, \tilde{\sigma}_a = \mathrm{rank}\, \sigma_a$

$\beta_a = \inf_D (\mathrm{rank}\, \tilde{\sigma}_a|_D)$, where D runs in the set of integral divisors of $\mathbf{P}^4$

$a_0 = 1+\sup\{a\in \mathbf{Z} \mid \alpha_a = \beta_a = \sum_{n\leq a} l(a)\}$ if it exists

$= +\infty$ if not

$a_1 = \inf\{a\in \mathbf{Z} \mid \alpha_a = \beta_a = \operatorname{rank}\mathcal{M}\}$.

Note that ([9],IV,prop.2.5) $a_0$-1, if it is finite, is the biggest integer such that the following equivalent facts are true (just correcting a misprint in their statement):

1) $\mathcal{M}_{\leq a}$ is a maximal sub-bundle, direct sum of line bundles, of $\mathcal{M}$

2) $\tilde{\sigma}_a$ is injective, with cokernel of constant rank on an open subset which contains the generic points of all divisors of $\mathbf{P}^4$

3) $\tilde{\sigma}_a$ is injective with cokernel without torsion

4) $\sigma_a$ is injective with cokernel without torsion.

Now we can define the key-function associated to $\mathcal{M}$ as

$q_{\mathcal{M}}(n) = l_M(n)$ if $n<a_0$

$q^*_{\mathcal{M}}(n) = \sum_{m\leq n} q(a) = \inf\{\alpha_n-1,\beta_n\}$ if not,

and we simply write q(n) if no confusion is possible.
The main properties of q are summarized in the following

THEOREM 3.7 ([9],IV, 3.1,3.4,3.7) *Let* $\mathcal{M}$, l, q *as above and* $r=\operatorname{rank}\mathcal{M}$.

a) *Let* $r:\mathbf{Z}\to\mathbf{N}$ *a function which is zero except for a finite number of values, and*

$$\tilde{u}: \bigoplus_{n\in\mathbf{Z}} \mathcal{O}_{\mathbf{P}}(-n)^{r(n)} \to \mathcal{M}$$

*an injection whose image is a maximal sub-bundle of* $\mathcal{M}$, *direct sum of line bundles. Then for every* $n\in\mathbf{Z}$ *it is* $\sum_{m\leq n} r(m) \leq \sum_{m\leq n} q_{\mathcal{M}}(m)$

b) *If* $\tilde{u}:\mathcal{K}= \bigoplus_{n\in\mathbf{Z}} \mathcal{O}_{\mathbf{P}}(-n)^{l(n)} \to \mathcal{M}$ *is the sheafification of the beginning of a minimal free resolution of* $M=H^0_*(\mathbf{P}^4,\mathcal{M})$, *then there exists a direct factor of* $\mathcal{K}$ *of the form* $\bigoplus_{n\in\mathbf{Z}} \mathcal{O}_{\mathbf{P}}(-n)^{q(n)}$, *which* $\tilde{\sigma}$ *identifies with a maximal sub-bundle of* $\mathcal{M}$.

c) *The maximum degree of a rank* r-1 *sub-bundle of* $\mathcal{M}$ *which is direct sum of line bundles is* $d(\mathcal{M}) = -\sum_{n\in\mathbf{Z}} nq(n)$; *every rank* r-1 *maximal sub-bundle of* $\mathcal{M}$ *of degree* $d(\mathcal{M})$ *which is direct sum of line bundles is isomorphic to* $\bigoplus_{n\in\mathbf{Z}} \mathcal{O}_{\mathbf{P}}(-n)^{q(n)}$ *and it is the image via* $\tilde{\sigma}$ *of a direct factor of* $\mathcal{K}$.

Now this theorem allows us to construct, for every even liaison class of surfaces in $\mathbf{P}^4$, a surface which is in the minimal shift, and hence to apply the whole LR-machinery.

Let $(M_1,M_2,b)$ be given as usual, and $b_2 \in \mathrm{Ext}^1(L_3,K_2)$ and $b_1 \in \mathrm{Ext}^1(L_2,K_1)$ be determined as before by b; they are represented by exact sequences

$$0 \to K_{\grave{a}} \to F \to L_3 \to 0$$

(this sequence was already used in §2 for the explicit construction of a surface in the liaison class) and

$$0 \to K_1 \to N \to L_2 \to 0.$$

We claim that these sequences fit into commutative diagrams

$$\text{(1)} \qquad \begin{array}{ccccccccc}
 & & & & 0 & & 0 & & \\
 & & & & \downarrow & & \downarrow & & \\
 & & & 0 \to & K_1 & \to & K_1 & \to & 0 \\
 & & & & \downarrow & & \downarrow & & \\
0 & \to & L_3 & \to & Q_3 \oplus K_1 & \to & N & \to & 0 \\
 & & \downarrow & & \downarrow & & \downarrow & & \\
0 & \to & L_3 & \xrightarrow{\alpha} & Q_3 & \to & L_2 & \to & 0 \\
 & & \downarrow & & \downarrow & & \downarrow & & \\
 & & 0 & & 0 & & 0 & &
\end{array}$$

and

$$\text{(2)} \qquad \begin{array}{ccccccccc}
 & & 0 & & 0 & & 0 & & \\
 & & \downarrow & & \downarrow & & \downarrow & & \\
0 & \to & K_2 & \to & F & \to & L_3 & \to & 0 \\
 & & \downarrow\chi & & \downarrow & & \downarrow\alpha & & \\
0 & \to & P_2 & \to & P_2 \oplus Q_3 & \to & Q_3 & \to & 0 \\
 & & \downarrow\psi & & \downarrow\theta & & \downarrow\gamma & & \\
0 & \to & K_1 & \to & N & \xrightarrow{\beta} & L_2 & \to & 0 \\
 & & \downarrow & & \downarrow & & \downarrow & & \\
 & & 0 & & 0 & & 0 & &
\end{array}$$

PROOF OF THE CLAIM.

(1) We start from $b_1 \in \mathrm{Ext}^1(L_2,K_1)$, represented by an extension

$$0 \to K_1 \to N \xrightarrow{\beta} L_2 \to 0.$$

One can lift $b_1$ to an element $c \in \mathrm{Ext}^1(Q_3,K_1)$ getting a commutative diagram

$$\begin{array}{ccccccccc}
0 & \to & K_1 & \to & V & \to & Q_3 & \to & 0 \\
 & & \downarrow \mathrm{id} & & \downarrow\alpha_2 & & \downarrow\gamma & & \\
0 & \to & K_1 & \to & N & \xrightarrow{\beta} & L_2 & \to & 0
\end{array}$$

where $V=\{(n,q) \in N \oplus Q_3 \mid \beta(n)=\gamma(q)\}$ is a submodule of $N \oplus Q_3$. Since $\mathrm{Ext}^1(Q_3,K_1)=0$, it follows that $V=Q_3 \oplus K_1$.

Now we apply the snake lemma to find an exact sequence

$$0 \to \mathrm{Ker}(\mathrm{id}_{K_1}) \to \mathrm{Ker}(\alpha_2) \to \mathrm{Ker}\gamma \to \mathrm{Coker}(\mathrm{id}_{K_1}) \to \mathrm{Coker}(\alpha_2) \to \mathrm{Coker}\gamma \to 0$$

where the first, fourth and sixth group are zero, and the third is isomorphic to $L_3$. From this it follows that $Ker(\alpha_2) = L_3$ and that $Coker(\alpha_2)=0$.

Note by the way that this is how $b_3 \in Hom(L_3,K_1)$ can be constructed, as a factor of this morphism $L_3 \to Q_3 \oplus K_1$.

(2) Start from the commutative square

$$\begin{array}{ccc} Q_3\oplus K_1 & \overset{(id,0)}{\to} & Q_3 \to 0 \\ \downarrow & & \downarrow \gamma \\ N & \overset{\beta}{\to} & L_2 \to 0 \\ \downarrow & & \downarrow \\ 0 & & 0. \end{array}$$

By composition with $Q_3\oplus P_2 \to Q_3\oplus K_1 \to 0$

$$q\oplus p \ \to \ q\oplus[\psi(p)]$$

one gets a commutative diagram

$$\begin{array}{ccc} Q_3\oplus P_2 & \overset{(id,0)}{\to} & Q_3 \to 0 \\ \downarrow\theta & & \downarrow \gamma \\ N & \overset{\beta}{\to} & L_2 \to 0 \\ \downarrow & & \downarrow \\ 0 & & 0 \end{array}$$

which is easily completed in the following one:

$$\begin{array}{ccccc} 0\to P_2 \to & & Q_3\oplus P_2 & \overset{(id,0)}{\to} & Q_3 \to 0 \\ \downarrow\psi & & \downarrow\theta & & \downarrow \gamma \\ 0\to K_1 \to & & N & \overset{\beta}{\to} & L_2 \to 0 \\ \downarrow & & \downarrow & & \downarrow \\ 0 & & 0 & & 0. \end{array}$$

Then the snake lemma gives the diagram

$$\begin{array}{ccccccc} & 0 & & 0 & & 0 & \\ & \downarrow & & \downarrow & & \downarrow & \\ 0\to & K_2 & \to & Ker\theta & \to & L_3 & \to 0 \\ & \downarrow\chi & & \downarrow & & \downarrow\alpha & \\ 0\to & P_2 & \to & P_2\oplus Q_3 & \to & Q_3 & \to 0 \\ & \downarrow\psi & & \downarrow\theta & & \downarrow\gamma & \\ 0\to & K_1 & \to & N & \overset{\beta}{\to} & L_2 & \to 0 \\ & \downarrow & & \downarrow & & \downarrow & \\ & 0 & & 0 & & 0 & . \end{array}$$

By construction, Ker$\theta$ is exactly the element individuated by $b_2 \in \mathrm{Ext}^1(L_3,K_2)$, and hence Ker$\theta$=F and we have the second diagram.

Note that $P_2 \xrightarrow{\Psi} K_1 \to 0$ and $Q_3 \xrightarrow{\gamma} L_2 \to 0$ are the beginning of minimal free resolutions of $K_1$ and $L_2$.

We claim that

LEMMA 3.8. $P_2 \oplus Q_3 \xrightarrow{\theta} N \to 0$ *is the beginning of a minimal free resolution* of N.

PROOF. If not, let $\theta'$: $V_0 \to N \to 0$ be the beginning of a minimal free resolution of N, let G=Ker$\theta'$, and let $V_0 \cong B \oplus E$, $P_2 \cong B \oplus C$, $Q_3 \cong E \oplus D$ (note that also $\theta$ splits through these isomorphisms). This would give (by composition) a surjection $E \to L_2 \to 0$, and this is a contradiction. hence D=0.

In the same way, the commutative diagram

$$\begin{array}{ccccc}
 & & 0 & & 0 \\
 & & \downarrow & & \downarrow \\
0 & \to & K_2 & \to & G \oplus C \\
 & & \downarrow & & \downarrow \\
0 & \to & B \oplus C & \to & B \oplus C \oplus Q_3 \\
 & & \downarrow \psi & & \downarrow \theta \\
0 & \to & K_1 & \to & N \\
 & & \downarrow & & \downarrow \\
 & & 0 & & 0
\end{array}$$

shows that $\psi'$: $B \to K_1$, the restriction of $\psi$, is surjective, hence a contradiction with the minimality of $\psi$: $P_2 \to K_1$ if $C \neq 0$; therefore the claim is proved.

To complete the picture, note that the sheafifications of N and F, let us denote them by $\tilde{N}$ and $\tilde{F}$, are locally free sheaves, and that $H^0_*(\mathbf{P}^4, \tilde{N})=N$.

THEOREM 3.9. *Let* $\mathcal{L}$ *be an even liaison class of surfaces of* $\mathbf{P}^4$, $(M_1,M_2,b)$ *its cohomological invariants and* N *be determined by* $b_1 \in \mathrm{Ext}^1(L_2,K_1)$.

*Then there exists an exact sequence*

$$0 \to \bigoplus_{n \in \mathbf{Z}} \mathcal{O}_{\mathbf{P}}(-n)^{q(n)} \to \tilde{N} \to \mathcal{J}_X(h) \to 0$$

*such that* X *is a surface in the minimal shift of* $\mathcal{L}$, *and* $h = \deg\tilde{N} - d(\tilde{N})$ (see 3.7).

PROOF. The existence of the sequence and of X is a consequence of theorem 3.7,b, as in [9], prop.IV.4.1. The minimality of X follows from the fact that if S is evenly linked to X, then it has a resolution of the form

$$0 \to \mathcal{P} \to \tilde{N} \oplus \mathcal{A} \to \mathcal{J}_S(h') \to 0,$$

with $\mathcal{P}$ and $\mathcal{A}$ direct sums of line bundles. Then as a consequence of theorem 3.7, with an argument similar to that used in [9], prop. IV.3.7, we get that $h' \geq h$, and hence X is in the minimal shift (see also [9],IV,4.3).

REMARK 3.10. In order to get a smooth surface in $\mathcal{L}$, it is enough to twists $\tilde{N}$ in such a way that $\tilde{N}(t)$ is globally generated. The zero locus of a set of rank($\tilde{N}$)-1 general sections of $\tilde{N}(t)$ is then a smooth surface (see [8]) belonging to $\mathcal{L}$.

## REFERENCES

[1] E. Ballico and G.Bolondi, Deficiency modules and specializations, Proc. Amer. Math. Soc. **108** 1 (1990), 43-48

[2] E. Ballico, G. Bolondi and J.C. Migliore:, The Lazarsfeld-Rao problem for two-codimensional subschemes of $\mathbf{P}^n$, Amer. J. Math. **113** (1991), 117-128.

[3] G. Bolondi and J.C. Migliore, The structure of an even liaison class, Trans. Amer. Math. Soc. **316** 1 (1989), 1-37

[4] G. Bolondi and J.C. Migliore, Configurations of linear projective subvarieties, In: *Algebraic curves and projective geometry*, Proceedings, Trento 1988, Lecture Notes in Mathematics **1389**, Springer-Verlag (1989), 19-32.

[5] G. Bolondi and J.C. Migliore, The Lazarsfeld-Rao property on an arithmetically Gorenstein variety, Preprint UTM 367 (1992), to appear in Manuscripta Mathematica

[6] W. Decker, L. Ein and F.O. Schreyer, Construction of surfaces in $\mathbf{P}^4$, Preprint 124, Bayreuth,Saarbrucken (1991)

[7] G. Horrocks, Vector bundles on the punctured spectrum of a local ring, Proc. London Math. Soc. (3) **14** (1964), 689-713

[8] S. Kleiman, Geometry on Grassmannians and applications to splitting bundles and smoothing cycles Publ. IHES **36** (1969), 282-298

[9] M. Martin-Deschamps and D. Perrin, *Sur la classification des courbes gauches* Astérisque, **184-185**, Soc. Math. Fr., Paris, 1990.

[10] C. Okonek, Flächen vom grad 8 in $\mathbf{P}^4$, Math. Z. **191** (1986), 207-233

[11] A.P. Rao, Liaison equivalence classes, Math. Ann. **258** (1981), 169-173.

DIPARTIMENTO DI MATEMATICA, UNIVERSITÀ DI TRENTO, I-38050 POVO (TRENTO), ITALY
*E-mail address* : IN%"Bolondi@ITNCISCA"

Contemporary Mathematics
Volume **162**, 1994

# The Class $\mathcal{C}$ is Not Stable By Small Deformations. II

F. CAMPANA

ABSTRACT. **We show that some rational threefolds admit arbitrarily small deformations which are no longer bimeromorphic to any compact Kähler manifoldsn thus contradicting the bimeromorphic analogue of Kodaira-Spencer's stability theorem.**

## § 0. Introduction.

Recall that a compact complex space $X$ is said to be in the class $\mathcal{C}$ if it is bimeromorphic to some compact complex Kähler manifold. Moishezon spaces are thus in $\mathcal{C}$.

The class $\mathcal{C}$ is stable under most natural operations : projective morphisms, product, meromorphic images and even the irreductible components of the cycle space $\mathcal{C}(X)$ of $X$ are compact and in $\mathcal{C}$ if so is $X$ ([C1]).

The stability theorem of Kodaira-Spencer ([K-S]) asserts that small deformations of compact Kähler manifolds $X$ are again Kähler.

For these reasons, it seems natural to expect the same property for the class $\mathcal{C}$. (*)

In [C3], we have shoown by a conterexample $X$ that this is false already in the smallest dimension (three) where non-Kähler manifolds of the class $\mathcal{C}$ exist. See also [L-P] for a similar proof.

We thus don't have any class enjoying both the geometric stability properties of $\mathcal{C}$ and the deformation stability of Kähler manifolds. Remark that, after [C2], the property of compactness of irreductible components of cycle spaces may also be lost by small deformations.

The $X$ considered here is even Moishezon : it is one of the twistor spaces recently constructed by $\mathcal{C}$. Lebrun ([L]) and subsequently by H. Kurke for $n = 3$ explicity ; he observes then that his construction can be extended to arbitrary $n$ (see [K]). Thus, $X$ is rational, with $\kappa(X, K_X^{-1}) = 3$. In particular, the Kodaira dimension $\kappa(X)$ of $X$ is $(-\infty)$.

---

(*) 1991 Mathematics Subject Classification : Primary 32G13 ; 14J30 ; 32L25; 32 J 17 ; 32J27. This paper is in final form, and no version of it will be submitted for publication elsewhere.

This naturally raises the question : are small deformations of $X$ again in $\mathcal{C}$ if $X$ is in $\mathcal{C}$ with $\kappa(X) \geq 0$ (or even : $\kappa(X) = \dim_{\mathbf{C}}(X)$) ? The answer is not necessarily expected to be yes.

Observe however that if $X$ is a Moishezon manifold of general type (i.e. $\kappa(X, K_X) = \dim_{\mathbf{C}}(X)$) of which the canonical bundle is numerically effective (see 1.0 below), then the small deformations of $X$ are again of this type. So the above question can be answered by yes in this case. (See 3.16 in [C3]). However, the proofs in [C3] and [L-P] heavily depended on the theory of twistor spaces.

We present here a new proof, independent from this theory. It is based on the ad hoc notion of T-space, which makes the proof very easy (almost by definition), and purely algebro-geometric. In particular, no differential geometric considerations are needed any longer. Hopefully, this simplification might also lead to new examples.

Actually, although our $X$ is Moishezon, it is not topologically projective (ie: not homeomorphic to any projective threefold), by [C-P].

## §1. Algebraic dimension of $T$-spaces.

**1.0. Notations :** By $Z$ or $X$ we denote a compact connected complex manifold, with $K_Z$, $K_Z^{-1}$ and $T_Z$ respectively its canonical, anticanonical and tangent bundles.

By $a(X)$, we denote the algebraic dimension of $X$. By $\kappa(X, L)$, where $L \in \mathrm{Pic}\,(\mathrm{X})$, we denote the Iitaka dimension of $L$, namely :

$$\kappa(X, L) := \limsup_{m \to +\infty} \left(\log h^0(X, L^{\otimes m}) / \log m\right).$$

Let then $\kappa^*(X) := \kappa(X, K_X^{-1})$ be the **anticanonical Kodaira** dimension of $X$.

For $L$ in Pic (X), let $\overline{L}$ be its **conjugate line bundle**, defined by the conjugate transition functions. A map $F : X \to X$ is said to be **antiholomorphic** if $F^* J_X = -J_X$.

**1.1. Définition :** *A compact connected smooth threefold $Z$ is said to be a $T$-space if :*

i) $H^i(Z, \mathcal{O}_Z) = 0 \quad (i = 1, 2)$.

ii) *$Z$ is rationally connected (ie : two generic points in $Z$ can be joined by some irreductible rational curve of $Z$).*

iii) *There exists an antiholomorphic involution $\sigma : Z \to Z$ such that : ($\forall L \in$ Pic (Z)) one has : $\sigma^*(\overline{L}) \cong L$ iff : $\exists\, p, q > 0$ integers such that : $L^{\otimes q} \cong K_Z^{\otimes p}$.*

**1.2. Example :** Any Moishezon twistor space is a $T$-space (see [C3] and references there).

**1.3. Remark :** From 1.1. i) and ii) it follows that : $\mathrm{Pic}\,(\mathrm{Z}) \cong \mathrm{H}^2(\mathrm{Z}, \mathbf{Z})$ and that $K_Z$ is not torsion. Thus, iii) means that $\sigma^*$ acts on $H^2(Z, \mathbf{Z})$ with a single anti-invariant direction. This is just a topological condition on $\sigma$.

One could replace 1.1. ii) by condition : 1.1. ii)' $Z$ is uniruled.

**1.4. Proposition :** *Let $Z$ be a $T$-space. Then :*

i) *$Z \in \mathcal{C}$ iff $Z$ is Moishezon iff all irreducible components of the Chow Scheme $\mathcal{C}(Z)$ of $Z$ are compact.*

ii) *$a(Z) = \kappa^*(Z)$, unless $a(Z) = 0$ and $\kappa^*(Z) = -\infty$.*

iii) *$Z \in \mathcal{C}$ iff $\kappa^*(Z) = 3$, in which case $\pi_1(Z) = 0$.*

**Proof :** i) as well as : $\pi_1(Z) = 0$ if $Z$ is in $\mathcal{C}$ are proved in [C2], using 1.1. ii). Also : ii) implies the first part of iii). Let us prove the remaining ii) : let $L \in \mathrm{Pic}\,(\mathrm{Z})$ be such that : $\kappa(Z, L) = a(Z)$. Put : $M := L \otimes \sigma^*(L)$. First : $a(Z) = \kappa(Z, M) \geq \kappa(Z, L) = a(Z)$, and so : $\kappa(Z, M) = a(Z) \geq 0$. We also obviously have : $\sigma^*(\overline{M}) \cong M$, since $\sigma^2 = 1_Z$. Thus : $M^{\otimes q} \cong K_X^{\otimes p}$ for some $p, q > 0$ integers. Thus : either $a(Z) = p = 0$ in which case :

$$\kappa^*(Z) \in \{-\infty, 0\}, \text{ or : } a(Z) \geq 1 \text{ and } p \neq 0.$$

So : $p < 0$, since $Z$ is uniruled. Thus : $a(Z) = \kappa(Z, M^{\otimes q}) = \kappa(Z, M) = \kappa(Z, K_Z^{\otimes p}) = \kappa(Z, K_Z^{-1}) = \kappa^*(Z)$ as desired.

**1.5. Proposition :** *Let $Z$ be a $T$-space. Assume that there exists a smooth member $D$ of $|K_Z^{-1/2}|$. Then :*

$$a(Z) \leq 1 + \kappa(D, K_D^{-1}),$$

*(where : $1 + (-\infty) = 0$ by convention, here, and $K_Z^{-1/2}$ is any $L$ in* Pic (Z) *with $L^{\otimes 2} \cong K_Z^{-1}$).*

**Proof :** By adjonction formula $K_{Z|D}^{-1/2} \cong K_D^{-1}$. The first terms of the cohomology sequence of :

$$0 \to K_Z^{-m/2} \to K_Z^{-(m+1)/2} \to K_D^{-(m+1)} \to 0 \quad (m \geq 0)$$

show that : $\kappa(Z, K_Z^{-1/2}) \leq 1 + \kappa(D, K_D^{-1})$. But, 1.3. ii) shows that : $a(Z) = \kappa(Z, K_Z^{-1/2})$. From this, the conclusion follows. Observe that in the situation of 1.5, we have :

**1.5.1. Remark :** If $Z$ is Moishezon and if $\kappa(Z, K_Z^{-1/2}) \geq 1$, then $D$ is a rational surface. Indeed: $D$ is then projective, with $\kappa(D, K_D^{-1}) \geq 0$.

**1.6. Definition :** *Let $D$ be a smooth rational surface with $K_D^2 = c_1(D)^2 \leq 0$. We say that $D$ is generic if: $\kappa(D, K_D^{-1}) \leq K_D^2$.*

Remark that generic surfaces $D$ exist for every value of $K_D^2 \leq 0$. More precisely, if $D$ is the blow-up of $\mathbf{P}_2(\mathbf{C})$ in $(9 - K_D^2) \geq 9$ points, the generic member of its Kuranishi family is generic in the sense of 1.6. Indeed : we just need to deform the blown-up points in such a way that they lie on no (resp. exactly one) cubic curve of $\mathbf{P}_3(\mathbf{C})$ if $K_D^2 < 0$ (resp. $K_D^2 = 0$).

**1.7. Corollary :** *Let $Z$ be a $T$-space. Assume that $|K_Z^{-1/2}|$ contains a smooth member $D$ which is a generic rational surface. Then : $a(Z) = 0$ if $c_1^3(Z) < 0$, and $a(Z) \leq 1$ if $c_1^3(Z) = 0$. In particular : $Z \notin \mathcal{C}$.*

**Proof :** By 1.5. we know that :

$$a(Z) \leq 1 + \kappa(D, K_D^{-1}) \leq 1 + K_D^2$$

(where $1 + (-\infty) = 0$).

But : $8.K_D^2 = c_1(Z)^3$. Hence the result.

**1.8. Remark :** The proof of 1.7. actually shows more, namely : $|K_Z^{-m/2}| = \{mD\}$ for any $m \geq 0$. This means : $Z$ contains only one effective irreductible divisor : $D$.

Our next task is to realize 1.7. by deforming a pair $(X_0, D_0)$ with the properties 1.7., except that $X_0$ is Moishezon, and so $D_0$ non-generic.

## §2. Deformations of $T$-spaces.

**2.1. Proposition :** *Let $Z$ be a $T$-space with $H^2(Z, \Theta_Z) = 0$. Let $\pi : Z \to S$ be the Kuraniski family of $Z := \pi^{-1}(0), 0 \in S$, and let $Z_s := \pi^{-1}(s), s \in S$. Then :*

1. *$\sigma$ extends to antiholomorphic involutions (also denoted $\sigma$) :*

$$\sigma : Z \to Z \text{ and } \sigma : S \to S, \quad \pi \text{ being} \quad \sigma\text{-equivariant.}$$

2. *The fix-point set $S^\sigma$ of $S$ is a totally real smooth submanifold of $S$ of half its dimension.*
3. *If $s \in S^\sigma$, then $Z_s$ is a $T$-space.*
4. *For $s, s'$ general in $S$ and in $S^\sigma$ respectively, one has : $a(Z_s) = a(Z_{s'})$.*

**Proof :** The assertion 1 and 2 follow directly from the versality of the Kuraniski family.

To show 3, just remark that $R^i\pi_*\mathcal{O}_Z = 0$ near $0 \in S$, that rational connectedness is an open condition, and that if $s \in S^\sigma$, then $\sigma$ (acting on $Z$) preserves $Z_s$, is involutive and antiholomorphic there, and has a single anti-invariant direction in $H^2(Z_s, \mathbf{Z}) = H^2(Z, \mathbf{Z})$.

Now to show 4, observe that $S^\sigma$ being Zariski dense in $S$, the rank of $K_Z^{-m}$ at $s$ and $s'$ coincide for every $m > 0$. Then, by 1.4., $\kappa^*(Z_s)$ determines $a(Z_s)$ for any $s \in S'$.

**2.2. Proposition :** *Let $Z$ be a compact connected threefold. Let $D \in |K_Z^{-1/2}|$ be smooth. Asume that :*

1. $H^2(Z,\Theta_Z) = H^2(D,\Theta_D) = 0$
2. $q(D) = P_g(D) = 0$
3. *The restriction map $H^1(Z,\Omega^1_Z) \to H^1(D,\Omega^1_D)$ is injective.*

*Then : any sufficiently small deformation of $D$ appears as a smooth member of $|K_{Z'}^{-1/2}|$ for $Z'$ an adequate deformation of $Z$.*

**Proof:** For relative deformation theory used here, we refer to [Ka]. Let $\Theta_{Z,D} :=$ Ker $(\alpha)$, where $\alpha : \Theta_Z \to N_{D|Z}$ is the natural map, with $N_{D|Z}$ the normal sheaf (on $Z$) to $D$ in $Z$.

The cohomology of the two exact sequences $c$ and $d$ below show that it is sufficient to show that :

**a.** $H^2(Z,\Theta_{Z,D}) = 0$
**b.** $H^1(Z,\Theta_{Z,D}) \to H^1(D,\Theta_D)$ is surjective.
**c.** $0 \to \Theta_Z(-D) \to \Theta_{Z,D} \to \Theta_D \to 0$
**d.** $0 \to \Theta_{Z,D} \to \Theta_Z \to N_{D|Z} \to 0$

So, let us show the assertions a and b : both of then follow from :

**e.** $H^2(\Theta_Z(-D)) = 0$,

which we now show : by Serre's duality :

$$h^2(Z,\Theta_Z(-D)) = h^1(Z,\Omega^1_Z(D+K_Z)) = h^1(Z,\Omega^1_Z(-D))$$

The cohomology of :

$$0 \to \mathcal{O}_D(-D) \to \Omega^1_{Z|D} \to \Omega^1_D \to 0$$

shows that (e) holds if so does :

**f.** $H^0(D,\Omega^1_{Z|D}) = 0$, by our assumption.

But (f) holds true since : $h^0(D,\Omega^1_D) = h^0(D,\mathcal{O}_D(-D)) = 0$. Indeed : $\mathcal{O}_D(-D) \cong K_D$.

**2.3. Theorem :** *Let $Z$ be a (Moishezon) $T$-space such that :*

1. $H^2(Z,\Theta_Z) = 0$
2. $\exists\, D \in |K_Z^{-1/2}|$ *; rational smooth.*
3. *The restriction map : $H^2(Z,\mathbf{R}) \to H^2(D,\mathbf{R})$ is injective.*
4. $c_1^3(Z) \leq 0$

*Then : any sufficiently small general deformation $Z'$ of $Z$ is such that : $a(Z') = 0$ (resp. $a(Z') \leq 1$) if $c_1^3(Z) < 0$ (resp. if $c_1^3(Z) = 0$). In particular : $Z' \notin \mathcal{C}$.*

**Proof :** $Z$ admits arbitrarily small deformations $Z'$ mapped to $S^\sigma$, containing $D' \in |K_{Z'}^{-1/2}|$ which is rational generic, by 2.2 (because $S^\sigma$ is Zariski dense in $S$). On the other hand, $c_1^3(Z) = 8c_1^2(D) \leq 0$. Then apply 1.9 to conclude.

We shall now show directly that Kurke-Lebrun twistor spaces fulfill all conditions in 2.3.

## §3. Kurke-Lebrun $T$-spaces

We give now the examples of Lebrun.

**3.1. Notations :** Let $Q := \mathbf{CP}_1 \times \mathbf{CP}_1$ ; for $(a,b) \in \mathbf{Z}^2$, let $\mathcal{O}_Q(a,b) \in \mathrm{Pic}\,(\mathrm{Q})$ inducing on either ruling of $Q$, $\mathcal{O}(a)$ and $\mathcal{O}(b)$ respectively.

Let $E$ be a vector bundle on $Q$ ; let $\pi : P := \mathbf{P}(E) \to Q$ be the natural projection, where $\mathbf{P}(E)$ is the set of lines of $E$ (we **don't** take Grothendieck's definition).

Let $\mathcal{O}_p(k), k \in \mathbf{Z}$ be the natural line bundle on $P$ associated to $E$ and $k$.

Thus :

$$\pi_* \mathcal{O}_p(k) \cong \begin{cases} 0 & k < 0 \\ \mathrm{Sym}^k(E^*) & k \geq 0 \end{cases}$$

where $E^* := \mathrm{Hom}\,(\mathrm{E}, \mathcal{O}_{\mathrm{p}})$.

For $(a,b,k) \in \mathbf{Z}^3$, let $\mathcal{O}_p(k)(a,b) :\cong \left[\mathcal{O}_p(k) \bigotimes_{\mathcal{O}_p} \pi^* \mathcal{O}_Q(a,b)\right]$, and $\pi^* E(k) :=$

$\pi^* E \bigotimes_{\mathcal{O}_p} \mathcal{O}_p(k)$.

Let $\sigma$ be the antiholomorphic involution on $Q$ sending $(u,v)$ to $(\overline{v};\overline{u})$ ; it induces a natural isomorphism, also denoted $\sigma$, between $\mathcal{O}_Q(a,b)$ and $\mathcal{O}_Q(b,a)$.

**3.2. Construction :** In the rest of this paper, we take : $E := \mathcal{O}_Q(n-1,1) \oplus \mathcal{O}_Q(1,(n-1) \oplus \mathcal{O}_Q$, where $n \in \mathbf{N}$ is a fixed nonnegative integer. We let $\sigma$ act in a natural way on $E$ and $\mathbf{P}(E) := P$.

Let $(f_1, \cdots, f_n) \in \Gamma(\mathcal{O}_Q(1,1))^n$ be such that their respective zero loci $(G_1, \cdots, G_n)$ are smooth conics meeting transversally.

The curve $G := G_1 + \cdots + G_n$ has thus $2\binom{n}{2}$ ordinary double points.

We define $h \in \mathrm{Hom}\,(\mathcal{O}_{\mathrm{p}}(-2), \pi^* \mathcal{O}_{\mathrm{Q}}(\mathrm{n},\mathrm{n}))$ by :

$$h(x,y,z) := x \otimes y - z \otimes z \otimes \Big(\bigotimes_{j=1}^{n} f_j(q)\Big),$$

where $(x,y,z) \in \mathcal{O}_p(-1) \subset \pi^* E$ and $q := \pi(x,y,z)$. More precisely : $x \in \pi^* L_1$; $y \in \pi^* L_2$, and $z \in \pi^* L_3$, where $l_1 = \mathcal{O}_Q(n-1,1)$ ; $L_2 = \mathcal{O}_Q(1,n-1)$, and $L_3 = \mathcal{O}_Q$.

Observe that, if all $f_i$ are Real (ie : $\sigma$-invariant), so is $h$, and so is $Z'$, which is by definition the zero locus of $h$ in $P$.

Thus : $Z'$ has exactly $2\binom{n}{2}$ singular points, all of them ordinary quadratic.

Let $S_1$ and $S_2$ respectively be the two surfaces of $Z'$ which are images of the sections $S_1, S_2$ of $\pi$ defined by : $S_1(q) := (0,y,0)$ and $S_2(q) := (x,0,0)$ with $x \neq 0$ and $y \neq 0$, $q \in Q$.

Both surfaces are contained in the smooth loci of $Z'$, and they have respectively normal bundles isomorphic to $\mathcal{O}_Q(1-n,-1)$ and $\mathcal{O}_Q(-1,1-n)$.

Let now $\eta : Z' \longrightarrow Z''$ be the contraction of $S_1$ and $S_2$ to smooth rational curves $C_1, C_2$ of normal bundles $\mathcal{O}_{C_1}(1-n) \oplus \mathcal{O}_{C_i}(1-n) \quad (i = 1,2)$.

Let now $\epsilon : Z^+ \longrightarrow Z'$ (resp. $\epsilon' : Z \longrightarrow Z''$) be small (compatible) resolutions of the quadratic singularities of $Z'$ and $Z''$. We thus have a commutative diagramm :

$$\begin{array}{ccc} Z^+ & \stackrel{\beta}{\longrightarrow} & Z \\ \epsilon \downarrow & & \downarrow \epsilon' \\ Z' & \stackrel{\eta}{\longrightarrow} & Z'' \end{array}$$

in which $\beta$ in the blow-up of $C = C_1 + C_2$. Remark that we can choose $\epsilon$ in $2^{\binom{n}{2}}$ ways so that $\sigma$ lifts to $Z^+$ equivariantly if all $f_j (j \leq n)$ are Real.

Assume now that all $f_j$ are Real, and that $\sigma$ lifts to $Z^+$.

**3.3. Lemma :**

i) $\beta^* K_Z^{-1} \cong \epsilon^* \mathcal{O}_{Z'}(2)(2,2)( :\cong \epsilon^* \mathcal{O}_P(2)(2,2)_{|Z'})$. *We put* : $K_Z^{-1/2} := \beta_* \epsilon^* \mathcal{O}_{Z'}(1)(1,1)$.

ii) *The generic member of* $|K_Z^{-1/2}|$ *is smooth, isomorphic to* $Q$ *blown-up in* $2n$ *points.*

iii) $c_1^3(Z) = 16(4-n)$.

**Proof :** Remark that $\mathcal{O}_{P|Z'}(1) = \mathcal{O}_{Z'}(S_1 + S_2)$. By adjonction formula, $K_{Z'}^{-1} = \mathcal{O}_{P|Z'}(1)(2,2)$. Now :

$$\begin{aligned} \beta^* K_Z = K_{Z^+} + S_1 + S_2 &= \epsilon^* \left( \mathcal{O}_{P|Z'}(1)(2,2) + S_1 + S_2 \right) \\ &= \epsilon^* \left( \mathcal{O}_{P|Z'}(2)(2,2) \right), \text{ which is i).} \end{aligned}$$

From this, we get : $|K_Z^{-1/2}| = \beta_* \epsilon^* \pi^* \left( |\mathcal{O}_Q(1,1)| \right)$.

Let now $D$ be the image of $C$, element of $|\mathcal{O}_Q(1,1)|$, by $\beta_* \epsilon^* \pi^*$.

We assume, as we can, that $C$ meets $G = G_1 + \cdots + G_n$ transversally at $2n$ distinct smooth points $(a_j, b_j) := (C \cup G_j)$, where $j \in J := \{1, \cdots, n\}$.

Let $D^+ := \epsilon^* \pi^* (C)$ : it is mapped isomorphically to $D$ by $\beta$.

Observe now that $\pi\epsilon : D^+ \longrightarrow C$ is a conic bundle with singular fibers $(F_j + F_j')$ (resp. $(F_j'' + F_j'''))$ over $a_j$ (resp. $b_j$). We choose $(F_j + F_j'')$ (resp. $(F_j' + F_j'''))$ to satisfy $x = 0$ (resp. $y = 0$).

From the above description, one easily concludes that $D$ is isomorphic to $Q$ blown-up in $2n$ points, because $C_1$ and $C_2$ are $(-n)$-curves of $D$.

(Remark that the blown-up points of $Q$ are in a very special position, easily described. We don't use this fact in the sequel).

Hence ii). But : $c_1^3(Z) = 8c_1^2(D) = 8.(8-2n)$ as claimed in iii).

We now describe explicit generators of $H^2(Z, \mathbf{Q})$.

For any $j \in \{1,\cdots,n\} := J$, let $D_j^+$ be the inverse image by $B$ of the irreducible component of $(\pi^*(G_j) \cap Z')$ defined by : $x = 0$. In other words : $D_j^+ := \epsilon^*((f_j = 0) \cap (x = 0))$. Let $D_j := \beta_*(D_j^+)$.

**3.4. Lemma :**
i) *The classes of* $K_Z^{-1/2}$ *and* $D_j$, $j \in J$, *form a basis of* $H^2(Z, \mathbf{Q})$.
ii) *The restriction map from* Pic (Z) *to* Pic (D) *is injective.*

**Proof :** We show first that the $(n+1)$ given classes are linearly independant over $\mathbf{Q}$.

This follows from the relations :

$$D_j.F_k = \begin{cases} 0 & \text{if } j = k \\ 1 & \text{if } j \neq k \end{cases}$$

$$D_j.F'_k = \begin{cases} 1 & \text{if } j = k \\ 0 & \text{if } j \neq k \end{cases}$$

$$K_Z.F_j = K_Z.F'_j = -2$$

for $j, k \in J$, for $F_j$ (resp. $F'_j$) defined as above and all intersection numbers being defined in $Z$.

To show the second assertion (3.4. ii)), we remark that $D_j$ restricts to $D$ as the union of $C_1$ and two $(-1)$-curves $F_j$ and $F''_j$ for $j \in J$, with $(F_j + F''_j)$ smooth and disjoint from $(F_k + F''_k)$ if $j \neq k$. Let $A = (C_1 + F_1 + F''_1)$. Then $(K_Z^{-1/2})_D$ is represented by : $A + \sigma(A)$. This shows the assertion, and hence 3.4. ii).

It remains to shwo that $H^2(Z, \mathbf{Q})$ is generated by the given classes. Let $f := \pi\epsilon : Z^+ \to Q$. Since $R^1 f_*\mathbf{Z} = 0$ and $f_*\mathbf{Z} \cong \mathbf{Z}$, we get, using Leray's spectral sequence for $f$ :

$$H^2(Z^+, \mathbf{Q}) \cong H^2(Q, \mathbf{Q}) \otimes H^0(Q, R^2F_*\mathbf{Q})$$

and so :

$$b_2(Z^+) = 2 + h^0(Q, R^2F_*\mathbf{Q}) = 2 + b_2(Z);$$

thus :

$$b_2(Z) = h^0(Q, R^2F_*\mathbf{Q})$$

For $q \in Q$, let $Z_q^+ := F^{-1}(q)$. Then :

$$(R^2F_*\mathbf{Z})_q \cong H^2(Z_q^+, \mathbf{Z}) \cong \begin{cases} \mathbf{Z} & \text{if } q \notin C \\ \mathbf{Z}^2 & \text{if } q \in (C_i/C_j) \quad (j \neq i) \\ \mathbf{Z}^3 & \text{if } q \in (C_i \cap C_j) \end{cases}$$

Now, because for each $j = 1, \cdots, n$, $F^{-1}(C_j)$ splits in two irreducible components, we see that :

$$h^0(Q, R^2F_*\mathbf{Q}) \leq n+1.$$

This concludes the proof.

Observing that : $\sigma(\overline{D}_j) = \beta_*(D'_j)$ $(\forall j \in J)$, with $D_{j'} := (y = F_j = 0)$, we get easily :

**3.5. Proposition :** *$Z$ is a Moishezon $T$-space ; moreover :*

i) $c_1^3(Z) = 16(4-n)$

ii) $|K_Z^{-1/2}|$ *contains smooth rational members $D$ such that :*

iii) *the restriction map : $H^2(Z,\mathbf{Q}) \to H^2(D,\mathbf{Q})$ is injective.*

**3.6. Note :** To fulfill all conditions of 2.3, it thus only remains to check that $H^2(Z,\Theta_Z) = 0$, and to choose $n \geq 4$.

**3.7. Proposition :** *Let $n \geq 3$. Then : $h^i(Z,T_Z) = 0$ for $i \geq 2$.*

**Proof :** It suffices to show :

$$h^i(Z^+, T_{Z^+}) = 0 \qquad (i \geq 2).$$

By [F], 3.4 and 4.1, we have :

$$h^i(Z^+, T_{Z^+}) \cong h^i(Z', T^0_{Z'}) \text{ for } i \geq 2.$$

We are thus reduced to show the same vanishing for $h^i(Z', T^0_{Z'})$.

We have an exact sequence :

3.7.a) $$0 \to T^0_{Z'} \to T_{P|Z'} \to \mathcal{O}_{Z'}(Z') \to T^1_{Z'} \to 0,$$

where $\mathcal{O}_{Z'}(Z') :\cong \mathcal{O}_p(Z')_{|Z'}$ and $T^1_{Z'}$ is concentrated over the singular points of $Z'$, where its stalk is $\mathbf{C}$. This sequence splits as :

3.7.a′) $$0 \to T^0_{Z'} \to T_{P|Z'} \to F \to 0,$$

3.7.a″) $$0 \to F \to \mathcal{O}_{Z'}(Z') \to T^1_{Z'} \to 0.$$

**3.8. Lemma :** *We have :*

i) $H^i(P, T_p(-Z)) = 0 \qquad (i \geq 0)$

ii) $H^i(P,T_p) \cong H^i(P, T_{P|Z}) \qquad (i \geq 0)$

iii) $H^i(P, \mathcal{O}_p(Z')) \cong H^i(Z', \mathcal{O}_{Z'}(Z')) \qquad (i \geq 1)$

iv) $H^i(P,T_p) = 0 \quad (i \geq 2)$ *and* $H^1(P,T_p) \cong H^1(P, \pi_*T_p) \cong H^1(Q, \mathrm{Hom}_0(\mathrm{E},\mathrm{E}))$, *where* $\mathrm{Hom}_0(\mathrm{E},\mathrm{E})$ := *sheaf of traceless endomorphisms of $E$.*

v) $H^i(P, \mathcal{O}_p(Z')) \cong H^i(P, \pi_* \mathcal{O}_p(Z')) \cong H^i(Q, \mathrm{Sym}^2(\mathrm{E}^*(\mathrm{n},\mathrm{n}))$.

**Proof :** Standard, using the Euler exact sequence :

$$0 \to \mathcal{O}_p \to (\pi^* E)(1) \to T_p \to \pi^* T_Q \to 0.$$

**3.9. Lemma :** *The map* $H^0(Z', \mathcal{O}_{Z'}(Z')) \to H^0(Z', T^1_{Z'})$ *induced from 3.7. a) is surjective, and the map induced from 3.7. a") :*

$$H^i(Z', F) \to H^i(Z', \mathcal{O}_{Z'}(Z')) \text{ are isomorphic, if } i \geq 1.$$

**Proof :** Obvious, because $Z'$ can be smoothed by just replacing $(\prod_{j=1}^n f_j)$ by a generic section of $\mathcal{O}_Q(n,n)$ in the definition of $h$ in 3.2.

**3.10. Corollary :**
i) $H^i(Z', T^0_{Z'}) = 0 \quad (i \geq 3)$
ii) $H^2(Z', T^0_{Z'}) = 0$ *iff the map*

$$\alpha_1 : H^1(Q, \mathrm{Hom}_0(\mathrm{E},\mathrm{E})) \to \mathrm{H}^1(\mathrm{Q}, \mathrm{Sym}^2(\mathrm{E}^*)(\mathrm{n},\mathrm{n}))$$

*defined below is surjective.*

**Proof :** Consider the following sequence and isomorphisms for $i \geq 1$ :

$$H^i(Z', T_{p|Z'}) \xrightarrow{\alpha_i} H^i(Z', F) \to H^{i+1}(Z', T^0_{Z'}) \to H^{i+1}(Z', T_{p|Z'})$$

$$H^i(Z', T_{P|Z'}) \cong H^i(P, T_p) \cong H^i(Q, \mathrm{Hom}_0(\mathrm{E},\mathrm{E}))$$

$$H^i(Z', F) \cong H^i(Z', \mathcal{O}_{Z'}(Z')) \cong H^i(P, \mathcal{O}_P(Z')) \cong H^i(Q, \mathrm{Sym}^2(\mathrm{E}^*)(\mathrm{n},\mathrm{n}))$$

$$H^{i+1}(Z', T_{P|Z'}) \cong H^{i+1}(P, T_p) \cong 0.$$

We define $\alpha_1$ by composition.

**3.11. Proposition :** *The map* $\alpha_1$ *of 3.10 is an isomorphism if* $n \geq 4$.

**Proof :** The map $T_{P|Z'} \to \mathcal{O}_{Z'}(Z')$ is induced by the derivative of the section $h \in \Gamma(P, \mathrm{Hom}(\mathcal{O}_{\mathrm{p}}(-2), (\mathrm{n},\mathrm{n}))$. Because the composite : $(H^1(P, T_{P|Z'}) \longleftarrow H^1(P, T_p) \longleftarrow H^1(P, \pi^* E(1)))$ is surjective, we need only to identify the vertical component of this derivative, which is a bilinear map on the fibers of $\mathcal{O}_p(-1)$ with values in $\pi^* \mathcal{O}_Q(n,n)$.

For $a \in P, \tilde{a} \in \mathcal{O}_p(-1)$ lying above $a$, and $t$ (resp. $l^*$) being elements of $E_q$ (resp. $E^*_q$), with $q = \pi(a)$, we get :

$$< dh_a, t \otimes l > (\tilde{a}) := \big(x(\tilde{a}) \otimes y(t) + x(t) \otimes y(\tilde{a})\big).l(\tilde{a}),$$

where $x$ (resp. $y$ ; resp. $z$) are the projections of $\pi^*E$ to its $\pi^*L_1$ (resp. $\pi^*L_2$ ; resp. $\pi^*L_3$) components.

Let $dh_* : \pi_*(\pi^*E(1)) \longrightarrow \pi_*(\mathcal{O}_p(2)(n,n))$ be the map induced from $(dh)$ at the level of $\pi_*$. An easy verification shows that if $t_i$ (resp. $l_j^*$) are germs of sections of $L_i$ (resp. $L_j^*$), then

$$dh_*(t_i \otimes l_j^*) = \begin{cases} 0 & \text{if } i = 3 \\ t_1 \otimes t_2 \otimes l_{i'}^* \otimes l_j^*) & \text{if } i \leq 2 \end{cases}$$

where $i'$ is the complement of $\{i\}$ in $\{1,2\}$, for some well-defined non-zero element $(t_{i'}^* \otimes l_{i'}^*)$. In particular, $dh_*$ induces a natural isomorphism from $L_i \otimes L_j^*$ to $L_1 \otimes L_2 \otimes L_{i'}^* \otimes L_j^*$.

The conclusion now follows from the fact that, for $n \geq 4$, we have :

$$H^1(Q, E \otimes E^*) \cong H^1(Q, L_1 \otimes L_2^*) \oplus H^1(Q, L_2 \otimes L_1^*),$$

allother terms vanishing ; and :

$$\begin{aligned} H^1(Q, \mathrm{Sym}^2(\mathrm{E}^*)(\mathrm{n},\mathrm{n})) \cong & H^1(Q, L_1 \otimes L_2 \otimes L_2^* \otimes L_3^*) \\ & \oplus H^1(Q, L_1 \otimes L_2 \otimes L_1^* \otimes L_3^*), \end{aligned}$$

all other terms vanishing. Hence the assertions 3.11, 3.7.

**3.12. Corollary :** *The assumptions of 2.3 are all satisfied by $Z$ if $n \geq 4$.*

**3.13. Corollary :** *If $n \geq 4$, $Z$ has arbitrarily small deformations which are not in the class $C$, although $Z$ is Moishezon. The general small deformations of $Z$ has algebraic dimension zero (resp. one) if $n \geq 5$ (resp. $n = 4$).*

## Bibliography

[C1] F. Campana, "Compacité et algébricité dans l'espace des cycles", Math. Ann. **251** (1980), 7-18.

[C2] F. Campana, "On twistor spaces in the class $C$", Journal of Diff. Geom.**33** (1991), 541-549.

[C3] F. Campana, "The class $C$ is not stable by small deformations", Math. Ann. **290** (1991), 19-30.

[CP] F. Campana, T. Peternell, "Rigidity theorems for Fano 3-folds", Preprint 1991.

[F] R. Friedman, "Simultaneous resolutions of threefold double points", Math. Ann. **274** (1986), 671-699.

[G] H. Grauert, "Ein theorem der analytischen Garbentheorie und die Modulraüme komplexer Strukturen", Publ. Math. IHES, **5** (1960).

[Ka] Y. Kawamata, "On deformations of compactifiable complex manifolds", Math. Ann. **235** (1978), 247-265.

[K-S] K. Kodaira, Spencer, "On deformations of complex analytic structures III", Ann. Math. **71** (1960), 43-76.

[K] H. Kurke, "Twistor spaces on $n$. $\mathbf{CP}_2$", Preprint Erlangen.

[L] C. Lebrun, "Explicit self-dual metrics on $(\sharp n\mathbf{CP}_2)$", Preprint Princeton.

[L-P] C. Lebrun, Y.S. Poon, "Twistors, Kähler manifolds and bimeromorphic geometry II", J. of the American Math. Society, **5** (1992), 317-325.

DEPARTEMENT DE MATHEMATIQUES
UNIVERSITE DE NANCY 1 - BP 239
F-54506 VANDOEUVRE CEDEX

Contemporary Mathematics
Volume **162**, 1994

# Surfaces with $K^2=\chi=2$ and special nets of quadrics in 3-space.

(F. Catanese - P. Cragnolini - P. Oliverio[1])

## 0 Introduction

Surfaces of general type with $K^2=\chi=2$ were first considered by F.Enriques in his book ([En], pages 316-321); later their geometry was investigated by the first named author and O. Debarre in the paper [Ca-De], where it was shown that, beyond the 16-dimensional irreducible component $\mathfrak{M}_E$ which is given by the closure of the family constructed by Enriques, there was a further (rational) component of their moduli space corresponding to surfaces with (2-) torsion. In particular it was conjectured in [Ca-De] that the moduli space should only consist of the above two irreducible components.

---

[1] This research has been partly supported by M.U.R.S.T., and by G.N.S.A.G.A. of C.N.R.
This paper is in final form and no version of it will be submitted for publication elsewhere
1991 Mathematics Subject Classification. Primary 14 J 29 Secondary 14 C 21, 14 M 12

The above surfaces are distinguished by the behaviour of their bicanonical map $\varphi$, which is a morphism to $\mathbb{P}^3$ with image either

0) a quadric cone (this occurs iff the surface S has 2-torsion);

i) a smooth quadric, iff S is simply connected and $\varphi$ is a Galois cover with group $(\mathbb{Z}/2)^2$ (here S belongs to a 11-dimensional subvariety of the Enriques component $\mathfrak{M}_E$);

ii) a K3 quartic symmetroid, iff S is simply connected and $\varphi$ is a double cover unramified in codimension 1 (again, here S belongs to a subvariety of the Enriques component, of dimension 12);

iii) a rational quartic with a double line, iff S is simply connected and $\varphi$ is a double cover ramified in codimension 1 (here S belongs to a 13-dimensional subvariety of the Enriques component);

iv) an octic surface $\Sigma$ everywhere tangent to a plane along a quartic curve $\Gamma$ having some smooth point, and which is the image of the canonical curve C of S (we call this case the <u>non-hyperelliptic birational case</u>, and we recall that if $\Sigma$ has ordinary singularities, then S belongs to the Enriques component);

v) an octic surface intersecting a plane in a plane conic Q counted with multiplicity 4, which is the image of the hyperelliptic canonical curve C (this will be referred to as the <u>hyperelliptic birational case</u>, and in [Ca-De], prop. 5.15, it was proven by an indirect argument that this locus intersects the Enriques component).

The main point in [Ca-De] was to show that, in each of the above cases, one could give a presentation of the canonical ring $\mathcal{R}(S)$ in terms of some matrix A of polynomials whose coefficients had to belong to some explicitly described locally closed set in some space of parameters; but, on the other hand, not for all cases it was possible to "solve" explicitly those polynomial equations (i.e., to show the unirationality of the given locally closed subvariety of the moduli space).

In particular, for class v), even explicit constructions were missing, except some relying on the fact that the double cover of a numerical Godeaux surface ($K^2 = \chi = 1$) with 2-torsion is a surface of this type (these latter surfaces were first shown to exist by F. Oort and C. Peters in [O-P], later a bigger subfamily was constructed by R. Barlow in [Bar]).

The main purpose of the present paper is to shed some light on the subvariety of the moduli space corresponding to surfaces in the class v), also in view of the determination of the moduli space of the numerical Godeaux surfaces with 2-torsion (this is conjectured to be irreducible, see [Re1], [Re2]).

On the other hand, the connection with nets of quadrics is not a peculiarity of the above family of surfaces, but a more general phenomenon.

In fact, given a canonical curve C on an algebraic surface S, the invertible sheaf $\mathcal{O}_C(K_S)$ is a thetacharacteristic on the curve C .

In our case this thetacharacteristic is ineffective and, since the bicanonical morphism $\varphi$ of S gives, by restriction to C , the canonical map of C, the direct image of $\mathcal{O}_C(K_S)$ gives a generalized thetacharacteristic on the schematic image of C (cf. [Ba], and also [Tju], [Bea], [Ca1] for extensions of

Dixon's theory ([Di]) of thetacharacteristics on plane curves).

We hope that our results , describing explicitly certain thetacharacteristics on non reduced plane quartics, may contribute to the understanding of thetacharacteristics on non reduced curves.

To explain in some detail our results, recall that in case v), the image $\varphi(C)$ is a plane conic, and the minimal resolution on $\mathbb{P}^2$ of $\mathcal{F} = \varphi_*(\mathcal{O}_C(K_S))$ is given by a symmetric $(4\times4)$ matrix $\alpha$ of linear forms in 3 homogeneous variables satisfying:

**a)** $\det(\alpha) = Q^2$, where $Q$ is a nonzero quadratic form;
**b)** the determinants of all the $3\times3$ minors of $\alpha$ are divisible by $Q$ (remark that condition b) implies a) unless $Q$ is a square).

In this paper we define a **special net of quadrics** in 3-space as a linear rational map $\alpha$ from $\mathbb{P}^2$ to the projective space of symmetric $4\times4$ matrices, satisfying the above properties a), b).

We shall denote by $\mathcal{S}$ the variety of special nets of quadrics, and we observe that one has a natural $\mathbb{P}GL(3)\times\mathbb{P}GL(4)$ action on this variety. The following theorem completely classifies the $\mathbb{P}GL(3)\times\mathbb{P}GL(4)$-orbits in $\mathcal{S}$ , and their incidence relationships.

## MAIN THEOREM

The variety $\mathcal{S}$ contains a finite number of $\mathbb{P}GL(3)\times\mathbb{P}GL(4)$-orbits, labelled $\mathcal{A},\mathcal{B}_1,\mathcal{B}_2,\mathcal{C}_1,\mathcal{C}_2,\mathcal{C}_3,\mathcal{C}_4,\mathcal{C}_5$ and of respective dimensions 19,18,17,18,17,16,15,11. These orbits satisfy moreover the following incidence relations ($\mathcal{A} \to \mathcal{B}$ means that the closure $\overline{\mathcal{A}}$ of $\mathcal{A}$ contains $\mathcal{B}$ )

$$\begin{array}{ccccccccc} \mathcal{G} & \to & \mathcal{B}_1 & \to & \mathcal{B}_2 & & & & \\ & \searrow & & \searrow & & \searrow & & & \\ & & \mathcal{C}_1 & \to & \mathcal{C}_2 & \to & \mathcal{C}_3 \to & \mathcal{C}_4 & \to \mathcal{C}_5 \end{array}$$

In particular the variety $\mathcal{S}$ is irreducible. Moreover, the singular set of $\mathcal{S}$ is exactly $\overline{\mathcal{B}}_2 = \mathcal{B}_2 \cup \mathcal{C}_3 \cup \mathcal{C}_4 \cup \mathcal{C}_5$ , and at the points of $\mathcal{C}_2$ the orbit closures $\overline{\mathcal{B}}_1, \overline{\mathcal{C}}_1$, are smooth subvarieties of $\mathcal{S}$ intersecting transversally in $\mathcal{C}_2$ .

The above specializations have the following nice geometrical explanations.

Case I. Fix two skew lines in $\mathbb{P}^3$ and consider the space $\mathcal{W}$ of quadrics containing them; $\mathcal{W}$ is a 3-dimensional projective subspace, and for a suitable choice of coordinates a quadric is in $\mathcal{W}$ iff an associated matrix has the form

$$A = \begin{pmatrix} 0 & 0 & a & b \\ 0 & 0 & c & d \\ a & c & 0 & 0 \\ b & d & 0 & 0 \end{pmatrix}$$

Thus on $\mathcal{W}$ det(A) is the square of the quadratic polynomial $\mathcal{Q} = ad - bc$ which divides the adjoint matrix $\Lambda^3 A$. Hence we get a special net of quadrics for any linear rational map $\alpha\colon \mathbb{P}^2 \to \mathcal{W}$ , such that $\alpha(\mathbb{P}^2)$ is not contained in the quadric $\mathcal{Q}$. We get the degenerations $\mathcal{G} \to \mathcal{B}_1 \to \mathcal{B}_2 \to \mathcal{C}_3 \to \mathcal{C}_5$ according to whether $\alpha(\mathbb{P}^2)$ is a plane transversal to, a plane tangent to, a line transversal to, a line tangent to, or a point not lying on $\mathcal{Q}$.

The next case II is a specialization of case I, since, given any pair of skew lines in $\mathbb{P}^3$, one can find a smooth quadric surface F containing them as two lines of one of the two rulings, and one can specialize the two lines in F till they become coincident.

<u>Case II</u>. Fix a line in $\mathbb{P}^3$ and a smooth quadric F containing it, and let $\Gamma$ be the subscheme of F given by the divisor on F consisting of the given line counted with multiplicity 2. Again, the quadrics containing $\Gamma$ form a 3-dimensional projective subspace $\mathcal{V}$, and for a suitable choice of coordinates their matrix A is of the form

$$A = \begin{pmatrix} a & b & d & 0 \\ b & 0 & 0 & 0 \\ d & 0 & c & -b \\ 0 & 0 & -b & 0 \end{pmatrix}$$

and $\det(A) = b^4$, while $b^2$ divides $\Lambda^3 A$.
The situation here is slightly different, since the rank of A is 4 outside of the plane $\{b=0\}$, 2 on the points of the plane which lie outside of the conic $C=\{b=ac-d^2=0\}$, and 1 on the points of C. Hence we get a special net of quadrics for any linear rational map $\alpha\colon \mathbb{P}^2 \to \mathcal{V}$, such that $\alpha(\mathbb{P}^2)$ is not contained in the plane $\{b=0\}$.

We get the degenerations $\mathcal{C}_1 \to \mathcal{C}_2 \to \mathcal{C}_3 \to \mathcal{C}_4 \to \mathcal{C}_5$ according to whether $\alpha(\mathbb{P}^2)$ is a plane transversal to C, a plane tangent to C, a line not meeting C, a line meeting C in 1 point, or a point not lying on the plane $\{b=0\}$.

In spite of this geometrical interpretation, the proof of the Main Theorem is very algebraic in nature, and relies on a careful analysis of the coherent sheaf $\mathcal{F}$ on $\mathbb{P}^2$ whose minimal resolution is given by the symmetric $(4\times4)$ matrix $\alpha$ of linear forms in 3 homogeneous variables.

A couple of words are needed concerning the existing literature on the subject: the much more general classification of nets of quadrics has been addressed by C.T.C. Wall in [Wa], but not completely carried through for obvious reasons of complexity. In fact (cf. [Wa], page 251) Wall encounters the orbit $\mathcal{Q}$ as the most degenerate case of his classification, inspired by Mumford's notion of stability, whereas for our classification,inspired by geometry, the orbit $\mathcal{Q}$ appears as the general case .

As to the contents of the paper, we shall prove in detail the main theorem, while we defer the reader to [Ca-De] for all the proofs concerning our previous very sketchy discussion; we simply notice that the result of the present paper is a first, but perhaps crucial step towards the determination of the canonical ring $\mathcal{R}(S)$ of surfaces in the class v). In fact, using our classification of special nets of quadrics, we are indeed able to prove the following result

**Theorem II**

Let S be a surface of general type with $K^2 = \chi = 2$ and belonging to the hyperelliptic birational locus in the moduli space: then, any associated special net of quadrics can only belong to an orbit of type $\mathcal{Q}$ or $\mathcal{B}_1$.

We plan to use the previous result in the following way: the coefficients of the matrix A' determining $\mathcal{R}(S)$ can be expanded according to powers of the coordinate $y_0$ corresponding to the plane containing the image of C, and our matrix $\alpha$ is just the degree 0 part.

In order to find A' given $\alpha$, one has to successively solve linear equations which determine the higher order term given the lower order ones, and we hope to perform this calculation with a nice computer program.

**Acknowledgements**: we would like to thank K. Ranestad for kindly sending us a copy of Wall's paper, and M. Reid for providing stimulus to carrying through a research plan of which the present paper is a part.

## 1 Notation and general setup

We shall work over the ground field $\mathbb{C}$ .

Given a $\mathbb{C}$-vector space V, we shall denote by $V^\vee$ its dual and by M(n,V) the $\mathbb{C}$-vector space of $n\times n$ matrices with coefficients in V. By $(\alpha)_{i,j}$ we denote the i,j-entry of the matrix $\alpha\in M(n,V)$.

$\mathbb{P}^5=\mathbb{P}(S^2(\mathbb{C}^3)^\vee)$ denotes the projective space of homogeneous polynomials of degree 2 in three variables.

By abuse of language, we shall often denote by Q the curve or surface $\{Q=0\}$ defined by a homogeneous polynomial Q.

$(\mathcal{O}_{\mathbb{P}^2,P},m_P)$ is the local ring of $P\in\mathbb{P}^2$, whereas $(\mathcal{O}_{Q,P},m_{Q,P})$ is the local ring of P as a point of the subscheme defined by Q .

$\mathfrak{N} = \mathbb{P}\{\alpha \in M(4,(\mathbb{C}^3)^\vee) \mid {}^t\alpha = \alpha\}$ will be the 29-dimensional complex projective space of "nets of quadrics", i.e., $4\times 4$ symmetric matrices whose entries are linear forms in $y_1, y_2, y_3$.

$\mathfrak{U}$ is the open set of $\mathfrak{N}$ consisting of those matrices whose determinant is not identically zero.

$\mathcal{S}' = \{(\alpha,Q) \in \mathfrak{U}\times\mathbb{P}^5 \mid \det(\alpha)=Q^2 \text{ and } Q \mid \Lambda^3\alpha\}$, where $Q \mid \Lambda^3\alpha$ means that Q divides the determinants of all the $3\times 3$ minors of $\alpha$.

It follows that $\mathcal{S}'$ is a closed quasiprojective subvariety of $\mathfrak{U}\times\mathbb{P}^5$.

Let $\varphi\colon \mathcal{S}'\to\mathbb{P}^5$ and $\psi\colon \mathcal{S}'\to\mathfrak{U}$ be defined by the restrictions of the natural projections of $\mathfrak{U}\times\mathbb{P}^5$ on the two factors.

Set $\mathcal{S}=\psi(\mathcal{S}')$; $\mathcal{S}$ is a closed quasiprojective subvariety of $\mathfrak{U}$, and it is precisely the variety of matrices satisfying the conditions a) and b) stated in the introduction. We remark that $\psi\colon \mathcal{S}'\to\mathcal{S}$ is in fact an isomorphism, since $\mathcal{S}'$ is the graph of the morphism $\pi\colon \mathcal{S}\to\mathbb{P}^5$, such that $\pi(\alpha)=Q$ where Q is the unique element of $\mathbb{P}^5$ s.t. $\det(\alpha)=Q^2$.

Let G be the group $\mathbb{P}GL(4,\mathbb{C})$, which acts on $\mathcal{S}$ by $g(\alpha)={}^tg\alpha g$ for $g\in G$, $\alpha\in\mathcal{S}$: here, if $\det(\alpha)=Q^2$ and $Q\mid\Lambda^3\alpha$, then $\det({}^tg\alpha g)=\det(g)^2Q^2=Q^2$ and $Q\mid\Lambda^3({}^tg\alpha g)$. We let G act trivially on $\mathbb{P}^5$.

Let $\Gamma$ be the group $\mathbb{P}GL(3,\mathbb{C})$ acting both on $\mathcal{S}$ and on $\mathbb{P}^5$ by projective coordinate change in the variables $y_1, y_2, y_3$: $\gamma(\alpha(y))=\alpha(\gamma^{-1}y)$ and $\gamma Q(y)=Q(\gamma^{-1}y)$ for $g\in\Gamma$, $\alpha\in\mathcal{S}$, $Q\in\mathbb{P}^5$.

The direct product group $G\times\Gamma$ acts therefore on the varieties $\mathcal{S}$ and $\mathbb{P}^5$ and the morphism $\pi: \mathcal{S} \to \mathbb{P}^5$ is $G\times\Gamma$-equivariant. Under the action of $G\times\Gamma$ (i.e. of $\Gamma$), $\mathbb{P}^5$ splits into 3 orbits: $\mathcal{A}'=\Gamma(y_1y_3-y_2^2)$ is the 5-dimensional open orbit corresponding to the smooth conics; $\mathcal{B}'=\Gamma(y_1y_2)$ is the 4-dimensional orbit corresponding to the reducible conics given by the union of two distinct lines; $\mathcal{C}'=\Gamma(y_1^2)$ is the 2-dimensional closed orbit corresponding to the non reduced conics given by a double line.

## 2 The normal form in the case where Q is a smooth conic.

Our aim is to study the $G\times\Gamma$-orbit structure of the variety $\mathcal{S}$. The first task is to find normal forms for its elements under the $G\times\Gamma$-action. From §1 we know that there is a surjective equivariant $G\times\Gamma$-morphism $\pi: \mathcal{S}\to\mathbb{P}^5$ and that $\mathbb{P}^5=\mathcal{A}'\cup\mathcal{B}'\cup\mathcal{C}'$, where $\mathcal{A}',\mathcal{B}',\mathcal{C}'$ are the $G\times\Gamma$-orbits corresponding respectively to smooth conics, two distinct lines and double lines. In this section we shall show that $\pi^{-1}(\mathcal{A}')$ consists of only one orbit.

So let $\alpha\in\pi^{-1}(\mathcal{A}')$. We know that $\alpha$ is a $4\times4$ matrix whose entries are linear forms in $y_1,y_2,y_3$ with complex coefficients and which satisfies the conditions a) and b) stated in §0: $\det(\alpha)=Q^2$ and $Q|\Lambda^3\alpha$, where $Q$ is a quadratic form in $y_1,y_2,y_3$ corresponding to a smooth conic.

We want to put $\alpha$ into a normal form up to $G\times\Gamma$-action. First we note that after a $\Gamma$-action we may assume that $Q=y_1y_3-y_2^2$. We remark that if $P\in\mathbb{P}^2$, then $rk(\alpha(P))=4$ if $P\notin Q$, while $rk(\alpha(P))\leq 2$ if $P\in Q$, because of condition b).

To study $\alpha$, we introduce the coherent sheaf $\mathcal{F}$ on $\mathbb{P}^2$ defined by the following exact sequence on $\mathbb{P}^2$:

$$2.1 \qquad 0 \to \mathcal{O}_{\mathbb{P}^2}(-2)^4 \xrightarrow{\alpha} \mathcal{O}_{\mathbb{P}^2}(-1)^4 \to \mathcal{F} \to 0$$

The support of $\mathcal{F}$ is the conic Q. For $P\in\mathbb{P}^2$, we denote by $\alpha_P$ the element of $M(4,\mathcal{O}_{\mathbb{P}^2,P})$ defined by $\alpha$ which gives the inclusion of stalks $\mathcal{O}_{\mathbb{P}^2}(-2)_P{}^4 \to \mathcal{O}_{\mathbb{P}^2}(-1)_P{}^4$.

We also denote by $m_P$ the maximal ideal of the local ring $\mathcal{O}_{\mathbb{P}^2,P}$ and by $(\mathcal{O}_{Q,P}, m_{Q,P})$ the local ring of P in Q.

**2.2 Lemma** Let $P\in Q$ . Then $rk(\alpha(P))=2$ , and up to an action of $\mathcal{G} = \mathbb{P}GL(4,\mathcal{O}_{\mathbb{P}^2,P})$ on $\alpha$, we may assume that $\alpha_P$ is in the following form:

$$\alpha_P = \begin{pmatrix} 1 & 0 & 0 & 0 \\ 0 & 1 & 0 & 0 \\ 0 & 0 & q & 0 \\ 0 & 0 & 0 & q \end{pmatrix}$$

where $q \in m_P$ is a local equation for $Q$ near $P$. In particular $\mathcal{F}_P \cong \mathcal{O}_{Q,P} \oplus \mathcal{O}_{Q,P}$. The statement of this lemma remains true even if the conic $Q$ is not smooth, provided $P$ is a smooth point of $Q$.

**Proof** Let $q \in m_P - m_P^2$ be a local equation of $Q$ in $P$. We may assume $\det(\alpha_P) = q^2$. Consider the symmetric matrix $\alpha_P(P) \in \mathbb{P}M(4,\mathbb{C})$ : up to an action of $G$ on $\alpha$, we may assume it to be diagonal with the main diagonal entries belonging to $\{0,1\}$. We claim that $\mathrm{rk}(\alpha_P(P)) = 2$: we saw before that $\mathrm{rk}(\alpha_P(P)) \leq 2$, and if $\mathrm{rk}(\alpha_P(P)) < 2$ then $\det(\alpha_P) = q^2 \in m_P^3$, a contradiction. It follows that $\alpha_P(P)$ is the following matrix:

$$\begin{pmatrix} 1 & 0 & 0 & 0 \\ 0 & 1 & 0 & 0 \\ 0 & 0 & 0 & 0 \\ 0 & 0 & 0 & 0 \end{pmatrix}$$

Since the entries of $\alpha_P(P)$ differ from those of $\alpha_P$ by elements of $m_P$, performing elementary operations (each row operation being followed by the transpose column operation, which corresponds to an action of $\mathcal{G} = \mathbb{P}GL(4, \mathcal{O}_{\mathbb{P}^2,P})$, we put $\alpha_P$ in the following form:

$$\alpha_P = \begin{pmatrix} I_2 & 0 \\ 0 & B \end{pmatrix}$$

where $B \in M(2, \mathcal{O}_{\mathbb{P}^2,P})$. From the condition $q | \Lambda^3 \alpha_P$ we see that $q|B$, thus $B = qC$, $C \in M(2, \mathcal{O}_{\mathbb{P}^2,P})$. Taking determinants, since $\det(\alpha_P) = q^2$, we get $\det(C) = 1$. Acting again with the group $\mathcal{G}$ we may transform

C into the $2\times 2$ identity matrix, yielding the result. Q.E.D.

**2.3 Corollary** $\mathcal{F}$ is a locally free sheaf of rank 2 on Q.

The normal form for $\alpha$ follows from the study of the sheaf $\mathcal{F}$. The smooth conic $Q=\{y_1y_3-y_2^2\}\subset\mathbb{P}^2$ is the image of the embedding j: $\mathbb{P}^1\to\mathbb{P}^2$ defined by $j(t_0,t_1)=(y_1,y_2,y_3)$, where:

**2.4** $$y_1=t_0^2\ ,\quad y_2=t_0t_1\ ,\quad y_3=t_1^2$$

Consider the locally free rank-2 sheaf $j^*\mathcal{F}$ on $\mathbb{P}^1$.

**2.5 Proposition** $j^*\mathcal{F}\cong\mathcal{O}_{\mathbb{P}^1}(-1)\oplus\mathcal{O}_{\mathbb{P}^1}(-1)$ .

**Proof** Since every finite rank locally free sheaf on $\mathbb{P}^1$ splits, we must have $j^*\mathcal{F}\cong\mathcal{O}_{\mathbb{P}^1}(m)\oplus\mathcal{O}_{\mathbb{P}^1}(n)$. From the long exact cohomology sequence attached to 2.1 we get:
$h^0(\mathbb{P}^1,j^*\mathcal{F})=0$ which implies $m<0$, $n<0$; we also obtain that $h^0(\mathbb{P}^1,j^*\mathcal{F}\otimes\mathcal{O}_{\mathbb{P}^1}(2))=4$, therefore $m=n=-1$. Q.E.D.

We shall now construct explicitly a symmetric minimal resolution of $\mathcal{F}$ as a $\mathcal{O}_{\mathbb{P}^2}$ module and show that any two such resolutions are isomorphic by a suitable G-action.
The coherent sheaf $\mathcal{F}$, via the Serre correspondence, is isomorphic to $M^\sim$ for M a suitable graded finite $\mathbb{C}[Y]$-module, where Y stands

for $y_1,y_2,y_3$. Since Q is the 2-uple embedding of $\mathbb{P}^1=\mathrm{Proj}\ \mathbb{C}[t_0,t_1]$, we have $Q\cong\mathrm{Proj}\ \mathbb{C}[t_0,t_1]^{(2)}$ (cf.[Ha]), where $\mathbb{C}[t_0,t_1]^{(2)}=\bigoplus_{d\geq 0}\mathbb{C}[t_0,t_1]_{2d}$ has the structure of $\mathbb{C}[Y]$-module given by 2.4 and is isomorphic to $\mathbb{C}[Y]/(y_1y_3-y_2^2)$. From 2.5 we see that we have an isomorphism of $\mathbb{C}[Y]$-modules:

2.6 $$M\cong\mathbb{C}[t_0,t_1](-1)^{(2)}\oplus\mathbb{C}[t_0,t_1](-1)^{(2)}$$

A minimal resolution of $\mathbb{C}[t_0,t_1](-1)^{(2)}$ as a $\mathbb{C}[Y]$-module is given by:

2.7 $$0\to\mathbb{C}[Y](-2)^2\to\mathbb{C}[Y](-1)^2\to\mathbb{C}[t_0,t_1](-1)^{(2)}\to 0$$

where the second and the third arrow are defined respectively by the following matrices:

2.8 $$\begin{pmatrix} y_3 & -y_2 \\ -y_2 & y_1 \end{pmatrix} \quad \begin{pmatrix} t_0 & t_1 \end{pmatrix}$$

We are now able to find the normal form for $\alpha$ in the case when Q is a smooth conic.

**2.9 Proposition** Let $\alpha\in\pi^{-1}(\mathcal{Q}')$; then there exists $(g,\gamma)\in G\times\Gamma$ such that $(g,\gamma)\alpha$ is the following matrix:

2.10 $$\begin{pmatrix} y_3 & -y_2 & 0 & 0 \\ -y_2 & y_1 & 0 & 0 \\ 0 & 0 & y_3 & -y_2 \\ 0 & 0 & -y_2 & y_1 \end{pmatrix}$$

We shall say that this normal form is of type A .

**Proof** A minimal resolution for M is obtained by 2.6 from the direct sum of two copies of 2.7. From 2.8 we see that the matrix $\alpha'$ giving the resolution is as in 2.10. Comparing with 2.1, since two minimal resolutions are isomorphic, we see that there do exist matrices $g_1, g_2$ such that $\alpha' = g_1 \alpha g_2$. Since the matrix $\alpha'$ in 2.10 is symmetric, we get $\alpha' = {}^t g_2 \alpha {}^t g_1$ and thus $(g_1^{-1}\ {}^t g_2)$ is a lifting of the identity of $\mathcal{F}$, thus $g_1 = {}^t g_2$. Q.E.D.

**2.11 Remark** The normal form 2.10 is G-equivalent to the following one:

2.12
$$\begin{pmatrix} 0 & 0 & y_3 & y_2 \\ 0 & 0 & y_2 & y_1 \\ y_3 & y_2 & 0 & 0 \\ y_2 & y_1 & 0 & 0 \end{pmatrix}$$

The corresponding net of quadrics has, as schematic base locus B, the union of two skew lines in $\mathbb{P}^3$: $\{x_0 = x_1 = 0\} \cup \{x_2 = x_3 = 0\}$.

Conversely, the base locus B determines a web of quadrics , associated to the same matrix .

## 3. The normal form when Q decomposes into two distinct lines.

We want now to find normal forms for $\alpha$ in the case $\alpha \in \pi^{-1}(\mathcal{B}')$, i.e., when the conic Q is the union of two distinct lines. After a suitable $\Gamma$-action we may assume that $Q = y_1 y_2$.

As in § 2, we define the sheaf $\mathcal{F}$ on Q by means of 2.1. Let $V = (0,0,1)$ be the unique singular point of Q. By 2.2, if $P \in Q$, $P \neq V$, $\mathrm{rk}(\alpha(P) = 2$ and $\mathcal{F}|_{Q-\{V\}}$ is a locally free sheaf on Q-{V} of rank 2. By condition b), $rk(\alpha(V))$ is $\leq 2$.

So we must distinguish between three cases.

**Case 1)** Suppose $rk(\alpha(V))=2$.
Then, if we consider $\alpha_V \in M(4,\mathcal{O}_{\mathbb{P}^2,V})$ and we argue as in 2.2, since $rk(\alpha_V(V))=2$, we get $\mathcal{F}_V \cong \mathcal{O}_{Q,V} \oplus \mathcal{O}_{Q,V}$ and $\mathcal{F}$ is locally free of rank 2 on Q. Let $\nu: Q^\nu \to Q$ be the normalization of Q. We have $Q^\nu = Q' \cup Q''$, where Q' and Q'' are two disjoint copies of $\mathbb{P}^1$. Consider the sheaf $\nu^*\mathcal{F}$ on $Q^\nu$.

**3.1 Proposition** $\nu^*\mathcal{F} \cong (\mathcal{O}_{Q'} \oplus \mathcal{O}_{Q'}(-1)) \oplus (\mathcal{O}_{Q''} \oplus \mathcal{O}_{Q''}(-1))$.

**Proof** Certainly $\nu^*\mathcal{F} \cong (\mathcal{O}_{Q'}(a') \oplus \mathcal{O}_{Q'}(b')) \oplus (\mathcal{O}_{Q''}(a'') \oplus \mathcal{O}_{Q''}(b''))$
From 2.1 we get:

**3.2** $h^0(\mathcal{F})=h^1(\mathcal{F})=0,\ h^0(\mathcal{F}(1))=4,\ h^0(\mathcal{F}(2))=8,\ h^1(\mathcal{F}(1))=0.$
From $0 \to \mathcal{O}_Q \to \nu_*\mathcal{O}_{Q^\nu} \to \mathbb{C} \to 0$ it follows:

**3.3**
$$0 \to \mathcal{F} \to \nu_*\nu^*\mathcal{F} \to \mathbb{C}^2 \to 0$$
and hence:

$h^0(\nu^*\mathcal{F})=2$ which implies $a',a'',b',b'' \leq 1$; if one is $=1$ then all the others are $<0$; otherwise two must be $=0$.

$h^1(\nu^*\mathcal{F})=0$ which implies by duality $a',a'',b',b'' \geq -1$.
Up to interchanging Q' and Q'', we have thus the following three possibilities for $(a',b',a'',b'')$:

$$(1,-1,-1,-1) \quad (0,0,-1,-1) \quad (0,-1,0,-1)$$

We are left to exclude the first two. It suffices to prove, setting $\mathcal{F}^\wedge = \mathcal{H}om_{\mathcal{O}_Q}(\mathcal{F},\mathcal{O}_Q)$, that $\mathcal{F} \cong \mathcal{F}^\wedge(-1)$, because then $\nu^*\mathcal{F} \cong (\nu^*\mathcal{F})^\wedge \otimes (\nu^*\mathcal{O}_Q(-1))$, which is not compatible with the first two cases.

To prove $\mathcal{F} \cong \mathcal{F}^\wedge(-1)$, we start from the exact sequence 2.1:

**3.4** $\quad 0 \to \mathcal{O}_{\mathbb{P}^2}(-2)^4 \to \mathcal{O}_{\mathbb{P}^2}(-1)^4 \to \mathcal{F} \to 0$

Taking $\mathcal{H}om_{\mathcal{O}_{\mathbb{P}^2}}(-,\mathcal{O}_{\mathbb{P}^2})$ and tensoring with $\mathcal{O}_{\mathbb{P}^2}(-3)$ we get:

**3.5** $\quad 0 \to \mathcal{O}_{\mathbb{P}^2}(-2)^4 \to \mathcal{O}_{\mathbb{P}^2}(-1)^4 \to \mathcal{E}xt^1(\mathcal{F},\mathcal{O}_{\mathbb{P}2}(-3)) \to 0$

The second arrows of 3.4 and 3.5 are given by $\alpha$ and ${}^t\alpha$ respectively. Since $\alpha$ is symmetric, we get $\mathcal{F} \cong \mathcal{E}xt^1(\mathcal{F},\mathcal{O}_{\mathbb{P}^2}(-3))$. But

$\mathcal{E}xt^1(\mathcal{F},\mathcal{O}_{\mathbb{P}^2}(-3)) \cong \mathcal{F}^\wedge \otimes \mathcal{E}xt^1(\mathcal{O}_Q,\omega_{\mathbb{P}^2}) \cong \mathcal{F}^\wedge \otimes \omega_Q \cong \mathcal{F}^\wedge(-1)$. Q.E.D.

Let $\nu^{-1}(V) = \{V',V''\}$ and $\mathcal{F}' = \mathcal{O}_{Q'} \oplus \mathcal{O}_{Q'}(-1))$, $\mathcal{F}'' = \mathcal{O}_{Q''} \oplus \mathcal{O}_{Q''}(-1)$. We want to reconstruct $\mathcal{F}$ from $\nu^*\mathcal{F} = \mathcal{F}' \oplus \mathcal{F}''$, simply by glueing (cf.3.3) $\mathcal{F}'_{V'} \otimes \mathbb{C}$ and $\mathcal{F}''_{V''} \otimes \mathbb{C}$, thus getting the informations we need to find the normal form of $\alpha$. The glueing between $\mathcal{F}'_{V'} \otimes \mathbb{C}$ and $\mathcal{F}''_{V''} \otimes \mathbb{C}$ is given by a matrix $A \in GL(2,\mathbb{C})$. An important restriction upon A comes from the above fact that $\nu^*\mathcal{F} \cong (\nu^*\mathcal{F})^\wedge(-1)$: we have an auto-duality $\nu^*\mathcal{F} \times \nu^*\mathcal{F} \to \mathcal{O}_{Q^\sim}(-1)$ given by: $(X,Y) \to {}^tXJY$, where $X = {}^t[x_0,x_1] \in \mathcal{O}_{Q^\sim} \oplus \mathcal{O}_{Q^\sim}(-1)$, $Y = {}^t[y_0,y_1] \in \mathcal{O}_{Q^\sim} \oplus \mathcal{O}_{Q^\sim}(-1)$ and where J is the following matrix:

$$J = \begin{pmatrix} 0 & 1 \\ 1 & 0 \end{pmatrix}$$

so that ${}^tXJY = x_0y_1 + x_1y_0$. The glueing must be compatible with this auto-duality, which gives the condition ${}^tAJA = J$. This equation has two solutions:

3.6 $\begin{pmatrix} a & 0 \\ 0 & 1/a \end{pmatrix}$ and $\begin{pmatrix} 0 & a \\ 1/a & 0 \end{pmatrix}$ where $a \in \mathbb{C}^*$

On these sets of matrices we can act via the automorphism group of $\nu^*\mathcal{F}$, which is $\{[a_{ij}] \mid 1 \leq i,j \leq 2;$ $a_{11}, a_{22} \in \mathbb{C}^*;\ a_{12} \in H^0(\mathcal{O}_{\mathbb{P}^1}(1));\ a_{21} = 0\}$. Up to this action, we have only two solutions: $I_2$ and J. But if the glueing matrix were $I_2$, then $\mathcal{F} \cong \mathcal{O}_Q \oplus \mathcal{O}_Q(-1)$ and $h^0(\mathcal{F}) = 1$, which is not the case, because $h^0(\mathcal{F}) = 0$: we conclude therefore that the glueing matrix must be J.

Since 2.1 gives a minimal resolution of the $\mathcal{O}_{\mathbb{P}^2}$-module $\mathcal{F}$ by means of the matrix $\alpha$, we want now to construct an explicit minimal resolution of $\mathcal{F}$, which compared with 2.1 will give the desired normal form of $\alpha$.

From 3.2 we know that $h^0(\mathcal{F}(1)) = 4$, $h^0(\mathcal{F}(2)) = 8$. Since the glueing matrix is J, $H^0(\mathcal{F}(1))$ is given by:

$$\{(\varphi, a, \psi, b) \in H^0((\mathcal{O}_{Q'}(1) \oplus \mathcal{O}_{Q'}) \oplus (\mathcal{O}_{Q''}(1) \oplus \mathcal{O}_{Q''}) \mid \varphi(0) = b,\ \psi(0) = a\}$$

The four generators of $H^0(\mathcal{F}(1))$ are thus:

3.7 $u_1 = (y_3, 0, 0, 1)$, $u_2 = (y_1, 0, 0, 0)$, $u_3 = (0, 1, y_3, 0)$, $u_4 = (0, 0, y_2, 0)$.

$H^0(\mathcal{F}(2))$ has 12 generators, namely: $\{y_i u_j\}_{1 \leq i \leq 3,\ 1 \leq j \leq 4}$. We expect thus 4 relations, which are in fact :

3.8 $y_2u_2 = 0$, $y_1u_4 = 0$, $y_1u_1 = y_3u_2$, $y_2u_3 = y_3u_4$

A minimal resolution of $\mathcal{F}(2)$ is then given by

3.9 $$0 \to \mathcal{O}_{\mathbb{P}^2}{}^4 \to \mathcal{O}_{\mathbb{P}^2}(1)^4 \to \mathcal{F}(2) \to 0$$

where the third arrow sends the four generators $(e_j)_{1 \le j \le 4}$ of $\mathcal{O}_{\mathbb{P}^2}(1)^4$ to $(u_j)_{1 \le j \le 4}$, while the second is defined by:

3.10 $$e_1 \to (0,y_2,0,0), \quad e_2 \to (0,0,y_2,-y_3),$$
$$e_3 \to (0,0,0,y_1), \quad e_4 \to (y_1,-y_3,0,0).$$

We have thus obtained the desired normal form for $\alpha$ in case 1):

**3.11 Proposition** Let $\alpha \in \pi^{-1}(\mathcal{B}')$ and suppose that $\mathrm{rk}(\alpha(V))=2$, where $V$ is the unique singular point of the conic $Q=\det(\alpha)$. Then there exists $(g,\gamma) \in G \times \Gamma$ such that $(g,\gamma)\alpha$ is the following matrix:

3.12 $$\begin{pmatrix} 0 & y_2 & 0 & 0 \\ y_2 & 0 & 0 & -y_3 \\ 0 & 0 & 0 & y_1 \\ 0 & -y_3 & y_1 & 0 \end{pmatrix}$$

We shall say that this normal form is of type $B_1$ .

**Proof** It suffices to exchange the role of $u_3$ and $u_1$ in 3.9 to get the symmetric matrix which is shown in 3.12 and which gives the minimal resolution 3.9. Comparing 2.1 and 3.9, by unicity of the minimal resolution of $\mathcal{F}$, since both $\alpha$ and 3.12 are symmetric, we can argue as in the proof of 2.9 to infer that we can transform $\alpha$ to 3.12 via a suitable G-action. Q.E.D.

**Case 2)** Suppose $\mathrm{rk}(\alpha(V))=1$. We are going to show that this case cannot occur.

We introduce in $\mathbb{P}^2$ affine coordinates $x_1=y_1/y_3$ and $x_2=y_2/y_3$; thus $Q=x_1x_2$ and the singular point

V of Q becomes the origin (0,0). We denote by $\alpha(x_1, x_2)$ again the matrix $(1/y_3)\alpha(y_1,y_2,y_3)$.

**3.13 Lemma** With the above assumptions, $Q=y_1y_2$, $\mathrm{rank}(\alpha(V))=1$, there exists a suitable action of the group $\mathcal{G}'=\mathbb{P}GL(4,\mathcal{O}'_{\mathbb{P}^2,V})$, (where $\mathcal{O}'_{\mathbb{P}^2,V}$ is the holomorphic local ring of the point V), which transforms the matrix $\alpha_V\in M(4,\mathcal{O}_{\mathbb{P}^2,V})$ to the following form:

$$\text{3.14}\qquad \begin{pmatrix} 1 & 0 & 0 & 0 \\ 0 & x_1 & 0 & 0 \\ 0 & 0 & x_2 & 0 \\ 0 & 0 & 0 & x_1x_2 \end{pmatrix}$$

**Proof** As in 2.2, we consider the matrix $\alpha_V(V)\in M(4,\mathbb{C})$; up to G-action, since its rank is 1, we may put it in the following form:

$$\begin{pmatrix} 1 & 0 & 0 & 0 \\ 0 & 0 & 0 & 0 \\ 0 & 0 & 0 & 0 \\ 0 & 0 & 0 & 0 \end{pmatrix}$$

It follows that $\alpha_V\in M(4,\mathcal{O}_{\mathbb{P}^2,V})$ may be brought by $\mathcal{G}=\mathbb{P}GL(4,\mathcal{O}_{\mathbb{P}^2,V})$-action in the following form:

$$\text{3.15}\qquad \begin{pmatrix} 1 & 0 \\ 0 & B \end{pmatrix}$$

with:

$$\text{3.16}\qquad B\in M(3,m_V),\quad \det(B)=(x_1x_2)^2,\quad x_1x_2|\Lambda^2 B.$$

Let $B=B'+B''$, with $B'$ a $3\times 3$ matrix of linear forms, $B''\in M(3,m_V{}^2)$ ($B'$ and $B''$ are both symmetric). Hence it follows: $\det(B')=0$ (being the homogeneous degree 3 part of $\det(B)$); $\Lambda^2 B \equiv \Lambda^2 B'$ (mod $m_V{}^3$) ; $x_1 x_2 | \Lambda^2 B'$.

*Claim1*: $\Lambda^2 B' \neq 0$. Otherwise, write $B' = x_1 A_1 + x_2 A_2$, $A_i \in M(3,\mathbb{C})$ symmetric. We must have $B' \neq 0$ (because $\det(B) \notin m_V{}^6$). If, for example, $A_1 \neq 0$, then, since $x_1 x_2 | \Lambda^2 B'$, $rk(A_1) = 1$, and after $\mathcal{G}$-action, $A_1$ would be:

3.17
$$\begin{pmatrix} 1 & 0 & 0 \\ 0 & 0 & 0 \\ 0 & 0 & 0 \end{pmatrix}$$

But then, if we assume $\Lambda^2 B' = 0$, also $A_2$, if $\neq 0$, would be of type 3.17, and $B'$ would be $A_1$ times a linear form in $x_1, x_2$. It would follow $\det(B) \in m_V{}^5$, contradicting 3.16.

*q.e.d. for claim 1.*

Thus we got a $2 \times 2$ minor of $B'$ with non zero determinant, which we may assume to equal $x_1 x_2$.

*Claim2:* we can find such a minor which is symmetric. Otherwise, the adjoint matrix $ad(B') = \Lambda^2 B'$ would have only zeros on the main diagonal. But $ad(B') \neq 0$ and $Q | ad(B')$. We see then $ad(ad(B')) \neq 0$, which contradicts, since $\det(B') = 0$, the identity $ad(ad(B')) = (\det(B'))^{n-2} B'$ (here $n=3$).

*q.e.d. for claim 2.*

After G-action, we may put this $2 \times 2$ minor, which we call $B'_{11}$, in the upper-left corner of $B'$. Let $B'_{11}$ be the following matrix: $\begin{pmatrix} a & b \\ b & c \end{pmatrix}$. Since $x_1 x_2$ is not a square, $a \neq 0$ and at least two of the linear forms a,b,c must be linearly independent. Say that a,b are independent: after G-action, we may assume that $a(x_1)$ depends only on $x_1$ and $b(x_2)$ depends only on $x_2$. But $a(x_1)c-(b(x_2))^2=x_1x_2$. It follows then that b=0, and the same can be obtained if a,b are not independent: then we may assume $a=x_1$, $c=x_2$. The matrix B' has thus gotten the following form:

$$B' = \begin{pmatrix} x_1 & 0 & e \\ 0 & x_2 & f \\ e & f & g \end{pmatrix}$$

But $x_1x_2 | \Lambda^2 B'$, and then $e=tx_1$, $f=sx_2$; after G-action we may assume $e=f=0$ and finally, since $\det(B')=0$, $g=0$. It follows that B has assumed the following form:

$$B = \begin{pmatrix} x_1+b_{11} & b_{12} & b_{13} \\ b_{12} & x_2+b_{22} & b_{23} \\ b_{13} & b_{23} & b_{33} \end{pmatrix}$$

with $b_{ij} \in m_V{}^2$. Since $\det(B)=(x_1x_2)^2$, we must have $b_{33}=x_1x_2+b$, $b \in m_V{}^3$. Since $\mathbb{C}\{x_1,x_2\}/(x_1+b_{11}) \cong \mathbb{C}\{x_2\}$, and analogously $\mathbb{C}\{x_1,x_2\}/(x_2+b_{22}) \cong \mathbb{C}\{x_1\}$, applying to B elementary operations, we see that B is equivalent in $M(3,\mathbb{C}\{x_1,x_2\})$ to a matrix $B^\wedge \in M(3,\mathbb{C}\{x_1,x_2\})$ of the form :

$$B^\wedge = \begin{pmatrix} x_1+b_{11} & f(x_2) & g(x_2) \\ f(x_2) & x_2+b_{22} & h(x_1) \\ g(x_2) & h(x_1) & x_1x_2+b \end{pmatrix}$$

with $f,g,h \in m_V{}^2 \mathcal{O}'_{\mathbb{P}^2,V}$, which verifies: $\det(B^\wedge)=(x_1x_2)^2$, $x_1x_2 | \Lambda^2 B^\wedge$. We observe that we may be more precise and specify that f,g,h are power series, while all the $b_{ij}$'s are in $\mathcal{O}_{\mathbb{P}^2,V}$.

**3.18** *Claim3:* we may bring $B^\wedge$ to the desired normal form

$$B^\wedge = \begin{pmatrix} x_1 & 0 & 0 \\ 0 & x_2 & 0 \\ 0 & 0 & x_1x_2 \end{pmatrix}$$

*Proof of claim3.*

From $x_1x_2 | (x_1+b_{11})h(x_1)-f(x_2)g(x_2)$ follows $fg=0$; then $x_1x_2 | (x_1+b_{11})h(x_1)$, whence, observing that the order of $x_1h(x_1)$ is strictly smaller than the order of $b_{11}h(x_1)$, $x_2 | h(x_1)$ and thus $h=0$. But then $x_1x_2 | g(x_2)(x_2+b_{22})$ thus, by the same argument as above, $x_1 | g(x_2)$, hence $g=0$.

Since $x_1x_2 | ((x_1+b_{11})(x_2+b_{22})-f(x_2)^2)$, we obtain that $(x_1+b_{11})b_{22}$ is congruent to 0 modulo $x_2$, thus $x_2 | b_{22}$. In particular, $x_2+b_{22}$ equals a unit times $x_2$, hence by elementary operations we may assume $f=0$, and then also $b_{22}=0$. In turn, therefore, we get $x_1 | b_{11}$, and we may assume $b_{11}=0$. But since

$\det(B^\wedge)=(x_1x_2)^2$, $x_1x_2 | b$, and we may multiply again the last row and column by a unit in order to have $b=0$.

*q.e.d. for claim 3.*

3.15 and 3.18 yield the proof of Lemma 3.13. Q.E.D.

Consider now on the conic $Q=\{x_1x_2=0\}$ the sheaf $\mathcal{F}$ defined as in 2.1 by the following exact sequence:

**3.19** $$0 \to \mathcal{O}_{\mathbb{P}^2}(-2)^4 \to \mathcal{O}_{\mathbb{P}^2}(-1)^4 \to \mathcal{F} \to 0$$

Let, as before, $\nu$: $Q^\nu \to Q$ be the normalization of Q.

We have $Q^\nu = Q' \cup Q''$, where Q' and Q'' are two

disjoint copies of $\mathbb{P}^1$. Let $\nu^{-1}(V)=\{V',V''\}$, $V'\in Q'$, $V''\in Q''$.

**3.20 Corollary** $\mathcal{F}_V \cong (\nu_*\mathcal{O}_{Q^\nu})_V \oplus \mathcal{O}_{Q,V}$. Moreover, there exists a rank-2 locally free sheaf $\mathcal{H}$ on $Q^\nu$ such that the following sequence is exact:

3.21 $$0 \to \mathcal{F} \to \nu_*\mathcal{H} \to \mathbb{C}_V \to 0$$

where $\mathbb{C}_V$ is the skycraper sheaf with stalk $\mathbb{C}$ on V.

**Proof** From lemma 3.13 and the G.AG.A. principle we see that $\mathcal{F}_V \cong (\nu_*\mathcal{O}_{Q^\nu})_V \oplus \mathcal{O}_{Q,V}$.
This isomorphism extends, since all sheaves are coherent, to a neighborhood U of V in Q: $\mathcal{F}|_U \cong (\nu_*\mathcal{O}_{Q^\nu})|_U \oplus \mathcal{O}_Q|_U$. If we restrict to $U-\{V\}$ we have a canonical isomorphism:

3.22 $$\mathcal{F}|_{U-\{V\}} \cong (\nu_*\mathcal{O}_{Q^\nu} \oplus \mathcal{O}_Q)|_{U-\{V\}} \cong \nu_*(\mathcal{O}_{Q^\nu} \oplus \mathcal{O}_{Q^\nu})|_{U-\{V\}}$$

To get $\mathcal{H}$, it suffices then to glue $\nu^*\mathcal{F}|_{Q^\nu-\{V',V''\}}$ with $(\mathcal{O}_{Q^\nu} \oplus \mathcal{O}_{Q^\nu})|_{\nu^{-1}(U)}$ using the isomorphism

3.23 $$\psi : \nu^*\mathcal{F}|_{\nu^{-1}(U)-\{V',V''\}} \to (\mathcal{O}_{Q^\nu} \oplus \mathcal{O}_{Q^\nu})|_{\nu^{-1}(U)-\{V',V''\}}$$

defined by 3.22. Q.E.D.

**3.24 Lemma** $\mathcal{H} \cong (\mathcal{O}_{Q'}(-1) \oplus \mathcal{O}_{Q'}(-1)) \oplus (\mathcal{O}_{Q''}(-1) \oplus \mathcal{O}_{Q''})$.

**Proof** Certainly $\mathcal{H} \cong (\mathcal{O}_{Q'}(a') \oplus \mathcal{O}_{Q'}(b')) \oplus (\mathcal{O}_{Q''}(a'') \oplus \mathcal{O}_{Q''}(b''))$. From the long cohomology sequence arising from 3.21, since $h^0(\mathcal{F})=h^1(\mathcal{F})=0$, we get: $h^0(\mathcal{H})=1$ (hence one among the integers a',b',a'',b'' is 0 and the others are $<0$), $h^1(\nu_*\mathcal{H})= h^1(\mathcal{H})=0$ ($\nu$ is affine) and hence all the above four integers are $\geq -1$. From this, up to interchanging Q' and Q", follows the desired assertion. Q.E.D.

**3.25 Lemma** There is a point $P_1 \in Q' - \{V\}$ such that

$$\mathcal{F} \cong \nu_*(\mathcal{O}_{Q^\nu}(-1)) \oplus \mathcal{O}_Q(-P_1) .$$

**Proof** From the exact sequence 3.21:

$$0 \to \mathcal{F} \to \nu_*\mathcal{H} \xrightarrow{p} \mathbb{C}_V \to 0$$

tensoring with $\mathcal{O}_Q(1)$ , we obtain

$$3.26 \quad 0 \to \mathcal{F}(1) \to \nu_*(\mathcal{O}_{Q'} \oplus \mathcal{O}_{Q''} \oplus \mathcal{O}_{Q'} \oplus \mathcal{O}_{Q''}(1)) \xrightarrow{p} \mathbb{C}_V \to 0 .$$

All the sheaves in the above exact sequence are generated by global sections, and we shall denote by $e_1$ the global generator of the first summand, respectively by $e_2$, $e_3$ the generators for the second and third summand.

Recall that both 3.21 and 3.26 are exact on global sections: in particular, the fact that 3.21 is exact on global sections implies that $H^0(\mathcal{O}_{Q''}(1))$ does not map to 0 under $H^0(p)$ in $\mathbb{C}_V \cong \mathbb{C}$. Therefore, we may choose two basis elements $y_0$, $y_1$ of $H^0(\mathcal{O}_{Q''}(1))$, mapping respectively to 1,0. Let $a,b,c \in \mathbb{C}_V \cong \mathbb{C}$ be the respective images of $e_1, e_2, e_3$ under $H^0(p)$. Replacing $e_2$ by $e_2 - by_0$, we may directly assume $b=0$.

Assume now $a=c=0$. Then, from 3.26 we infer that $\mathcal{F}(1) \cong \mathcal{O}_{Q'} \oplus \mathcal{O}_{Q''} \oplus \mathcal{O}_{Q'} \oplus \mathcal{O}_{Q''}$, contradicting (cf. 3.20) $\mathcal{F}_V \cong (\nu_*\mathcal{O}_{Q^\nu})_V \oplus \mathcal{O}_{Q,V}$ (notice that the two sheaves are not isomorphic, since the rank of the stalks at V are 4, respectively 3).

Hence, after replacing $e_1$, $e_3$ by suitable linear combinations (with constant coefficients), we may assume $a=0$, $c=1$ .
Hence $\mathcal{F}(1)\cong \nu_*(\mathcal{O}_{Q'}\oplus\mathcal{O}_{Q''})\oplus\mathcal{G}(1)$, where $\mathcal{G}(1)$ is the subsheaf of $\nu_*(\mathcal{O}_{Q'}\oplus\mathcal{O}_{Q''}(1))$ given by the kernel of the restriction of p to the latter summands. The desired assertion follows now immediately. Q.E.D.

**3.27 Corollary** When Q decomposes into two distinct lines, the case 2), namely $rk(\alpha(V))=1$, where V is the intersection point of the two components, cannot occur.

**Proof** By lemma 3.25 $\mathcal{F}\cong \nu_*(\mathcal{O}_{Q^\nu}(-1))\oplus\mathcal{O}_Q(-P_1)$,

but this is impossible, since $\mathcal{F}\cong\mathcal{F}\hat{}(-1)$, hence this relation would hold true also for the invertible summand $\mathcal{L}=\mathcal{O}_Q(-P_1)$: we would have $\mathcal{L}\cong\mathcal{L}\hat{}(-1)$, what is obviously absurd by degree considerations. Q.E.D.

**Case 3)** Suppose $rk(\alpha(V))=0$. Since $\alpha(V)=\alpha(0,0,1)=0$, it follows therefore that $\alpha(y_1,y_2,y_3)=\alpha(y_1,y_2)=y_1C+y_2D$ with $C,D\in M(4,\mathbb{C})$ symmetric.

We have: $y_1y_2|\Lambda^3\alpha$, and hence C and D have rank $\leq 2$; $\det(\alpha)=(y_1y_2)^2$ implies that $\alpha(1,1)=C+D$ is invertible, whence $Im(C)+Im(D)=\mathbb{C}^4$, so that $rk(C)=rk(D)=2$. Thus there exist vectors $v_1,v_2,v_3,v_4$ such that $C=v_1{}^tv_1+v_2{}^tv_2$, $D=v_3{}^tv_3+v_4{}^tv_4$; by our assumptions $v_1,v_2,v_3,v_4$ are a basis of $\mathbb{C}^4$, whence, by G-action, we have reached the following normal form:

3.28
$$\begin{pmatrix} y_1 & 0 & 0 & 0 \\ 0 & y_1 & 0 & 0 \\ 0 & 0 & y_2 & 0 \\ 0 & 0 & 0 & y_2 \end{pmatrix}$$

We shall call this the normal form of type $B_2$ .

## 4. The normal form when Q is a double line.

We want now to find a normal form for $\alpha$ when $\alpha \in \pi^{-1}(\mathfrak{C}')$, i.e., when the conic Q is a double line. After a suitable $\Gamma$-action, we may assume that $Q = y_1^2$.

4.1 **Proposition** Let $\alpha'$ be a matrix representative for $\alpha \in \mathcal{S}$. Then the following two facts are equivalent:

i) $Q = y_1^2$ (i.e., more precisely, a) $\det(\alpha') = y_1^4$; b) $y_1^2 | \Lambda^3 \alpha'$);

ii) there exist $P \in GL(4,\mathbb{C})$ , $A, B \in M(4,\mathbb{C})$ such that the following conditions hold:

1) $\det(P) = \pm 1$;

2) ${}^tA = A$ , ${}^tB = B$ , $A^2 = B^2 = AB + BA = 0$ ;

3) ${}^tP\alpha'P = y_1 I_4 + y_2 A + y_3 B$ .

**Proof** Notice first of all that, setting $C = y_2 A + y_3 B$, 2) is equivalent to ${}^tC = C$, $C^2 = 0$.

ii) $\Rightarrow$ i): let P,A,B satisfy 1),2),3).

**a)**: since $C^2 = 0$, the characteristic polynomial of C equals $y_1^4$, and thus $y_1^4 = \det(y_1 I_4 + C) = \det(\alpha')$.

**b)**: we have to show that $y_1^2 | \Lambda^3(y_1 I_4 + C)$. This condition is invariant under conjugation, hence we may assume that C is in Jordan normal form. If $\mathrm{rank}(C) \le 1$, then $\Lambda^2 C = 0$ and we are done. Since $C^2 = 0$, the only other possibility for the normal form of C is the following one:

4.2
$$\begin{pmatrix} 0 & 1 & 0 & 0 \\ 0 & 0 & 0 & 0 \\ 0 & 0 & 0 & 1 \\ 0 & 0 & 0 & 0 \end{pmatrix}$$

which can be checked by direct verification.
i) ⇒ ii): Let $\alpha'$ be such that **a)** and **b)** hold. Suppose $\alpha' = y_1 M + y_2 A' + y_3 B'$, with $M, A', B' \in M(4,\mathbb{C})$ symmetric; since $\det(\alpha') = y_1^4$ we see that $\det(M) = 1$ and hence we can find $P \in GL(4,\mathbb{C})$, $\det(P) = \pm 1$, such that ${}^tPMP = I_4$. So we have: ${}^tP\alpha'P = y_1 I_4 + y_2 A + y_3 B$ with $A, B \in M(4,\mathbb{C})$ symmetric. It suffices again to show $C^2 = 0$. Since $\det(y_1 I_4 + C) = y_1^4$, we see that the minimal polynomial of C divides $y_1^4$. Since $y_1^2 | \Lambda^3(y_1 I_4 + C)$ it follows that $\mathrm{rk}(C) \le 2$ and that the Jordan type of C is 4.2, hence $C^2 = 0$. Q.E.D.

Consider the given matrix $\alpha \in \pi^{-1}(\mathfrak{C}')$. By 4.1, there do exist $A, B \in M(4,\mathbb{C})$ such that after $G \times \Gamma$-action we have:

$$\alpha = y_1 I_4 + y_2 A + y_3 B$$

and ${}^tA=A$, ${}^tB=B$, $A^2=B^2=AB+BA=0$. Denote by $\langle A,B\rangle$ the $\mathbb{C}$-subspace of dimension $\leq 2$ of $M(4,\mathbb{C})$ spanned by A,B.

Let $(z,w)={}^tzw$ be the standard bilinear symmetric form on $\mathbb{C}^4$, and denote by $\perp$ the orthogonal complement with respect to this form. Let $\mathcal{Q}$ be the smooth quadric $\{z\in\mathbb{P}^3|(z,z)=0\}$.
Its group of automorphisms is the complex orthogonal group: $\mathbb{P}O(4)=\{P\in\mathbb{P}GL(4,\mathbb{C})|\ {}^tPP=I_4\}$.

We are going to prove, through a series of remarks, lemmas, and intermediary propositions, the following

## 4.10 Proposition

Any "special net" $\alpha=y_1I_4+y_2A+y_3B\in\pi^{-1}(\mathcal{C}')$ can be put, by a suitable $G\times\Gamma$ action, in one of the following normal forms:

$$C_1=\left(y_1I_4+\begin{pmatrix} y_2S & 0\\ 0 & y_3S\end{pmatrix}\right),\quad S=\begin{pmatrix}1 & i\\ i & -1\end{pmatrix};$$

(denoting by $\mathcal{C}_1$ the corresponding orbit, $\alpha\in\mathcal{C}_1\Leftrightarrow Q$ is a double line; $\dim\langle A,B\rangle=2$, and in $\mathbb{P}\langle A,B\rangle$ there are 2 matrices of rank 1)

$$C_2=\left(y_1I_4+\begin{pmatrix} y_3S & y_2S\\ y_2S & 0\end{pmatrix}\right)$$

($\alpha\in\mathcal{C}_2\Leftrightarrow Q$ is a double line; $\dim\langle A,B\rangle=2$, and in $\mathbb{P}\langle A,B\rangle$ there is 1 matrix of rank 1)

$$C_3=\left(y_1I_4+\begin{pmatrix} y_2S & 0\\ 0 & y_2S\end{pmatrix}\right)$$

($\alpha \in \mathcal{C}_3 \Leftrightarrow$ Q is a double line; dim<A,B>=1, and in $\mathbb{P}$<A,B> there are no matrices of rank 1)

$$C_4 = (y_1 I_4 + \begin{pmatrix} y_2 S & 0 \\ 0 & 0 \end{pmatrix})$$

($\alpha \in \mathcal{C}_4 \Leftrightarrow$ Q is a double line; dim<A,B>=1, and in $\mathbb{P}$<A,B> there is one matrix of rank 1)

$C_5 = (y_1 I_4)$ .

We proceed now to the proof of 4.10.

**4.3 Remark** The following hold true:

**a)** $\forall$ C$\in$<A,B>, $C^2=0$, ${}^tC=C$, rk(C)$\leq$2;
**b)** if C$\in$<A,B> and rk(C)=2, then

$$W_C = \mathrm{Ker}(C) = \mathrm{Im}(C) = (\mathrm{Ker}(C))^\perp = (\mathrm{Im}(C))^\perp$$

and this means that $\mathbb{P}$Ker(C)=$\mathbb{P}$Im(C) is a line of the ruling of $\mathbb{Q}$;
**c)** if C$\in$<A,B> and rk(C)=1, then:

$$(\mathrm{Ker}(C))^\perp = \mathrm{Im}(C) \subset \mathrm{Ker}(C) = (\mathrm{Im}(C))^\perp$$

and this means that $\mathbb{P}\mathrm{Im}(C) = \mathbb{P}(\mathrm{Ker}(C))^\perp$ is a point in $\mathbb{Q}$ and $\mathbb{P}\mathrm{Ker}(C) = \mathbb{P}(\mathrm{Im}(C))^\perp$ is its tangent plane.

**4.4 Lemma** Assume that $\forall$ C$\in\mathbb{P}$<A,B>, rk(C)=1 . Then, dim<A,B> = 1.

**Proof** If A and B were not proportional, then (cf. 4.3.c)) Ker(B) and Ker(A) would differ, hence there would be vectors v in Ker(B)-Ker(A), respectively w in Ker(A)- Ker(B). But then the two vectors (A+B)(v) and (A+B)(w) would not be proportional, hence rk(A+B)=2, a contradiction. Q.E.D.

Assume dim<A,B>=2: by 4.4 there exists C$\in$<A,B> s.t. rk(C)=2. Moreover, we notice that there are at most two C's in $\mathbb{P}\langle A,B\rangle$ with rank equal to 1 (those are defined by equations of degree 2). Hence we get a rational map of $\mathbb{P}^1$ to the variety of lines contained in $\mathbb{Q}$ (to C we associate the line $\mathbb{P}\mathrm{Ker}(C)$). We claim that indeed this rational map is constant. This follows from :

**4.5 Lemma** Assume $rk(A)=rk(B)=2$, then $\mathrm{Ker}(B)=\mathrm{Ker}(A)$.

**Proof** This is obvious if A,B are proportional. Otherwise, a preliminary remark is that, since $\mathbb{P}^1$ is irreducible, the lines belong to only one of the two rulings of $\mathbb{Q}$. Whence, those subspaces are skew. But then A maps Ker(B) isomorphically to $\mathrm{Im}(A)=\mathrm{Ker}(A)$ and similarly B maps Ker(A) isomorphically to Ker(B). On the other hand, the equations $A^2=B^2=AB+BA=0$ imply that $ABA=0$, a contradiction. Q.E.D.

**4.6 Remark** Assume that $\mathrm{rank}(C)=2$ and set $W=\mathrm{Ker}(C)=\mathrm{Im}(C)$ . Then C factors through:

the canonical projection ($V=\mathbb{C}^4$) $\pi: V\to V/W$,
an isomorphism $C'': V/W\to W$,
and the inclusion $i: W\to V$.
Moreover, C" is symmetrical in the following sense: the standard bilinear map on V, since $W=W^{\perp}$, induces a perfect duality between V/W and W, and the symmetry of C" simply reads off as

4.7 $$(C''(u), u') = (C''(u'), u) .$$

Taking dual bases in V/W and W, C" is given by a symmetric $(2\times 2)$ matrix, hence our job is now reduced to recalling (without proof) the standard classification of pencils of quadratic forms in 2 homogeneous variables.

**4.8 Proposition** Up to change of coordinates , the only pencils of symmetric $(2\times 2)$ matrices are:

$$\begin{pmatrix} y_3 & y_2 \\ y_2 & 0 \end{pmatrix} \quad \begin{pmatrix} y_2 & 0 \\ 0 & y_3 \end{pmatrix}$$

To end our classification in case $\dim\langle A,B\rangle = 2$, it suffices to remark that the group $\mathbb{P}O(4)$ acts transitively on points and lines of Q. We may thus let $W=\langle \eta,\beta\rangle$, where ${}^t\eta=[1,i,0,0]$, ${}^t\beta=[0,0,1,i]$. It follows easily that each C has the following form:

$$4.9 \quad C=\begin{pmatrix} a & ia & b & ib \\ ia & -a & ib & -b \\ b & ib & c & ic \\ ib & -b & ic & -c \end{pmatrix}=\begin{pmatrix} aS & bS \\ bS & cS \end{pmatrix}; \text{ where } S=\begin{pmatrix} 1 & i \\ i & -1 \end{pmatrix}.$$

Remark also that $\operatorname{rk}\left(\begin{pmatrix} aS & bS \\ bS & cS \end{pmatrix}\right)=\operatorname{rk}\begin{pmatrix} a & b \\ b & c \end{pmatrix}$.

The cases $\dim\langle A,B\rangle=1,0$ follow easily. The proof of 4.10 is thus completed.

## 5. Adjacencies and dimensions of the orbits of $\mathcal{S}$.

In the next proposition, we summarize 2.9, 3.11, 3.28, 4.10, giving a list of the normal forms obtained up to now. In the sequel, we shall sometimes adopt different normal forms, essentially because we shall fix $x_0x_1-x_2x_3$ as quadratic form in $\mathbb{C}^4$ (the lines on this quadric are more easily parametrized).

**5.1 Proposition** Under the action of the group $G\times\Gamma$, the variety $\mathcal{S}$ splits into 8 orbits:

$$\mathcal{A} = (G\times\Gamma)\begin{pmatrix} y_3 & y_2 & 0 & 0 \\ y_2 & y_1 & 0 & 0 \\ 0 & 0 & y_3 & y_2 \\ 0 & 0 & y_2 & y_1 \end{pmatrix},$$

$\alpha\in\mathcal{A} \Leftrightarrow Q$ is a smooth conic;

$$\mathcal{B}_1 = (G\times\Gamma)\begin{pmatrix} 0 & y_2 & 0 & 0 \\ y_2 & 0 & 0 & y_3 \\ 0 & 0 & 0 & y_1 \\ 0 & y_3 & y_1 & 0 \end{pmatrix},$$

$\alpha\in\mathcal{B}_1 \Leftrightarrow Q$ decomposes into 2 distinct lines and the rank of $\alpha$ at the intersection is 2;

$$\mathcal{B}_2 = (G\times\Gamma)\begin{pmatrix} y_1 & 0 & 0 & 0 \\ 0 & y_1 & 0 & 0 \\ 0 & 0 & y_2 & 0 \\ 0 & 0 & 0 & y_2 \end{pmatrix},$$

$\alpha\in\mathcal{B}_2 \Leftrightarrow Q$ decomposes into 2 distinct lines and the rank of $\alpha$ at the intersection is 0;

$$\mathcal{C}_1 = (G\times\Gamma)(y_1 I_4 + \begin{pmatrix} y_2 S & 0 \\ 0 & y_3 S \end{pmatrix}), \quad S = \begin{pmatrix} 1 & i \\ i & -1 \end{pmatrix};$$

$$\mathcal{C}_2 = (G\times\Gamma)(y_1 I_4 + \begin{pmatrix} y_3 S & y_2 S \\ y_2 S & 0 \end{pmatrix});$$

$$\mathcal{C}_3 = (G\times\Gamma)(y_1 I_4 + \begin{pmatrix} y_2 S & 0 \\ 0 & y_2 S \end{pmatrix});$$

$$\mathcal{C}_4 = (G\times\Gamma)(y_1 I_4 + \begin{pmatrix} y_2 S & 0 \\ 0 & 0 \end{pmatrix}); \qquad \mathcal{C}_5 = (G\times\Gamma)(y_1 I_4).$$

$\alpha \in \mathcal{C}_i \Leftrightarrow Q$ is a double line. Moreover, if we consider the $G\times\Gamma$-equivariant morphism $\pi: \mathcal{S} \to \mathbb{P}^5$, we have: $\pi^{-1}(\mathcal{A}')=\mathcal{A}$; $\pi^{-1}(\mathcal{B}')=\mathcal{B}_1 \cup \mathcal{B}_2$; $\pi^{-1}(\mathcal{C}')=\mathcal{C}_1\cup\mathcal{C}_2\cup\mathcal{C}_3\cup\mathcal{C}_4\cup\mathcal{C}_5$.

**5.2 Remark** As the reader will already have noticed, only in cases $\mathcal{A}$, $\mathcal{B}_1$, $\mathcal{C}_1$, $\mathcal{C}_2$ we actually have a net. Case $\mathcal{C}_5$ corresponds to the case of constant maps $\alpha$, and therefore (9 being the dimension of the projective space of symmetric $4\times 4$ matrices, 2 being the dimension of the dual space of $\mathbb{P}^2$) we get:

**5.3 Proposition** The orbit $\mathcal{C}_5$ has dimension 11.

**5.4 Proposition** The base locus of the net determined by $\alpha$ consists of one of the following subschemes of a smooth quadric $\mathcal{Q}$ in the net:

i) case $\mathcal{A}$: two distinct lines of a ruling of $\mathcal{Q}$;

ii) case $\mathcal{B}_1$: two distinct lines of a ruling of $\mathcal{Q}$ plus a line of the other ruling;

iii) case $\mathcal{C}_1$: a line of $\mathcal{Q}$ counted with multiplicity 2;

iv) case $\mathcal{C}_2$: a line of $\mathcal{Q}$ counted with multiplicity 2, plus a line of the other ruling.

**Proof** The base locus of the net is easily computed. In cases $\mathcal{C}_1$ and $\mathcal{C}_2$, it is more convenient to change coordinates in $\mathbb{P}^3$: instead of $z=(z_0,z_1,z_2,z_3)$, we shall use coordinates $x=(x_0,x_1,x_2,x_3)$ such that

**5.5** $x_0=z_0+i\,z_1,\quad x_1=z_0-i\,z_1,\quad -x_2=z_2+i\,z_3,\quad x_3=z_2-i\,z_3.$

We can therefore assume $\mathcal{Q}$ to be the quadric

5.6 $$\mathcal{Q} = \{ x \mid x_0 x_1 = x_2 x_3 \}$$

A basis of the net is completed by $x_0^2$, $x_2^2$ in case $\mathcal{C}_1$ and by $x_0 x_2$, $x_0^2$ in case $\mathcal{C}_2$. Q.E.D.

**5.7 Proposition** The family $\mathcal{Q}$ has dimension 19.

**Proof** The variety of pairs of 2 skew lines in $\mathbb{P}^3$ has dimension 8. Let us fix such a pair, say, $x_0 = x_1 = 0$, $x_2 = x_3 = 0$.

Then the set $\mathcal{W}$ of quadrics containing these lines is a 3-dimensional projective space. A quadric is in $\mathcal{W}$ iff its matrix has the form:

$$A = \begin{pmatrix} 0 & 0 & a & b \\ 0 & 0 & c & d \\ a & c & 0 & 0 \\ b & d & 0 & 0 \end{pmatrix}$$

Thus on $\mathcal{W}$ det(A) is the square of the quadratic polynomial $\mathcal{Q} = ad - bc$ which divides $\Lambda^3 A$. Hence we get a special net of quadrics for any linear rational map $\alpha \colon \mathbb{P}^2 \to \mathcal{W}$, such that $\alpha(\mathbb{P}^2)$ is not contained in the quadric $\mathcal{Q}$. We can thus conclude that, fixing the two lines, we get a family of dimension 11. Q.E.D.

**5.8 Proposition** The family $\mathcal{C}_4$ has dimension 15.

**Proof** The "special nets" in the family $\mathcal{C}_4$ are rational linear maps $\alpha \colon \mathbb{P}^2 \to \mathcal{P}$, where $\mathcal{P}$ is a pencil generated by a smooth quadric in $\mathbb{P}^3$ and a tangent plane counted with multiplicity 2. It will be proven that the dimension is 15 if we show that the variety of those pencils $\mathfrak{P}$ has dimension equal to 10. Now, the variety of quadrics has dimension 9, each quadric has a 2-dimensional family of tangent planes; since each such pencil $\mathfrak{P}$ contains a unique quadric of rank 1, we have a 1-dimensional family of such pairs yielding the same pencil. Q.E.D.

**5.9 Definition** We shall say that the orbit $\mathfrak{D}$ is a specialization of the orbit $\mathfrak{F}$, or that $\mathfrak{D}$ is adjacent to $\mathfrak{F}$, if $\mathfrak{D}$ is contained in the closure of $\mathfrak{F}$. Notice that $\mathfrak{D}$ is adjacent to $\mathfrak{F}$ iff there is a point of $\mathfrak{D}$ in the closure of $\mathfrak{F}$. Moreover, the adjacency relation is transitive.

**5.10 Proposition** There exist specializations:

$$\mathfrak{A} \to \mathfrak{B}_1 \to \mathfrak{B}_2 \to \mathfrak{C}_3 \to \mathfrak{C}_5$$

**Proof** As in 5.7, consider the space $\mathfrak{W}$ of quadrics containing two fixed skew lines, and the family of linear rational maps $\alpha$: $\mathbb{P}^2 \to \mathfrak{W}$, such that $\alpha(\mathbb{P}^2)$ is not contained in the locus $\mathfrak{Q}=\{ad-bc=0\}$ of degenerate quadrics of $\mathfrak{W}$. Recall that, for Q in $\mathfrak{W}$, the rank of Q is 2 if Q belongs to $\mathfrak{Q}$, and is otherwise equal to 4. We get cases $\mathfrak{A}$, $\mathfrak{B}_1$, $\mathfrak{B}_2$, $\mathfrak{C}_3$, $\mathfrak{C}_5$, according to whether $\alpha(\mathbb{P}^2)$ is a plane transversal to $\mathfrak{Q}$, a plane tangent to $\mathfrak{Q}$, a line transversal to $\mathfrak{Q}$, a line tangent to $\mathfrak{Q}$, a point not in $\mathfrak{Q}$. Q.E.D.

**5.11 Proposition** There exist specializations:

$$\mathcal{C}_1 \to \mathcal{C}_2 \to \mathcal{C}_3 \to \mathcal{C}_4 \to \mathcal{C}_5$$

**Proof** Fix again the smooth quadric $Q=\{x_0x_1=x_2x_3\}$ in $\mathbb{P}^3$, and let $\Gamma$ be the subscheme of Q given by the divisor on Q consisting of the line $x_0=x_2=0$ counted with multiplicity 2 (cf. the proof of 5.4, the base locus in case $\mathcal{C}_1$ is given by $x_0^2=x_2^2=x_0x_1-x_2x_3=0$). Again, the quadrics containing $\Gamma$ form a 3-dimensional projective space $\mathcal{V}$, in fact they are the quadrics whose matrix A is of the form:

$$A=\begin{pmatrix} a & b & d & 0 \\ b & 0 & 0 & 0 \\ d & 0 & c & -b \\ 0 & 0 & -b & 0 \end{pmatrix}$$

and we remark that $\det(A)=b^4$, while $b^2 | \Lambda^3 A$. Here, the rank of A is 4 outside of the plane $\{b=0\}$, 2 on the points of the plane which lie outside of the conic $C=\{b=ac-d^2=0\}$, and 1 on the points of C. Hence we get a special net of quadrics for any linear rational map $\alpha\colon \mathbb{P}^2 \to \mathcal{V}$, such that $\alpha(\mathbb{P}^2)$ is not contained in the plane $\{b=0\}$. We get the degenerations $\mathcal{C}_1 \to \mathcal{C}_2 \to \mathcal{C}_3 \to \mathcal{C}_4 \to \mathcal{C}_5$, according to whether $\alpha(\mathbb{P}^2)$ is a plane transversal to C, a plane tangent to C, a line not meeting C, a line meeting C in 1 point, or a point not in the plane $\{b=0\}$. Q.E.D.

**5.12 Proposition** The dimensions of the orbits are: $\dim(\mathcal{A}) = 19$, $\dim(\mathcal{B}_1) = \dim(\mathcal{C}_1) = 18$, $\dim(\mathcal{B}_2) = \dim(\mathcal{C}_2) = 17$, $\dim(\mathcal{C}_3) = 16$, $\dim(\mathcal{C}_4) = 15$, $\dim(\mathcal{C}_5)=11$.

**Proof** The dimensions of the orbits $\mathfrak{A}$, $\mathfrak{C}_4$, $\mathfrak{C}_5$, have been computed in prop. 5.7, 5.8, 5.3. The proof is completed by the following argument: observe first that an orbit can only specialize to an orbit of lower dimension, hence we have shown that the dimension of the orbits are at most as indicated. But if strict inequality would hold, we would obtain the contradiction $\dim(\mathfrak{C}_4)<15$, since in 5.13 we shall see that $\mathfrak{A}\to\mathfrak{C}_1$. Q.E.D.

**5.13 Proposition** There exist specializations $\mathfrak{A}\to\mathfrak{C}_1$, respectively $\mathfrak{B}_1\to\mathfrak{C}_2$, whereas there do not exist specializations $\mathfrak{C}_j\to\mathfrak{B}_i$ (for any j,i), respectively $\mathfrak{B}_i\to\mathfrak{C}_i$ (for i =1,2).

**Proof** The non existence of specializations $\mathfrak{C}_j\to\mathfrak{B}_i$ is obvious since $\pi$ is continuous and the variety $\mathfrak{C}'$ of double lines is closed. $\mathfrak{B}_i$ cannot specialize to $\mathfrak{C}_i$ since, by proposition 5.12, those orbits have the same dimension.

In order to show the specializations indicated above, we shall consider the following 2-parameter family:

Let Q be the quadric $x_0x_1-x_2x_3=0$, and, for $\lambda,\mu\in\mathbb{C}$, take the following two quadrics:

**5.14** $$Q_1=\{\mu x_3x_1+x_3^2=0\},$$

$$Q_2=\{\mu\lambda x_0x_2+\lambda x_0^2+\mu x_0x_1+x_0x_3=0\}.$$

We shall consider, for $\lambda,\mu\in\mathbb{C}$, the net spanned by $Q,Q_1,Q_2$.

**5.15 Claim**: we have a net in $\mathfrak{C}_2$ for $\lambda=\mu=0$, in $\mathfrak{C}_1$ for $\mu=0$, $\lambda\neq0$, in $\mathfrak{B}_1$ for $\mu\neq0$, $\lambda=0$, in $\mathfrak{A}$ for $\lambda\neq0$, $\mu\neq0$.

**Proof of the claim**: it suffices to compute the base locus of the net, which is a subscheme of Q. Hence we shall perform the computations using the isomorphism of Q with $\mathbb{P}^1 \times \mathbb{P}^1$, given as follows:

**5.16** $x_0 = u_0 v_0$, $x_1 = u_1 v_1$, $x_2 = u_0 v_1$, $x_3 = u_1 v_0$.

Since $Q_1 = x_3(\mu x_1 + x_3) = u_1^2 v_0(\mu v_1 + v_0)$, while

$Q_2 = x_0(\mu\lambda x_2 + \lambda x_0 + \mu x_1 + x_3) = u_0 v_0(\mu v_1 + v_0)(u_1 + \lambda u_0)$,

the base locus is given by $v_0(\mu v_1 + v_0) = 0$, unless $\lambda = 0$ or $\lambda = \mu = 0$, cases in which the base locus becomes respectively $u_1 v_0(\mu v_1 + v_0) = 0$, $u_1 v_0^2 = 0$. We can now conclude by proposition 5.4. Q.E.D. for 5.15 and 5.13

## 6. Smoothness and normal crossings structure for the orbits of the variety $\mathcal{S}$.

In this section, we shall prove that $\mathcal{S} - (\mathcal{B}_2 \cup \mathcal{C}_3 \cup \mathcal{C}_4 \cup \mathcal{C}_5)$ is smooth, and that at every point of $\mathcal{C}_2$ the orbit closures $\overline{\mathcal{B}}_1, \overline{\mathcal{C}}_1$, are smooth subvarieties of $\mathcal{S}$ intersecting transversally in $\mathcal{C}_2$.

Since smoothness is an open property, and $\mathcal{C}_2$ is a single orbit, for the first statement it will suffice to exhibit a point of $\mathcal{C}_2$ which is smooth on $\mathcal{S}$, since it will imply that all points of the orbits whose closure contains $\mathcal{C}_2$ are smooth.

**6.1 Lemma** If $\alpha \in \mathcal{C}_2$, then the dimension of the Zariski tangent space $T\mathcal{S}_\alpha$ of the variety $\mathcal{S}$ at $\alpha$ is 19.

**Proof** Let $\alpha(y_1,y_2,y_3)=y_1x_0^2+2y_2x_0x_2+2y_3(x_0x_3-x_1x_2)$. As in 5.15, using the isomorphism 5.16, we see that the base locus of the net $\alpha$ is given on the smooth quadric $x_0x_3-x_1x_2$ by $\{u_0^2v_0^2=u_0u_1v_0^2=0\}$; hence, by 5.4, $\alpha\in\mathfrak{C}_2$. The net $\alpha$ corresponds to the following matrix A:

$$A=\begin{pmatrix} y_1 & 0 & y_2 & y_3 \\ 0 & 0 & -y_3 & 0 \\ y_2 & -y_3 & 0 & 0 \\ y_3 & 0 & 0 & 0 \end{pmatrix}$$

Let $C=A+tB$, where $B=(b_{ij})$ is a $4\times4$ symmetric matrix of linear forms $b_{ij}$ in $y_1,y_2,y_3$, and let $D=(D_{ij})=\Lambda^3C$.

In the ring $\mathbb{C}[y_1,y_2,...,y_3,t]/(t^2)$ we have $\det(C)=(y_3^2+tQ')^2$, where $Q'=y_3b_{14}-y_3b_{23}-y_1(b_{44}/2)+y_2b_{24}$. The Zariski tangent space $T\mathcal{S}_\alpha$ is the set of morphisms of Spec $\mathbb{C}[t]/(t^2)$ to $\mathcal{S}$ mapping the closed point to $\alpha$. It can be identified to the quotient by the 1-dimensional subspace generated by A of the vector space of matrices B for which in the above ring there exists an element R such that R divides $D_{ij}$ for $1\le i\le j\le 4$ and $\det(C)=R^2$. It follows that $R=(y_3^2+tQ')u$, u being a unit, whence we get the condition that $y_3^2+tQ'|D_{ij}$ for $1\le i\le j\le 4$. We now calculate the $D_{ij}$'s and find the subspace of matrices B for which $y_3^2+tQ'|D_{ij}$.

$D_{11}=tD'_{11}$ with $D'_{11}=-y_3^2b_{44}$; imposing $(y_3^2+tQ')|D_{11}$, we get $(y_3^2+tQ')t\eta=tD'_{11}$, where $\eta$ is an unknown linear form in the y's. This equation is equivalent to $y_3^2|D'_{11}$, which is always satisfied and gives no condition on B.

$D_{12}=tD'_{12}$ with $D'_{12}=y_3^2b_{34}-y_3y_2b_{44}$; as before, we get $y_3^2|y_3^2b_{34}-y_3y_2b_{44}$ which is equivalent to $\mathbf{y_3|b_{44}}$, which gives two linear conditions on B. Denote by $\beta_{44}$ the complex number $b_{44}/y_3$.

$D_{13}=tD'_{13}$ with $D'_{13}=y_3^2b_{24}$; as for $D_{11}$, no condition on B.

$D_{14}=-y_3^3+tD'_{14}$, $D'_{14}=-y_3^2b_{44}+2y_3^2b_{23}-y_3y_2b_{24}$; as before, we get $(y_3^2+tQ')(-y_3+t\eta)=-y_3^2+tD'_{14}$, which is equivalent to $y_3^2|y_3Q'+D'_{14}$. But, (mod $y_3^2$), since $y_3|b_{44}$, we have $y_3Q'+D'_{14}=y_3y_2b_{24}-y_3y_2b_{24}=0$: no condition on B.

$-D_{22}=tD'_{22}$, $D'_{22}=y_3^2b_{33}+y_2^2b_{44}-2y_3y_2b_{34}$; we get $y_3^2|D'_{22}$, which is equivalent to $\mathbf{y_3|(y_2\beta_{44}-2b_{34})}$, and defines two linear conditions on B.

$D_{23}=y_3^3+tD'_{23}$, $D'_{23}=-y_3^2b_{23}+2y_3^2b_{14}+y_3y_2b_{24}-y_3y_1b_{44}$; $(y_3^2+tQ')(y_3+t\eta)=y_3^3+tD'_{23}$, $y_3^2|-y_3Q'+D'_{23}$. But, (mod $y_3^2$), since $y_3|b_{44}$, we have $-y_3Q'+D'_{23}=-y_3y_2b_{24}+y_3y_2b_{24}=0$: no condition on B.

$D_{24}=y_3^2y_2+tD'_{24}$, $D'_{24}=y_3y_2b_{14}-y_3y_2b_{23}+y_3^2b_{13}-y_3y_1b_{34}+ +y_2^2b_{24}$; $(y_3^2+tQ')(y_2+t\eta)=y_3^2y_2+tD'_{24}$;

$y_3^2 | -y_2Q'+D'_{24}$; we get $y_3 | -y_2(\beta_{44}/2)+b_{34}$, which is the same condition as $D_{22}$.

$D_{33}=-ty_3^2b_{22}$: no condition.

$D_{34}=tD'_{34}$, $D'_{34}=-y_3^2b_{12}-y_3y_2b_{22}+y_3y_1b_{24}$; we get

$y_3^2 | D'_{34}$, equivalent to $\mathbf{y_3 | (y_2b_{22}-y_1b_{24})}$, which gives three linear conditions on B. Denote by $\lambda$ the linear form $(y_2b_{22}-y_1b_{24})/y_3$.

$-D_{44}=y_3^2y_1+tD'_{44}$, $D'_{44}=b_{11}y_3^2-2y_3y_1b_{23}+y_2^2b_{22}+2y_3y_2b_{12}$;

$(y_3^2+tQ')(y_1+t\eta)=y_3^2y_1+tD'_{44}$; we get $y_3^2 | -y_1Q'+D'_{44}$ which is the same as

$\mathbf{y_3 | (-y_1(b_{14}+b_{23})+y_1^2(\beta_{44}/2)+y_2\lambda+2y_2b_{12})}$, and gives three linear conditions on B. Since the associated matrix is triangular, we see that the $2+2+3+3=10$ linear conditions equivalent to $B \in T\mathcal{S}_\alpha$ are independent so that $\dim(T\mathcal{S}_\alpha)=\dim(T\mathfrak{U}_\alpha)-10=19$, because $\mathfrak{U}$ is smooth. Q.E.D.

**6.2 Lemma** If $\alpha \in \mathcal{B}_2$, then $\dim(T\mathcal{S}_\alpha)=21$.

**Proof** Let $\alpha=2y_1x_1x_2+2y_2x_0x_3$; then, by 5.4, $\alpha \in \mathcal{B}_2$. The net $\alpha$ corresponds to the following matrix A:

$$A=\begin{pmatrix} 0 & 0 & 0 & y_2 \\ 0 & 0 & y_1 & 0 \\ 0 & y_1 & 0 & 0 \\ y_2 & 0 & 0 & 0 \end{pmatrix}$$

As in 6.1, let $C=A+tB$, $B={}^tB$ a matrix of linear forms $b_{ij}$, and let $D=(D_{ij})=\Lambda^3C$. In $\mathbb{C}[y_1,y_2,y_3,t]/(t^2)$,

$\det(C)=(y_1y_2+tQ')^2$, $Q'=y_1b_{14}+y_2b_{23}$. Then $B\in T\mathcal{S}_\alpha$ iff $(y_1y_2+tQ')|D_{ij}$ for $1\le i\le j\le 4$.

$D_{11}=tD'_{11}$, $D'_{11}=-y_1^2b_{44}$; then $(y_1y_2+tQ')|tD'_{11}$ is equivalent to $y_2|b_{44}$. In the same way, $D_{22}$ = $-ty_2^2b_{33}$, $D_{33}=-ty_2^2b_{22}$, $D_{44}=-ty_1^2b_{11}$, give respectively $y_1|b_{33}$, $y_1|b_{22}$, $y_2|b_{11}$. So far, we have 8 independent linear conditions on B. It is easy to check that the other minors $D_{ij}$ give no conditions; it follows $\dim(T\mathcal{S}_\alpha)=\dim(T\mathcal{U}_\alpha)-8=21$. Q.E.D.

For $(a,b)\in\mathbb{C}^2$, consider the following family of "special nets": $\alpha(a,b)(y_1,y_2,y_3)=2y_1Q_1+2y_2Q_2+2y_3(x_0x_3-x_1x_2)$, where

$$Q_1=x_0^2+x_0x_2+ax_0x_1+ax_0x_3,$$

$$Q_2=x_2^2+(1-b)x_0x_2+ax_2x_3+a(1-b)x_0x_3$$

Using the isomorphism 5.16, we have $Q_1=v_0(v_0+av_1)u_0(u_0+u_1)$, $Q_2=v_0(v_0+av_1)u_1((1-b)u_0+u_1)$; then, looking at the base locus, by 5.4 we see that in $\mathbb{C}^2$

**6.3** $\alpha^{-1}(\mathcal{B}_1^-)=\{b=0\}$, $\alpha^{-1}(\mathcal{C}_1^-)=\{a=0\}$, $\alpha^{-1}(\mathcal{C}_2^-)=\{a=b=0\}$ as sets.

In particular, $\alpha(0)\in\mathcal{C}_2$.

**6.4 Lemma** The regular map $\alpha:\mathbb{C}^2\to\mathcal{S}$ meets transversally the orbit $\mathcal{C}_2$ at $\alpha(0)$, i.e. $\operatorname{Im}(d\alpha_0)\oplus T\mathcal{C}_{2,\alpha(0)}=T\mathcal{S}_{\alpha(0)}$.

**Proof** By 6.1 and 5.12, it will suffice to produce two functions in $\mathcal{O}_{\mathcal{S},\alpha(0)}$ vanishing on $\mathcal{C}_2$ and such that their pull-back to $\mathbb{C}^2$ spans the maximal ideal (a,b) of $\mathcal{O}_{\mathbb{C}^2,0}$. The family $\alpha$ is associated to the following matrix:

$$M=\begin{pmatrix} 2y_1 & ay_1 & y_1+(1-b)y_2 & y_3+ay_1+a(1-b)y_2 \\ ay_1 & 0 & -y_3 & 0 \\ y_1+(1-b)y_2 & -y_3 & 2y_2 & ay_2 \\ y_3+ay_1+a(1-b)y_2 & 0 & ay_2 & 0 \end{pmatrix}$$

Then, (mod $y_2$), $\det(M)=(y_3(y_3+ay_1))^2$. One of the conditions that define the orbit $\mathcal{C}_2$ is that there

exist a linear form in the y's whose $4^{th}$ power equals det(M): so we get the equation $a=0$. Looking at the classification 5.1, we see that an other condition defining $\mathcal{C}_2$ is that in $\mathbb{P}\langle A,B\rangle$ there is only 1 matrix of rank 1, A and B being, in the above notation, the $4\times4$ matrices associated to the quadrics $Q_1$ and $Q_2$. Consider the following $2\times2$ minor of M

$$\begin{pmatrix} 2y_1 & y_1+(1-b)y_2 \\ y_1+(1-b)y_2 & 2y_2 \end{pmatrix}$$

Its determinant is $-y_1^2-(1-b)^2y^2+(2+2b)y_1y_2$; since the discriminant is 16b, we get the equation $b=0$. Q.E.D.

**6.5 Theorem** The variety $\mathcal{S}-\overline{\mathcal{B}}_2$, is smooth; more precisely, $\mathrm{Sing}(\mathcal{S})=\overline{\mathcal{B}}_2=\mathcal{B}_2\cup\mathcal{C}_3\cup\mathcal{C}_4\cup\mathcal{C}_5$. Moreover, at every point of $\mathcal{C}_2$ the orbit closures $\overline{\mathcal{B}}_1$, $\overline{\mathcal{C}}_1$, $\overline{\mathcal{C}}_2$ are smooth and with respective ideals $(\lambda)$, $(\mu)$, $(\lambda,\mu)$ for a suitable pair of functions $\lambda,\mu$, which can be completed to a system of parameters .

**Proof** Since the singular locus $\mathrm{Sing}(\mathcal{S})$ of the variety $\mathcal{S}$ is closed, the first assertion follows from the fact that $\dim\mathcal{S}=19$ and 6.1, 6.2, 5.10, 5.11, 5.13. The second assertion follows then from 6.3 and the proof of 6.4. Q.E.D.

**6.6 Remark** Moreover, one can prove that the dimension of the Zariski tangent space of the variety $\mathcal{S}$ at the points of $\mathcal{B}_2 \cup \mathcal{C}_3 \cup \mathcal{C}_4$ is 21.

From propositions 5.1, 5.3, 5.7, 5.8, 5.10, 5.11, 5.12, 5.13, and theorem 6.5 follows the main theorem stated in the introduction.

MAIN THEOREM

The variety $\mathcal{S}$ contains a finite number of $\mathbb{P}GL(3)\times\mathbb{P}GL(4)$-orbits, labelled $\mathcal{A},\mathcal{B}_1,\mathcal{B}_2,\mathcal{C}_1,\mathcal{C}_2,\mathcal{C}_3,\mathcal{C}_4,\mathcal{C}_5$ and of respective dimensions 19,18,17,18,17,16,15,11. These orbits satisfy moreover the following incidence relations ( $\mathcal{A} \rightarrow \mathcal{B}$ means that the closure $\overline{\mathcal{A}}$ of $\mathcal{A}$ contains $\mathcal{B}$ )

$$\begin{array}{ccccccccc}
\mathcal{A} & \rightarrow & \mathcal{B}_1 & \rightarrow & \mathcal{B}_2 & & & & \\
 & \searrow & & \searrow & & \searrow & & & \\
 & & \mathcal{C}_1 & \rightarrow & \mathcal{C}_2 & \rightarrow & \mathcal{C}_3 & \rightarrow \ \mathcal{C}_4 & \rightarrow \ \mathcal{C}_5
\end{array}$$

In particular the variety $\mathcal{S}$ is irreducible. Moreover $\mathcal{S}-(\mathcal{B}_2\cup\mathcal{C}_3\cup\mathcal{C}_4\cup\mathcal{C}_5)$ is smooth, and at the points of $\mathcal{C}_2$ the orbit closures $\overline{\mathcal{B}}_1, \overline{\mathcal{C}}_1$, are smooth subvarieties of $\mathcal{S}$ intersecting transversally in $\mathcal{C}_2$ .

On the other hand, we omit the full analysis of the nonsingularity of the orbit closures , in view of the fact that, as we are going to show in the following section, only actual nets (i.e. of projective dimension 2) occur when dealing with surfaces of general type with $K^2 = \chi = 2$.

## 7. Surfaces of general type with $K^2 = \chi = 2$ of the hyperelliptic birational class.

We recall that (cf. [Ca-De], page 37), if we are in the hyperelliptic birational case, letting $\varphi$ be the bicanonical morphism $\varphi\colon S \to \mathbb{P}^3$ the sheaf $\mathcal{F} = \varphi_*(\mathcal{O}_S(K_S))$ admits a minimal free resolution of length 1

**7.1** $$0 \to \mathcal{L}_1 \to \mathcal{L}_0 \to \mathcal{F} \to 0$$

with $\mathcal{L}_1$, $\mathcal{L}_0$ direct sums of invertible sheaves on $\mathbb{P}^3$, and the homomorphism $\mathcal{L}_1 \to \mathcal{L}_0$ being given by a matrix of the following form :

**7.2** $$A' = \begin{pmatrix} y_0G+\det(a) & y_0q_1 & y_0q_2 & y_0q_3 & y_0q_4 & Q+y_0L \\ q_1 & & & & & y_0 \\ q_2 & & (a) & & & 0 \\ q_3 & & & & & 0 \\ q_4 & & & & & 0 \\ Q+y_0L & y_0 & 0 & 0 & 0 & 0 \end{pmatrix}$$

where $Q = Q(y_1, y_2, y_3)$ is a quadratic form, $L = L(y_0, y_1, y_2, y_3)$ is a linear form, $G = G(y_0, y_1, y_2, y_3)$ is a cubic form, $(a) = (\alpha) + y_0(a')$ is a symmetric $4\times 4$ matrix of linear forms ((a') is thus a matrix of constants), and $(\alpha) = (\alpha)(y_1, y_2, y_3)$ gives a "special net of quadrics" with determinant $= Q^2$.

Setting $y_0 = w^2$, one obtains the equations of the canonical model X of S in the weighted projective space $\mathbb{P}(1,2,2,2,3,3,3,3,4)$, with coordinates $(w, y_1, y_2, y_3, v_1, v_2, v_3, v_4, v_5)$ as follows:
let A be the following (symmetric) matrix

$$7.3 \quad A = \begin{pmatrix} y_0G+\det(a) & wq_1 & wq_2 & wq_3 & wq_4 & Q+y_0L \\ wq_1 & & & & & w \\ wq_2 & & (a) & & & 0 \\ wq_3 & & & & & 0 \\ wq_4 & & & & & 0 \\ Q+y_0L & w & 0 & 0 & 0 & 0 \end{pmatrix}$$

and let v be the transposed vector of $(v_0, v_1, v_2, v_3, v_4, v_5)$ where we have set $v_0 = 1$.

If $B = \Lambda^5(A)$ is the adjoint matrix of A, and $B_{ij}$ is the i,j-th entry of B, then the following so called Rank Conditions hold for the matrix A :

$$7.4 \qquad B_{ij} = \Sigma_{k=0,..5}\, \lambda_{ij,k}(w, y_1, y_2, y_3)\, B_{0k} \qquad (i,j=1,..5)$$

Then (cf. loc. cit.) the equations which describe the canonical ring $\mathcal{R}$ of S (i.e., $X = \mathrm{Proj}(\mathcal{R})$) are:

$$7.5 \qquad Av = 0,$$
$$v_i v_j = \Sigma_{k=0,..5}\, \lambda_{ij,k}(w, y_1, y_2, y_3)\, v_k \quad (i,j=1,..5).$$

**7.6 Proposition** If a "special net" $\alpha$ is obtained from a surface of general type with $K^2 = \chi = 2$, then it is actually a net, i.e., it belongs only to an orbit of type $\mathcal{A}, \mathcal{B}_1, \mathcal{C}_1$ or $\mathcal{C}_2$.

**Proof** If a "special net" $\alpha$ is obtained from a surface of general type S with $K^2=\chi=2$, then the canonical model X of S is defined by equations 7.5.

Now, if the image of $\alpha$: $\mathbb{P}^2 \to \mathbb{P}'$ = Projective space of symmetric $4\times4$ matrices, has projective dimension $\leq 1$, then there is a point P of $\mathbb{P}^2$ for which the matrix $\alpha(P)$ is zero.

In particular, there are linear forms y,y' in $(y_1,y_2,y_3)$ such that A vanishes for $w=y=y'=0$ (in fact, by our normal forms, one can choose $w=y_1=y_2=0$).

Since the bicanonical mapping is finite, setting $w=y_1=y_2=0$, we get a finite subscheme Z of X, and we can assume that also setting $w=y_1=0$, we get a finite subscheme Z' of X.

Inspecting equations 7.5, we shall show that the length of Z is at least 6, contradicting the fact that Z is a subscheme of Z' and that Z' has length 4 (being a complete intersection of a canonical curve with a bicanonical curve, and since $K^2=2$). Let us in fact look at the relations 7.4 (which hold in the ring of polynomials $\mathbb{C}[w,y_1,y_2,y_3]$) .

Passing to the quotient ring $\mathbb{C}[y_1,y_2,y_3]$ by setting $w=0$, and denoting by a" the homomorphic image of an element a in the quotient ring, we obtain

7.7 $$B''_{ij} = \Sigma_{k=0,..5}\, \lambda''_{ij,k}(y_1,y_2,y_3)\, B''_{0k} \qquad (i,j=1,..5).$$

On the other hand, B" is the adjoint matrix of the matrix A"

$$\text{7.8} \qquad A'' = \begin{pmatrix} \det(\alpha)=Q^2 & 0 & 0 & 0 & 0 & Q \\ 0 & & & & & 0 \\ 0 & & (\alpha) & & & 0 \\ 0 & & & & & 0 \\ 0 & & & & & 0 \\ Q & 0 & 0 & 0 & 0 & 0 \end{pmatrix}$$

whence $B''_{0k}=0$ for $k\neq 5$, whereas $B''_{05}=Q^3$, $B''_{55}=Q^4$, $B''_{i5}=0$ for $i\neq 0,5$. Also, $B''_{ij}=Q^2\beta_{ij}$ $(i,j=1,..,4)$, $\beta$ being the adjoint matrix of $\alpha$. Hence $\lambda''_{ij,k}(y_1,y_2,y_3)=0$, for $k\neq 5$, while $\lambda''_{55,5}(y_1,y_2,y_3)=Q$, $\lambda''_{ij,5}(y_1,y_2,y_3)=\beta_{ij}/Q$. Now, the polynomials $\beta_{ij}$ involve only the variables $y_1,y_2$, and are divisible by $Q(y_1,y_2)$, hence the $\lambda''_{ij,k}$ also involve only the variables $y_1,y_2$.
Therefore the ideal of the scheme Z, generated by $w,y_1,y_2$ and the equations 7.5, is also generated by $w,y_1,y_2$ and the $v_iv_j$'s. It follows that Z is contained in the affine piece $y_3\neq 0$ and that its ring is isomorphic as a $\mathbb{C}$-vector space to the vector space generated by $1,v_1,...,v_5$. The length of Z is 6 and this is the desired contradiction. Q.E.D.

**7.9 Proposition** If a "special net" $\alpha$ is obtained from a surface of general type with $K^2=\chi=2$, then it cannot belong to an orbit of type $\mathfrak{C}_1$ or $\mathfrak{C}_2$.

**Proof** We shall follow the same reasoning we have adopted in the course of proving 7.6, using the particular nets whose matrices $\alpha$ are as follows:

$$\begin{pmatrix} y_2 & y_1 & 0 & 0 \\ y_1 & 0 & 0 & 0 \\ 0 & 0 & 0 & y_1 \\ 0 & 0 & y_1 & y_3 \end{pmatrix} \qquad \begin{pmatrix} 0 & y_1 & 0 & y_2 \\ y_1 & 0 & 0 & 0 \\ 0 & 0 & 0 & y_1 \\ y_2 & 0 & y_1 & y_3 \end{pmatrix}$$

The respective non-zero $\beta_{ij}$'s, $i \le j$, are (up to sign)

for the first matrix: $\beta_{12} = y_1^3$, $\beta_{22} = y_1^2 y_2$, $\beta_{33} = y_1^2 y_3$, $\beta_{34} = y_1^3$;

for the second matrix: $\beta_{12} = y_1^3$, $\beta_{23} = y_1^2 y_2$, $\beta_{33} = y_1^2 y_3$, $\beta_{34} = y_1^3$.

Let us consider again the subscheme Z of X obtained by setting $w = y_1 = y_2 = 0$. As before we get $\lambda''_{ij,k}(y_1, y_2, y_3) = 0$, for $k \neq 5$, while $\lambda''_{55,5}(y_1, y_2, y_3) = Q$, $\lambda''_{ij,5}(y_1, y_2, y_3) = \beta_{ij}/Q$, and here $Q = y_1^2$.

The ideal of the scheme Z, which is generated by $w, y_1, y_2$ and the equations 7.5, is also generated by $w, y_1, y_2$, $v_4$, $v_3^2 + v_5$, and $v_i v_j$ for $(i,j) \neq (3,3)$. It follows that Z is contained in the affine piece $y_3 \neq 0$ and that its ring is isomorphic as a $\mathbb{C}$-vector space to the vector space generated by $1, v_1, v_2, v_3, v_3^2$. The length of Z is 5 and this is the desired contradiction. Q.E.D.

Propositions 7.6 and 7.9 yield immediately theorem II stated in the introduction

## Theorem II

Let S be a surface of general type with $K^2 = \chi = 2$ and belonging to the <u>hyperelliptic birational locus</u>

in the moduli space: then, if $\alpha$ is an associated special net of quadrics, then it can only belong to an orbit of type $\mathcal{A}$ or $\mathcal{B}_1$.

## References

[Ba] W. Barth, "Moduli of vector bundles on the projective plane", Inv. Math. **42**, 63-91 (1977).

[Bar] R. Barlow, "Some new surfaces with $p_g = 0$", Duke Math. Jour. Vol. **51**, No. 4 (1984), 889-904.

[Bea] A. Beauville, "Variétés de Prym et jacobiennes intermediaires", Annales de l' E.N.S., **10**, 309-391 (1977).

[Bo] E. Bombieri, "Canonical models of surfaces of general type", Inst. Hautes Études Sci. Publ. Math. **42** (1973), 171-219

[BPV] W. Barth, C. Peters and A. Van de Ven, "Compact complex surfaces", Ergebnisse der Math., Springer Verlag, 1984

[Ca1] F. Catanese, "Babbage's conjecture, contact of surfaces, symmetric determinantal varieties and applications", Inv. Math. **63**, 433-465 (1981).

[Ca2] F. Catanese, "Commutative algebra methods and equations of regular surfaces", in Alg. Geom. Bucharest 1982, L.N.M. **1056** (1984), 68-111.

[Ca3] F. Catanese, "Canonical rings and "special surfaces"", Proc. Symposia in Pure Math., **46** (I), 1987, 175-194.

[Ca-De] F. Catanese, O. Debarre, "Surfaces with $K^2 = 2$, $p_g = 1$, $q = 0$", Crelle's Journal reine angew. Math. **395** (1989), 1-55.

[Di] A.C. Dixon, "Note on the reduction of a ternary quantic to a symmetrical determinant", Proc. Camb. Phil. Soc. **11**, 350-351 (1902).

[En] F. Enriques, "Le superficie algebriche", Zanichelli (Bologna) 1949.

[Ha] R. Hartshorne, Algebraic geometry, G.T.M. 52 Springer, 1976

[O-P] F. Oort, C. Peters, "A Campedelli surface with torsion group $\mathbb{Z}/2$", Indag. Math. **43** (1981), 399-407.

[Re1] M. Reid, "Surfaces with $p_g=0$, $K^2=1$", J. Fac. Sc. Univ. Tokyo S. IA **25** (1978) 75-92.

[Re2] M. Reid, "Infinitesimal view of extending a hyperplane section", Proc. of the conference "Hyperplane sections", L'Aquila 1988, S.L.N. **1437**, 214-286.

[Tju] A. N. Tjurin, "On intersections of quadrics", Russian Math. Surveys 30: 6, 51-105 (1975).

[Wa] C.T.C. Wall, "Nets of quadrics, and theta-characteristics of singular curves", Phil. Trans. Royal Soc. London **289** A 1357 (1978) 229-269.

Authors' addresses:
F. Catanese , P.Cragnolini
Dipartimento di Matematica
Universita' di Pisa,
Via Buonarroti 2, 56100 Pisa
(Italy)

P.Oliverio
Dipartimento di Matematica
Universita' della Calabria
Arcavacata di Rende
87036 Rende (Cosenza)
(Italy).

Contemporary Mathematics
Volume **162**, 1994

# Non-isomorphic curves of genus four with isomorphic (non-polarized) jacobians

CIRO CILIBERTO AND GERARD VAN DER GEER

ABSTRACT. We show that there exist non-isomorphic Riemann surfaces of genus four with isomorphic jacobians.

In this note we shall prove that there exist non-isomorphic Riemann surfaces of genus 4 for which the jacobian varieties are isomorphic (as non-polarized abelian varieties).

To put this result in perspective, we remark that for $g = 2$ and $g = 3$ the image of $\mathcal{M}_g(\mathbb{C})$ under Torelli is dense in $\mathcal{A}_g(\mathbb{C})$. Therefore, it is easy to show that there are abelian surfaces and abelian threefolds which possesses two inequivalent principal polarizations which make them a jacobian (by using abelian varieties with multiplication by a totally real field of degree $g$ which possess a totally positive unit which is not a square). It would be nice to have an explicit example of a pair of non-isomorphic curves of genus two given by an equation $y^2 = f_i$ $(i = 1, 2)$ with isomorphic jacobians (as abelian varieties). The question remains whether there are simple abelian varieties of dimension $g > 4$ carrying two principal polarizations which make them a jacobian.

Our investigation was inspired by a paper of Comessatti, see [C].

1) Let $K$ be a totally real field of degree $g$ over $\mathbb{Q}$ and let $O_K$ be its ring of integers. We let

$$\delta^{-1} = \{x \in K\colon \mathrm{Tr}(xy) \in \mathbb{Z} \text{ for all } y \in O_K\}$$

1991 *Mathematics Subject Classification.* Primary14H10,14H40 ; Secondary 14K10.
*Key words and phrases.* Curves of genus 4, jacobian variety, abelian varieties.
This paper is in final form and no version of it will be submitted for publication elsewhere

be the inverse different of $K$. We consider the $O_K$-module $L = O_K \oplus \delta^{-1}$. The set $P$ of $O_K$-linear alternating forms on $L$ is in 1-1 correspondence with $(\wedge^2_{O_K} L)^\vee \cong O_K$. The element $1 \in O_K$ corresponds to the alternating form $R$ given by $\langle (a,b),(c,d)\rangle = \mathrm{Tr}_{K/\mathbb{Q}}(ad-bc)$. This alternating form is unimodular.

2) We can construct a complex torus of dimension $g$ with real multiplication, that is, with an injective homomorphism $O_K \to \mathrm{End}(A)$ by putting a complex structure on $L \otimes_{\mathbb{Z}} \mathbb{R}$. These complex structures are parametrized by

$$GSp(L)\backslash GSp(L\otimes_{\mathbb{Z}}\mathbb{R})/k,$$

where $k$ is the centralizer of one complex structure and where $GSp$ stands for $O_K$-linear maps which preserve the symplectic form up to a scalar. We get a polarized abelian variety if we fix a non-degenerate (= non-zero) alternating form $F$ from $P$ and take a complex structure $I$ such that $F$ becomes a Riemann form, i.e. if $F(Ix,x)>0$ for all $x \in L\otimes\mathbb{R}, x\neq 0$. The set of such complex structures is not empty and can be identified with

$$Sp(L)\backslash Sp(L\otimes_{\mathbb{Z}}\mathbb{R})/k, \qquad (*)$$

where $k$ is a maximal compact subgroup.

We set

$$X_L = Sp(L)\backslash Sp(L\otimes_{\mathbb{Z}}\mathbb{R})/k.$$

This can be given the structure of a complex analytic variety.

3) Let $A$ be an abelian variety with real multiplication by $O_K$. We have a natural injective map $\nu$

$$\nu\colon O_K \to NS(A),$$

where $NS$ is the Néron-Severi group of $A$. The image of 1 corresponds to a principal polarization in $NS(A)$. Let $\mathcal{L}_\alpha$ be a line bundle representing the class $\nu(\alpha) \in NS(A)$. In the case treated in 2) its Chern class is represented by the alternating form $\mathrm{Tr}(\alpha(ad-bc))$. This form satisfies the positivity condition for polarizations if $\alpha$ is totally positive and the polarization is principal if $\alpha$ is a unit of $O_K$, cf [L].

Let $\epsilon \in O_K^*$ be a unit. It acts as an automorphism on the abelian variety $A$ and thus acts on the Neron-Severi group $\subset H^2(A,\mathbb{Z}) = \wedge^2 H^1(A,\mathbb{Z})$ by multiplication by $\epsilon^2$. It sends a principal polarization to a principal polarization. If the group of totally positive units $U_K^+$ is larger than the group of squares $U_K^{(2)}$ of units of $O_K$ then there exist two inequivalent principal polarizations on $A$, represented by $\mathcal{L}_1$ and $\mathcal{L}_\epsilon$ with $\epsilon \in U_K^+\backslash U_K^{(2)}$.

4) We consider again the situation of 1). By identifying the pair $(L,R)$ with the standard symplectic space $(\mathbb{Z}^{2g}, B(x,y) = \sum_{i=1}^g x_i y_{g+i} - x_{g+i}y_i))$ we obtain a homomorphism of algebraic groups

$$Sp(2,K) \to Sp(2g,\mathbb{Q})$$

and an associated immersion of symmetric spaces

$$\mathcal{H}_1^g \to \mathcal{H}_g,$$

where $\mathcal{H}_i = \{\tau \in M_i(\mathbb{C}): \tau^t = \tau,\ \mathrm{Im}\,(\tau) > 0\}$ is the Siegel upper half space of degree $i$. This induces a generically finite morphism

$$\phi: X_L \to Sp(2g, \mathbb{Z})\backslash\mathcal{H}_g .$$

Let $\mathcal{A}_g$ be the moduli space of principally polarized abelian varieties of dimension $g$. We have

$$A_g(\mathbb{C}) \cong Sp(2g, \mathbb{Z})\backslash\mathcal{H}_g.$$

5) From now on $K$ will be a totally real quartic field which contains no quadratic subfields and possesses a totally positive unit which is not a square.

Then the constructions of 2) and 4) gives us a morphism

$$\phi: X_L \to Sp(8, \mathbb{Z})\backslash\mathcal{H}_4 = \mathcal{A}_4(\mathbb{C}).$$

By changing the polarization $\mathrm{Tr}_{K/\mathbb{Q}}(ad-bc)$ to $\mathrm{Tr}_{K/\mathbb{Q}}\epsilon(ad-bc)$ we find a second morphism

$$\phi': X_{L'} \to Sp(8, \mathbb{Z})\backslash\mathcal{H}_4 = \mathcal{A}_4(\mathbb{C}),$$

where $L' = O_K \oplus \delta^{-1}$ with $\mathrm{Tr}_{K/\mathbb{Q}}\epsilon(ad-bc)$.

As usual, $\mathcal{M}_g$ denotes the moduli space of curves of genus $g$. Let

$$j: \mathcal{M}_4(\mathbb{C}) \to \mathcal{A}_4(\mathbb{C})$$

be the morphism which associates to a curve of genus 4 its jacobian.

There exists a subgroup $\Gamma \subset Sp(2, O_K)$ which is of finite index in $Sp(L)$ and in $Sp(L')$. The quotient $Y = \Gamma\backslash\mathcal{H}_1^4$ is a finite cover both of $X_L$ and of $X'_L$. Composing with the maps $\phi$ and $\phi'$ we obtain two generically finite morphisms

$$\psi: Y \to \mathcal{A}_4(\mathbb{C}) \qquad \psi': Y \to \mathcal{A}_4(\mathbb{C}).$$

They can be extended to morphisms of Satake compactifications $\overline{Y} \to \overline{\mathcal{A}}_4(\mathbb{C})$.

Since the closure of the image of $j$ in the Satake compactification $\overline{\mathcal{A}}_4(\mathbb{C})$ is an ample divisor (the zero divisor of a Siegel modular form) and since $\overline{Y} - Y$ consists of finitely many points we find

$$\dim(\psi^{-1}(\mathcal{M}_4) \cap \psi'^{-1}(\mathcal{M}_4)) \geq 2.$$

A point $y$ of $Y$ corresponds to an abelian fourfold (with some additional structure) which carries two principal polarizations (represented by $\psi(y)$ and $\psi'(y)$) which make the abelian fourfold a jacobian if $\psi(y)$ and $\psi'(y)$ belong to $j(\mathcal{M}_4)$. Therefore we find families of dimension $\geq 2$ of abelian varieties which carry two principal polarizations which make them a jacobian. However we want to exclude that such a family consists of jacobians of reducible curves only. For this we must check that the locus in $X_L$ (or in $X'_L$) of polarized abelian varieties that are products of polarized abelian varieties has dimension $\leq 1$. But this follows immediately from the following lemma.

**Lemma.** *Let $K$ be a totally real quartic field which contains no quadratic subfields and let $A$ be a 4-dimensional abelian variety with multiplication by $K$. Then either $A$ is simple or $A$ is isogenous to the fourth power of an elliptic curve.*

**Proof.** If $B$ is a simple abelian variety of dimension $< 4$ then $\mathrm{E}nd^0(B) = \mathrm{E}nd(B) \otimes \mathbb{Q}$ cannot contain $K$, as follows from the description of endomorphism rings in [M]. From this it follows that if $A$ is not simple at least one simple isogeny factor is contained twice in $A$.

If $E$ is an elliptic curve then $M_2(\mathrm{E}nd(E))$ cannot contain $K$, since this would imply that $K$ is quadratic over $\mathbb{Q}$ or over an imaginary quadratic field.

Suppose that $A$ is isogenous to $B^2$ with $B$ simple. Then we have an injection $K \to M_2(L)$ with $L = \mathrm{E}nd^0(B)$. This implies that elements of $K$ are of degree $\leq 2$ over $L$. Hence $L \neq \mathbb{Q}$. If $L$ is real quadratic this implies that $L \subset K$, a contradiction. If $L$ is a division quaternion algebra then $K$ is quadratic over a quadratic subfield of $L$, and the same holds if $L$ is a CM-field of degree 4. This is impossible.

If $A$ is isogenous to $E^2 \times B$ with $\dim(E) = 1$ and $B$ simple then we must have $K \subset M_2(\mathrm{E}nd^0(E))$ which was excluded above. Similarly, one excludes the case that $A$ is isogenous to $E^2 \times E'^2$ with $E$ and $E'$ non-isogenous elliptic curves.

Finally, if $A$ is isogenous to $E^3 \times E'$ and $E'$ is not isogenous to $E$, then $K$ is of degree $\leq 3$ over a CM-field or over $\mathbb{Q}$, and this is again impossible. □

6) It remains to show that there exist totally real quartic fields without quadratic subfields and with a totally positive unit which is not a square. We give an example of such a field communicated to us by J. Schmettow.

Let $f \in \mathbb{Q}[x]$ be the irreducible polynomial $x^4 - x^3 - 5x^2 + 4x + 3$. Then $K = \mathbb{Q}(\alpha)$ with $\alpha$ a root of $f$ is such a field. Indeed, the field is totally real and has discriminant 7537 which is a prime, hence there are no quadratic subfields. We have a system of fundamental units:

$$\epsilon_1 = -2 + \alpha, \epsilon_2 = -4 + \alpha^2, \epsilon_3 = 2 + 4\alpha - \alpha^3.$$

Then $\epsilon_3$ is totally positive and not a square since it is a fundamental unit.

## References

[C] A. Comessatti : Sulle superficie di Jacobi semplicemente singolari. *Mem. della Società Italiana delle Scienze (detta dei XL)*, **21**(3), 45-71 (1919).

[L] H. Lange: Abelian varieties with several principal polarizations. *Duke Math. Journal* **55**, 617-628 (1987).

[M] D. Mumford: Abelian varieties. Oxford University Press, 1974.

Dipartimento di Matematica, Università di Roma II, Via Fontanile di Carcaricola, 00173 Roma, Italy

*E-mail address*: ciliberto@mat.utovrm.it

Faculteit Wiskunde en Informatica, Universiteit van Amsterdam, Plantage Muidergracht 24, 1018 TV Amsterdam, The Netherlands

*E-mail address*: geer@fwi.uva.nl

Contemporary Mathematics
Volume **162**, 1994

# Deformations of symmetric products of curves

BARBARA FANTECHI

ABSTRACT. A well-known theorem by Kempf says that if $C$ is a non-hyperelliptic curve, then the deformation functors for $C$ and for any symmetric product $C^{(r)}$ are isomorphic.

In this paper we prove that they are isomorphic precisely when the curve $C$ has genus $\geq 3$, giving also a different proof for Kempf's result.

## 0. Introduction

Let $X$ be a smooth projective variety over an algebraically closed field of characteristic 0, $r$ a positive integer; denote by $X^r$ the product of $r$ copies of $X$ and by $X^{(r)}$ its quotient by the natural action of the symmetric group. A deformation of $X$ naturally induces a deformation of $X^{(r)}$. A natural question is to ask when all small deformations of $X^{(r)}$ arise in this way.

Easy considerations show that if $X$ is of general type and has dimension $\geq 3$ the answer to this question is indeed yes (cfr [**F2**]). For surfaces of general type this is also true (see [**F1**]).

If $X$ is a curve $C$ the symmetric products are smooth varieties, and high enough symmetric products are naturally projective bundles over the Jacobian variety of the curve. The question was answered in the affirmative by Kempf in [**K**] under the hypothesis that the curve is not hyperelliptic.

The aim of this paper is to sharpen this result. We will prove the following

THEOREM. *Let $C$ be a smooth curve of genus $g \geq 2$, $r \geq 2$ an integer. Then the natural map $Def(C) \to Def(C^{(r)})$ is an isomorphism if and only if $g \geq 3$.*

In particular if $g \geq 3$ the deformations are unobstructed. The result is proved by comparing the deformations of the symmetric product with the deformations of the product, and is related to the more general question of studying deformations of quotients of smooth varieties via the action of finite groups.

---

1991 *Mathematics Subject Classification*. Primary 14D15, 32G05.

The author is a member of GNSAGA of CNR and was partially supported by the European Science Program contract no. SCI-0398-C(A).

This paper is in final form and no version of it will be submitted for publication elsewhere.

## 1. Preliminaries and notation

**(1.1).** If $C$ is a smooth curve of genus $g \geq 2$ and $m$ is an integer, it is well known that the natural maps

$$H^0(C, K_C) \otimes H^0(C, mK_C) \to H^0(C, (m+1)K_C)$$

are surjective if $m \geq 3$; surjectivity for $m = 2$ holds if and only if $g \geq 3$.

**(1.2).** If $Y$ is a projective variety we will denote by $Def(Y)$ the functor of infinitesimal deformations, and by $Def'(Y)$ the functor of locally trivial infinitesimal deformations. If $Y$ is smooth $Def(Y) = Def'(Y)$ has as projective hull the germ of the Kuranishi family; moreover in this case $Def(Y)$ has as tangent space (corresponding to first order deformations) $H^1(Y, \theta_Y)$ and as obstruction space $H^2(Y, \theta_Y)$.

If $G$ is a finite group acting on $Y$ and $Def(Y)$ is prorepresentable we can consider the subfunctor $Def(Y)^G$ of $G$-invariant deformations (cfr [**C1**], §7 or [**F2**]); if $Y$ is smooth and $X = Y/G$ we have a natural map $Def(Y)^G \to Def'(X)$.

**(1.3).** Let $X$ be a variety, $B$ a subscheme defined by the ideal $\mathcal{I}_B$ (if B is a divisor, we will also write $\mathcal{O}_X(-B)$) instead of $\mathcal{I}_B$). We define $\theta_X(-\log B)$ to be the subsheaf of $\theta_X$ given by derivations sending the ideal $\mathcal{I}_B$ to itself. We also define $N'_{B/X}$, the equisingular normal sheaf (see [**C2**]), by the exact sequence

$$0 \to \theta_X(-\log B) \to \theta_X \to N'_{B/X} \to 0. \tag{1.3.1}$$

**(1.4).** If $r : Y \to X$ is a finite map of varieties and $\mathcal{F}$ is a sheaf on $Y$, we have natural isomorphisms $H^i(X, r_*(\mathcal{F})) \cong H^i(Y, \mathcal{F})$ as the higher direct image functors are zero; if moreover $r$ is the quotient by the action of a finite group $G$ and $\mathcal{F}$ is a $G$-sheaf on $Y$, we also have isomorphisms $H^i(X, r_*^G(\mathcal{F})) \cong H^i(Y, \mathcal{F})^G$.

If in addition $Y$ is smooth and $B \subset X$ is the branch locus of $r$, then

$$r_*^G(\theta_Y) \cong \theta_X(-\log B)$$

(see [**C2**]). In this case the natural mapping $Def(Y)^G \to Def'(X)$ can be described by the maps $H^i(Y, \theta_Y)^G \cong H^i(X, \theta_X(-\log B)) \to H^i(X, \theta_X)$ (for $i = 1, 2$) induced by the exact sequence (1.3.1).

**(1.5).** If $X$ is a variety, $\mathcal{F}$ a sheaf on $X$ and $Z$ a locally closed subscheme of $X$, we will denote by $H^i_Z(X, \mathcal{F})$ the cohomology groups with support on $Z$ (cfr [**G**] for definitions) and by $\mathcal{H}^i_Z(X, \mathcal{F})$ the corresponding cohomology sheaves.

**(1.6).** If we have a mapping $f : X \to Y$ between germs of complex spaces such that $df$ is an isomorphism and $X$ is smooth, then $f$ is an isomorphism. This follows from the fact that as $df$ is injective it induces an isomorphism between $X$ and its image, so $\dim Y \geq \dim X$. On the other hand surjectivity implies that the dimension of the tangent space of $Y$ equals $\dim X$. Therefore the dimension of $Y$ is the same of its tangent space, hence $Y$ is smooth and $f$ is an isomorphism.

We will use this elementary result in the following way; given a mapping between two deformation functors which is an isomorphism on tangent spaces, unobstructedness of the first functor implies that the map is an isomorphism.

## 2. Proof of the main theorem

**(2.1).** Let $C$ be a smooth curve of genus $g \geq 2$, $r \geq 2$ an integer. We denote by $C^r$ the product of $r$ copies of $C$; $\pi : C^r \to C^{(r)}$ will be the quotient by the natural action of $S_r$.

We will denote by $\Delta_{ij} \subset C^r$ the locus where the $i$-th and $j$-th factor coincide, and by $B \subset C^{(r)}$ the branch locus of $\pi$. We will also consider the factorization

$$C^r \xrightarrow{f} C^2 \times C^{(r-2)} \xrightarrow{g} C^{(r)}$$

of $\pi$ given by first dividing by the subgroup $S_{r-2}$ of $S_r$ of permutations leaving the first two coordinates unchanged.

Let $D := f(\Delta_{12})$, $Z := f((\Delta_{12} \cap \Delta_{13})$ (or empty if $r = 2$). $D$ (resp. $Z$) is smooth and naturally isomorphic to $C \times C^{(r-2)}$ (resp. $C \times C^{(r-3)}$). We denote by $p_1 : D \to C$ the projection to the first factor.

LEMMA 2.2. *If the genus of $C$ is at least 2, the natural map $(Def(C))^r \to Def(C^r)$ is an isomorphism; it induces (via the multidiagonal embedding) an isomorphism $Def(C) \to Def(C^r)^{S_r}$. In particular both $Def(C^r)$ and $Def(C^r)^{S_r}$ are unobstructed.*

PROOF. First order deformations of $C^r$ are parametrized by $H^1(\theta_{C^r})$. By Künneth we have

$$H^1(\theta_{C^r}) \cong \bigoplus_{i \neq j} H^0(C_i, \theta_{C_i}) \otimes H^1(C_j, \mathcal{O}_{C_j}) \oplus \bigoplus_i H^1(C_i, \theta_{C_i})$$

where $C_i \cong C$ is the $i$-th factor of the product; as the genus is $\geq 2$ $H^0(C_i, \theta_{C_i}) = 0$, hence the natural map $(Def(C))^r \to Def(C^r)$ is an isomorphism on tangent spaces; as $(Def(C))^r$ is unobstructed it is an isomorphism by (1.6).

Moreover by Künneth decomposition $H^0(C^r, \theta_{C^r}) = 0$, hence $Def(C^r)$ is prorepresentable. The tangent space to $Def(C^r)^{S_r}$ is the $S_r$-invariant subspace of $\oplus_i H^1(C_i, \theta_{C_i})$, hence the natural map $Def(C) \to Def(C^r)^{S_r}$ is an isomorphism on tangent spaces; again the unobstructedness of $Def(C)$ implies that it is an isomorphism.

CLAIM 2.3. There is a natural isomorphism

$$N'_{B/C^{(r)}} \cong g_*(p_1^*(2\theta_C) \otimes \mathcal{O}_D(Z)).$$

This claim will be proved in the next section.

LEMMA 2.4. *$H^i(B, N'_{B/C^{(r)}})$ is canonically isomorphic to $H^i(C, 2\theta_C)$ for $i = 0, 1$.*

PROOF. By (1.4) together with claim (2.3) we get

$$H^i(B, N'_{B/C^{(r)}}) \cong H^i(D, p_1^*(2\theta_C \otimes \mathcal{O}_D(Z)) \qquad \text{for all } i$$

Now for $i = 0, 1$ it follows from Künneth decomposition that $H^i(D, p_1^*(2\theta_C)) \cong H^i(C, 2\theta_C)$.

If $r = 2$, then $Z$ is empty and we are done. Otherwise consider the exact sequence

$$0 \to p_1^*(2\theta_C) \to p_1^*(2\theta_C) \otimes \mathcal{O}_D(Z) \to p_1^*(2\theta_C) \otimes \mathcal{O}_D(Z) \otimes \mathcal{O}_Z \to 0.$$

As $p_1^*(2\theta_C) \otimes \mathcal{O}_D(Z) \otimes \mathcal{O}_Z \cong p_1^*(3\theta_C) \otimes \mathcal{O}_Z$, it suffices now to prove that

$$H^0(Z, p_1^*(3\theta_C) \otimes \mathcal{O}_Z) = 0 \tag{2.4.1}$$

and that the natural map

$$H^1(Z, p_1^*(3\theta_C) \otimes \mathcal{O}_Z) \to H^2(D, p_1^*(2\theta_C)) \tag{2.4.2}$$

is injective. Again we have that for $i = 0, 1$

$$H^i(Z, p_1^*(3\theta_C) \otimes \mathcal{O}_Z) \cong H^i(C, 3\theta_C)$$

which immediately yields (2.4.1). As

$$H^2(D, p_1^*(2\theta_C)) \cong H^1(C, 2\theta_C) \otimes H^1(C, \mathcal{O}_C),$$

we see that the dual of (2.4.2) is the map

$$H^0(C, 3K_C) \otimes H^0(C, K_C) \to H^0(C, 4K_C)$$

which is surjective by (1.1). Hence (2.4.2) is injective and we are done.

(2.5) PROOF OF THE MAIN THEOREM. By (2.2), the theorem is equivalent to proving that the natural map $Def(C^r)^{S_r} \to Def(C^{(r)})$ is an isomorphism. As the former is unobstructed, it is enough to prove it is an isomorphism on tangent spaces by (1.6). By (1.4) the natural map of tangent spaces fits into an exact sequence

$$\begin{aligned} &H^0(B, N'_{B/C^{(r)}}) \to H^1(C^r, \theta_{C^r})^{S_r} \to H^1(C^{(r)}, \theta_{C^{(r)}}) \to \\ &\to H^1(B, N'_{B/C^{(r)}}) \to H^2(C^r, \theta_{C^r})^{S_r}. \end{aligned}$$

Therefore the injectivity of the map on tangent spaces follows from the isomorphism of lemma 2.4. The surjectivity of the map on tangent spaces is equivalent to injectivity of $H^1(B, N'_{B/C^{(r)}}) \to H^2(C^r, \theta_{C^r})^{S_r}$. From lemma 2.4 $H^1(B, N'_{B/C^{(r)}})$ is naturally isomorphic to $H^1(C, 2\theta_C)$. On the other hand

$H^2(C^r, \theta_{C^r})^{S_r}$ is naturally isomorphic to $H^1(C, \theta_C) \otimes H^1(C, \vee_C)$. In fact, by Künneth decomposition together with the assumption $g \geq 2$, we have

$$H^2(C^r, \theta_{C^r}) \cong \bigoplus_{i \neq j} H^1(C_i, \theta_{C_i}) \otimes H^1(C_j, \mathcal{O}_{C_j});$$

as $S_r$ acts transitively on the couples $i \neq j$ the result follows.

Passing to the dual shows that the requested injectivity is equivalent to the surjectivity of the induced map

$$H^0(C, 2K_C) \otimes H^0(C, K_C) \to H^0(C, 3K_C).$$

The theorem now follows from (1.1).

REMARK 2.6. One might ask oneself whether it is possible that $Def(C) \to Def(C^{(r)})$ is at least bijective if $g(C) = 2$, if not an isomorphism. This is not true if $r = 2$; in this case, fixing a point $P_0 \in C$, the natural map $C^{(2)} \to Jac(C)$ (given by $(P, Q) \mapsto \mathcal{O}_C(2P_0 - P - Q)$) is the blow up of a point, hence deformation theory of $C^{(2)}$ is the same as that of $Jac(C)$; this has however a 4-parameter space of deformations which is unobstructed, while $\mathcal{M}_2$ has dimension 3. We remark that the "extra" deformations are in general non-algebraic, as e.g. in the case of $K3$ surfaces.

## 3. Proof of claim 2.3

**(3.1).** Using hypothesis and notations of (1.3), we remark that $N'_{B/X}$ is supported on $B$, and we have a map

$$N'_{B/X} \otimes \mathcal{I}_B \to \mathcal{O}_B$$

induced by the natural map

$$\theta_X \otimes \mathcal{I}_B \to \mathcal{O}_X \qquad \varphi \otimes f \mapsto \varphi(f).$$

In case $X$ is smooth and $B$ is a hypersurface, the above map is an injection and gives a natural isomorphism

$$N'_{B/X} \otimes \mathcal{I}_B \cong \mathcal{J} \tag{3.1.1}$$

where $\mathcal{J} \subset \mathcal{O}_B$ is the jacobian ideal of $B$ (i.e. the ideal locally generated by the partial derivatives of the function defining $B$). In particular if $B$ is smooth $N'_{B/X}$ is isomorphic to the normal sheaf $N_{B/X}$ of $B$ in $X$.

**(3.2).** We introduce some more notation. Let $Z_1 := f(\Delta_{13})$ (or empty if $r = 2$), $Z_2 := f(\Delta_{23})$ (or empty if $r = 2$), $T := f(\Delta_{34})$ (or empty if $r = 2, 3$). It is easy to verify that

$$g^*(B) = 2D + 2Z_1 + 2Z_2 + T. \tag{3.2.1}$$

We have $Z = D \cap Z_1 = D \cap Z_2$; let $W = T \cap D$. Then it is easy to see that $\mathcal{O}_D(-Z) = \mathcal{O}_{C^2 \times C^{(r-2)}}(-Z_i) \otimes \mathcal{O}_D$ for $i = 1, 2$, and that $\mathcal{O}_D(-W) = \mathcal{O}_{C^2 \times C^{(r-2)}}(-T) \otimes \mathcal{O}_D$.

$Z_1$ and $Z_2$ are smooth varieties, isomorphic to $C^2 \times C^{(r-3)}$.

LEMMA 3.3. *If $\mathcal{J} \subset \mathcal{O}_B$ is the jacobian ideal, then*

$$\mathcal{J} = g_*\mathcal{O}_D(-3Z - T).$$

PROOF. The map $\mathcal{J} \to g_*\mathcal{O}_D(-3Z-T)$ is given by considering the composite map

$$\mathcal{J} \to \mathcal{O}_B \to g_*\mathcal{O}_D$$

and proving that the image is $g_*\mathcal{O}_D(-3Z - T)$.

It is enough to check this up to codimension 1 in $B$. In fact, if $B_1$ is a subvariety in $B$ of codimension $\geq 2$ (hence of codimension $\geq 3$ in $C^{(r)}$), by depth considerations we must have

$$\mathcal{H}^i_{B_1}(\theta_{C^{(r)}}) = \mathcal{H}^i_{B_1}(\theta_{C^{(r)}}(-\log B)) = 0$$

for $i \leq 2$. Therefore $\mathcal{H}^i_{B_1}(N'_{B/C^{(r)}}) = 0$ for $i = 0, 1$, hence if $j : U \to B$ is the inclusion of the complement of $B_1$ we have $N'_{B/C^{(r)}} = j_*(j^*(N'_{B/C^{(r)}}))$. The same must then hold for $\mathcal{J}$, and it obviously holds for the (push forward of) any line bundle.

The result is clear on the smooth locus. It is therefore enough to check the result near the points $(Q_1, \ldots, Q_r)$ in $B$ such that either

(i) $Q_1 = Q_2 = Q_3$ and no other equality holds or
(ii) $Q_1 = Q_2$ and $Q_3 = Q_4$ and no other equality holds.

In case (i) the point is the image of a unique point in $D$, lying in $Z$ but not in $W$; one can choose local coordinates $(t, u_1, \ldots, u_{r-2})$ on $D$ and $(x, y, v_1, \ldots, v_{r-2})$ in $C^{(r)}$ such that

$$g(t, u_1, \ldots, u_{r-2}) = (t^2, t^3, u_1, \ldots, u_{r-2})$$

and $Z$ has equation $t = 0$. Then $B$ has equation $x^3 = y^2$, the jacobian ideal is $(x^2, y)$ and its image in $g_*(\mathcal{O}_D)$ is $(t^3, t^4)$, which is $g_*\mathcal{O}_D(-3Z - T)$.

In case (ii) the point has two inverse images in $D$, and a similar local computation proves the result.

(3.4) PROOF OF CLAIM 2.3.. By (3.2.1) we know that

$$g^*\mathcal{O}_{C^{(r)}}(-B) \otimes \mathcal{O}_D = N_{D/C^2 \times C^{(r-2)}} \otimes \mathcal{O}_D(-4Z - W).$$

Now $N_{D/C^2 \times C^{(r-2)}}$ is naturally isomorphic to $p_1^*(2\theta_C)$; combining (3.1.1) and lemma (3.3) proves the claim.

ACKNOWLEDGEMENTS. I want to thank Fabrizio Catanese and the referee for useful suggestions.

## REFERENCES

[C1] Catanese, F., *Moduli of Algebraic Surfaces*, Theory of Moduli, LNM 1337 (E. Sernesi, ed.), Springer, Berlin-Heidelberg-New York, 1988.

[C2] ———, *Everywhere non-reduced moduli spaces*, Invent. Math. **98** (1989), 293–310.

[F1] Fantechi, B., *Deformations of Hilbert schemes of points on a surface* (1993), preprint University of Trento.

[F2] ———, *Thesis*, in preparation.

[G] Grothendieck, A., *Local Cohomology, LNM 41*, Springer, Berlin-Heidelberg-New York, 1967.

[K] Kempf, G., *Deformations of symmetric products*, Riemann Surfaces and Related Topics, Ann. Math. Studies, vol. 97, 1980, pp. 319–341.

DIPARTIMENTO DI MATEMATICA, UNIVERSITÀ DI TRENTO, I - 38050 POVO (TN), ITALY

*E-mail address*: fantechi@itncisca.bitnet

Contemporary Mathematics
Volume **162**, 1994

# $\mathbb{Q}$–acyclic Surfaces and their Deformations

HUBERT FLENNER AND MIKHAIL ZAIDENBERG

ABSTRACT. In this paper the classification and deformation theory of $\mathbb{Q}$–acyclic surfaces is studied. If $V$ is a smooth compactification of an open algebraic surface $X$ and $D = V \setminus X$ is a simple normal crossing divisor then we consider deformations $(\mathcal{V}, \mathcal{D})$ of $(V, D)$ over a base space $S$ such that the number of components of $\mathcal{D}_s$, $s \in S$, is locally constant on $S$. If $X$ is a $\mathbb{Q}$–acyclic surface, then also the nearby fibers $X_s = V_s - D_s$ are $\mathbb{Q}$–acyclic. The number of moduli of $(V, D)$ is given by $h^1(\Theta_V\langle D\rangle)$. We compute this number for the minimal compactifications of $\mathbb{Q}$–acyclic surfaces of logarithmic Kodaira dimension $\overline{k}(X) \leq 1$ and also for many examples with $\overline{k}(X) = 2$. Moreover we show that the deformations are unobstructed in all these cases. For the calculations the classification of such surfaces is needed, which we review and refine in various ways. As an application we obtain e. g. that $\mathbb{C}^3$ admits uncountably many exotic algebraic structures.

## 0. Introduction

This paper grew out of an attempt to understand the classification theory and especially the deformations of $\mathbb{Q}$–acyclic surfaces, i.e. smooth complex algebraic surfaces, whose reduced homology with coefficients in $\mathbb{Q}$ vanishes. For open surfaces there exist many classification results, see especially the fundamental paper of Fujita [**13**] and also the work of Gurjar, Miyanishi, Ramanujam, Shastri, Sugie, tom Dieck, Petrie, the second author and many others.

One of the first remarkable results is due to Ramanujam [**26**]: If $X$ is a contractible algebraic surface, which is simply connected at infinity, then $X$ is algebraically isomorphic to the affine plane. He also gave the first example of a contractible surface $X$ not isomorphic to $\mathbb{C}^2$. Moreover, he observed that for

1991 *Mathematics Subject Classification.* Primary 14J10; Secondary 14J25.
This paper is in final form and no version of it will be submitted for publication elsewhere.
This research was supported by the SFB 170 Göttingen and the MPI Bonn. We thank both institutions for their hospitality.

such a surface, $X \times \mathbb{C}$ is diffeomorphic to $\mathbb{C}^3$, thus exhibiting an exotic algebraic structure on affine 3–space. As an application of our calculations we will show that there are even uncountable many algebraic structures on $\mathbb{C}^3$, see (6.10).

The deformations of open algebraic surfaces $X$, which we will consider in this paper, are logarithmic deformations of a compactification $(V, D)$ of $X = V \setminus D$, where $V$ is a smooth compact algebraic surface and the divisor $D$ at infinity has at most simple normal crossings; $(V, D)$ is then called an SNC–*pair* for short. A *logarithmic deformation* of $(V, D)$ over $S$ is a deformation $\mathcal{V} \to S$ of $V$ and a deformation $\mathcal{D} \to S$ of $D$ together with an $S$–embedding $\mathcal{D} \subseteq \mathcal{V}$ such that $\mathcal{D}$ is an SNC–divisor relative $S$. Such deformations have been studied by Kawamata [**20**] and, in a more general context, by the first author [**12**]. The number of moduli is given by $h^1(\Theta_V\langle D\rangle)$, and $H^2(V, \Theta_V\langle D\rangle)$ is the space of obstructions for extending infinitesimal deformations, see (1.1). In general, the deformations depend not only on the open part $X$, but also on the compactification, even if we require $(V, D)$ to be minimal, i.e. $V$ contains no rational $-1$–curve, which meets at most two other irreducible components of D. The simplest example is $\mathbb{C}^2$, where we will compute the possible number of moduli for the minimal compactifications in section 3.

However, if $\overline{k}(X)$, the logarithmic Kodaira dimension of $X$, is $\neq -\infty$ , then the number of moduli, as well as the basis of the semiuniversal deformation, do not depend on the minimal compactification (which is not unique, in general!), and thus they are invariants of the open part $X = V \setminus D$, see (6.3).

In the case of $\mathbb{Q}$–acyclic surfaces of logarithmic Kodaira dimension $\leq 1$ we are able to give a very precise picture of these deformations: The basis of the semiuniversal deformation is always smooth, and even $h^2(\Theta_V\langle D\rangle) = 0$, see (6.5), (6.6) and (6.13). In order to determine the number of moduli, it will be necessary to give a classification of these surfaces. In principle, this was done by Fujita [**13**], Miyanishi–Tsunoda [**23**] and Gurjar–Miyanishi [**16**], but for our purposes we need a description in a more explicit form, at least in the case of logarithmic Kodaira dimension $-\infty$ and 1. To have a simple picture it is convenient to consider the surfaces up to a process called half–point detachment in [**13**]: If $(V, D)$ is an SNC–pair and $E \subseteq V$ is a smooth rational $-1$–curve with $ED \leq 1$ not contained in $D$, then the pair $(V', D')$, where $V \to V'$ is the blowing down of $E$ and $D' = \pi(D)$, is called the *half–point detachment* of $(V, D)$. The (iterated) half–point detachment of a $\mathbb{Q}$–acyclic surface leads to a surface, which we call *pre–$\mathbb{Q}$–acyclic*. A pair $(V, D)$ is called *strictly minimal*, if it is minimal and admits no half–point detachment. In section 4 resp. 5 we give a quite explicit description of the strictly minimal SNC–pairs $(V, D)$, which are pre–$\mathbb{Q}$–acyclic and of logarithmic Kodaira dimension $-\infty$ resp. 1. As an application we obtain a complete list of the possible number of moduli of such pairs in logarithmic Kodaira dimension 1, see (6.8). For $\overline{k} = 0$ we have $h^i(\Theta_V\langle D\rangle) = 0$ for all $i$. In particular, such surfaces are rigid, see (6.12). Finally, for $\overline{k} = -\infty$ it is, at least in principle, possible to determine the number of moduli, since $\chi(\Theta_V\langle D\rangle)$ can be computed from the dual graph of $D$, see (1.3), and $h^2(\Theta_V\langle D\rangle) = 0$. Hence, by calculating the number of vector fields one obtains the number of moduli.

For logarithmic Kodaira dimension 2 there is up to now no classification of $\mathbb{Q}$– or even $\mathbb{Z}$–acyclic surfaces. Most of the known examples can be obtained from line configurations in the projective plane, as described by tom Dieck and Petrie [**5**]. We will verify in section 6 that the $\mathbb{Q}$–acyclic surfaces of logarithmic Kodaira dimension 2, arising in this way, are always rigid, and even $h^i(\Theta_V\langle D\rangle) = 0$ for all $i$, see (6.13). We do not know, whether this holds for $\mathbb{Q}$–acyclic surfaces in Kodaira dimension 2, in general. Even the equality $\chi(\Theta_V\langle D\rangle) = 0$, which by (1.3) is equivalent to $K_V(K_V + D) = 0$, we are unable to prove.

We will give a rough outline of the contents of the various sections: In section 1 we start by defining the logarithmic deformations of the SNC–pairs $(V, D)$, and describe the tangent space and obstruction space of the semiuniversal deformation, see (1.3). In (1.5), (1.6), resp. (1.7) we study the behaviour of the deformation space under blowing up in a point of $D$, resp. deleting components of $D$. The aim of section 2 is to describe normal forms of vector fields and their algebraic integral curves on minimal rational surfaces. The material in this section is elementary, but necessary for the explicit computations in the subsequent paragraphs.

In section 3 this will be applied to logarithmic deformations of minimal compactifications of $\mathbb{C}^2$. We describe, at least in principle, all possibilities for the number of moduli. For this we deal first with the case that $V$ is a minimal rational surface, cf. (3.1). We give a complete list of possibilities for $(V, D)$ and the associated number of moduli, which in general can be positive (in contrast to Theorem 3 in [**25**]). The general case is reduced to this by using the classification of minimal compactifications of $\mathbb{C}^2$ due to Ramanujam [**26**] and Morrow [**25**], see (3.8).

In section 4 we first describe, how to obtain form a minimal $\mathbb{Q}$–acyclic SNC–pair $(V, D)$ a strictly minimal pre–$\mathbb{Q}$–acyclic surface $(V', D')$ by using half–point detachment and blowing down of certain $-1$–curves. The main aim of this section is the explicit description of all pre–$\mathbb{Q}$–acyclic surfaces with $\overline{k} = -\infty$, see (4.10) (cf. also [**23**]). For this we use, similarly as Fujita [**13**], the existence of cylinderlike open subsets of affine surfaces with $\overline{k} = -\infty$, which are connected at infinity, see [**22**], [**29**]. We have included a simplified proof of this basic result in (4.11), following the lines of Russell [**29**].

In section 3 we have seen, that the compactification $(V, D)$ of $X = \mathbb{C}^2$ can have many moduli, while the affine part remains constant in the deformation. It is easy to see that this is no longer true for arbitrary $\mathbb{Q}$–acyclic surfaces with $\overline{k} = -\infty$. In (4.16) we even construct a family $(V_t, D_t), t \in \mathbb{C}$, such that, for $t \neq t'$, the affine surfaces $X_t = V_t - D_t$ and $X_{t'}$ are nonisomorphic, whereas the boundaries are all isomorphic.

In an appendix to section 4 we describe the structure of birational maps of pairs $(V, D) \to (V', D')$, which are biregular on the open parts $V \setminus D \cong V' \setminus D'$, in terms of modifications of the dual graphs of $D$ and $D'$.

Section 5 is mainly devoted to the description of pre–$\mathbb{Q}$–acyclic surfaces with $\overline{k} = 1$. Following **[19]**, each of them has a $\mathbb{C}^\star$–fibration and so, more generally, we describe pre–$\mathbb{Q}$–acyclic strictly minimal $\mathbb{C}^\star$–triples $(V, D, p : V \to \mathbb{P}^1)$, see (5.7) and (5.8). We also show, which $\mathbb{C}^\star$–triples lead to surfaces with logarithmic Kodaira dimension $-\infty, 0, 1$ resp., see (5.11), (5.14).

In section 6 we study deformations of surfaces of logarithmic Kodaira dimension $\geq 0$. First we show, that the logarithmic deformations do not depend on the choice of the minimal SNC–completion, cf. (6.1). Moreover, for $\overline{k} \leq 1$ the deformations are unobstructed. In (6.8) we give a complete list for the possible values of $h^i(\Theta_V\langle D\rangle), 0 \leq i \leq 2$, where $(V, D, p)$ is a strictly minimal pre–$\mathbb{Q}$–acyclic $\mathbb{C}^\star$–triple. As an application we show, that there are families $(V_t, D_t)$ such that $X_t = V_t \setminus D_t$ are topologically contractible pairwise nonisomorphic surfaces with $\overline{k} = 1$, see (6.11). This leads to uncountable many exotic algebraic structures on $\mathbb{C}^3$, as mentioned above. We conclude this section by discussing the examples of $\mathbb{Q}$–acyclic surfaces, arising from the line configurations in **[5]**.

We add a few remarks about the notations used in this paper. For an algebraic variety $V$ we denote by $\Theta_V$ the sheaf of vector fields on $V$, and by $\Theta_V\langle D\rangle$ the sheaf of vector fields tangent to $D$, where $D$ is a divisor in $V$. By $V^{(n)}$ we indicate the normalization of $V$. If $V$ is complete and $\mathcal{E}$ is a coherent sheaf on $V$, then we set $h^i(\mathcal{E}) := \dim_{\mathbb{C}}(H^i(V, \mathcal{E}))$ , and $\chi(\mathcal{E})$ will be the Euler characteristic of $\mathcal{E}$. For a smooth $V$, $P_n(V)$ will denote the n–th plurigenus of $V$, and $p_g(V) = P_1(V)$, $p_a(V)$ its geometric resp. arithmetic genus. If $V \to S$ is a morphism, then $L^\bullet_{V/S}$ will be the relative cotangent complex. Assume that $V$ is a smooth surface and $D$ is a divisor with simple normal crossings in $V$ such that 2 irreducible components of $D$ have at most one intersection point. Such pairs $(V, D)$ will be called shortly *SNC–pairs.* The pair (V,D) will be called *minimal,* if there is no rational (–1)–curve in $D$, which meets at most 2 other components of D. If, moreover, the open part $X := V \setminus D$ admits, up to isomorphism, only one minimal SNC–compactification, then $(V, D)$ is said to be *absolutely minimal.* Let $\Gamma := \Gamma(D)$ be the weighted dual graph of $D$. We say that $D$ is a *rational tree* if all components $D_i$ of $D$ are smooth rational curves and $\Gamma$ is connected and has no cycles. For a vertex $e$ of $\Gamma$ the number of edges from $e$ is called the *valence* of $e$. A vertex of valence 1 resp. $\geq 3$ is also called an *end* resp. a *branching point* of $\Gamma$. The algebraic varieties, which we consider in this paper, are all assumed to be defined over the field of complex numbers. For simplicity we do not distinguish in our notation between the algebraic variety and its associated analytic space. Usually we work in the algebraic category, but when speaking about the cohomology in constant sheaves, or about deformations, we deal always with the underlying analytic spaces.

## 1. Deformations of completions of open surfaces: Generalities

Let $(V, D)$ be an SNC–pair such that $X = V \setminus D$ is $\mathbb{Q}$–acyclic. Let

$$\pi : (\mathcal{V}, \mathcal{D}) \to S$$

be a deformation of the pair $(V, D)$ , i. e. $\mathcal{V}, \mathcal{D}$ are flat over $S$ with fibers $V, D$ resp. over some distinguished point $s_0$ of $S$. Assume that for all fibers $\mathcal{V}(s)$ and $\mathcal{D}(s)$, the complements $X(s) = \mathcal{V}(s) \setminus \mathcal{D}(s)$ are $\mathbb{Q}$–acyclic. Then the number of components of $\mathcal{D}(s)$ is necessarily constant on $S$, since it is equal to the rank of $H^2(V, \mathbb{Q})$. We will therefore consider the following type of deformations, which are described more generally for manifolds of arbitrary dimension:

Let $V$ be a compact complex manifold of dimension $n$ and $D = \bigcup_{i=1}^{r} D_i$ a divisor with simple normal crossings. Then we consider deformations of the map

$$f : D^{(n)} := \dot{\bigcup_{i=1}^{r}} D_i \to V$$

in the sense of [**12**], [**20**]: a deformation of $f$ over the base space $S$ consists of flat morphisms $\mathcal{D} \to S$ and $\mathcal{V} \to S$ together with an $S$–morphism $F : \mathcal{D} \to \mathcal{V}$ such that F induces f over some distinguished point $s_0$ of $S$. We will call these deformations also *logarithmic deformations* of the pair $(V, D)$. By [**12**] the infinitesimal deformations of $f$ are described by $T^1_f(\mathcal{O}_V)$. A description of this vector space is given by the following lemma.

LEMMA 1.1. *There are canonical isomorphisms* $T^i_f(\mathcal{O}_V) \cong H^i(V, \Theta_V\langle D\rangle)$.

PROOF. The groups $T^\star_f(\mathcal{O}_V)$ are the Ext–groups of the pair of complexes

$$(L^\bullet_V, L^\bullet_{D^{(n)}}),$$

see [**12**], and there is a natural triangle

$$\begin{array}{ccc} & \mathcal{M}^\bullet = R\,\mathcal{H}om_{(V,D^{(n)})}((L^\bullet_V, L^\bullet_{D^{(n)}}), (\mathcal{O}_V, \mathcal{O}_{D^{(n)}})) & \\ \nearrow & & \searrow \\ R\,\mathcal{H}om_{D^{(n)}}(L^\bullet_{D^{(n)}/V}, \mathcal{O}_{D^{(n)}}) & \xleftarrow[\psi]{+1} & R\,\mathcal{H}om_V(L^\bullet_V, \mathcal{O}_V) \end{array}$$

in the derived category. But $L^\bullet_{D^{(n)}/V} \cong \bigoplus_i \mathcal{N}^\vee_{D_i/V}, L^\bullet_V = \Omega^1_V$, and $\psi$ is induced by the canonical map $\Theta_V \to \bigoplus \mathcal{N}_{D_i/V}$. Since this map is surjective with kernel $\Theta_V\langle D\rangle$, it follows that $\mathcal{M}^\bullet$ is quasi–isomorphic to $\Theta_V\langle D\rangle$, which implies the lemma. $\square$

In the following, we will call $h^1(\Theta_V\langle D\rangle) := \dim_{\mathbb{C}} H^1(\Theta_V\langle D\rangle)$ the *number of moduli of the pair* $(V, D)$. If

$$\begin{array}{ccc} \mathcal{D} & \longrightarrow & \mathcal{V} \\ & \searrow \swarrow & \\ & S & \end{array}$$

is the versal deformation of $(V, D)$, then $S$ is smooth if $H^2(\Theta_V\langle D\rangle) = 0$. The pair $(V, D)$ will be called *rigid* if its number of moduli is zero. Observe that $(V, D)$ has a versal deformation by [**12**] (8.5) . The set of isomorphism classes of logarithmic deformations over a complex space germ $(S, 0)$ will also be denoted by $\mathrm{Def}_{V,D}(S, 0)$. Obviously, $\mathrm{Def}_{V,D}$ is a functor on the category of complex space germs.

Let now $(V, D)$ be an SNC–pair such that $X = V \setminus D$ is a $\mathbb{Q}$–acyclic surface. In order to be able to compute the number of moduli it will be useful to have a simple formula for the Euler characteristic of $\Theta_V\langle D\rangle$. Moreover, it is important to understand the behaviour of logarithmic deformations under blowing up of the boundary D. For this we start with the following well known lemma.

LEMMA 1.2. *A rational tree is a curve of arithmetical genus* $p_a(D) = 0$.

PROOF. This is easily seen by an induction on the number $r$ of components. If $r = 1$, then there is nothing to prove. For the induction step, choose a decomposition $D = D' \cup C$, where $D'$ is a rational tree with $(r-1)$ components and $C \cong \mathbb{P}^1$. From the natural sequence

$$0 \to \mathcal{O}_D \to \mathcal{O}_{D'} \oplus \mathcal{O}_C \to \mathbb{C} \to 0$$

we obtain $p_a(D') + p_a(C) = p_a(D)$. This proves (1.2). □

LEMMA 1.3. *Let* $(V, D)$ *be an SNC–pair and assume that* $X := V \setminus D$ *is* $\mathbb{Q}$–*acyclic. Let* $D_1, \ldots, D_r$ *be the irreducible components of* $D$. *Then the following holds:*

(0) $D$ *is a rational tree.*
(1) $\chi(\Theta_V) = 10 - 2r$.
(2) $K_V^2 = 10 - r$.
(3) $c_2 := c_2(V) = 2 + r$.
(4) $\chi(\bigoplus \mathcal{N}_{D_i/V}) = \sum(D_i^2 + 1) = (\sum D_i^2) + r$.
(5) $\chi(\Theta_V\langle D\rangle) = 10 - 3r - \sum D_i^2 = (K_V + D)^2 + 2 = -D^2 - r + 8 = K_V(K_V + D)$.

PROOF. (0) follows from **[13]** (2.6) and (2.8). Since $V \setminus D$ is $\mathbb{Q}$-acyclic, we have $h^2(V, \mathbb{Q}) = r$, $h^1(V, \mathbb{Q}) = h^3(V, \mathbb{Q}) = 0$ and $h^0(V, \mathbb{Q}) = h^4(V, \mathbb{Q}) = 1$. This gives (3). By Noether's formula

$$12 \cdot \chi(\mathcal{O}_V) = K_V^2 + c_2.$$

Since $\chi(\mathcal{O}_V) = 1$, this shows (2). (4) is obvious, since all the curves $D_i$ are rational. (1) follows easily from the Riemann–Roch theorem

$$\chi(\mathcal{E}) = \frac{1}{2} c_1(\mathcal{E})(c_1(\mathcal{E}) - K_V) - c_2(\mathcal{E}) + s \cdot \chi(\mathcal{O}_V),$$

which holds for every locally free sheaf $\mathcal{E}$ of rank $s$ on $V$. Finally, the first equality in (5) follows from (1) and (4). Since $D$ is a rational tree, we have $K_V D = -D^2 - 2$ by (1.2) and the adjunction formula. Hence

$$\begin{aligned} (K_V + D)^2 &= K_V^2 + 2K_V D + D^2 \\ &= 10 - r - 2D^2 - 4 + D^2 \\ &= 6 - r - D^2, \end{aligned}$$

which shows the last two equalities of (5). Since the dual graph $\Gamma$ of $D$ is a rational tree, we have

$$1 = e(\Gamma) = r - \frac{1}{2} \sum_{i \neq j} D_i D_j.$$

Hence

$$D^2 = \sum D_i^2 + \sum_{i \neq j} D_i D_j = \sum D_i^2 + 2r - 2,$$

concluding the proof. □

*Remark* 1.4. (1) By (5) of the preceding lemma, the following conditions are equivalent:

(i) $\chi(\Theta_V\langle D\rangle) = 0$,

(ii) $K_V(K_V + D) = 0$,

(iii) $D^2 = 8 - r$.

(2) Let $(V, D)$ be a minimal SNC–pair. Assume that $X = V \setminus D$ is $\mathbb{Q}$-acyclic and of logarithmic Kodaira dimension $\overline{k} = 2$. In all the examples of such surfaces known to us, the three conditions of (1) are satisfied. We do not know, whether this holds for any such surface.

Now we will study the behaviour of $\chi(\Theta_V\langle D\rangle)$ under blowing up $\pi : V' \to V$ of $V$ in a point $P$ of $D$. Set $D' = \pi^{-1}(D)_{red}$, and denote by $E \cong \mathbb{P}^1 \subseteq V'$ the exceptional curve of $\pi$.

LEMMA 1.5. (1) $\pi^\star(\Theta_V) = \Theta_{V'}\langle E\rangle(E)$.

(2) $\pi_\star(\Theta_{V'}) = \mathfrak{m}_P \cdot \Theta_V$, *where* $\mathfrak{m}_P \subseteq \mathcal{O}_V$ *denotes the maximal ideal sheaf of* $P$. *Moreover,* $R^1\pi_\star(\Theta_{V'}) = 0$. *In particular,*

$$\chi(\Theta_{V'}) = \chi(\Theta_V) - 2.$$

(3) *Assume that* $P$ *is a smooth point of* $D$. *Then* $R^1\pi_\star(\Theta_{V'}\langle D'\rangle) = 0$, *and there is an exact sequence*

$$0 \to \pi_\star(\Theta_{V'}\langle D'\rangle) \to \Theta_V\langle D\rangle \to \mathbb{C} \to 0.$$

*Thus* $\chi(\Theta_{V'}\langle D'\rangle) = \chi(\Theta_V\langle D\rangle) - 1$. *Moreover,* $h^2(\Theta_V\langle D\rangle) = h^2(\Theta_{V'}\langle D'\rangle)$.

(4) *Assume that* $P$ *is not a smooth point of* $D$, *i.e.* $P$ *is on two components* $D_1, D_2$ *of* $D$. *Then* $\pi_\star(\Theta_{V'}\langle D'\rangle) = \Theta_V\langle D\rangle, R^1\pi_\star(\Theta_{V'}\langle D'\rangle) = 0$ *and* $\Theta_{V'}\langle D'\rangle = \pi^\star(\Theta_V\langle D\rangle)$. *In particular,* $h^i(\Theta_{V'}\langle D'\rangle) = h^i(\Theta_V\langle D\rangle)$ *for all* $i \geq 0$.

(5) $h^2(\Theta_V\langle D\rangle)$ *only depends on the open surfaces* $X = V \setminus D$ *and not on the SNC–compactification* $(V, D)$.

PROOF. (1) is well known and easily verified by an explicit computation. For the proof of (2) we consider the exact sequence

$$0 \to \pi^\star(\Theta_V)(-E) \to \Theta_{V'} \to \mathcal{N}_{E/V'} \cong \mathcal{O}_{\mathbb{P}^1}(-1) \to 0\ ,$$

which follows from (1). Taking the direct image sheaves we obtain $\mathfrak{m}_p \cdot \Theta_V \cong \pi_\star(\Theta_{V'})$ and $R^1\pi_\star(\Theta_{V'}) = 0$. This proves (2). In order to show (4), one verifies first, that $\pi^\star(\Theta_V\langle D\rangle) = \Theta_{V'}\langle D'\rangle$. This is easily done by an explicit computation: Choose local coordinates $(x, y)$ around $P$ in $V$ such that $D_1 = \{x = 0\}, D_2 = \{y = 0\}$ in a neighbourhood of $P$. Let $\hat{D}_i$ be the proper transform of $D_i$, and $P_i$ the intersection point of $\hat{D}_i$ and $E$. Then there are coordinates $(u, v)$ on $V'$ around $P_1$ such that $x = uv, y = v$. The vector field $x\frac{\partial}{\partial x}$ is mapped onto $u\frac{\partial}{\partial u}$ and $y\frac{\partial}{\partial y}$ onto $v\frac{\partial}{\partial v} - u\frac{\partial}{\partial u}$, which are local generators of $\Theta_{V'}\langle D'\rangle$ at $P_1$. Now the other formulas of (4) follow easily from this, since $R^i\pi_\star(\pi^\star(\Theta_V\langle D\rangle)) = R^i\pi_\star(\mathcal{O}_{V'}) \otimes \Theta_V\langle D\rangle$ for all i.

For the proof of (3) it is sufficient to show the existence of the exact sequence and $R^1\pi_\star(\Theta_{V'}\langle D'\rangle) = 0$. This is of local nature, and so we may assume that there is a curve $C$ meeting $D$ in $P$ transversally. Set $\tilde{D} = D \cup C, \tilde{D}' = \pi^{-1}(\tilde{D})_{red}$, and denote by $\hat{C}$ the proper transform of $C$. It is easily seen that $\mathcal{N}_{\hat{C}/V'} \cong \mathcal{N}_{C/V}(-P)$. Taking the direct image of the canonical exact sequence

$$0 \to \Theta_{V'}\langle\tilde{D}'\rangle \to \Theta_{V'}\langle D'\rangle \to \mathcal{N}_{\hat{C}/V'} \to 0,$$

we get a diagram

$$\begin{array}{ccccccccc}
0 & \to & \pi_\star(\Theta_{V'}\langle\tilde{D}'\rangle) & \to & \pi_\star(\Theta_{V'}\langle D'\rangle) & \xrightarrow{p} & \pi_\star(\mathcal{N}_{\hat{C}/V'}) & \to & 0 \\
 & & \cong\downarrow\alpha & & \beta\downarrow & & \downarrow\gamma & & \\
0 & \to & \Theta_V\langle\tilde{D}\rangle & \to & \Theta_V\langle D\rangle & \to & \mathcal{N}_{C/V} & \to & 0
\end{array}.$$

Observe that $p$ is surjective and $\alpha$ is an isomorphism because of (4). Now a simple diagram chasing shows that coker $(\beta) \cong$ coker $(\gamma) = \mathcal{N}_{C/V}/\mathcal{N}_{C/V}(-P) \cong \mathbb{C}$ as desired. Clearly, (5) follows immediately from (3) and (4). □

Finally we will study, how the logarithmic deformations behave under blowing up.

PROPOSITION 1.6. *Let $(V, D)$ be an SNC–pair, and assume that $(V', D')$ is obtained form $(V, D)$ by blowing up a point $x$ of $D$. Let $(S, 0)$ resp. $(S', 0)$ be the basis of the semiuniversal logarithmic deformation of $(V, D)$ resp. $(V', D')$.*

(1) *If $x$ is a nonsmooth point of $D$, then $(S, 0)$ and $(S', 0')$ are canonically isomorphic.*

(2) *Assume that $x$ is a smooth point of $D$. Then $(S, 0)$ and $(S', 0')$ are isomorphic if $H^1(\Theta_V\langle D\rangle) = H^1(\Theta_{V'}\langle D'\rangle)$, and $(S', 0)$ is canonically isomorphic to $(S \times \mathbb{C}, 0)$ if $H^1(\Theta_V\langle D\rangle) \neq H^1(\Theta_{V'}\langle D'\rangle)$.*

PROOF. Let $(\mathcal{V}', \mathcal{D}') \to (\Sigma, 0)$ be any deformation of $(V', D')$, where $\mathcal{D}' = \bigcup_{i=1}^{r} \mathcal{D}'_i$ and $\mathcal{D}'_1 \to (\Sigma, 0)$ is the deformation of the exceptional $(-1)$–curve $E \subseteq D'$. By **[28]** Theorem 2, $\mathcal{D}'_1$ can be contracted to give a deformation $(\mathcal{V}, D) \to (\Sigma, 0)$. Hence we get a canonical map of deformation functors

$$\mathrm{Def}_{V',D'} \to \mathrm{Def}_{V,D}.$$

By (1.1) an obstruction theory for $\mathrm{Def}_{V,D}$ is given by $H^2(V, \Theta_V\langle D\rangle)$. The canonical map $H^1(\Theta_{V'}\langle D'\rangle) \to H^1(\Theta_V\langle D\rangle)$ is surjective and $H^2(\Theta_V\langle D\rangle) \cong H^2(\Theta_{V'}\langle D'\rangle)$ by the preceding lemma. Hence $\mathrm{Def}_{V',D'} \to \mathrm{Def}_{V,D}$ is a smooth morphism of functors, and so by **[27]** (2.10) $(S', 0) \cong (S, 0) \times (\mathbb{C}^k, 0)$ for some $k$. Clearly, $k$ is the difference of the dimensions of the tangent spaces of $S'$ resp. $S$ at 0, and so is equal to $h^1(\Theta_{V'}\langle D'\rangle) - h^1(\Theta_V\langle D\rangle)$. This concludes the proof of (1.6). □

For the next proposition we consider the following situation. Let $(V, D)$ be an SNC–pair, and assume that $E \subseteq D$ is a smooth rational curve. Let $D^\star := D - E$. Denote by $S$ resp. $S^\star$ the basis of the logarithmic semiuniversal deformation of $(V, D)$ resp. $(V, D^\star)$. Since every logarithmic deformation of $(V, D)$ can also be considered as a logarithmic deformation of $(V, D^\star)$, we get a map

$$\varphi : S \to S^\star.$$

Moreover, let

$$\gamma^i : H^i(\Theta_V\langle D\rangle) \to H^i(\Theta_V\langle D^\star\rangle)$$

be the natural map. Then we have:

PROPOSITION 1.7. (1) *If* $E^2 \geq -1$, *then* $\varphi$ *is smooth. Moreover,* $\gamma^2$ *is bijective and* $\gamma^1$ *is surjective.*

(2) *If* $E^2 \leq -1$, *then* $\varphi$ *is an embedding. Moreover,* $\gamma^0$ *is bijective and* $\gamma^1$ *is injective.*

(3) *If* $E^2 = -1$, *then* $\varphi$ *and all* $\gamma^i$ *are isomorphisms.*

PROOF. There is a canonical exact sequence

$$0 \to \Theta_V\langle D\rangle \to \Theta_V\langle D^\star\rangle \to \mathcal{N}_{E/V} \to 0.$$

Now $\mathcal{N}_{E/V} = \mathcal{O}_E(E^2)$, and so $h^1(\mathcal{N}_{E/V}) = 0$ iff $E^2 \geq -1$ and $h^0(\mathcal{N}_{E/V}) = 0$ iff $E^2 \leq -1$. This implies immediately the assertions about $\gamma^i$.

Assume that $E^2 \geq -1$. Since $\gamma^1$ is surjective, we get that the tangent map of $\varphi$ is surjective. Since $\gamma^2$ is bijective, the obstruction groups for extending deformations are also the same. Hence by **[27]** (2.8) $\varphi$ is smooth.

If $E^2 \leq -1$, then $\gamma^1$ is injective and so the tangent map of $\varphi$ is injective, too. Hence $\varphi$ is an embedding. Finally, if $E^2 = -1$, then $\varphi$, being an embedding and smooth, is an isomorphism. □

## 2. Normal forms and integral curves of vector fields on minimal rational surfaces

We first treat vector fields on $\mathbb{P}^2$. More generally, a vector field on $\mathbb{P}^n = \mathbb{P}(V)$, $\dim V = n+1$, is given by a derivation of degree 0 of the symmetric algebra

$$\delta : S(V) \to S(V)\ ,$$

and so corresponds in a unique way to an endomorphism of $V$, modulo scalar operators. So, we can choose a basis of $V$ such that $\delta \mid V : V \to V$ has Jordan normal form and the eigenvalue of the largest Jordan bloc is zero. For $\mathbb{P}^2$ we obtain the possible normal forms

$$\begin{pmatrix} 0 & 1 & 0 \\ 0 & 0 & 1 \\ 0 & 0 & 0 \end{pmatrix}, \quad \begin{pmatrix} 0 & 1 & 0 \\ 0 & 0 & 0 \\ 0 & 0 & \lambda \end{pmatrix}, \quad \begin{pmatrix} 0 & 0 & 0 \\ 0 & \lambda_1 & 0 \\ 0 & 0 & \lambda_2 \end{pmatrix}.$$

Calculating these vector fields in affine coordinates $u = x_1/x_0, v = x_2/x_0$, we get

LEMMA 2.1. *After a suitable automorphism of* $\mathbb{P}^2$, *any vector field* $\delta$ *on* $\mathbb{P}^2$ *is one of the following:*

(i) $\delta = \frac{\partial}{\partial u} + u\frac{\partial}{\partial v}$,

(ii) $\delta = \frac{\partial}{\partial u} + \lambda v\frac{\partial}{\partial v}$ $(\lambda \in \mathbb{C})$,

(iii) $\delta = \lambda_1 u\frac{\partial}{\partial u} + \lambda_2 v\frac{\partial}{\partial v}$ $(\lambda_1, \lambda_2 \in \mathbb{C})$.

Now we will list all the possible irreducible integral curves of such vector fields.

LEMMA 2.2. *Let $l_\infty$ denote the line $\{x_0 = 0\}$. Then the possible irreducible integral curves are given by* TABLE *1 below.*

TABLE 1

| | Vector field | Integral curves |
|---|---|---|
| 1) | $\frac{\partial}{\partial u} + u\frac{\partial}{\partial v}$ | a) $l_\infty$<br>b) $v = c + \frac{u^2}{2}$, $c \in \mathbb{C}$ |
| 2) | $\frac{\partial}{\partial u} + \lambda v\frac{\partial}{\partial v}$, $\lambda \neq 0$ | a) $l_\infty$<br>b) $v = 0$ |
| 3) | $\frac{\partial}{\partial u}$ | a) $l_\infty$<br>b) $v = c, c \in \mathbb{C}$ |
| 4) | $\lambda_1 u\frac{\partial}{\partial u} + \lambda_2 v\frac{\partial}{\partial v}$<br>$\lambda_2 \neq 0, \frac{\lambda_1}{\lambda_2} = \lambda \notin \mathbb{Q}$ | a) $l_\infty$<br>b) $u = 0$<br>c) $v = 0$ |
| 5) | $\lambda_1 u\frac{\partial}{\partial u} + \lambda_2 v\frac{\partial}{\partial v}$<br>$\lambda_2 \neq 0, \lambda = \lambda_1/\lambda_2 \in \mathbb{Q}_{>0}$<br>$\lambda = \frac{s}{r}, r, s \in \mathbb{N}, (r, s) = 1$ | a) $l_\infty$<br>b) $u^r + c\, v^s = 0, c \in \mathbb{C}$<br>c) $v = 0$<br>d) $u = 0$ |
| 6) | $\lambda_1 u\frac{\partial}{\partial u} + \lambda_2 v\frac{\partial}{\partial v}$<br>$\lambda_2 \neq 0, \lambda = \lambda_1/\lambda_2 \in \mathbb{Q}_{<0}$<br>$\lambda = -\frac{s}{r}, r, s \in \mathbb{N}, (r, s) = 1$ | a) $l_\infty$<br>b) $1 - cu^r v^s = 0, c \in \mathbb{C}$<br>c) $u = 0$<br>d) $v = 0$ |
| 7) | $\lambda_1 u\frac{\partial}{\partial u}, \lambda_1 \neq 0$ | a) $l_\infty$<br>b) $v - c = 0$, $c \in \mathbb{C}$<br>c) $u = 0$ |

PROOF. The families of integral curves, which are mentioned in cases 1,3,5,6 and 7, fill out the whole projective plane. Therefore other curves can only be contained in the zero sets of the respective vector fields. But it is easily checked that in these cases the irreducible components of the zero sets appear already in the list. For the proof of 2) assume that the vector field $\delta = \frac{\partial}{\partial u} + \lambda v\frac{\partial}{\partial v} (\lambda \neq 0)$ is tangent to $\{g = 0\}$, where $g(u, v) = \sum_j g_j(u)v^j$ is an irreducible polynomial.

Then $\delta(g) = \epsilon(u, v) \cdot g$ for some polynomial $\epsilon(u, v)$. Comparing the degrees in $u$ and $v$, we obtain that $\epsilon$ is constant and

$$\frac{\partial g_j}{\partial u} + j\lambda g_j = \epsilon g_j.$$

The only solutions of this equation are $c_j \exp\left((\epsilon - j\lambda)u\right)$. Hence $c_j = 0$ except for possibly one $j$, and $\epsilon = j\lambda$ in this case. But then $g = c_j \cdot v^j$, forcing $j = 1$ because of the irreducibility of $g$.

Finally assume that $\delta = \lambda_1 u\frac{\partial}{\partial u} + \lambda_2 v\frac{\partial}{\partial v}$ as in (4) is tangent to $\{g = 0\}$, where $g = \sum g_j(u)v^j$ is irreducible as above. We may assume that $\lambda_1 = 1$ and $\lambda = \lambda_2 \notin \mathbb{Q}$. Then as above $\delta(g) = \epsilon g$ for some constant $\epsilon$, and so

$$u\frac{\partial g_j}{\partial u} + j\lambda g_j = \epsilon g_j.$$

All solutions of this equation are the functions $g_j(u) = c_j u^{\epsilon - j\lambda}$. Hence, if $c_j \neq 0$, then $\epsilon - j\lambda \in \mathbb{N}_0$, and since $\lambda \notin \mathbb{Q}$, this implies that $c_j \neq 0$ for at most one index $j$. This gives $g = g_j v^j$ and again $j = 1$, $g = c_1 v$ or $j = 0$, $g = c_0 u$ by the irreducibility of $g$. □

Next we treat the case $\mathbb{P}^1 \times \mathbb{P}^1$.

LEMMA 2.3. *After a suitable automorphism, every vector field on $\mathbb{P}^1 \times \mathbb{P}^1$ is one of the following:*

(i) $\frac{\partial}{\partial u} + \frac{\partial}{\partial v}$,

(ii) $\frac{\partial}{\partial u} + \lambda v\frac{\partial}{\partial v}$, $\lambda \in \mathbb{C}$,

(iii) $\lambda_1 u\frac{\partial}{\partial u} + \lambda_2 v\frac{\partial}{\partial v}$, $\lambda_1, \lambda_2 \in \mathbb{C}$.

PROOF. The reasoning at the beginning of this section shows that up to an automorphism the only vector fields on $\mathbb{P}^1$ are $\frac{\partial}{\partial u}$ and $\lambda u\frac{\partial}{\partial u}$, where $u = x_1/x_0$ is the affine coordinate function and $\lambda \in \mathbb{C}$. So, by the Künneth formula the lemma follows. □

LEMMA 2.4. *Besides the two lines $\{u = \infty\}$ and $\{v = \infty\}$ in affine coordinates of $\mathbb{P}^1 \times \mathbb{P}^1$, the integral curves to the vector fields in* (2.3) *are the following:*

(i) *If $\delta = \frac{\partial}{\partial u} + \frac{\partial}{\partial v}$, then $\{u - v = c\}, c \in \mathbb{C}$, are the irreducible integral curves.*

(ii) *In the remaining cases* (ii) *and* (iii) *of* (2.3), *the description is as in* (2.2), 2 – 7.

PROOF. This follows as in Lemma (2.2). □

Now we will investigate vector fields on the other minimal rational surfaces. Let $\Sigma(n)$, $n \geq 1$, be the Hirzebruch surface

$$\text{Proj }(S(\mathcal{O}_{\mathbb{P}^1} \oplus \mathcal{O}_{\mathbb{P}^1}(-n))) \xrightarrow{\pi} \mathbb{P}^1,$$

where $\pi$ is the canonical projection. On $\mathbb{P}^1$ we consider the affine coordinate $u = x_1/x_0$. On $\Sigma(n)$ there is a unique section $C_\infty$ of $\pi$ with $C_\infty^2 = -n$, which is given by the projection $\mathcal{O}_{\mathbb{P}^1} \oplus \mathcal{O}_{\mathbb{P}^1}(-n) \to \mathcal{O}_{\mathbb{P}^1}(-n)$. Denote by $F_\infty$ the fibre of $\pi$ over $u = \infty$. On $\mathbb{C} = \mathbb{P}^1 \setminus \{u = \infty\}$ we have that

$$S(\mathcal{O}_{\mathbb{C}} \oplus \mathcal{O}_{\mathbb{C}}(-n)) = \mathcal{O}_{\mathbb{C}}[y_0, y_1],$$

and $C_\infty$ is given by $y_0 = 0$. Hence $u$ and $v = y_1/y_0$ constitute an affine chart of $\Sigma(n) \setminus (F_\infty \cup C_\infty)$. In the next lemma we describe automorphisms, vector fields and sections of $\pi$ with respect to this affine chart.

LEMMA 2.5. (1) *The automorphisms of $\Sigma(n), n \geq 1$, preserving $F_\infty$, are given by*

$$(u, v) \to (a + bu, cv + P(u)),$$

*where $P \in \mathbb{C}[u]$ is a polynomial of degree at most $n$ and $bc \neq o$.*

(2) *The vector fields, which are tangent to $F_\infty$, are given by*

$$(a + bu)\frac{\partial}{\partial u} + (cv + P(u))\frac{\partial}{\partial v},$$

*where $a, b, c \in \mathbb{C}$ and $P$ is as above.*

(3) *The sections $C$ of $\pi$ are given by the equations $\{q_0(u) \cdot v - q_1(u) = 0\}$, where $q_0(u)$ and $q_1(u)$ have no common zero. If $a = \max\{\deg q_0(u), \deg(q_1(u)) - n\}$, then $C \cdot C_\infty = a$, $C^2 = n + 2a$ and $C \cdot C_0 = n + a$, where $C_0 = \{v = 0\}$. If $a = 0$, then there is an automorphism mapping $C$ onto $C_0$.*

PROOF. In order to prove (1) we first observe the following general fact. Let $X$ be a scheme and $\varphi : X \to X$ be an automorphism. Assume that $\mathcal{E}$ is a locally free sheaf of rank $r$ on $X$. Then the automorphisms $\Phi$ of $\mathbb{P}(\mathcal{E})$ with $\varphi\pi = \pi\Phi$, where $\pi : \mathbb{P}(\mathcal{E}) \to X$ is the canonical projection, are in a $1 - 1$ correspondence with the elements of

$$\text{Iso }(\mathcal{E}, \varphi^\star(\mathcal{E}))/\mathbb{C}^\star,$$

where Iso$(\dots)$ denotes the set of $\mathcal{O}_X$-linear isomorphisms, on which $\mathbb{C}^\star$ acts in an obvious way. Hence, if $\varphi : \mathbb{P}^1 \to \mathbb{P}^1$ is an automorphism leaving $u = \infty$ fixed, then the automorphisms $\Sigma(n) \xrightarrow{\Phi} \Sigma(n)$ with $\pi\Phi = \varphi\pi$ are in a $1 - 1$ correspondence with the elements of

$$\text{Iso }(\mathcal{O}_{\mathbb{P}^1} \oplus \mathcal{O}_{\mathbb{P}^1}(-n), \mathcal{O}_{\mathbb{P}^1} \oplus \mathcal{O}_{\mathbb{P}^1}(-n))/\mathbb{C}^\star.$$

Such a $\varphi$ is given by $u \mapsto a+bu$, $b \neq 0$, and an isomorphism $\mathcal{O}_{\mathbb{P}^1} \oplus \mathcal{O}_{\mathbb{P}^1}(-n)$ into itself is, up to a scalar, given by a matrix $\begin{pmatrix} 1 & P \\ 0 & c \end{pmatrix}$, where $c \neq 0$ and $P = P(x_0, x_1)$ is a homogeneous polynomial of degree $n$. With respect to the affine chart $(u, v)$, this automorphism is the one described by (1). In order to complete the proof of (1), it is sufficient to show that every automorphism of $\Sigma(n)$ into itself maps fibers onto fibers. But this follows from the well known fact that any effective irreducible curve in $\Sigma(n)$ is linearly equivalent to $C_\infty, F$ or $aF + bC_\infty$, where $a/b \geq n$ and $F$ is a fibre of $\pi$. Hence, if $F'$ is a irreducible curve of selfintersection 0, it must be a fibre.

The proof of (2) follows immediately from (1), since the space of vector fields, which are tangent to $F_\infty$, is the Lie algebra associated to the Lie group of automorphisms preserving $F_\infty$.

In order to show (3) we remark that by the universal property of $\mathbb{P}(\mathcal{O}_{\mathbb{P}^1} \oplus \mathcal{O}_{\mathbb{P}^1}(-n))$ there is a $1-1$ correspondence of the sections of $\pi$ with the surjective maps $\mathcal{O}_{\mathbb{P}^1} \oplus \mathcal{O}_{\mathbb{P}^1}(-n) \to \mathcal{O}_{\mathbb{P}^1}(a)$. Such a map is given by two homogeneous polynomials $\tilde{q}_0(x_0, x_1)$ and $\tilde{q}_1(x_0, x_1)$ without common factors of degrees $a$ and $n+a$, respectively. In affine coordinates, the equation of this section obviously is $q_1(u) - q_0(u)v = 0$, where $q_i(u) = \tilde{q}_i(1, u)$. Since $\tilde{q}_0$ and $\tilde{q}_1$ have no common factor, the same is true for $q_0$ and $q_1$, and $a = \max\{\deg q_0, \deg q_1 - n\}$ because of $(\tilde{q}_1(0,1), \tilde{q}_1(0,1)) \neq (0,0)$. It is easily checked that for a fixed $a$ such a section coincide with the zero set of an element of $H^0(\mathcal{O}_{\Sigma(n)}(C_\infty + (a+n)F))$. Hence $C^2 = 2a + n,\ CC_\infty = a,\ CC_0 = n + a.$ $\square$

The next lemma gives the normal forms of vector fields on $\Sigma(n),\ n \geq 1$ :

LEMMA 2.6. *After a suitable automorphism every vector field on the Hirzebruch surface $\Sigma(n), n \geq 1$, is in affine coordinate $(u, v)$ given by one of the following:*

(i) $\lambda_1 u \frac{\partial}{\partial u} + \lambda_2 v \frac{\partial}{\partial v},\ \lambda_1, \lambda_2 \in \mathbb{C}$,

(ii) $P(u) \frac{\partial}{\partial v},\ P \in \mathbb{C}[u],\ \deg P \leq n$,

(iii) $\lambda_1 u \frac{\partial}{\partial u} + (\lambda_2 v + cu^\lambda) \frac{\partial}{\partial v},\ c, \lambda_1, \lambda_2 \in \mathbb{C},\ \lambda_1 \neq 0$ *and* $\lambda = \lambda_2/\lambda_1 \in \mathbb{N}_0,\ 0 \leq \lambda \leq n$,

(iv) $\frac{\partial}{\partial u} + \lambda v \frac{\partial}{\partial v},\ \lambda \in \mathbb{C}$,

(v) $\frac{\partial}{\partial u} + cu^n \frac{\partial}{\partial v},\ c \in \mathbb{C}^\star$.

PROOF. With respect to suitable coordinates every vector field on $\mathbb{P}^1$ is either $\partial/\partial u$ or $\lambda_1 u \cdot \partial/\partial u$. Hence, if $\delta$ is a vector field on $\Sigma(n)$, we may assume that $\delta(u) = 1$ or $\delta(u) = \lambda_1 u$.

First we consider the case $\delta(u) = \lambda_1 u$, i.e. $\delta = \lambda_1 u \frac{\partial}{\partial u} + (\lambda_2 v + P(u)) \frac{\partial}{\partial v}$, where $P(u) = \sum_{j=0}^{n} b_j u^j$, see (2.5). Then, if $v' = v + Q(u)$, we get

$$\delta(v') = (\lambda_1 u Q'(u) + P(u) - \lambda_2 Q(u)) + \lambda_2 v'.$$

If $\lambda_1 = 0$ and $\lambda_2 \neq 0$, we can set $Q(u) := P/\lambda_2$ and arrive at the normal form (i). If $\lambda_1 \neq 0$ and $\lambda = \lambda_2/\lambda_1$ is not in $\mathbb{N}_0$, then

$$uQ'(u) = -\frac{P(u)}{\lambda_1} + \lambda Q(u)$$

has a solution, namely $Q(u) = \sum_j \frac{1}{\lambda_1(\lambda - j)} b_j u^j$. If $\lambda_1 \neq 0$ and $\lambda \in \mathbb{N}_0$, then at least

$$uQ'(u) = -\frac{P(u)}{\lambda_1} + \lambda Q(u) + \frac{b_\lambda}{\lambda_1} u^\lambda$$

has a solution of degree $\leq n$, and so

$$\delta(v') = \lambda_2 v' + b_\lambda u^\lambda.$$

In any case, we arrive at (i) or (iii). Finally, assume that $\delta = \frac{\partial}{\partial u} + (\lambda v + P(u))\frac{\partial}{\partial v}$ and consider again $v' = v + Q(u)$. Then

$$\delta(v') = (Q'(u) + P(u) - \lambda Q(u)) + \lambda v'.$$

If $\lambda = 0$, then the equation $Q'(u) + P(u) = b_n u^n$ has a solution $Q(u)$ of degree $\leq n$, and so $\delta(v') = b_n u^n$ in this case. If $\lambda \neq 0$, then there is a unique polynomial $Q$ of degree $\leq n$, satisfying $Q'(u) + P(u) - \lambda Q(u) = 0$, as is easily seen from the set of equations $(\nu + 1)c_{\nu+1} + b_\nu - \lambda c_\nu = 0$, $0 \leq \nu \leq n, c_{n+1} = 0$. Hence $\delta(v') = \lambda v'$ and we have the normal form (iv). $\square$

Now we can give a list of integral curves of the vector fields listed in (2.6).

LEMMA 2.7. *Besides $F_\infty$ and $C_\infty$, the vector fields listed in* (2.6) *have the irreducible integral curves as listed in* TABLE 2 *below.*

PROOF. The same arguments as in the proof of (2.2) yield 3) and 5). Assume that $\delta$ is as in 4), and consider an integral curve $\{g = 0\}$, where $g = \sum g_j(u)v^j$. Then $\delta(g) = \epsilon \cdot g$. Comparing the degrees in $u$ and $v$ on both sides, we get $\epsilon \in \mathbb{C}$. Moreover,

$$\epsilon g_j(u) = \lambda_1 u g_j'(u) + j\lambda_2 g_j(u) + cu^\lambda (j+1) g_{j+1}(u).$$

Let $\nu_0 = \deg_v(g)$. Then $g_{\nu_0+1} = 0$, and $(\epsilon/\lambda_1 - \nu_0\lambda) g_{\nu_0}(u) = u g_{\nu_0}'(u)$ has the solutions $g_{\nu_0}(u) = au^{\epsilon/\lambda_1 - \nu_0\lambda}$, $a \in \mathbb{C}$, which is polynomial only in the case $\epsilon/\lambda_1 \in \mathbb{N}_0, \epsilon/\lambda_1 \geq \nu_0\lambda$. If $\nu_0 = 0$, then $\epsilon/\lambda_1 = 1$, since otherwise the curve is not irreducible. Hence $\{g = 0\} = \{u = 0\}$. Now assume $\nu_0 > 0$. Then

$$(\epsilon/\lambda_1 - (\nu_0 - 1)\lambda) g_{\nu_0 - 1}(u) = u g_{\nu_0 - 1}'(u) + \frac{c}{\lambda_1} u^\lambda \nu_0 a u^{\epsilon/\lambda_1 - \nu_0\lambda}.$$

TABLE 2

| | Vector field | Integral curves |
|---|---|---|
| 1) | $\lambda_1 u \frac{\partial}{\partial u} + \lambda_2 v \frac{\partial}{\partial v}$ | see (2.2), cases 4) – 7) |
| 2) | $\frac{\partial}{\partial u} + \lambda v \frac{\partial}{\partial v}$ | see (2.2), cases 2) – 3) |
| 3) | $P(u)\frac{\partial}{\partial v}, P(u) \neq 0$ | all fibers $F_c = \pi^{-1}(c), c \in \mathbb{P}^1$ |
| 4) | $\lambda_1 u \frac{\partial}{\partial u} + (\lambda_2 v + cu^\lambda)\frac{\partial}{\partial v}$<br>$c \in \mathbb{C}^\star, \lambda_1 \neq 0$<br>$\lambda = \lambda_2/\lambda_1 \in \mathbb{N}_0, 0 \leq \lambda \leq n$ | u=0 |
| 5) | $\frac{\partial}{\partial u} + \lambda u^n \frac{\partial}{\partial v},\ \lambda \in \mathbb{C}^\star$ | $v = c + \frac{\lambda}{n+1} u^{n+1}, c \in \mathbb{C}$ |

But this equation has the only solutions

$$g_{\nu_0 - 1}(u) = u^{\epsilon/\lambda_1 - \lambda(\nu_0 - 1)}(b + \log u^{-d}),\ d = \frac{c}{\lambda_1}\nu_0,\ b \in \mathbb{C}\ ,$$

which obviously is not polynomial, since $d \neq 0$. □

*Remark* 2.8. Let $C$ be an integral curve of some nonzero vector field $\eta$ on $\Sigma(n)$, $n \geq 0$, such that $\eta$ is tangent to $F_\infty$. Then there is even an automorphism of $\Sigma(n)$, which maps $F_\infty$ into itself such that under this automorphism $C$ and $\eta$ get the normal forms described in (2.7) and (2.4). This is easily seen by the proofs.

## 3. Minimal compactifications of $\mathbb{C}^2$

By a theorem of Fujita [**14**] (3.2) it is known that every $\mathbb{Z}$–acyclic surface $X$ of logarithmic Kodaira dimension $-\infty$ is isomorphic to $\mathbb{C}^2$. In this section we will consider deformations of minimal compactifications $(V, D)$ of $\mathbb{C}^2$. Let us recall that, by a theorem of Ramanujam [**26**], the dual graph $\Gamma(D)$ of $D$ is linear. For a classification of the possible graphs we refer to [**26**] and [**25**]. By [**21**] it is known that $V$ is rational (see also [**25**]).

First we will give a complete list of all those compactifications $(V, D)$, where $V = \mathbb{P}^2$ or $V = \Sigma(n)$, $n \geq 0$. We also compute in all these cases the numbers $h^i(\Theta_V\langle D\rangle)$. We will describe the possible divisors $D$ in $\Sigma(n)$, $n \geq 0$, by their equations with respect to the affine chart $(u, v)$, which is described after Lemma (2.4). With respect to these coordinates, we denote by $F_\infty$ resp. $C_\infty$ the curves $\{u = \infty\}$ and $\{v = \infty\}$. Here $F_\infty$ is the fibre of $\Sigma(n) \xrightarrow{\pi} \mathbb{P}^1$ over $\infty$, and $C_\infty$ is the unique curve of $\Sigma(n)$ with $C_\infty^2 = -n$, if $n \geq 1$.

THEOREM 3.1. *Up to automorphisms, the possible minimal compactifications* $(V, D)$ *of* $\mathbb{C}^2$ *with* $V = \mathbb{P}^2$ *or* $\Sigma(n)$, *and the dimensions* $h^i := h^i(\Theta_V\langle D\rangle)$, *are given by* TABLE 3 *below.*

TABLE 3

| | V | D | $\Gamma(D)$ | $(h^0, h^1, h^2)$ |
|---|---|---|---|---|
| (I) | $\mathbb{P}^2$ | $\mathbb{P}^1 = \{x_0 = 0\}$ | $\circ$ 1 | $(6, 0, 0)$ |
| (II) | $\Sigma(n)$<br>$n \neq 1$ | $F_\infty \cup C_\infty$ | $0 \quad k$<br>$\circ\!\!-\!\!\!-\!\!\circ$<br>$k = -n$ | $(4 - k, 0, 0)$ |
| (III) | $\Sigma(n)$<br>$n \geq 0$ | $F_\infty \cup C, C \equiv C_\infty + (n+a)F$,<br>$a \geq 2$, such that there is no automorphism mapping $F_\infty \cup C$ to one of the pairs described in IV, V, VI | $0 \quad k$<br>$\circ\!\!-\!\!\!-\!\!\circ$<br>$k = n + 2a$ | $(0, k - 4, 0)$ |
| (IV) | $\Sigma(n)$<br>$n \geq 0$ | $F_\infty \cup C$, where $C = \{v = u^{n+a}\}$<br>$a \geq 2$, or $C = \{u^a v = 1\}$,<br>where $a \geq 1, n \geq 1$ | $0 \quad k$<br>$\circ\!\!-\!\!\!-\!\!\circ$<br>$k = n + 2a$ | $(1, k - 3, 0)$ |
| (V) | $\Sigma(n)$<br>$n \geq 0$ | $F_\infty \cup C$ , where $C = \{v = u^{n+1}\}$ | $0 \quad k$<br>$\circ\!\!-\!\!\!-\!\!\circ$<br>$k = n + 2$ | $(2, k - 2, 0)$ |
| (VI) | $\Sigma(n)$<br>$n \geq 1$ | $F_\infty \cup C_0$ , where $C_0 = \{v = 0\}$ | $0 \quad k$<br>$\circ\!\!-\!\!\!-\!\!\circ$<br>$k = n$ | $(3, k - 1, 0)$ |

Here $\Gamma(D)$ denotes the dual graph of $D$. For the proof we need the following two lemmas.

LEMMA 3.2. *Let* $(\Sigma(n), D)$ *be an SNC–compactification of some* $\mathbb{Z}$*–acyclic surface* $X$. *Then* $D = F \cup C$, *where* $F$ *is a fibre of* $\pi$ *and* $C$ *is a section, and so* $X$ *is isomorphic to* $\mathbb{C}^2$.

PROOF. Assume that $D = D_1 \cup D_2$, where $D_1 \equiv aF + bC_\infty, D_2 \equiv cF + dC_\infty$ with $a, b, c, d \in \mathbb{N}_0$. The condition that $X$ is acyclic is equivalent to the unimodularity of the matrix $\left(\begin{smallmatrix} a & b \\ c & d \end{smallmatrix}\right)$. After interchanging $D_1$ and $D_2$ if necessary, we may assume that $ad - bc = 1$. The condition $D_1 D_2 = 1$ gives $1 = ad + bc - bdn$. After replacing $ad$ by $1 + bc$ in this equation, we get $2bc = bdn$. If $D_2 \not\equiv C_\infty$, then $c \neq 0$ and $0 \leq D_2 C_\infty = c - nd$. This is only possible if $bc \leq 0$, i.e. $b = 0$.

But then $D_1 = F$ and $D_2 = cF + C_\infty$ , and so $D_2$ is a section. If $D_2 = C_\infty$, then $c = 0, d = 1$ and $D_1 = F + bC_\infty$ , and from $D_1D_2 = 1$ we get $bn = 0$. Hence for $n \geq 1$, $D_1 = F$ is a fibre, whereas for $n = 0$, after interchanging the factors of $\mathbb{P}^1 \times \mathbb{P}^1$, $D_2$ will become a fibre and $D_1$ a section. □

LEMMA 3.3. *Let $(V, D)$ be an SNC–compactification of $\mathbb{C}^2$. Then the group $H^2(\Theta_V\langle D\rangle)$ vanishes.*

PROOF. Since $H^2(\Theta_{\mathbb{P}^2}\langle l\rangle) = 0$ for a line $l$, this follows immediately from (1.5) (5). □

PROOF OF THEOREM (3.1). In the case of $V = \mathbb{P}^2$ the theorem is evident. So assume from now on that $V = \Sigma(n)$, $n \geq 0$. By (3.2) $D = F \cup C$, where $F$ is a fibre and $C$ a section of $\pi$. After an automorphism we may assume that $F = F_\infty$.

If $C = C_\infty$ and $n \geq 1$, then every vector field on $\Sigma(n)$ is tangent to $C_\infty$, since every automorphism of $\Sigma(n)$ maps $C_\infty$ onto itself. From the description in (2.5) (2) we get $h^0(\Theta_{\Sigma(n)}\langle F_\infty \cup C_\infty\rangle) = n + 4$, whereas $\chi(\Theta_{\Sigma(n)}\langle F_\infty \cup C_\infty\rangle) = n + 4$, see (1.3) (5), and $h^2(\Theta_{\Sigma(n)}\langle F_\infty \cup C_\infty\rangle) = 0$ by (3.3). Since the compactification is assumed to be minimal, the case $C = C_\infty$, $n = 1$ is not possible. If $n = 0$ and $C = C_\infty$, then (2.3) easily shows that $h^0(\Theta_{\mathbb{P}^1\times\mathbb{P}^1}\langle D\rangle) = 4$, and as above $\chi(\Theta_{\mathbb{P}^1\times\mathbb{P}^1}\langle D\rangle) = 4$, $h^2(\Theta_{\mathbb{P}^1\times\mathbb{P}^1}\langle D\rangle) = 0$. This proves our assertion in the case $D = F_\infty \cup C_\infty$.

So, assume now that $D = F_\infty \cup C$, where $C$ is a section of $\Sigma(n)$ with $k = C^2 \geq 1$. The curve $C$ is linearly equivalent to $C_\infty + (n+a)F$ for some $a \geq 0$, and so $k = n+2a$. By (1.3) (5) $\chi(\Theta_{\Sigma(n)}\langle D\rangle) = 4 - k$ and by (3.3) $h^2(\Theta_{\Sigma(n)}\langle D\rangle) = 0$. If $h^0(\Theta_{\Sigma(n)}\langle D\rangle) = 0$, then $h^1(\Theta_{\Sigma(n)}\langle D\rangle) = k - 4$ leads to case III by exclusion. Observe that in the case $a \leq 1$ the curve $C$ is given by an equation $q_0(u)v - q_1(u) = 0$ with $a = max\ \{\deg(q_0), \deg(q_1) - n\}$, see (2.5) (3), and thus after a suitable automorphism (cf. (2.5) (1)) we arrive at one of the cases described in IV – VI.

Assume from now on that $h^0(\Theta_{\Sigma(n)}\langle D\rangle) \neq 0$. Then after a suitable change of coordinates we have the following possibilities, see (2.4), (2.5), (2.7) and (2.8):

(a) $C = \{v = c\}, c \in \mathbb{C}$,

(b) $C = \{cv - u^r = 0\}, c \in \mathbb{C}^\star$,

(c) $C = \{1 - cu^r v = 0\}, c \in \mathbb{C}^\star$,

where $r \geq 1$. In the case (a), after applying the automorphism $(u, v) \mapsto (u, v - c)$, we may assume that $c = 0$. Since $C^2 \geq 1$, the case $n = 0$ cannot occur anymore. From the description in (2.5) (2) we obtain that the only vector fields tangent to $F_\infty \cup C$ are given by

$$(\alpha + \beta u)\frac{\partial}{\partial u} + \gamma v\frac{\partial}{\partial v}\ .$$

This can be seen as follows. If

$$\delta = (\alpha + \beta u)\frac{\partial}{\partial u} + (\gamma v + P(u))\frac{\partial}{\partial v}$$

in a vector field tangent to $v = 0$, then $\delta(v) \equiv 0$ modulo $v$, and so $P = 0$. Hence $h^0(\Theta_{\Sigma(n)}\langle D\rangle) = 3$. This shows the theorem in case VI.

Next we consider a curve $C$ as in (b). Again we can assume that $c = 1$. If $r \leq n$, then, after the automorphism $(u,v) \mapsto (u, v - u^r), C$ becomes the zero section, and again we are in case VI. If $r > n \geq 1$, then by (2.5) (3) $C^2 = n + 2(r - n)$, and so $r = n + a, a \geq 1$. If $a = 1$, then a basis of the space of vector fields tangent to $F_\infty \cup C$ is given by

$$\frac{\partial}{\partial u} + (n+1)u^n\frac{\partial}{\partial v}, \qquad u\frac{\partial}{\partial u} + (n+1)v\frac{\partial}{\partial v},$$

as is easily seen from the description in (2.5), (2). So we are in case V. If $a > 1, n \geq 1$, then $H^0(\Theta_{\Sigma(n)}\langle D\rangle)$ is generated by

$$u\frac{\partial}{\partial u} + (n+a)v\frac{\partial}{\partial v},$$

i.e. $h^0 = 1$, and this is described in IV. If $r > n = 0$, then the computations are similar using (2.3) instead of (2.5). We omit the details. Finally, we examine the case (c), where $C$ is given by the equation $1 - cu^r v = 0$; again we may restrict ourselves to the case $c = 1, r \geq 1$. First assume that $n \geq 1$. By (2.5) (3) $r = a$. Similarly as above, a basis of $H^0(\Theta_{\Sigma(n)}\langle D\rangle)$ is given by

$$u\frac{\partial}{\partial u} - av\frac{\partial}{\partial v},$$

and $h^0 = 1$. If $n = 0$, then the change of coordinates $(u,v) \mapsto (u, 1/v)$ maps $F_\infty \cup C$ onto $F_\infty \cup \{v = u^r\}$, which was already treated. This completes the proof of the theorem. □

*Remark* 3.4. (1) Let $C \subseteq \Sigma(n)$ be a section, given by the equation $g(u,v) = q_0(u)v - q_1(u) = 0$, and $a = \max(\deg(q_0), \deg(q_1) - n)) \geq 2$. Then $F_\infty \cup C$ is as in case III of the theorem, iff its equation is none of the following:

(1) $g(u,v) = \alpha v(u-c)^a + \beta(u-c)^a P(u) + \gamma$, where $\alpha, \gamma \in \mathbb{C}$, $c, \beta \in \mathbb{C}$ and $P$ is a polynomial of degree $\leq n$,

(2) $g(u,v) = \alpha v + \beta(u-c)^{n+a} + P(u)$, where $P$ is as above and $\beta \in \mathbb{C}^\star, \alpha \in \mathbb{C}$,

(3) $n = 0, g(u,v) = v(\alpha + \beta(u-c)^a) + \gamma + \delta(u-c)^a$, where $\det\left(\begin{smallmatrix}\alpha & \beta \\ \gamma & \delta\end{smallmatrix}\right) \neq 0$.

This is easily seen by applying automorphisms to the equations, which occur in IV,V.

(2) Among the cases described in (3.1), the only rigid ones are up to automorphism those listed in TABLE 4 below. This follows from the description in (3.1) and the preceding remark.

TABLE 4

| Case | |
|---|---|
| I | line in $\mathbb{P}^2$ |
| II | $F_\infty \cup C_\infty \subseteq \Sigma(n),\ n \neq 1$ |
| III | $F_\infty \cup \{uv = u^2 + 1\} \subseteq \mathbb{P}^1 \times \mathbb{P}^1$ |
| IV | $F_\infty \cup \{uv = 1\} \subseteq \Sigma(1)$ |
| V | $F_\infty \cup \{u = v\} \subseteq \mathbb{P}^1 \times \mathbb{P}^1$ |
| VI | $F_\infty \cup \{v = 0\} \subseteq \Sigma(1)$ |

(3) Assume that $(V, D)$ is an SNC–pair. Then there is an exact sequence

$$0 \to H^0(\Theta_V\langle D\rangle) \to H^0(\Theta_V) \xrightarrow{\alpha} \bigoplus_i H^0(\mathcal{N}_{D_i/V}) \to$$

$$\xrightarrow{\beta} H^1(\Theta_V\langle D\rangle) \xrightarrow{\gamma} H^1(\Theta_V) \to \bigoplus_i H^1(\mathcal{N}_{D_i/V}) \to H^2(\Theta_V) \to \cdots\ ,$$

which is the cohomology sequence of

$$0 \to \Theta_V\langle D\rangle \to \Theta_V \to \bigoplus_i \mathcal{N}_{D_i/V} \to 0\ .$$

Here the $D_i$ are the components of $D$. From this one can see, in which way the moduli $h^1(\Theta_V)$ arise. In our case we have $H^2(\Theta_V) = 0$. For instance, in case III we have $\bigoplus H^1(\mathcal{N}_{D_i/V}) = 0$ and $H^0(\Theta_V\langle D\rangle) = 0$. Hence $\gamma$ is surjective, i.e. all deformations of $\Sigma(n)$ come from deformations of $(\Sigma(n), D)$. We remark that the deformations of $\Sigma(n)$ are in a $1-1$ correspondence with the extensions

$$0 \to \mathcal{O}_{\mathbb{P}^1}(-n) \to \mathcal{E} \to \mathcal{O}_{\mathbb{P}^1} \to 0.$$

Moreover, deforming the equation of $C$ up to automorphisms, one obtains additional parameters, which correspond to Im $\beta$.

(4) Let $(\overline{X}, D)$ and $(\overline{X}', D')$ be SNC–pairs, which are minimal compactifications of the same $\mathbb{Z}$–acyclic surface $X$. Assume that there is an isomorphism $D \to D'$ such that the corresponding self–intersection numbers of the irreducible components are the same. Assume further that $\overline{X}$ and $\overline{X}'$ are isomorphic. Then it does not necessarily follow that the pairs are isomorphic. For instance, if $\overline{X} = \mathbb{P}^1 \times \mathbb{P}^1$, and if $(\overline{X}, D)$ is a compactification of $X = \mathbb{C}^2$, which is nonrigid (see (3.1), cases III – V), then for any small deformation $(\overline{X}', D)$ of $(\overline{X}, D)$ we have $\overline{X}' = \mathbb{P}^1 \times \mathbb{P}^1, D \cong D', \Gamma(D) \cong \Gamma(D')$ as weighted graphs, but $(\overline{X}, D) \ncong (\overline{X}', D')$, as the Fischer–Grauert type theorem of [**3**] shows.

3.5. We will now describe how one can, at least in principle, compute the number of moduli for any minimal compactification $(V, D)$ of $\mathbb{C}^2$. For this we need the description of possible graphs $\Gamma(D)$, which one can find in [**25**]. In order to reproduce this list, we will abbreviate the subgraph of the form

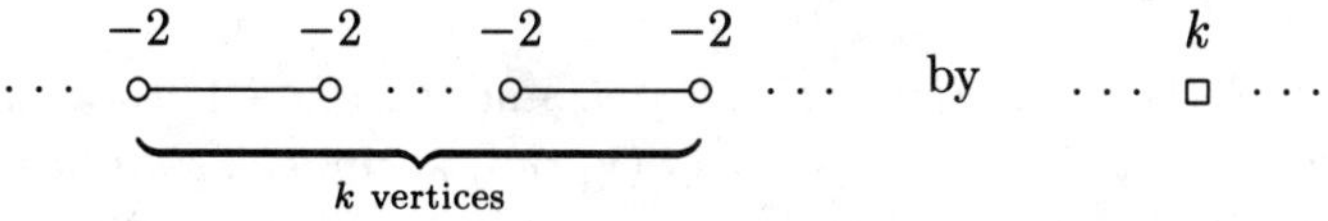


Then the possible graphs are:

(a)

(b)

, $n \neq -1$

(c)

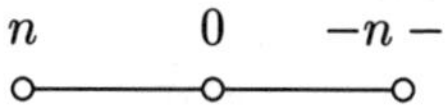


, $n > 0$

(d)

, $n, L > 0$

(e)

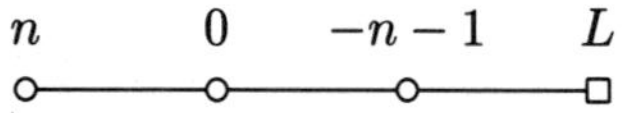


, $n, L > 0$

(f)

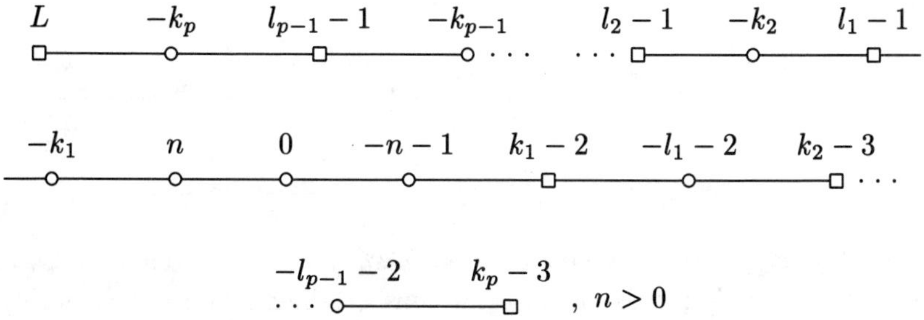


, $n > 0$

(g)

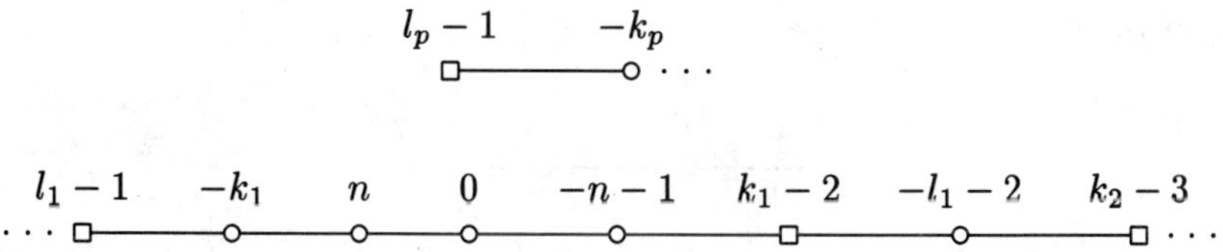

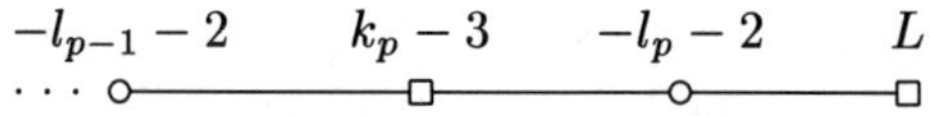


, $n>0$

For the conditions on $p, k_j, l_j$ we refer to **[25]**; but different from loc. cit. we assume that (c) – (e) are not special cases of (f) and (g).

3.6. Let $(V, D)$ be an SNC–pair, $P \in D$ and $\pi : V' \to V$ the blowing up of $V$ at $P$. Set $D' = \pi^{-1}(D)$. We will call $(V', D')$ an *inner modification* or *inner blowing up* of $(V, D)$ if $P$ is in two components of $D$. Let $\Gamma(D)$ resp. $\Gamma(D')$ be the corresponding graph and $P \in D_1 \cap D_2$, where $D_1, D_2 \subseteq D$ are irreducible components. Then the graph $\Gamma(D')$ is obtained from $\Gamma(D)$ by the operation

For inner modifications $\pi : (V', D') \to (V, D)$ we have $h^i(\Theta_{V'}\langle D'\rangle) = h^i(\Theta_V\langle D\rangle)$, $i = 0, 1, 2$, as follows from (1.5). The operation inverse to an inner blowing up will be called an *inner blowing down.*

LEMMA 3.7. *Each of the graph (c) – (g) can be transformed by a sequence of inner blowing ups and blowing downs to one of the graphs:*

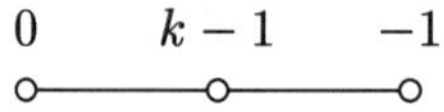


, $k \geq 1$ .

PROOF. If a graph $\Gamma$ contains a subgraph of the form (c), then one can perform the following inner blowing ups and downs:

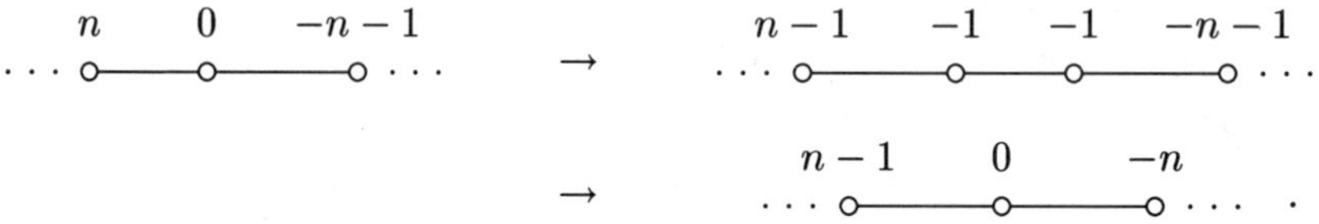


We will call this operation an *elementary transformation* and denote it by $\mathrm{el}(\Gamma)$. Moreover, by applying these transformations or their inverse successively, we have operations $el^m(\Gamma)$ for any $m \in \mathbb{Z}$. This operation corresponds in an obvious way to a birational transformation $el^m(V, D)$, if $\Gamma = \Gamma(D)$. If there is a subgraph of $\Gamma$ of the form

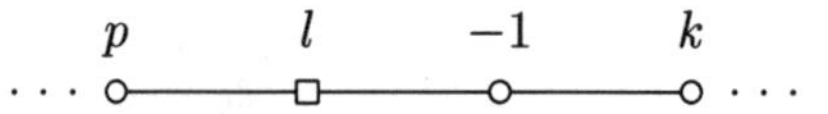


,

then we can blow down the $(-1)$ vertices successively and arrive at the graph

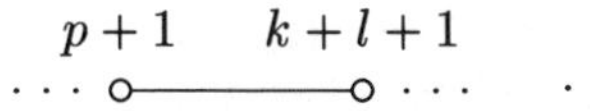


We denote this operation by $bk(l)$ ("box killing"). Similarly, we can perform the operation

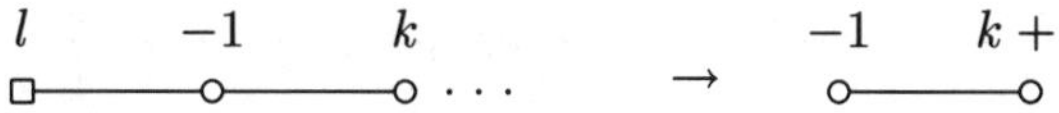


which will be denoted by $ebk(l)$ ("end box killing").

Applying $el^n$ to the graph in (c), we obtain the graph 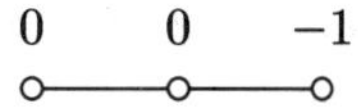

. In case (d) and (e), after applying $el^{n+1}$ resp. $el^n$ and then $ebk(L)$, we arrive at 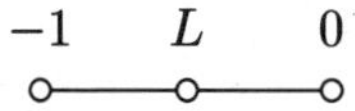

. For the graphs (f) and (g) we remark first that we obtain type (g) from (f) by $el^{2n+1}$. Moreover, performing in (f) in case $p \geq 2$ the transformations $el^n$ and $bk(k_1)$, we arrive at a new graph of smaller length, which has again, up to a transformation $el^m$ for some $m$, type (f). If $p = 1$ in (f), then we can perform $el^m$ and after that $ebk(k_1)$ to obtain the graph

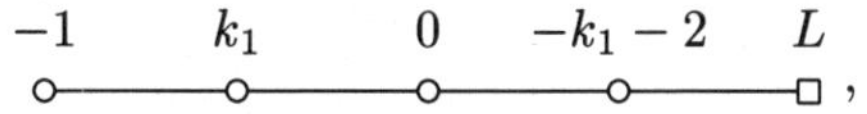


which is the blowing up of a graph of type (d) or (e), and so can be transformed similarly as above. □

To complete the picture for the possible number of moduli for minimal compactifications of $\mathbb{C}^2$, we have to calculate the $h^i$ for the graphs occurring in (3.4).

PROPOSITION 3.8. *Let $(V, D)$ be a compactification of $\mathbb{C}^2$ such that $(V, D)$ is the blowing up of $(\Sigma(n), F_\infty \cup C)$ in some point $P$ of $C \setminus F_\infty$. Assume that $C^2 = k = n + 2a \geq 1$, where $a \geq 0$. Then the numbers $h^i := h^i(\Theta_V\langle D\rangle)$ are given by* TABLE 5 *below, where the numbering refers to the different cases in Theorem* (3.1).

PROOF. From the exact sequence in (1.5) (3) we get that $h^0 \leq h^0(\Theta_{\Sigma(n)}\langle C_\infty \cup F\rangle) \leq h^0 + 1$. Moreover, we know that $\chi(\Theta_V\langle D\rangle) = \chi(\Theta_{\Sigma(n)}\langle F_\infty \cup C\rangle) - 1$ and $h^2 = 0$. This implies the proposition in case III. If, in one of the remaining cases, there is a vector field in $H^0(\Theta_{\Sigma(n)}\langle F_\infty \cup C\rangle)$, which is nonzero in $P$, then $h^0(\Theta_{\Sigma(n)}\langle F_\infty \cup C\rangle) = h^0 - 1$ and $h^1(\Theta_{\Sigma(n)}, F_\infty \cup C) = h^1$. But it can be easily checked that in cases IV (a), V and VI there is a vector field in $H^0(\Theta_{\Sigma(n)}\langle F_\infty \cup$

TABLE 5

| Case | Point, which has been blown up | $(h^0, h^1, h^2)$ |
|---|---|---|
| III | $P \in C \setminus F_\infty$ any point | $(0, k-3, 0)$ |
| IV (a) | $P \in C \setminus (F_\infty \cup F_0)$, where $F_0$ is the fibre of $\pi$ over $u = 0$ of $\mathbb{P}^1$ | $(0, k-3, 0)$ |
| IV (b) | $P \in C \cap F_0$ | $(1, k-2, 0)$ |
| V | $P \in C \setminus F_\infty$ any point | $(1, k-2, 0)$ |
| VI | $P \in C \setminus F_\infty$ any point | $(2, k-1, 0)$ |

$C\rangle)$, which is nonzero in $P$ , whereas in IV (b) every such vector field vanishes at $P$. □

*Remark* 3.9. If one of the graphs of type (c) – (g) is transformed into the graph 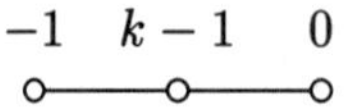

, then in case (c) we have $k = 1$, whereas for (d), (e) we have $k = L + 1$ and for (f), (g) $k = L + 2$. This can be seen from the proof in (3.4).

### 4. $\mathbb{Q}$–acyclic surfaces of logarithmic Kodaira dimension $-\infty$

For the description of $\mathbb{Q}$–acyclic surfaces with $\bar{k} \leq 1$ it is convenient to introduce the half–point attachment process, see [**13**].

4.1. Let $(V, D)$ be an SNC–pair and $x \in D$ a smooth point. Denote by $V'$ the blowing up of $V$ at $x$ and let $D'$ be the proper transform of $D$ in $V'$. Then $(V', D')$ is called *the half–point attachment of* $(V, D)$ *at* $x$, and conversely $(V, D)$ *the half–point detachment of* $(V', D')$. An SNC–pair $(V, D)$ will be called *strictly minimal*, if $(V, D)$ is minimal, not a half–point attachment of some other SNC–pair, and $V \setminus D$ contains no $(-1)$ curve, i.e. exceptional curve of the first kind. Obviously, these last two conditions are equivalent to the non–existence of smooth rational curves $E$ in $V$ such that $E^2 = -1,\ ED \leq 1$. Clearly, any SNC–pair can be transformed into a strictly minimal one by iterated blowing downs of components of $D$, $(-1)$–curves in $V \setminus D$ and half–point detachments. Observe that by [**13**] (6.23) (see also [**29**] (2.7)) the logarithmic Kodaira dimension of $V \setminus D$ does not change by passing to the strictly minimal model.

DEFINITION 4.2. An SNC–pair $(V, D)$ will be called *pre–$\mathbb{Q}$–acyclic* if $D$ is a rational tree, $b_1(V) = 0$ and if the components $D_i$ of $D$ generate $h^2(V, \mathbb{Q})$.

LEMMA 4.3. (1) (cf.[**13**] (2.4)) *If* $(V, D)$ *is pre–*$\mathbb{Q}$*–acyclic, then* $X = V \setminus D$ *is affine.*

(2) (cf. [**26**], [**13**]) $(V, D)$ *is* $\mathbb{Q}$*–acyclic iff the following conditions are satisfied:*

(i) $D$ *is a rational tree.*

(ii) *The components of* $D$ *are a basis of* $H^2(V, \mathbb{Q})$.

(iii) $b_1(V) = 0$.

PROOF. (2) follows easily form the exact sequence of local cohomology groups

$$\ldots \to H^i(X, \mathbb{Q}) \to H^{i+1}_D(V, \mathbb{Q}) \to H^{i+1}(V, \mathbb{Q}) \to \ldots$$

and Poincaré duality $H^i_D(V, \mathbb{Q}) \simeq H_{4-i}(D, \mathbb{Q})$. □

LEMMA 4.4. (1) *Let* $(V, D)$ *be a* $\mathbb{Q}$*–acyclic surface and* $(V', D')$ *a strictly minimal model of* $(V, D)$. *Then* $(V', D')$ *is pre–*$\mathbb{Q}$*–acyclic.*

(2) *Conversely, if* $(V', D')$ *is pre–*$\mathbb{Q}$*–acyclic, then there exists a* $\mathbb{Q}$*–acyclic pair* $(V, D)$ *such that* $(V', D')$ *is obtained from* $(V, D)$ *by a succession of blowing downs of components of* $D$ *and half–point detachments.*

PROOF. It is clear that $b_1(V)$ and the condition that $D$ is a rational tree, do not change under blowing ups in points of $D$ and also not by half–point attachments. So for (1) it is sufficient to remark that after half–point detachment the $[D_i]$ will remain to form generators of $H^2$. For the proof of (2) we will show that for a pre–$\mathbb{Q}$–acyclic pair $(V', D')$, which is not $\mathbb{Q}$–acyclic, it is possible to find a half–point attachment $(V'', D'')$, which is again pre–$\mathbb{Q}$–acyclic. Indeed, let $D' = D'_1 \cup \ldots \cup D'_r$ be the irreducible components and assume that

$$\sum \alpha_\rho [D'_\rho] = 0$$

is a nontrivial relation in $H^2(V, \mathbb{Q})$. Choose $D'_\rho$ with $\alpha_\rho \neq 0$, and let $V'' \xrightarrow{\pi} V'$ be the blowing up in a smooth point of $D'_\rho$. Let $D''_i$ be the proper transform of $D'_i$ on $V''$ and $E \subseteq V''$ the exceptional curve of $\pi$. Taking $\pi^\star$ we get from the relation that

$$\sum_i \alpha_i [D''_i] + \alpha_\rho [E] = 0.$$

Hence the $[D''_i]$ generate $H^2(V'', \mathbb{Q})$, and clearly $b_2(V'') = b_2(V') + 1$. Repeating this process, we get a pre–$\mathbb{Q}$–acyclic pair $(V, D)$ with rk $(H^2(V, \mathbb{Q})) = r$ and which therefore is $\mathbb{Q}$–acyclic, see Lemma 4.3 (2). □

*Remarks.* (1) In (4.4) (a) the strictly minimal model $(V', D')$ of $(V, D)$ is not uniquely determined. This is easily seen from a blowing up of a Hirzebruch surface.

(2) The proof shows that, in order to obtain a $\mathbb{Q}$–acyclic surface from a pre–$\mathbb{Q}$–acyclic one, we have to perform successively comb–attachments (which are introduced below) in points of the support of nontrivial relations $\sum a_i[D'_i] = 0$ in Pic $(V')$.

DEFINITION 4.5. Let $(V', D') \xrightarrow{x} (V, D)$ be a birational morphism of SNC–pairs. $(V', D')$ is called *a comb–attachment of* $(V, D)$, if the following conditions are satisfied:

(1) $V'$ is a succession of blowing ups in points over a given point $x \in D$.

(2) The exceptional set of $V' \to V$ contains only one $(-1)$–curve $E$; furthermore $ED' = 1$ and $\pi^{-1}(D)_{red} = D' + E$.

For instance, the half–point attachment is a special case of such a comb–attachment. In this case $x \in D$ is a smooth point and $E = \pi^{-1}(x)$. In general, any comb–attachment is a succession of blowing ups

$$(V', D') = (V_n, D_n) \to \ldots \to (V_1, D_1) \to (V_0, D_0) = (V, D) ,$$

where $V_{i+1}$ is obtained from $V_i$ by blowing up a point on the $(-1)$–curve $E_i \subseteq D_i, i \leq n-1$, which is exceptional in the preceding step and where $(V', D') \to (V_{n-1}, D_{n-1})$ is a half–point attachment.

In the next lemma we will determine the structure of the dual graph of $D'$ from the structure of $D$. For this we introduce the following notation.

4.6. A weighted graph of the type as shown in FIGURE 1 will be abbreviated

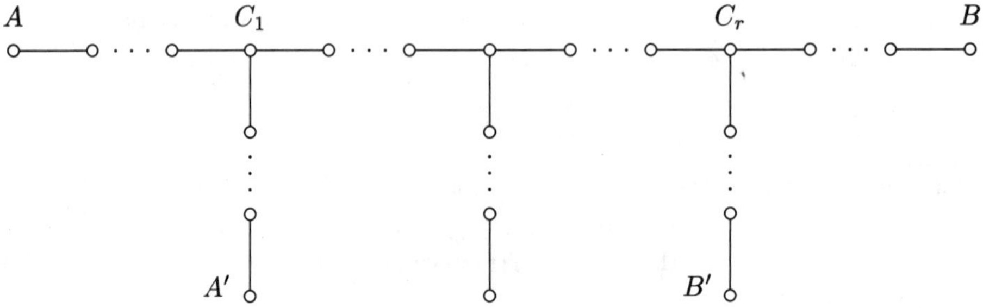


FIGURE 1.

by a box and denoted by ⊞ . Here $C_1, \ldots, C_r$ are the branching points and $r \geq 0$. If we write

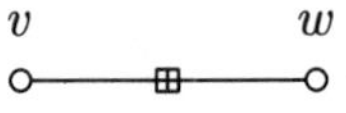

,

then we mean by this the graph, which is obtained from ⊞ by connecting $v$, $w$ with $A$ resp. $A'$. If we write

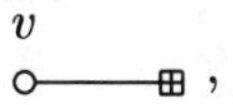


then this denotes the graph, in which $v$ is connected with some vertex on ⊞ , which is lying on the linear branch between $A$ and $C_1$ (excluding $C_1$). A box ⊞ will be called *a comb*, if there is a vertex $\Gamma$ on the linear branch between $B$ and $C_r$, such that by connecting a $(-1)$–vertex to $\Gamma$ we obtain a graph ⊞ – – – $\overset{-1}{\circ}$ which is contractible of discriminant $\pm 1$ in the sense of (A.2)(2).

In particular, this implies that the intersection matrix of a comb is negative definite. For technical reasons, we allow ⊞ to be the empty box. In this case $\cdots\circ$——⊞——$\circ\cdots$ means $\cdots\circ$——$\circ\cdots$ , similarly for $\cdots\circ$——⊞ and ⊞——$\circ\cdots$ .

LEMMA 4.7. *Let $\pi : (V', D') \to (V, D)$ be a comb–attachment at $x \in D$.*

(1) *If $x \in D_i$ is smooth, then the graph $\Gamma(D')$ is obtained from the graph of $D$ by attaching a comb at $D_i$:*

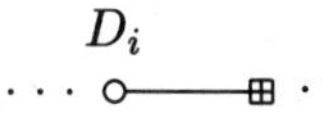


(2) *If $x \in D_i \cap D_j, i \neq j$, then the graph $\Gamma(D')$ is obtained from the graph of $D$ by replacing the edge joining $D_i$ , $D_j$ by a comb connected with $D_i$ and $D_j$ :*

PROOF. More generally we will show: If $\pi : U' \to U$ is a sequence of blowing ups over the point $x \in D$ such that the exceptional divisor of $\pi$ contains only one rational $(-1)$–curve $E$, then the graph $\Gamma(\pi^*(D))$ has the form

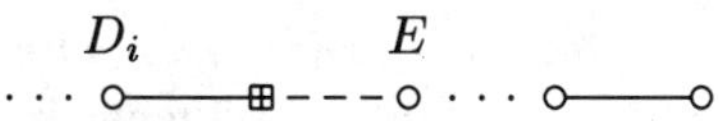


in case (1) resp.

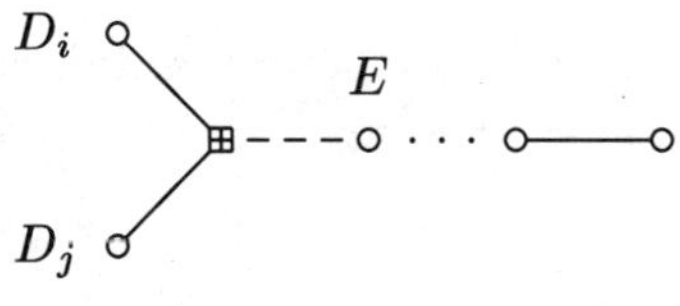

in case (2), where the box together with the linear branch on the right hand side is contractible with discriminant $\pm 1$. Indeed, if $\pi$ is a comb–attachment, then $E$ must be an end of $\Gamma(\pi^\star(D))$ , and so the lemma will follow from this.

We proceed by induction on the number $n$ of blowing ups in $\pi$. The case $n = 1$ is obvious (here ⊞ is the empty box). If $n > 1$, we blow down $E$ and obtain a new modification $V'' \xrightarrow{\pi''} V$ such that $\pi''^{-1}(x)$ has only $(n-1)$ components. If $\pi''^{-1}(x)$ would have two $(-1)$–curves, then they would intersect, and blowing down one of them gives a curve of selfintersection 0, contradicting the contractibility condition. Hence there is only one $(-1)$– curve in $\pi''^{-1}(x)$. Applying the induction hypothesis it is easily seen that $\Gamma(D)$ has the required form. □

We will now give examples of (pre–) $\mathbb{Q}$–acyclic surfaces of logarithmic Kodaira dimension $-\infty$ :

*Example* 4.8. Let $\pi : \Sigma(n) \to \mathbb{P}^1$ be the Hirzebruch surface and $B \subseteq \Sigma(n)$ a section. Then, for any fibers $F_0, \ldots, F_s$ of $p$,

$$(\Sigma(n), \;\; B \cup F_0 \cup \ldots \cup F_s) \; , \; s \geq 0 \; ,$$

is pre–$\mathbb{Q}$–acyclic and the complement $X = \Sigma(n) \setminus (B \cup F_0 \cup \ldots \cup F_s) \cong \mathbb{C} \times (\mathbb{C} \setminus \{x_1, \ldots, x_s\})$ is of logarithmic Kodaira dimension $-\infty$.

PROPOSITION 4.9. *Let* $(\Sigma(n), B \cup F_0 \cup \ldots \cup F_s)$ *be as above. Let* $p : (V, D) \to (\Sigma(n), B \cup F_0 \cup \ldots \cup F_s)$ *be comb–attachments in points* $x_j \in F_j \setminus B$, $j = 1, \ldots, r$ , $r \leq s$. *Then* $X := V \setminus D$ *is pre–*$\mathbb{Q}$*–acyclic and* $\bar{k}(X) = -\infty$. *Furthermore,* $X$ *is* $\mathbb{Q}$*–acyclic, if* $r = s$.

PROOF. That $\bar{k}(X) = -\infty$ follows from the existence of the fibration $X \to \mathbb{C}$ with generic fibers isomorphic to $\mathbb{C}$, induced by the fibration $\varphi := \pi \circ p : V \to \mathbb{P}^1$ with rational generic fibers. In order to show that $X$ is pre–$\mathbb{Q}$–acyclic, observe that $H^2(\Sigma(n), \mathbb{Q})$ is generated by $B, F_0, \ldots, F_s$ with the relations $F_0 - F_\sigma \approx 0, \sigma = 1, \ldots, s$. If we perform a comb–attachment in $F_\sigma \setminus B$, then the relation $F_0 - F_\sigma$ will be killed on the resulting surface, see (4.4) (2) and its proof. Hence $X$ will be pre–$\mathbb{Q}$–acyclic and $\mathbb{Q}$–acyclic, if $r = s$. □

*Remark.* The dual graph $\Gamma(D)$ has the form as indicated in FIGURE 2, where $\hat{F}_i$ denotes the proper transform of $F_i$ and $\hat{F}_0^2 = 0$ (see (4.7) (1)). If $s \geq 2$, then $\Gamma(D)$ is minimal.

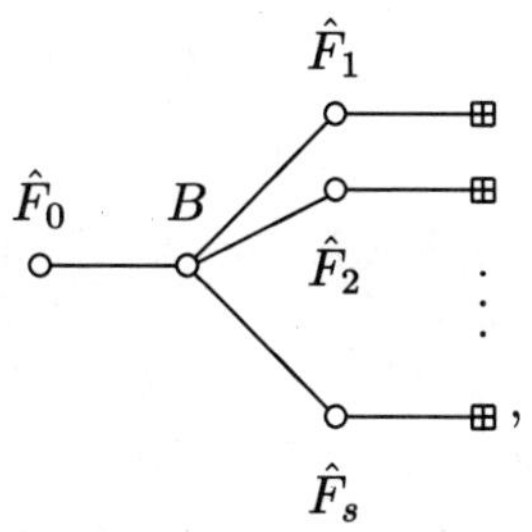


FIGURE 2.

THEOREM 4.10. *Let $X$ be a pre-$\mathbb{Q}$ acyclic surface of logarithmic Kodaira dimension $-\infty$. Then there is an SNC–completion $(V, D)$ of $X$, which is one of the pairs described in* (4.9).

The proof is based on:

THEOREM 4.11. (Miyanishi–Sugie; Russell, see **[24]**, **[22]**, **[29]**) *Let $(V, D)$ be an SNC–pair, where $D$ is connected and $q = h^1(\mathcal{O}_V) = 0$. Set $X := V \setminus D$. Then the following conditions are equivalent.*

(1) $\overline{k}(X) = -\infty$.

(2) *After blowing up $V$ suitably in points of $D$, there is a commutative diagram*

$$\begin{array}{ccc} V & \xrightarrow{\ \pi\ } & \Sigma(n) \\ & {\scriptstyle p}\searrow \quad \swarrow{\scriptstyle \bar{p}} & \\ & \mathbb{P}^1 & \end{array}$$

*such that $\pi$ is birational and the generic fibre $F$ of $p$ satisfies the inequality $FD \leq 1$.*

For the proof we will need three lemmas:

LEMMA 4.12. *Let $V$ be a smooth projective surface with $q(V) = p_g(V) = 0$.*

a) *Let $C$ be a reduced effective divisors on $V$. Then $\mid K_V + C \mid = \phi$ iff $C$ is a union of rational trees.*

b) *Let $C_1, C_2$ be connected reduced effective divisors on $V$. If $\mid K_V + C_1 + C_2 \mid = \phi$, then $C_1 C_2 \leq 1$.*

For the proof see **[24]** (2.1.3) and **[29]** (2.1), (2.2).

LEMMA 4.13. *Let $D$ be an effective divisor on a rational surface $V$ with $k(K + D) = -\infty$. Then, for any divisor $\Gamma$ on $V$, there is an $m \geq 1$ such that $\mid \Gamma + m(K + D) \mid = \phi$.*

For the proof, see [**14**] (2.5).

LEMMA 4.14. *Let* $(V, D)$ *be a minimal SNC–pair. Assume that one of the following assumptions is satisfied:*

(a) *The dual graph of* $\Gamma$ *contains a non–admissible outer chain, see* (A.2), *i.e. there is a rational curve* $E$ *with* $E^2 \geq 0$*, lying on an outer chain of* $\Gamma$.

(b) *There is a smooth rational curve* $E$ *in* $V$ *with* $E^2 \geq 0$ *and* $ED \leq 1$.

*Then* (4.11) (2) *holds. In particular,* $X := V \setminus D$ *has logarithmic Kodaira dimension* $-\infty$.

PROOF. If (a) is satisfied, then after elementary transformations we can assume that there is an irreducible component $E$ of $D$ with $E^2 = 0$, which is an end–vertex of $\Gamma$. By [**1**](V.4.3), $E$ is the nonsingular fibre of a $\mathbb{P}^1$–fibration $p : V \to \mathbb{P}^1$. Since such a fibration factors through some Hirzebruch surface, we obtain (4.11) (2). If (b) is satisfied, then again after blowing up $E$ we can assume that $E^2 = 0$. Thus the same arguments as in the first case can be applied. □

PROOF OF (4.11) We give a simplified version of the proof in [**29**]. Assume that (2) is satisfied. Then obviously $X$ contains a set of the form $B \times \mathbb{C}$, where $B$ is a curve, and so $\overline{k}(X) \leq \overline{k}(B \times \mathbb{C}) = -\infty$.

For the converse implication we will show that there is a curve $\Gamma$ on $V$ satisfying one of the assumptions of (4.14), see also [**29**] (2.3) and (2.5). We remark that by Castelnuovo's criterion $V$ is a rational surface, since $q = 0$ and $2K \leq 2(K + D)$, and so $P_2(V) = 0$. First we observe that $D$ is a union of rational trees by Lemma (4.12). If the Picard number $\rho(V) \leq 2$, then $V$ is a Hirzebruch surface or $\mathbb{P}^2$. If $D$ is irreducible, then $D^2 \geq 0$ or $D \subseteq \Sigma(n)$ is a section of the ruling $\Sigma(n) \to \mathbb{P}^1$, and so the assumptions of (4.14) are fulfilled. For a reducible $D$ the dual graph has at least two end vertices, and so there is an end vertex with nonnegative selfintersection.

From now on we assume that $\rho(V) \geq 3$. By the remark in (4.1), we can also suppose that $(V, D)$ is strictly minimal, i.e. there is no smooth rational curve $C$ with $C^2 = -1$ and $CD \leq 1$. First we will treat the case that $K(K + D) \leq 0$. We claim that in this case we can find an effective divisor $F = \sum F_i$, such that $FK < 0$, $\mid F + K + D \mid = \phi$. In fact, let $E \subseteq V$ be a smooth rational curve with $E^2 = 0$. By Lemma (4.13) we can find $m \in \mathbb{N}$ such that $\mid E + (m-1)(K+D) \mid \neq \phi$ and $\mid E + m(K + D) \mid = \phi$. Then $F \in \mid E + (m-1)(K + D) \mid$ has the properties above, since $EK = -2$, and so

$$FK = EK + (m-1)(K + D)K \leq -2.$$

Because of $\mid (F + D)_{red} + K \mid = \phi$, the divisor $(F + D)_{red}$ is a union of rational trees, see Lemma (4.12)(a). If $F_i \subseteq F$ is a component, then also $\mid F_i + K + D \mid = \phi$, and so by the connectedness of $D$ and Lemma (4.12)(b) $F_i D \leq 1$. Because of

$FK < 0$, there is a component $F_i$ with $F_iK < 0$, i.e. $F_i^2 \geq -1$. Since $(V, D)$ is strictly minimal, $F_i^2 \geq 0$. Thus condition (b) of (4.14) is fulfilled, and so we are done.

In the second case we assume that $K(K + D) > 0$. By the Riemann–Roch inequality we have

$$h^0(-D-K) + h^0(D+2K) \geq 1 + \frac{1}{2}(D+K)(D+2K) = (D+K)K > 0.$$

Since $D + 2K \leq 2(D + K)$ , we get $h^0(D + 2K) = 0$ and so $-D - K$ is effective. Let $F \in \mid -D - K \mid$ and consider a smooth rational $(-1)$–curve $E$ on $V$. Since $D$ is connected, $\mid E + D + K \mid = \phi$ would imply $ED \leq 1$ by Lemma (4.12)(b), contradicting the strict minimality of $(V, D)$. Hence $\mid E + D + K \mid \neq \phi$, and so we have $E = F + G$ for some $G \in \mid E + D + K \mid$. Since $\mid E \mid = \{E\}$ , this shows that $G = 0$ , and so $F = E$ is irreducible. Hence $F = E$ is the only smooth rational $(-1)$–curve on $V$.

Now let $C$ be a curve with $C^2 < 0$ and $C \not\subseteq D \cup E$. Then again $\mid C+D+K \mid \neq \phi$ would imply $C = E + G$ for some $G \in \mid C + D + K \mid$, contradicting $\mid C \mid = \{C\}$. Hence $\mid C+D+K \mid = \phi$, and so $C$ is a smooth rational curve and $C^2 \leq -2$, since $E$ was already shown to be the only $(-1)$–curve on $V$. Moreover, $C$ is disjoint from $E + D$ because of the inequalities

$$0 \geq -CE = C(K+D) = CD - 2 - C^2 \geq CD \geq 0.$$

Let $\pi : V \to \Sigma(n)$ be a blowing down to some Hirzebruch surface. Since $E$ is the only smooth rational $(-1)$–curve, and any other exceptional curve, not contained in $D \cup E$, is disjoint from $E + D$, all the exceptional set of $\pi$ must be connected and contained in $E + D$.

We have $ED = 2$ , since $E = -K - D$ and $D$ is a rational tree. If $D$ is irreducible, then $V$ is obtained from $\Sigma(n)$ by at most 2 blowing ups. If there is only one blowing up, then there must be a second $(-1)$ curve (the proper image of the fibre), which is a contradiction. If there are two blowing ups, then $E \cup D$ is the exceptional set for $\pi$ and so $ED = 1$, which is also impossible. Hence $D$ must be reducible.

If $B \subseteq D$ is an irreducible component, then $E + B + K = -(D - B)$ and so $\mid E + B + K \mid = \phi$. Thus $EB \leq 1$ by Lemma (4.12)(b). Moreover $v_D(B) = B(D - B)$, which is

$$-EB - BK - B^2 = -EB + 2 = \begin{cases} 2 & \text{iff } B \cap E = \phi, \\ 3 & \text{iff } E \subseteq D \text{ and } B = E, \\ 1 & \text{else.} \end{cases}$$

Hence, if $E \not\subseteq D$, then $D$ has a linear graph and $E$ intersects the end curves transversally. Thus $E$ and $D$ are either as in FIGURE 3 (1) or (2). If $E \subseteq D$, then $v_D(E) = 3$, $E$ meets every end curve of the graph of $D$, and so the only possible graph is as in FIGURE 4. Let $F_0 \subseteq \Sigma(n)$ be the fibre of the ruling $p : \Sigma(n) \to \mathbb{P}^1$ containing the point $x_0 = \pi(E)$, and let $\hat{F}_0 = \pi^{-1}(F_0)$. Then

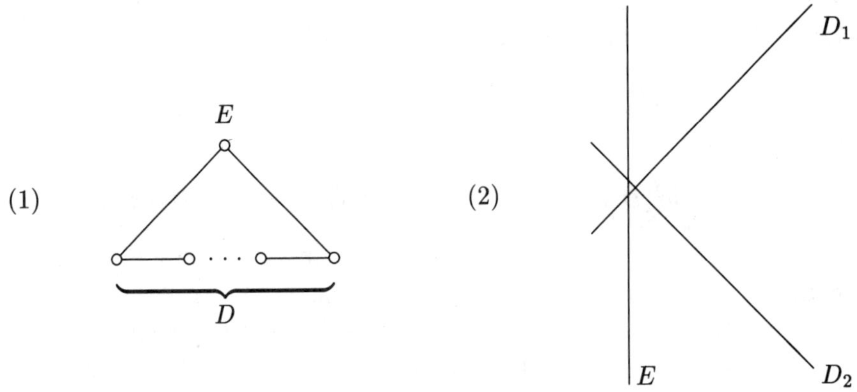


FIGURE 3.

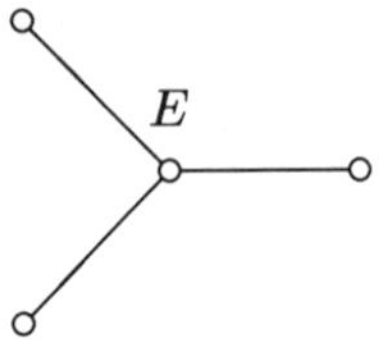


FIGURE 4.

$\hat{F}_0$ is reducible, consists of at least 3 components (otherwise there would be two $(-1)$–curves in $V$), and is a rational tree contained in $D \cup E$. Hence (2) cannot occur.

Let $F$ be the generic fibre of $\overline{p} : \Sigma(n) \to \mathbb{P}^1$. Then $F \simeq \hat{F} = \pi^{-1}(F)$ does not meet $E$, and so

$$\hat{F}D = \hat{F}(-E-K) = 2.$$

If $n \geq 1$ , then the proper transform $\hat{C}_\infty$ of the unique curve $C_\infty \subseteq \Sigma(n)$ of negative selfintersection is contained in $D$. For $n = 0$ the proper preimage $\hat{C}_\infty$ of the horizontal section $C_\infty$ of $\overline{p}$ containing $x_0$ also has negative selfintersection and so belongs to $D$. In any case, $D$ contains a section $\hat{C}_\infty$ of negative selfintersection, and because of $\hat{F}D = 2$ also must contain the proper transform $\hat{C}_0$ of another section $C_0$ of $\overline{p}$. The center of the second blowing up must be the intersection point of the exceptional curve of the first blowing up with the proper transform of $F_0$, since otherwise the resulting surface $V$ would contain at least two $(-1)$–curves. This means that $\hat{C}_0^2 \geq C_0^2 - 1$. For $n \geq 1$ this proves the existence of a component as in (4.14), since $C_0^2 \geq n \geq 1$. If $n = 0$, then either $x_0 \notin C_0$ , in which case $\hat{C}_0^2 = C_0^2 \geq 0$, or $x_0 \in C_0$ , in which case $\hat{C}_0^2 = C_0^2 - 1 \geq 2 - 1 = 1$. This completes the proof of (4.11). □

PROOF OF (4.10): Let $(V, D)$ be an SNC–completion of $X$. After blowing up $V$ in points of $D$, there is a diagram as in (4.11) (2). We may assume that $D$ contains no $(-1)$–curve, which is mapped onto a point on $\Sigma(n)$ under $\pi$. Let

$F_{b_1}, \dots F_{b_s}$ denote the singular fibers of $p$. Since for the generic fibre $F$ we have $FD = 1$, there must be a section $C \subseteq V$ of $p$, which is contained in $D$, and all other components of $D$ must be contained in fibers of $p$. Let $k$ be the number of components of the exceptional set of $\pi$ and $r$ the number of components of $D$. Since $X$ is pre–$\mathbb{Q}$–acyclic, we have $k + 2 \leq r$, and $D$ contains at least one full fibre $F_0$ of $p$ and possibly other nonsingular fibers, say $F_1, \dots, F_t$. The union of the singular fibers $F^\star = F_{b_1} \cup \dots \cup F_{b_s}$ has $k + s$ components, and in each $F_{b_i}$ there is at least one $(-1)$–curve. Hence there can be at most one $(-1)$–curve $E_i$ in each fibre $F_{b_i}$, and $D$ contains all other irreducible components of $F^\star$. So

$$D = F_0 + \dots + F_t + C + \sum_{j=1}^{s} ((F_{b_j})_{red} - E_j) .$$

This shows that $(V, D)$ is obtained from $(\Sigma(n),\ C \cup F_0 \cup \dots \cup F_t \cup \overline{F}_{b_1} \cup \dots \cup \overline{F}_{b_s})$ by comb–attachments in points $x_j \in \overline{F}_{b_j}$, where $\overline{F}\dots = \pi(F\dots)$. It is easily seen that we can also achieve $x_j \in \overline{F}_{b_j} \setminus C$, since otherwise we can perform elementary transformations of $\Sigma(n)$ at $x_j$. □

*Remark.* Performing elementary transformations at $F_0$, we can even get a representation as in (4.9) with $\Sigma(n) = \mathbb{P}^1 \times \mathbb{P}^1$, $C^2 = 0$.

As an application of (4.10) we will show that for certain $\mathbb{Q}$–acyclic surfaces $X$ at least the dual graph $\Gamma(D)$ of the boundary of minimal compactification $(V, D)$ is an invariant:

COROLLARY 4.15. *Let $(V, D)$ be as in (4.2), and assume that the curve $B$ from the remark after (4.9) is a branching point of the dual graph $\Gamma(D)$, i.e. $s \geq 2$. Let $(V', D')$ be another minimal SNC–completion of $X = V \setminus D$. Then the weighted dual graph $\Gamma(D)$ and $\Gamma(D')$ are equal up to a changing of the weight of $B$. Conversely, if $\Gamma$ is any (minimal) weighted graph, which is obtained from $\Gamma(D)$ by changing the weight of $B$, then $\Gamma = \Gamma(D')$ for some (minimal) SNC–completion $(V', D')$ of $X$. Moreover, the birational map $V - - \to V'$, induced by the identity on $X$, is biholomorphic outside $\hat{F}_0$, where $\hat{F}_0$ is as in the remark following* (4.9).

PROOF. The proof follows from (A.2) of the appendix to this section and the following simple observation: *If $\Gamma$ is the one–point weighted graph with weight 0 and $\Gamma'$ is minimal and it is obtained from $\Gamma$ by a sequence of blowing ups and downs, then* $\Gamma = \Gamma'$. This is easily seen by representing $\Gamma$ as the dual graph of a fibre of $\Sigma(0) \to \mathbb{P}^1$. Then $\Gamma'$ corresponds to the dual graph of a fibre of some modification $\Sigma' \to \mathbb{P}^1$ of $\Sigma(0) \to \mathbb{P}^1$. But such a fibre is irreducible iff $\Gamma'$ is minimal, see e.g. [**2**]. □

We remark that any minimal SNC–pair $(V', D')$ as in the corollary is obtained from the construction described in (4.9), as follows from the proof of (4.10).

*Example* 4.16. We will give an example of two SNC–pairs $(V, D)$ and $(V', D')$ such that $D \simeq D'$ with weights, but $X \not\simeq X'$, where $X = V \setminus D, X' = V' \setminus D'$ are $\mathbb{Q}$–acyclic surfaces of logarithmic Kodaira dimension $-\infty$. Moreover, $(V, D)$ and $(V', D')$ give the canonical representation of $X$, resp. $X'$ over the same Hirzebruch surface $\Sigma(n)$. Let $\Sigma = \Sigma(n)$, $n \geq 0$, be a fixed Hirzebruch surface and let $F_\infty, C_0, C_\infty$ be as in sect. 3. In the fibers $F_0, F_1$ of $\Sigma \xrightarrow{\pi} \mathbb{P}^1$ over points $0, 1 \in \mathbb{P}^1$ we perform the following blowing ups: Let $q_0 : \Sigma_0 \to \Sigma$ be the monoidal transformation in the two points $p_0 := F_0 \cap C_0$ and $p_1 := F_1 \cap C_0$. Blowing up again in the singular points of $q_0^{-1}(F_0)$ and $q_1^{-1}(F_1)$, we obtain a surface $q_1 : \Sigma_1 \to \Sigma_0$ with exceptional set $E_0 + E_1$ over 0 resp. 1. Now $\Sigma_1$ can be blown up in two points $x_0 \in E_0$, $x_1 \in E_1$, which are nonsingular on $q_1^{-1}q_0^{-1}(F_0)$ resp. $q_1^{-1}q_0^{-1}(F_1)$, to get a surface $V(x_0, x_1)$. Denote by $q : V(x_0, x_1) \to \Sigma$ the projection. The divisor $q^{-1}(F_i)$ can be decomposed $q^{-1}(F_i) = \hat{F}_i + \hat{E}_i + \hat{G}_i + H_i$, where $\hat{F}_i$ is the proper transform of the fibre $F_i$, $H_0 + H_1$ is exceptional for the last blowing up, and $\hat{E}_0 + \hat{E}_1$ resp. $\hat{G}_0 + \hat{G}_1$ is the proper transform of the exceptional locus for $q_1$ resp. $q_0$. Let $\hat{C}_\infty$ and $\hat{F}_\infty$ be the strict transform of $C_\infty$ and $F_\infty$. The divisor

$$D = \hat{F}_\infty + \hat{C}_\infty + \hat{F}_0 + \hat{F}_1 + \hat{E}_1 + \hat{G}_0 + \hat{G}_1$$

has dual graph as shown in FIGURE 5 which, as well as the complex structure

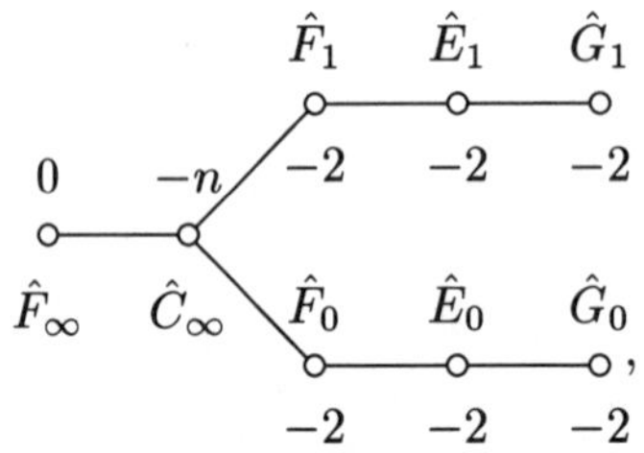


FIGURE 5.

of $D$, is independent of the position of the points $x_0, x_1$. We claim that among the $\mathbb{Q}$–acyclic surfaces $X(x_0, x_1) := V(x_0, x_1) \setminus D$ there is a 1–parameter family of pairwise nonisomorphic ones. Set $V = V(x_0, x_1)$, $V = V(x_0', x_1')$, $X := V \setminus D$, $X' = V' \setminus D'$, and assume that there is an isomorphism $X \xrightarrow{\gamma} X'$. We will show that in this case $x_0/x_0' = x_1/x_1'$ or $x_1/x_0' = x_0/x_1'$. This has an invariant sense, if we regard the intersection of $\hat{E}_i$ with $\hat{F}_i$ and $\hat{G}_i$ as the points $\infty$ and 0 respectively.

By (4.15) $\gamma$ can be extended to give an isomorphism $\hat{\gamma} : V \setminus \hat{F}\infty \to V \setminus \hat{F}_\infty$, which maps $q^{-1}(F_0 \cup F_1)$ into $q^{-1}(F_0 \cup F_1)$ and $\hat{C}_\infty$ into $\hat{C}_\infty$. Blowing down the surfaces, we get an automorphism $\beta : \Sigma \setminus F_\infty \to \Sigma \setminus F_\infty$ mapping $F_0 \cup F_1$ resp. $C_\infty$ into itself. After interchanging the positions of the points $x_0, x_1$, we may assume that $\beta(F_0) = F_0$, $\beta(F_1) = F_1$, and so $\beta \mid C_\infty$, having three fixed points, is the identity. As in (2.5) (1), $\beta$ is the map

$$(x, y) \overset{\beta}{\mapsto} (x, \lambda y + P(x))$$

for some polynomial $P \in \mathbb{C}[X]$, where we have used the affine chart $(x, y)$ as described in the discussion preceding (2.5). With respect to the coordinates $(x', y')$ with $x = x'y'$, $y' = y$ of the blowing up at (0,0), we get

$$\beta(x', y') = \left( \frac{x'y'}{\lambda y' + P(x'y')}, \ \lambda y' + P(xy') \right),$$

which is holomorphic on the blown up surface only if $P(x) = x\tilde{P}(x)$, in which case $\beta$ can be rewritten as $(x := x', \ y := y')$

$$\beta(x, y) = \left( \frac{x}{\lambda y + \tilde{P}(xy)x}, \ \lambda y + P(xy) \right).$$

Blowing up again in (0,0) we get, with respect to new coordinates $x' = \frac{x}{y}$, $y' = y$, that

$$\beta(x', y') = \left( \frac{x'}{(\lambda + \tilde{P}(x'y'^2)x'y')^2}, \ \lambda y' + P(x'y'^2) \right).$$

On the curve $E_0$, which is just given by $y' = 0$, the induced automorphism is $\beta(x') = \frac{x'}{\lambda^2}$. In order to obtain an isomorphism $\hat{\beta} = \hat{\gamma} : V \setminus F_\infty \to V' \setminus F_\infty$ as above, the morphism $\beta$ must map $x_0$ onto $x_0'$. Observe that $x_0, x_0' \notin \{0, \infty\}$. Similarly, by the same calculation for the blowing up over $F_1$, $\beta$ must also map $x_1$ onto $x_1'$. But this is only possible if $x_0/\lambda^2 = x_0'$, $x_1/\lambda^2 = x_1'$, or $x_0/x_0' = x_1/x_1'$.

*Remark* 4.17. It will be shown in sect. 6 that the basis of the versal logarithmic deformation of a pair $(V, D)$ is smooth, if $\overline{k}(V \setminus D) = -\infty$.

## Appendix A. A graph–theoretic lemma

In this appendix, we consider graphs $\Gamma$ with weights in $\mathbb{Z} \cup \{-\infty\}$. For such a graph we have the blowing up and blowing down process as usual. A blowing up of a vertex will be called *outer* and a blowing up of an edge will be called *inner* blowing up. If $\Gamma'$ is obtained from $\Gamma$ by a sequence of blowing ups, then we say that $\Gamma'$ *dominates* $\Gamma$ and write shortly $\Gamma' \to \Gamma$. If $v \in \Gamma$ is a vertex, then $v$ can be considered as a vertex in any $\Gamma'$ dominating $\Gamma$, so we will consider the set of vertices of $\Gamma$ as a subset of the set of vertices of $\Gamma'$. In analogy with the case of modifications of surfaces, we will say that $w \in \Gamma'$ is *contracted* (in $\Gamma$), if $w \notin \Gamma$.

If $\Gamma' \to \Gamma$ is a simple blowing up, then the new created $(-1)$–vertex $e$ is also called the *exceptional* vertex. If $\Delta \subseteq \Gamma$ is a subset of the set of vertices of $\Gamma$, then we set $\tilde{\Delta} := \Delta \cup \{e\}$, if a vertex of $\Delta$, or an edge incident with $\Delta$, is blown up, and $\tilde{\Delta} := \Delta$ otherwise. Iterating this we can define $\tilde{\Delta}$ for any sequence of blowing ups $\Gamma' \to \Gamma$. We will call $\tilde{\Delta}$ the *total transform* of $\Delta$ in $\Gamma'$. We say that two graphs $\Gamma, \Gamma'$ are *birationally equivalent* ($\Gamma \sim \Gamma'$ for short), if $\Gamma$ is obtained from $\Gamma'$ by a sequence of blowing ups and downs. We call $\Gamma$ *minimal*, if there is no possible blowing down in $\Gamma$; we call it *absolutely minimal*, if any graph birationally equivalent to $\Gamma$ dominates $\Gamma$, or, in other words, if any minimal graph, which is birationally equivalent to $\Gamma$, is isomorphic to $\Gamma$. For a connected graph $\Gamma$ and a vertex $v$, the connected components of $\Gamma \setminus \{v\}$ will be called *the branches of* $\Gamma$ *at* $v$. *A branching point* $v$ is a vertex with at least three branches.

Let $B = B(\Gamma)$ be the set of all branching points and vertices of weight $-\infty$. The connected components of $\Gamma \setminus B$ will be called *chains.* A chain will be called an *outer chain* if it is the branch at some point $b \in B$ and otherwise an *inner chain.*

*Remark* A.1. (1) If $\Gamma_1$ and $\Gamma_1$ are simple blowing ups of $\Gamma$, then $\Gamma_1, \Gamma_2$ can be dominated by a third one:

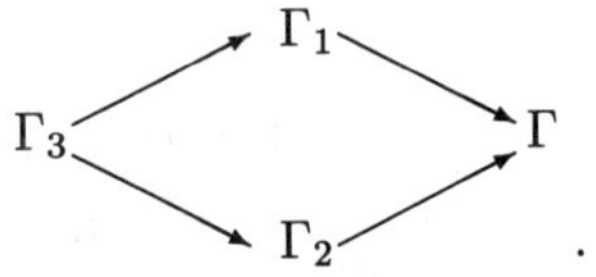

Using this simple observation it is easily seen that, if $\Gamma$ and $\Gamma'$ are birationally equivalent, then there exists a diagram

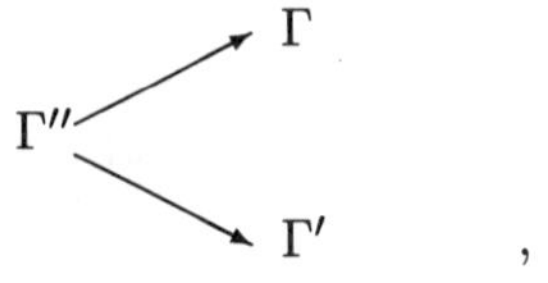

i.e. $\Gamma$ and $\Gamma'$ are dominated by some $\Gamma''$.

(2) A graph is called *contractible*, if it dominates the one point graph with the weight $-1$. If $\Gamma'$ is birationally equivalent to a contractible graph $\Gamma$, then $\Gamma'$ is well known to be contractible, too (indeed, both of them are realizable as dual graphs of exceptional curves of the first type in smooth surfaces).

A chain is called *admissible* if the weights of all its vertices are $\leq -2$. For any weighted graph, we consider the following three subsets of the set of vertices:

$\Gamma$ (1) is the union of $B = B(\Gamma)$ and all vertices of admissible chains.

$\Gamma$ (2) is the set of all vertices of inner non admissible chains.

$\Gamma$ (3) is the set of all vertices of outer non admissible chains.

Then we will show:

PROPOSITION A.2. *Let $\Gamma, \Gamma'$ be two weighted minimal graphs which are birationally equivalent. Then $\Gamma'$ can be obtained from $\Gamma$ by a sequence of blowing ups and downs, which only take place in the total transforms of $\Gamma(2) \cup \Gamma(3)$, and where the outer blowing ups and downs are only performed over $\Gamma(3)$. More precisely, there is a graph $\Gamma_0$, dominating both $\Gamma$ and $\Gamma'$, such that*

$$\Gamma(1) = \Gamma'(1),$$
$$\tilde{\Gamma}(2) = \tilde{\Gamma}'(2),$$
$$\tilde{\Gamma}(3) = \tilde{\Gamma}'(3),$$

*considered as subsets of $\Gamma_0$, and every branching point of $\Gamma_0$, which is not in $B(\Gamma)$, is contained in $\tilde{\Gamma}(3) = \tilde{\Gamma}'(3)$.*

PROOF. By (A.1), (1) there is a graph $\Gamma_0$ dominating both $\Gamma$ and $\Gamma'$. Blowing down successively those at most linear $(-1)$–vertices of $\Gamma_0$, which are not in $\Gamma \cup \Gamma'$, we can assume:

($\star$) $\Gamma_0 \setminus (\Gamma \cup \Gamma')$ contains no $(-1)$–vertex, which is at most linear in $\Gamma_0$. We will show that then

(1) $B(\Gamma) = B(\Gamma_0) \cap \Gamma = B(\Gamma_0) \cap \Gamma' = B(\Gamma')$,

(2) $B(\Gamma_0) \setminus B(\Gamma) \subseteq \tilde{\Gamma}(3)$,

(3) $\Gamma(1) = \Gamma'(1)$, $\tilde{\Gamma}(2) = \tilde{\Gamma}'(2)$, $\tilde{\Gamma}(3) = \tilde{\Gamma}'(3)$.

Obviously, (1) – (3) imply the proposition.

For the proof of (1), let $w \in B(\Gamma_0) \cap \Gamma'$, and assume that $w \notin B(\Gamma)$. Let $\Delta_1, \ldots, \Delta_r$ be the branches of $w$ in $\Gamma_0$. Since $w \notin B(\Gamma)$, at least $(r-2)$ of these branches, say $\Delta_3, \ldots, \Delta_r$ are blown down in $\Gamma$. So, each $\Delta_i$ is contractible and contains at least one at most linear $(-1)$–vertex $e_i, i = 3, \ldots, r$. Since $w$ is not contracted by blowing down to $\Gamma'$, the necessary blowing downs can be performed separately in the branches $\Delta_1, \ldots, \Delta_r$. By the minimality of $\Gamma'$, the vertices $e_i \in \Delta_i$ are blown down in $\Gamma'$ as well as in $\Gamma$, contradicting ($\star$). This proves that $B(\Gamma) \supseteq B(\Gamma_0) \cap \Gamma'$. By symmetry, $B(\Gamma') \supseteq B(\Gamma_0) \cap \Gamma$, and since trivially $B(\Gamma), B(\Gamma') \subseteq B(\Gamma_0)$, we get

$$B(\Gamma_0) \cap \Gamma \supseteq B(\Gamma) \supseteq B(\Gamma_0) \cap \Gamma' \supseteq B(\Gamma') \supseteq B(\Gamma_0) \cap \Gamma,$$

showing (1).

For the proof of (2), let $w \in B(\Gamma_0) \setminus B(\Gamma)$. By (1), $w$ is contracted in $\Gamma$ as well as in $\Gamma'$. Let $\Delta_1, \ldots \Delta_r$ denote again the branches of $\Gamma_0$ at $w$, where as above $\Delta_3, \ldots \Delta_r$ are blown down in $\Gamma$, i.e. $\Delta_3, \ldots \Delta_r$ are contractible. By symmetry also $(r-2)$ of the branches are blown down in $\Gamma'$. Because of ($\star$) there is no branch contracted both in $\Gamma$ and $\Gamma'$. Hence $r \leq 4$ and at least one of the branches $\Delta_1, \Delta_2$, say $\Delta_2$, is blown down to $\Gamma'$. Again by ($\star$), every $(-1)$–vertex of the contractible graph $\Delta_2$ (resp. $\Delta_3$) is not contracted in $\Gamma$ (resp. $\Gamma'$), and because of the minimality of $\Gamma$ and $\Gamma'$ it must have a weight $\geq 0$ in $\Gamma$ (resp. $\Gamma'$). On the other hand, $\Delta_2$ being a tree, also the blowing down $\overline{\Delta}_2$ of $\Delta_2$ (i.e. $\Delta_2 \cap \Gamma$) in $\Gamma$

is a tree, which cannot contain a point of $B(\Gamma) = B(\Gamma')$, since $\Delta_2 \cap B(\Gamma') = \phi$. Hence $\overline{\Delta}_2$, being a branch at some point of $\Gamma$, is contained in some outer chain, which is not admissible, since it contains a vertex of weight $\geq 0$. Thus $\overline{\Delta}_2 \subset \Gamma(3)$ and so $\Delta_2 \subset \tilde{\Gamma}(3)$. It follows, that also $w \in \tilde{\Gamma}(3)$. This completes the proof of (2).

In the next step we show that there is a one to one correspondence between the chains of $\Gamma$ and $\Gamma'$. For this it is sufficient to show that the map $\Lambda \overset{\gamma}{\mapsto} \tilde{\Lambda}$ gives a 1–1–correspondence between the connected components of $\Gamma \setminus B(\Gamma)$ and $\Gamma_0 \setminus B(\Gamma)$, since we can then associate to $\Lambda$ the chain $\Lambda'$ in $\Gamma'$, if $\tilde{\Lambda}' = \tilde{\Lambda}$ in $\Gamma_0$. Clearly, the map $\Lambda \overset{\gamma}{\mapsto} \tilde{\Lambda}$ is injective. If it would not be surjective, then there would be a connected component $\Delta$ of $\Gamma_0 \setminus B(\Gamma)$, which is completely in $\Gamma_0 \setminus \Gamma$ and so contains an exceptional $(-1)$–vertex $e_0$. But blowing down to $\Gamma'$, all blowing downs can be performed separately in the connected components of $\Gamma_0 \setminus B(\Gamma)$, since $B(\Gamma) = B(\Gamma')$. Hence, if no vertex of $\Delta$ is blown in $\Gamma'$, all weights of $\Delta$ are the same in $\Gamma'$ and $\Gamma_0$, contradicting the minimality of $\Gamma'$. If a vertex of $\Delta$ is blown down in $\Gamma'$, then the first in the sequence of blowing down from $\Gamma_0$ to $\Gamma'$ is a $(-1)$–vertex blown down both in $\Gamma$ and $\Gamma'$, contradicting $(\star)$. Hence it follows that $\Lambda \overset{\gamma}{\to} \tilde{\Lambda}$ is bijective.

Let $\Lambda$ be a chain in $\Gamma$ and $\Lambda'$ the associated chain in $\Gamma'$. In order to show (3) it remains to prove that

(a) $\Lambda$ is an inner chain iff $\Lambda'$ is inner. Moreover, in this case $\tilde{\Lambda}$ is a chain of $\Gamma_0$,

(b) $\Lambda$ is admissible iff $\Lambda'$ is admissible, and in this case $\Lambda = \Lambda' = \tilde{\Lambda} = \tilde{\Lambda}'$.

For (a) we remark, that if $\Lambda$ is inner, then by (2) $\tilde{\Lambda} = \tilde{\Lambda}'$ has no branching point, and so it must be a chain of $\Gamma_0$. Clearly, then also $\Lambda'$ is inner in $\Gamma'$, since $B(\Gamma) = B(\Gamma')$.

In order to prove (b), observe that an at most linear $(-1)$–vertex $v$ of $\tilde{\Lambda}$ has to be contracted in $\Lambda$, since otherwise $v$ would have weight $\geq -1$ in $\Gamma$. Hence $\Lambda' = \tilde{\Lambda}$, because otherwise $\tilde{\Lambda}$ would contain a $(-1)$ vertex $v$, which would be blown down both in $\Gamma$ and $\Gamma'$. By the minimality of $\Gamma', \tilde{\Lambda} = \Lambda'$ contains no $(-1)$ vertex, and so $\tilde{\Lambda} = \Lambda' = \Lambda$. □

We give two simple geometric applications.

COROLLARY A.3. (compare **[15]** Lemma 2) *Let $(V, D)$ be an SNC pair. Assume that for every rational component $E$ of $D$ with $E^2 \geq -1$ we have $v_D(E) := (D - E)E \geq 3$. Then $(V, D)$ is absolutely minimal.*

PROOF. Let $\Gamma = \Gamma(D)$ be the weighted dual graph of $D$, where we assign to a component $E$ the weight $E^2$ if it is rational, and $-\infty$ otherwise. Then with the notations above, $\Gamma = \Gamma(1)$, and so by (A.2) every minimal graph, birationally equivalent to $\Gamma$, is $\Gamma$ itself. This easily implies (A.3). □

COROLLARY A.4. *Let $(V, D)$ be an SNC–pair. Assume that the dual weighted graph $\Gamma = \Gamma(D)$ of $D$ has only admissible outer chains (where as above we assign to a curve of genus $\geq 1$ the weight $-\infty$ in $\Gamma$). Then every birational transformation of $(V, D)$ can be realized by performing only inner modifications of* $\Gamma$.

PROOF. This follows immediately from (A.2), since $\Gamma(3) = \phi$ in this case. □

*Remark* A.5. The assumptions in (A.4) are satisfied if $\bar{k}(X) \geq 0$, where $X = V \setminus D$, see (4.14).

*Remark* A.6. Assume that $B(\Gamma) \neq \phi$, and that $\Lambda$ is a nonadmissible outer chain. Let $\Gamma_0$ and $\Gamma'$ be as in (A.2), and assume that $(\star)$ in the proof of (A.2) is satisfied. Then $\tilde{\Lambda}$ contains at most one branching point of $\Gamma_0$. If there is such a branching point $w \in B(\Gamma_0) \cap \tilde{\Lambda}$, then $\Gamma_0$ has 3 branches $\Delta_1, \Delta_2, \Delta_3$ at $w$, and two of them, say $\Delta_1, \Delta_2$, are linear, contractible and contain at most one $(-1)$–vertex, see FIGURE 6.

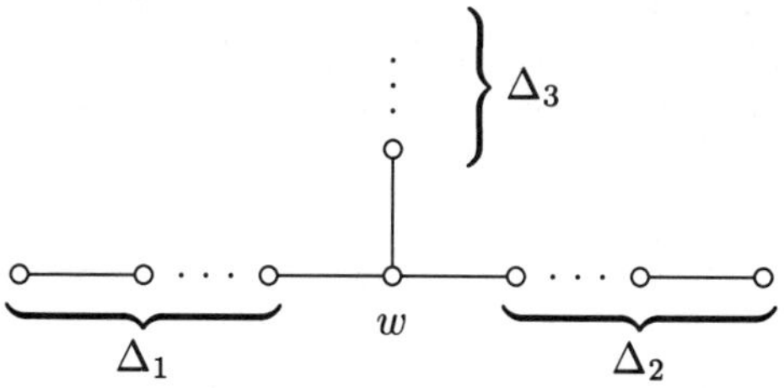


FIGURE 6.

PROOF. Let $w \in B(\Gamma_0) \cap \tilde{\Lambda}$ and $\Delta_1, \ldots, \Delta_r$ be the branches of $\Gamma_0$ at $w$. Let $\Delta_3$ be the branch containing $B(\Gamma)$. As is shown in the proof of (A.2), $(r-2)$ of the $\Delta_i$ are contractible and contracted in $\Gamma$, and also $(r-2)$ of the $\Delta_i$ are contractible and contracted in $\Gamma'$, but by $(\star)$ none of the $\Delta_i$ is contracted in both $\Gamma$ and $\Gamma'$. Since $\Delta_3$ clearly is not contractible, this shows that $r = 3, \Delta_1$ and $\Delta_2$ are contractible and one of them, say $\Delta_1$, is contracted in $\Gamma$.

Next we show that the $\Delta_i$ are linear and contain at most one $(-1)$–vertex. In fact, let $v \in \Delta_1$ be a further branching point with branches $\Delta'_1, \Delta'_2, \Delta'_3$ in $\Gamma_0$, and assume that $w \in \Delta'_3$. By what was shown above, $\Delta'_1$ and $\Delta'_2$ are contractible, and one of them, say $\Delta'_1$, is contracted in $\Gamma'$. But $\Delta'_1$ contains an at most linear $(-1)$–vertex $e'$, which is contracted in $\Gamma'$ and also in $\Gamma$, since $\Delta'_1 \subseteq \Delta_1$. This is a contradiction to ($\star$), and so the branches $\Delta_i$ are linear. Every $(-1)$–vertex of $\Delta_1$ resp. $\Delta_2$ is by ($\star$) not contracted in $\Gamma'$ resp. $\Gamma$, and so must get a weight $\geq 0$ there. Let $v$ be the $(-1)$–vertex of $\Delta_1$ nearest to $w$, and denote by $\Delta^\star$ the branch of $\Gamma_0$ at $v$ not containing $w$ (possibly $\Delta^\star = \phi$), see FIGURE 7. Then

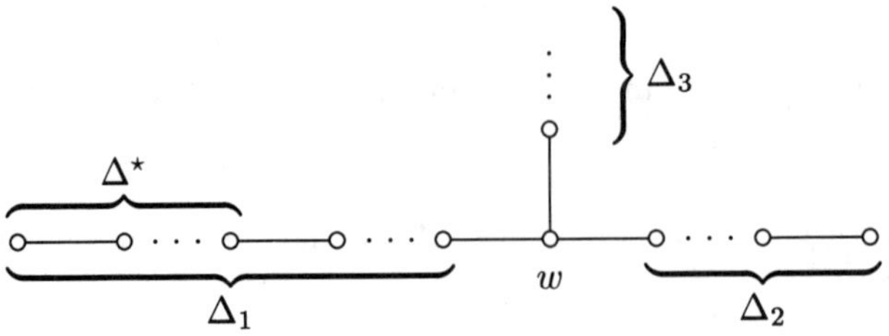


FIGURE 7.

blowing down to $\Gamma'$, the blowing down in $\Delta^\star$ can be performed separately. Hence if some vertex in $\Delta^\star$ would be blown down to $\Gamma'$, then there would be a vertex in $\Delta^\star$ of weight $(-1)$ in $\Gamma_0$, contradicting ($\star$). Hence $\Delta^\star$ contains no further $(-1)$–vertex, i.e. $v$ is the unique $(-1)$ vertex in $\Delta_1$.

Now it follows easily that there is at most one branching point of $\Gamma_0$ in $\tilde{\Lambda}$. Indeed, let $b \in B(\Gamma)$ be the vertex incident with $\Delta_3$. Applying the considerations above to the branching point $w \in B(\Gamma_0) \cap \tilde{\Lambda}$ nearest to $b$, it follows that $w$ is unique. $\square$

## 5. $\mathbb{Q}$–acyclic surfaces of logarithmic Kodaira dimension 1

Let $(V, D)$ be a minimal SNC–pair such that $\bar{k}(X) = 1$, where as usual $X = V \setminus D$. We recall

THEOREM 5.1. ([**19**]) *Let $K_V + D = H + N$ be the Zariski decomposition of $K_V + D$. Then for $t$ sufficiently large the linear system $\mid tH \mid$ is base point free and defines a fiber space structure $p : V \to C$ over a curve $C$, such that the generic fibre $F$ is either elliptic with $FD = 0$, or $F$ is rational with $FD = 2$.*

Assume from now on that $X$ is also pre–$\mathbb{Q}$–acyclic. Then $X$ is affine and so contains no complete curves, see [**13**] (2.4). So in this case $F$ is rational. Moreover, since $q(V) = 0$ we also have $q(C) = 0$ and hence $C = \mathbb{P}^1$ is rational, too. In particular, $V$ is a rational surface and so dominates some Hirzebruch surface.

5.2. In the following, we consider triples $(V, D, p)$, where $(V, D)$ is an SNC–pair, $p : V \to \mathbb{P}^1$ is a $\mathbb{P}^1$–fibration, such that the generic fibre meets $D$ in 2 points. We will call $(V, D, p)$ shortly a *$\mathbb{C}^\star$–triple.*

A curve $C \subseteq V$ is called *vertical*, if $\dim p(C) = 0$, and *horizontal*, if $p(C)$ is 1–dimensional. We will call $(V, D, p)$ of *twisted type*, if there is only one irreducible component $C$ of $D$, which is horizontal. Obviously, $CF = 2$ for a fibre $F$, i.e. $p : C \to \mathbb{P}^1$ is $2:1$. We call $(V, D, p)$ of *untwisted type*, if there are 2 horizontal components in $D$ (also called of *sandwich–type* in [**13**]).

LEMMA 5.3. *Let $(V, D, p)$ be a $\mathbb{C}^\star$–triple, and denote by $D^h \subseteq D$ the sum of horizontal components. Then there is a diagram*

$$\begin{array}{ccc} V & \xrightarrow{\ \pi\ } & \Sigma(n) \\ & {\scriptstyle p}\searrow \qquad \swarrow{\scriptstyle \overline{p}} & \\ & \mathbb{P}^1 & \end{array}$$

*such that $D^h \xrightarrow{\pi} \overline{D}^h := \pi(D^h)$ is an isomorphism.*

PROOF. Clearly, there exists a diagram

$$\begin{array}{ccc} V & \xrightarrow{\ \pi\ } & \Sigma(n) \\ & {\scriptstyle p}\searrow \qquad \swarrow{\scriptstyle \overline{p}} & \\ & \mathbb{P}^1 & \end{array}$$

We will show that, after replacing $\pi$ possibly by the contraction onto a new Hirzebruch surface, we can achieve that $D^h \xrightarrow{\pi} \overline{D}^h$ is an isomorphism. Since $D^h \to \overline{D}^h$ is birational, this map is an isomorphism over every regular point of $\overline{D}^h$. First assume that $z \in \overline{D}^h$ is singular, but $D^h$ is regular in $\pi^{-1}(z)$. Let $\sigma : \hat{\Sigma} \to \Sigma(n)$ be the blowing up of $\Sigma(n)$ in $z$. Assume that $p(z) = x$, denote the proper transform of $\overline{F} = \overline{p}^{-1}(x)$ resp. $\overline{D}^h$ on $\hat{\Sigma}$ by $\hat{F}$ resp. $\hat{D}^h$, and let $E$ denote the exceptional curve for $\sigma$. Since $D^h$ is smooth over $z$, the morphism $\pi$ must factor through $\sigma$. Moreover, the multiplicity of $\overline{D}^h$ in $z$ must be 2 because of $\overline{D}^h\overline{F} = 2$. Hence $\sigma^\star\left(\overline{D}^h\right) = \hat{D}^h + 2E$, and so $2 = \sigma_\star\left(\hat{F}\right)\overline{D}^h = \hat{F} \cdot \sigma^\star\left(\overline{D}^h\right) = \hat{F}\hat{D}^h + 2$, i.e. $\hat{F}\hat{D}^h = 0$. This allows us to contract the curve $\hat{F}$, and on the new Hirzebruch surface the image of $D^h$ is isomorphic to $\hat{D}^h$. After a finite number of steps we obtain that $\overline{D}^h$ is smooth in $z$.

Finally assume that $z \in \overline{D}^h$ is singular, that also $D^h$ is singular in $\pi^{-1}(z)$ and that $\pi : D^h \to \overline{D}^h$ is not an isomorphism. This is only possible in the case that $D^h$ consists of two horizontal components $C_0, C_1$, which meet transversally in a point $z^\star$ over $z$, whereas $\overline{C}_0 = \pi(C_0)$ and $\overline{C}_1 = \pi(C_1)$ have a contact of order

$\geq 2$. With the same argument as above, it is possible to perform elementary transformations until also $\overline{C}_0$ and $\overline{C}_1$ meet transversally. This finishes the proof of (5.3). □

In order to give a fairly explicit description of those strictly minimal $\mathbb{C}^*$–triples, which are pre–$\mathbb{Q}$–acyclic, we need some notation.

DEFINITION 5.4. Let $(V, D, p)$ be a $\mathbb{C}^*$–triple. Assume that $x \in \mathbb{P}^1$ and set $F = p^{-1}(x)_{red}$, $F' := F \cap D$. We say that $F'$ has *broken chain type*, if

(1) $F$ has a linear dual graph,
(2) $F$ contains a unique $(-1)$–curve say $G$, and $F = F' + G$,
(3) $F'$ has two connected components $F_1'$ and $F_2'$, each of which meets $D - F'$ in one point, and this intersection point lies on an irreducible component of $F_i'$, which is an end component of $F_i'$.

This implies that the dual graph of $D + G$ looks as follows:

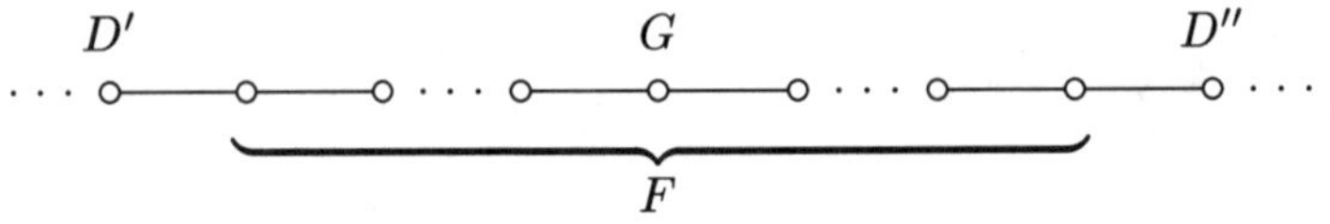

,

where $D', D''$ are (not necessarily distinct) irreducible components of $D$.

*Remark.* If $D^h \subseteq \Sigma(n)$ is a horizontal divisor, such that $D^h$ is smooth over a given point $x \in \mathbb{P}^1$ and intersects $\overline{F} = \overline{p}^{-1}(x)$ in two points, it is easy to construct such broken chain type fibre over $x$. Indeed, let $V$ be the surface obtained from $\Sigma(n)$ by a sequence of blowing ups, where at the first step we blow up a point of $\overline{F} \cap D^h$, at the second step the intersection point of the resulting two $(-1)$–curves in the fibre, and in each further step a point of the preceding $(-1)$–curve, which is an intersection point in the full fibre. Then finally we have only one $(-1)$–curve, and, by deleting this curve from the full fibre, arrive at a fibre of broken chain type.

LEMMA 5.5. *Let $(V, D, p)$ be a strictly minimal $\mathbb{C}^*$–triple, where $D$ is connected. Let $F = p^{-1}(x)$ be a singular fibre, such that $F \cap D^h$ consists of two different points, where $D^h$ is the sum of horizontal components of $D$. Then $F' = F \cap D$ has broken chain type.*

PROOF. Using (5.3) and its proof, we find a sequence of blowing downs of $(-1)$–curves in the fibre $F$:

$$V = V_n \stackrel{\sigma_n}{\to} V_{n-1} \stackrel{\sigma_{n-1}}{\to} \dots \stackrel{\sigma_1}{\to} V_0$$

such that, if $\pi_k := \sigma_1 \circ \dots \circ \sigma_k$, then

(1) $D^h \to \pi_n\left(D^h\right)$ is an isomorphism, and

(2) $\pi_n(F)$ is irreducible.

Let $p_k : V_k \to \mathbb{P}^1$ denote the morphism induced by $p$, and set $F_{(k)} := p_k^{-1}(x)$, $D_{(k)} := (\sigma_{k+1} \circ \ldots \circ \sigma_n)(D)$. Since $D^h$ and $F$ meet in two different points, by (1) the same is true for $\pi_n\left(D^h\right)$ and $F_{(0)}$, i.e. $\pi_n\left(D^h\right) \cup F_{(0)}$ is a divisor with normal crossings. Assume first that at the step $k$, a smooth point $z$ of $D_{(k)} \cup F_{(k)}$ is blown up. Then the preimage of $z$ in $V$ is a rational tree $T$ with a $(-1)$ curve, say $E$, meeting at most two other curves of $D \cup F$. Since $(V, D)$ is strictly minimal, $E$ is not contained in $D$ and intersects $D$ in two points. From the connectedness of $D$ we obtain that $D+E$ would contain a cycle. But such a cycle would lie in $T$, since $T$ and the proper transform of $D_{(k)}$ meet in just one point. Thus we get a contradiction.

Hence in every step, only singular points of $D_{(k)} \cup F_{(k)}$ are blown up. From this it follows easily by induction that the dual graph of $D^h \cup F$ is a chain

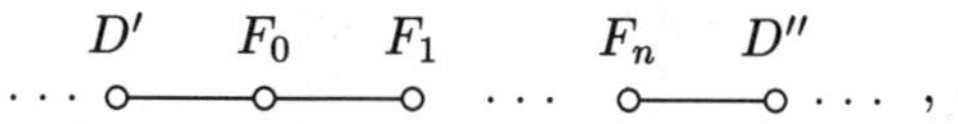


where $D', D''$ are the (not necessarily distinct) components of $D^h$. Since $D$ is connected, the components of $F$, not contained in $D$, must form a linear chain $F_j \cup F_{j+1} \ldots \cup F_k$. But because of the strict minimality, every $(-1)$–curve must be contained in this segment, and the segment can only have length 1. Hence, there is a unique $(-1)$–curve $G = F_j$ in $F$, and this curve is the only curve of $F$ which is not in $D$. If $j = 0$ or $j = n$, then blowing down successively it is easily seen that also the other end component of $F$ is a $(-1)$–curve. Hence $j \neq 0, n$, and so $F' = F \cap D$ has broken chain type. □

LEMMA 5.6. *Let $(V, D, p)$ be a strictly minimal $\mathbb{C}^*$–triple which is of twisted type, where $D$ is connected. Denote by $C$ the horizontal curve, and assume that the ramification points of $C \xrightarrow{p} \mathbb{P}^1$ are over $0, \infty \in \mathbb{P}^1$. Assume that $F_\infty = p^{-1}(\infty)$ is a singular fibre. Then $F_\infty \subseteq D$, and $F_\infty \cup C$ has dual graph as in* FIGURE *8.*

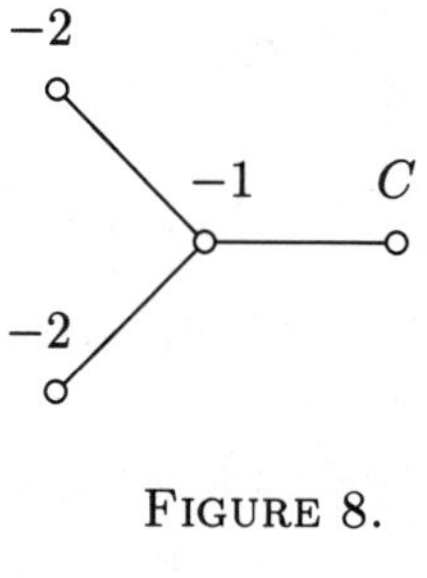


FIGURE 8.

PROOF. Using again (5.3) and its proof, we find a sequence of blowing downs of $(-1)$–curves in the fibre $F = F_\infty$

$$V = V_n \overset{\sigma_n}{\to} V_{n-1} \to \ldots \overset{\sigma_1}{\to} V_0$$

such that, with $\pi_k := \sigma_1 \circ \ldots \circ \sigma_k$,

(1) $\overline{C} = \pi_n(C)$ is smooth,
(2) $\pi_n(F)$ is irreducible, and
(3) $\sigma_1 \circ \sigma_2 : V_2 \to V_0$ is the minimal resolution of the intersection point of $\pi_n(F)$ and $\overline{C}$, which has dual graph as shown in FIGURE 9, where $C_{(k)} = (\sigma_{k+1} \circ \ldots \circ \sigma_n)(C)$, $F_{(k)} := (\sigma_{k+1} \circ \ldots \circ \sigma_n)(F)$.

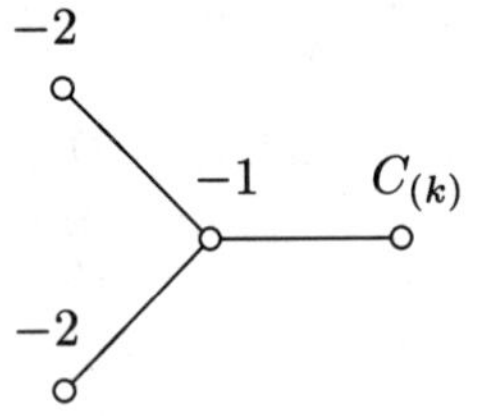

,

FIGURE 9.

With the same argument as in the proof of (5.5) we see, that in each step $k \leq 2$, only intersection points of curves in $F_{(k)} \cup C_{(k)}$ can be blown up. Hence, if there is a further blowing up, then $F$ contains a $(-1)$–curve $G$, which meets at most two other components of $F \cup C$, as is easily seen by induction. By the strict minimality of $(V, D)$ we have $G \not\subseteq D$, and $G$ meets two components of $D$, which must be contained in $F \cup C$. But then $G$ would disconnect the tree $F \cup C$, and so in particular $D$ would be disconnected, contradicting our assumption. Hence $n = 2$ and $F_\infty \cup C$ has the desired form. □

THEOREM 5.7. *Let $(V, D, p)$ be a strictly minimal $\mathbb{C}^*$–triple, which is pre–$\mathbb{Q}$–acyclic. Assume that $(V, D, p)$ is of twisted type. Then there is a diagram*

$$(\star) \qquad \begin{array}{ccc} V & \xrightarrow{\ \pi\ } & \Sigma(n) \\ & {\scriptstyle p}\searrow \quad \swarrow{\scriptstyle \bar{p}} & \\ & \mathbb{P}^1 & \end{array},$$

*and $(V, D, p)$ is one of the following types* A1 *or* A2*:*

**A1** $D = C + F'_1 + \ldots + F'_s + F_\infty$, *where*

(1) *$C$ is horizontal, $p : C \to \mathbb{P}^1$ is 2–sheeted and has $0, \infty \in \mathbb{P}^1$ as ramification points,*
(2) *$F'_1, \ldots F'_s$ are of broken chain type over points $x_1, \ldots x_s \notin \{0, \infty\}$,*
(3) *$\overline{C} = \pi(C)$ is smooth,*
(4) *$F_\infty = p^{-1}(\infty)$ is a full fibre, and $F_\infty \cup C$ has the dual graph as in* (5.6).

**A2** *$D = C + F_1' + \ldots + F_s' + F_0 + F_\infty$, where $C, F_1', \ldots, F_s', F_\infty$ are as in* A1, *and where for $F_0 = p^{-1}(0)$ also* (4) *holds.*

It follows that in the case of A2 the divisor D and its dual graph are as shown in FIGURE 10 resp. FIGURE 11. In the case A1 the diagrams are similar with the evident changes.

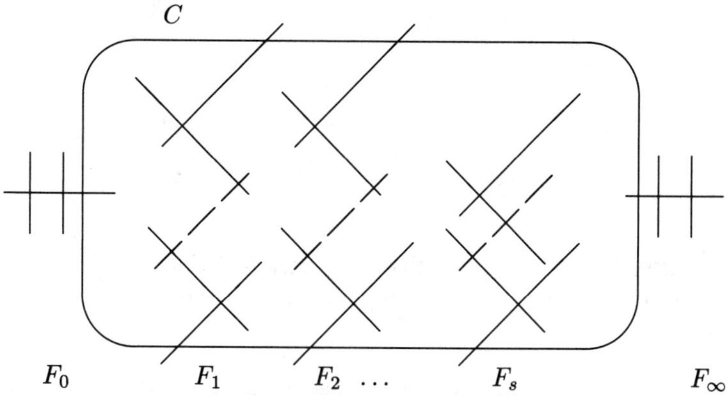


FIGURE 10. The divisor D in case A2.

PROOF. If $C$ is the horizontal component of $D$, then $C \to \mathbb{P}^1$ is a $2:1$ covering and so has two branching points, which we may assume to lie over $0, \infty \in \mathbb{P}^1$. By (5.3) there is a diagram ($\star$) such that (3) holds. Since $D$ is an SNC–divisor without cycles, it cannot contain a nonsingular fibre of $p$. From (5.5) we obtain that for all singular fibers $F_\sigma = p^{-1}(x_\sigma), x_\sigma \in \mathbb{P}^1 \setminus \{0, \infty\}$, the intersection $F_\sigma' = F_\sigma \cap D$ has broken chain type.

Lemma (5.6) implies that if $F_\infty$ resp. $F_0$ is singular, then this fibre has the structure described in (4). So it remains to prove that at least one of these fibers is singular. Assume that this is not the case. Let $k_\sigma$ be the number of components of $F_\sigma$, $\sigma \in \{1, \ldots, s\}$. Then the Picard number of $V$ is

$$\rho = \sum_1^s (k_\sigma - 1) + 2,$$

and $F_\sigma' = F_\sigma \cap D$ has $(k_\sigma - 1)$ components, since it is of broken chain type. Hence $D = \cup F_\sigma' \cup C$ has $1 + \sum (k_\sigma - 1) < \rho$ components, contradicting the assumption that $(V, D)$ is pre–$\mathbb{Q}$–acyclic. □

*Remarks.* (1) If $\overline{C} \subseteq \Sigma(n)$ is a curve as in (5.7), then necessarily $n \leq 1$ and $\overline{C} \sim 2C_\infty + (n+1)\overline{F}$. This is easily seen from the adjunction formula $(K_{\Sigma(n)} + \overline{C})\overline{C} = -2$ and $C \cdot C_\infty \geq 0$. In particular, $\overline{C}^2 = 4$.

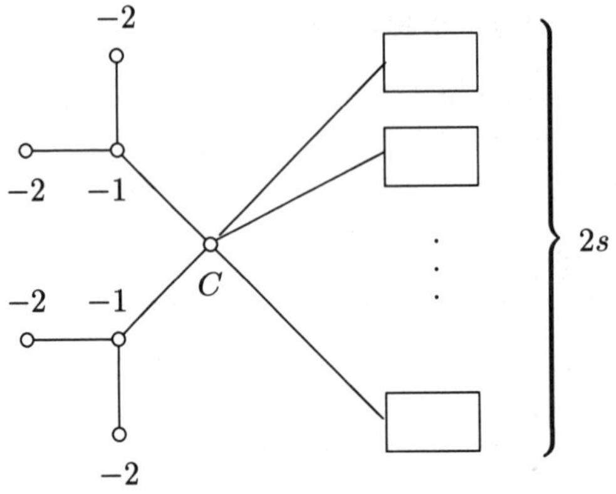


FIGURE 11. The dual graph of $D$ in case A2. The boxes denote the connected components of the fibers $F_1', \ldots, F_s'$.

(2) If we only assume that $D$ is a (connected) rational tree and $(V, D, p)$ is a strictly minimal $\mathbb{C}^\star$–triple, then $(V, D, p)$ is one of the types A1 or A2, or $D$ can be written as a sum $C + F_1' + \ldots + F_s'$, where again (1), (2) and (3) in (5.7) hold.

THEOREM 5.8. *Let $(V, D, p)$ be a strictly minimal $\mathbb{C}^\star$–triple, which is pre–$\mathbb{Q}$–acyclic. Assume that $(V, D, p)$ is of untwisted type with horizontal sections $C_0, C_1 \subseteq D$. Then there is a diagram*

$$\begin{array}{ccc} V & \xrightarrow{\ \pi\ } & \Sigma(n) \\ & {\scriptstyle p}\searrow \quad \swarrow{\scriptstyle \bar{p}} & \\ & \mathbb{P}^1 & \end{array},$$

*and $(V, D, p)$ is one of the following types* B1 *or* B2 *(see also* FIGURES 12–15*):*

**B1** $D = C_0 + C_1 + F_\infty + F_1' + \ldots + F_s'$, *where*

(1) $\overline{C}_1 = \pi(C_i)$ *is smooth,* $i = 1, 2$,

(2) $F_1', \ldots, F_s'$ *are of broken chain type over points* $x_1, \ldots, x_s \in \mathbb{P}^1 \setminus \{\infty\}$,

(3) $F_\infty = p^{-1}(\infty)$ *is a nonsingular fibre,*

(4) $\overline{C}_0 \cap \overline{C}_1 = \phi$.

**B2** $D = C_0 + C_1 + F_1' + \ldots + F_s'$, *where* (1) *and* (2) *above hold and* (3), (4) *are replaced by the condition*

(3') $C_0, C_1$ *and also* $\overline{C}_0$ *and* $\overline{C}_1$ *meet transversally in just one point.*

PROOF. By (5.3) there is a blowing down to some Hirzebruch surface, such that (1), (4) resp. (1), (3') are satisfied in the case that $C_0 \cap C_1 = \phi$ resp.

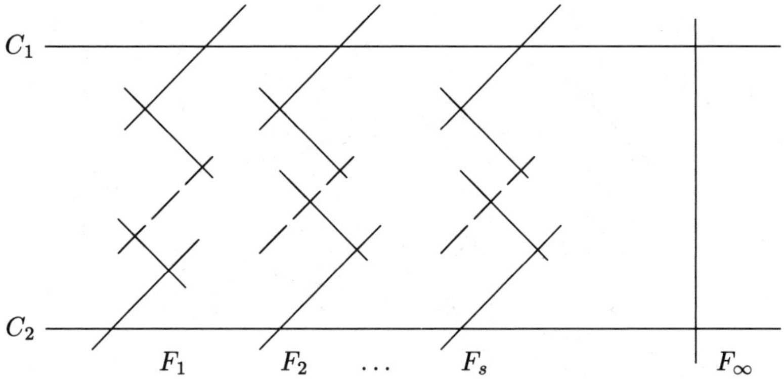


FIGURE 12. The divisor $D$ in case B1.

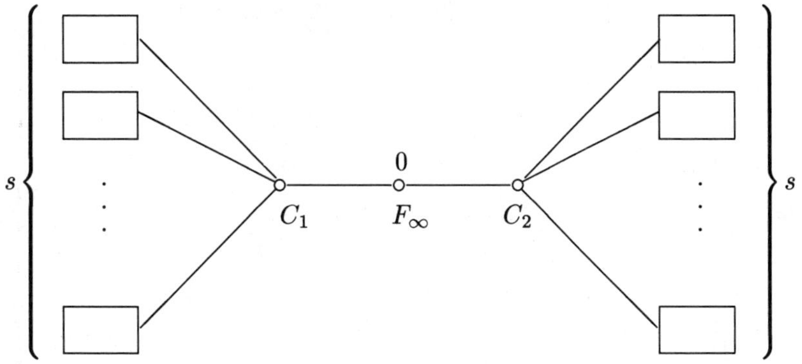


FIGURE 13. The dual graph of $D$ in case B1.

$\neq \phi$. By (5.5) the other singular fibers are of broken chain type. Assume first that $C_0 \cap C_1 = \phi$. Then because of the connectedness of $D$, there must be a nonsingular fibre in $D$, which we may assume to be $F_\infty = p^{-1}(\infty)$. Since $D$ is without cycles, there can be no further nonsingular fibre in $D$, and so we are done in this case. Finally, if $C_0 \cap C_1 \neq \phi$, then $D$ cannot contain a nonsingular fibre, since otherwise it would have either a cycle, or a triple point. This finishes the proof of (5.8). □

*Remarks.* (1) In (5.8) we could normalize the situation a little bit further. E.g., in B1 we could also require that $n = 0$, i.e. that we have a blowing down onto $\Sigma(0) = \mathbb{P}^1 \times \mathbb{P}^1$, since we can perform elementary transformation at $F_\infty$. Observe, that for $n \geq 1$ one of the curves $\overline{C}_0, \overline{C}_1$ must have negative selfintersection, because of $\overline{C}_0 \cap \overline{C}_1 = \phi$. Similarly, in B2 one of the curves $\overline{C}_0, \overline{C}_1$ is easily seen to have selfintersection $\leq 0$, except in the case $n = 1$, where also $\overline{C}_0 \sim F + C_\infty, \overline{C}_1 \sim F + C_\infty$ is possible. Here $C_\infty \subseteq \Sigma(n), n \geq 1$, is as usual the unique curve with negative selfintersection and $F$ is a fibre in $\Sigma(n)$.

(2) In (5.8) the assumption that $(V, D, p)$ is pre–$\mathbb{Q}$–acyclic can also be replaced

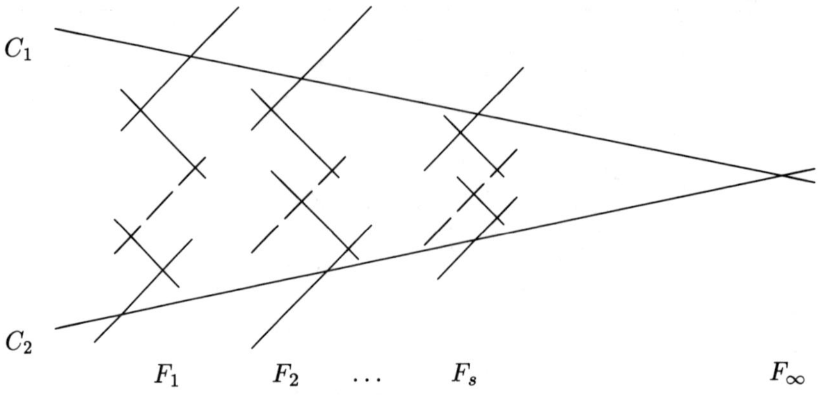


FIGURE 14. The divisor D in case B2.

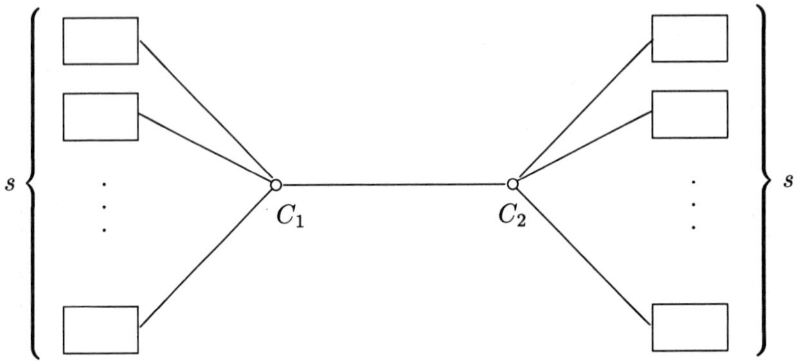


FIGURE 15. The dual graph of $D$ in case B2.

by the weaker condition, that $D$ is a (connected) rational tree, as is seen by the proof.

We will show next that (5.7) and (5.8) can also be used to construct many examples of pre–$\mathbb{Q}$–acyclic surfaces:

PROPOSITION 5.9. *Let $\pi : V \to \Sigma(n)$ be a smooth modification of $\Sigma(n)$ and $D \subseteq V$ an SNC–divisor. Let $p : V \to \mathbb{P}^1$ be the composition of $\pi$ with the ruling $\overline{p} : \Sigma(n) \to \mathbb{P}^1$. Then the following holds:*

(1) *If we are in case* A1 *or* A2 *in* (5.7) *or* B1 *in* (5.8), *then $(V, D, p)$ is a pre–$\mathbb{Q}$–acyclic $\mathbb{C}^*$–triple.*
(2) *If* B2 *in* (5.8) *holds, let $G_\sigma \subseteq F_\sigma = p^{-1}(x_\sigma)$ be the unique $(-1)$–curve in $F_\sigma$, and denote by $k_\sigma$ resp. $a_\sigma$ the coefficient of $G_\sigma$ in $F_\sigma$ resp. $\pi^*\left(\overline{C}_0\right) - \pi^*\left(\overline{C}_1\right)$. Let $\overline{C}_0 - \overline{C}_1 \sim \gamma\overline{F}$, $\overline{F}$ being a generic fibre of $\Sigma(n)$. Then $(V, D, p)$ is pre–$\mathbb{Q}$–acyclic iff*

$$\gamma \neq \sum_{\sigma=1}^{s} \frac{a_\sigma}{k_\sigma} .$$

PROOF. We have to show that the components of $D$ generate $\mathrm{Pic}(V) \otimes \mathbb{Q}$. The Picard group of $V$ is obviously generated over $\mathbb{Q}$ by a generic fibre $F$ of $p$, any horizontal irreducible curve and the components of the exceptional set of $\pi$. Let $\equiv$ denote the equality of $\mathbb{Q}$–divisors modulo linear equivalence and modulo those $\mathbb{Q}$–divisors, which are generated by components of $D$. Then in case A1 we have for $\sigma \in \{1, \dots, s\}$:

(0) $$k_\sigma G_\sigma \equiv F_\sigma \sim F_\infty \equiv 0 ,$$

if $G_\sigma$ is the unique $(-1)$–curve in $F_\sigma$ and $k_\sigma$ is its multiplicity in $F_\sigma$. Hence in this case $\mathrm{Pic}(X) \otimes \mathbb{Q} = 0$, since the exceptional curves, $F_\infty$ and $C$ are generators of $\mathrm{Pic}(V) \otimes \mathbb{Q}$. That $(V, D, p)$ is pre–$\mathbb{Q}$–acyclic in case A2 follows immediately from case A1, and for B1 the proof is similar.

Finally assume that (B2) holds. Then, for all $\sigma, \tau = 1, \dots, s$

(1) $$k_\sigma G_\sigma \equiv F_\sigma \sim F_\tau \equiv k_\tau G_\tau .$$

Moreover,

(2) $$\pi^\star\left(\overline{C}_0\right) - \pi^\star\left(\overline{C}_1\right) \equiv \sum_{1}^{s} a_\sigma G_\sigma ,$$

and on the other hand,

(3) $$\pi^\star\left(\overline{C}_0\right) - \pi^\star\left(\overline{C}_1\right) \sim \gamma F \equiv \gamma k_1 G_1 .$$

Replacing $G_\sigma$ by $\frac{k_1}{k_\sigma} G_1$, we obtain from (2), (3) that

(4) $$k_1 \left( \gamma - \Sigma \frac{a_\sigma}{k_\sigma} \right) \cdot G_1 \equiv 0 .$$

Hence, if $(V, D, p)$ is pre–$\mathbb{Q}$–acyclic, then $\gamma \neq \Sigma \frac{a_\sigma}{k_\sigma}$. If, conversely, $\gamma \neq \Sigma \frac{a_\sigma}{k_\sigma}$, then $G_1 \equiv 0$ by (4), hence $G_\sigma \equiv 0$ for all $\sigma$ by (1), and so $(V, D, p)$ is pre–$\mathbb{Q}$–acyclic. □

*Remark* 5.10. (1) In general, $(V, D, p)$ is not strictly minimal, since a horizontal curve in $D$ can happen to be a smooth rational $(-1)$–curve, which meets at most two other components of $D$.

(2) Looking more closely at the proof of (5.9), it is easy to calculate the order of the finite group

$$\mathrm{Pic}(X) = \mathrm{Pic}(V) / \sum \mathbb{Z}[D_i] .$$

In case (A1), the subgroup $G'$ of $\mathrm{Pic}(V)$, generated by $F_\infty$, the exceptional curves and $C$, has index 2 in $\mathrm{Pic}(V)$, and because of (0) the subgroup $\sum \mathbb{Z}[D_i]$ has index $\prod_1^s k_\sigma$ in $G'$. Hence in this case $\mathrm{Pic}(X)$ has $2 \cdot \prod_1^s k_\sigma$ elements. The same reasoning shows that

$$\sharp(\mathrm{Pic}(X)) = 2 \cdot \prod_1^s k_\sigma \quad \text{for} \quad A2,$$

$$\sharp(\mathrm{Pic}(X)) = \prod_1^s k_\sigma \quad \text{for} \quad B1,$$

whereas in case B2, because of (1) and (2), (3), the number of elements in $\mathrm{Pic}(X)$ is equal to the absolute value of the determinant

$$\begin{vmatrix} -k_1 & k_2 & 0 & \dots \\ -k_1 & 0 & k_3 & \dots \\ \dots\dots & \dots & \dots & \dots \\ -k_1 & \dots & 0 & k_s \\ -\gamma k_1 + a_1 & a_2 & \dots & a_s \end{vmatrix},$$

which is equal to

$$(-1)^s \left(\prod_1^s k_\sigma\right) \quad \left(-\gamma + \sum_{\sigma=1}^s \frac{a_\sigma}{k_\sigma}\right).$$

(3) The surfaces in A1 and B2 are obviously $\mathbb{Q}$–acyclic. In order to obtain from type A2 a minimal $\mathbb{Q}$–acyclic surface, one has to perform a comb–attachment in a point of $F_0 \cup F_\infty$, since $F_0 - F_\infty$ is, up to a multiple, the only relation in $\mathrm{Pic}(V)$, see Remark (2) following (4.4). Similarly, if we are in type B1, then it is easily seen that among the components of $D$ the relations are generated by

$$\Delta := \pi^\star(\overline{C}_0 - \overline{C}_1) - \gamma F_\infty,$$

if $\overline{C}_0 - \overline{C}_1 = \gamma \cdot \overline{F}_\infty$. Hence we must perform in this case a comb–attachment in a point of $\mathrm{supp}(\Delta)$. Observe, that $\gamma \neq 0$ since $\overline{C}_0 \cap \overline{C}_1 \neq \phi$.

We will now determine the logarithmic Kodaira dimensions of the surfaces in A1, A2, B1, B2.

THEOREM 5.11. *Let $(V, D, p)$ be a strictly minimal $\mathbb{C}^\star$–triple with $s$ broken chain type fibers, and suppose that $\pi : V \to \Sigma(n)$ is a blowing down of $V$, according to (5.7) resp. (5.8). Denote by $G_\sigma, \sigma = 1, \dots, s$, the unique $(-1)$–curve in the broken chain type fibre $F'_\sigma$, and let $k_\sigma$ be the multiplicity of $G_\sigma$ in $F_\sigma = p^{-1}(x_\sigma)$. Set*

$$\delta = \delta(V, D, p) := \begin{cases} s - \frac{1}{2} - \Sigma \frac{1}{k_\sigma} & \textit{if A1 holds,} \\ s - \Sigma \frac{1}{k_\sigma} & \textit{if A2 holds,} \\ s - 1 - \Sigma \frac{1}{k_\sigma} & \textit{if B1 or B2 holds.} \end{cases}$$

*Then*

$$\overline{k}(X) = \begin{cases} -\infty & \textit{iff} \ \ \delta < 0, \\ 0 & \textit{iff} \ \ \delta = 0, \\ 1 & \textit{iff} \ \ \delta > 0. \end{cases}$$

For the proof we will need the following two lemmas, which are well known.

LEMMA 5.12. *Let $\pi : V' \to V$ be the blowing up of $V$ in $x \in V, E = \pi^{-1}(x)$ and $D \subseteq V$ a reduced divisor with multiplicity $m$ in $x$. Set $D' = \pi^{-1}(D)_{red}$. Then*

$$\pi^\star(K_V + D) = K_{V'} + D' + (m-2)E.$$

PROOF. This follows immediately form the formulas $\pi^\star(D) = D' + (m-1)E$ and $K_{V'} = \pi^\star(K_V) + E$. $\square$

LEMMA 5.13. *Let $V$ be a normal variety of dimension $\leq 2$, and $D, E$ two divisors on $V$, such that $E > 0$ is exceptional and $\mid D \mid \cap \mid E \mid = \phi$. Then $H^0(\mathcal{O}_V(D+E)) = H^0(\mathcal{O}_V(D))$. In particular, $k(D+E) = k(D)$, where $k$ denotes the Kodaira dimension.*

PROOF. Let $\pi : V \to \overline{V}$ be the blowing down of $E$. Then $\pi_\star(\mathcal{O}_V(D+E)) = \pi_\star(\mathcal{O}_V(D))$. Taking sections we obtain $H^0(\mathcal{O}_V(D+E)) = H^0(\mathcal{O}_V(D))$. The equality for the Kodaira dimension is immediate from this. $\square$

PROOF OF (5.11) Obviously, it is sufficient to prove that $\overline{k}(X) = k(\delta \cdot F)$, where $F \subseteq V$ is the generic fibre. Assume first that we are in case A1, see (5.7). Let $C_\infty \subseteq \Sigma(n)$ be a horizontal section with $C_\infty^2 \leq 0$ (which is unique if $n \geq 1$), and $\overline{F} \subseteq \Sigma(n)$ the generic fibre. Then

$$\overline{K} := K_{\Sigma(n)} \ = \ -2C_\infty - (n+2)\overline{F},$$

and

$$\overline{C} = 2C_\infty + (n+1)\overline{F},$$

see the remark following (5.7). Hence

(1) $$\overline{K} + \overline{C} + \overline{F}_\infty + \overline{F}_1 + \ldots + \overline{F}_s \sim s\overline{F}.$$

Denote the curves in $F_\infty \cup C \subseteq V$ as follows:

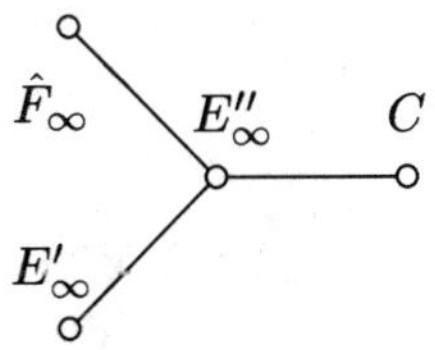

Here $\hat{F}_\infty$ is the proper transform of $\overline{F}_\infty$. Locally around $F_\infty, V$ is obtained by two blowing ups, the first of which is performed in the singular point of $\overline{C} \cup \overline{F}_\infty$, which is of multiplicity 2, whereas the second one is performed in the resulting triple point of the preimage of $\overline{C} \cup \overline{F}_\infty$ under the first blowing up.

Using (5.12) repeatedly, we obtain

$$\pi^\star(\overline{K} + \overline{C} + \overline{F}_\infty + \overline{F}_1 + \ldots + \overline{F}_s) = K + D + \sum_1^s G_\sigma + E''_\infty. \tag{2}$$

Together with (1) this implies that

$$K + D \sim \left(s - \frac{1}{2} - \sum_1^s \frac{1}{k_\sigma}\right) F + \sum_1^s \left(\frac{1}{k_\sigma} F_\sigma - G_\sigma\right) + \left(\frac{1}{2} F_\infty - E''_\infty\right).$$

Now $\left(\frac{1}{k_\sigma} F_\sigma - G_\sigma\right)$ resp. $\frac{1}{2} F_\infty - E''_\infty$ is a sum of components of $F_\sigma$ different from $G_\sigma$ resp. $E''_\infty$. Hence $\sum_1^s \left(\frac{1}{k_\sigma} F_\sigma - G_\sigma\right)$ and $\left(\frac{1}{2} F_\infty - E''_\infty\right)$ are exceptional and disjoint from the generic fibre. So, we get from (5.13) that

$$\overline{k}(X) = k(K + D) = k\left(\left(s - \frac{1}{2} - \Sigma \frac{1}{k_\sigma}\right) F\right) = k(\delta \cdot F).$$

In the case A2 the same reasoning shows that

$$\overline{K} + \overline{C} + \overline{F}_\infty + \overline{F}_0 + \overline{F}_1 + \ldots + \overline{F}_s \sim (s+1)F, \tag{3}$$

and

$$\pi^\star\left(\overline{K} + \overline{C} + \overline{F}_\infty + \overline{F}_0 + \ldots + \overline{F}_s\right) = K + D + \sum_1^s G_\sigma + E''_\infty + E''_0, \tag{4}$$

if $E''_0$ is the $(-1)$ curve in the fibre $F_0$. Again, we get as above from (3), (4) that

$$K + D \sim \left(s - \sum_1^s \frac{1}{k_\sigma}\right) F + \sum_1^s \left(\frac{1}{k_\sigma} F_\sigma - G_\sigma\right) + \left(\frac{1}{2} F_\infty - E''_\infty\right) + \left(\frac{1}{2} F_0 - E''_\infty\right),$$

and so $\overline{k}(X) = k(\delta F)$. Now assume that we are in case B1. Then $\overline{C}_i = C_\infty + \gamma_i \overline{F}$, and because of $\overline{C}_0 \overline{C}_1 = 0$ we have $\gamma_0 + \gamma_1 = n$. Hence

$$\overline{K} + \overline{C}_0 + \overline{C}_1 + \overline{F}_\infty + \overline{F}_1 + \ldots + \overline{F}_s \sim (s-1)\overline{F}, \tag{5}$$

and

$$\pi^\star\left(\overline{K} + \overline{C}_0 + \overline{C}_1 + \overline{F}_\infty + \overline{F}_1 + \ldots + \overline{F}_s\right) = K + D + \sum_1^s G_\sigma. \tag{6}$$

Thus we obtain

$$K + D \sim \left(s - 1 - \sum_1^s \frac{1}{k_\sigma}\right) F + \sum_1^s \left(F_\sigma - \frac{1}{k_\sigma} G_\sigma\right),$$

which again shows that $\overline{k}(X) = k(\delta F)$. If B2 holds, then $\overline{C}_0 \cdot \overline{C}_1 = 1$ implies $\gamma_0 + \gamma_1 = n + 1$, where $\overline{C}_i = C_\infty + \gamma_i \overline{F}$ as above. Hence

$$\overline{K} + \overline{C}_0 + \overline{C}_1 + \overline{F}_1 + \ldots + \overline{F}_s \sim (s-1)\overline{F} \tag{7}$$

and

$$\pi^\star\left(\overline{K} + \ldots + \overline{F}_s\right) = K + D + \Sigma G_\sigma, \tag{8}$$

which shows the assertion as in the previous case. □

For instance, with (5.11) it is an easy exercise to write down all strictly minimal pre-$\mathbb{Q}$-acyclic $\mathbb{C}^\star$-triples $(V, D, p)$ of logarithmic Kodaira dimension 0.

COROLLARY 5.14. *Each strictly minimal pre-$\mathbb{Q}$-acyclic $\mathbb{C}^\star$-triples $(V, D, p)$ with $\overline{k}(X) = 0$, where $X := V \setminus D$, is one of the following:*

(1) *Type* A1, $s = 1, k_\sigma = 2$.

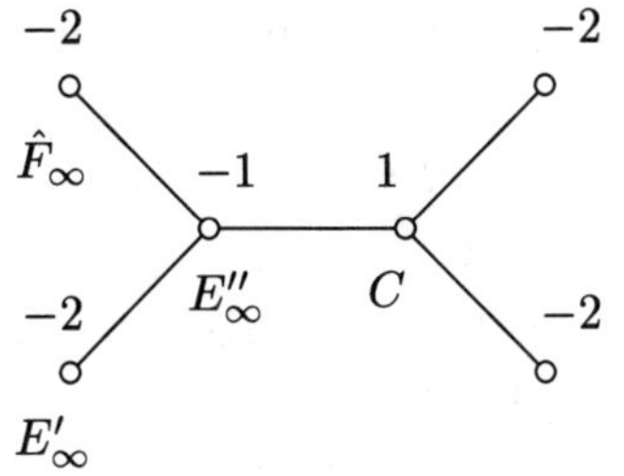


*Here $\hat{F}_\infty \cup E'_\infty \cup E''_\infty = F_\infty$, and the two other $(-2)$–curves are in $F_1$.*

(2) *Type* A2, $s = 0$.

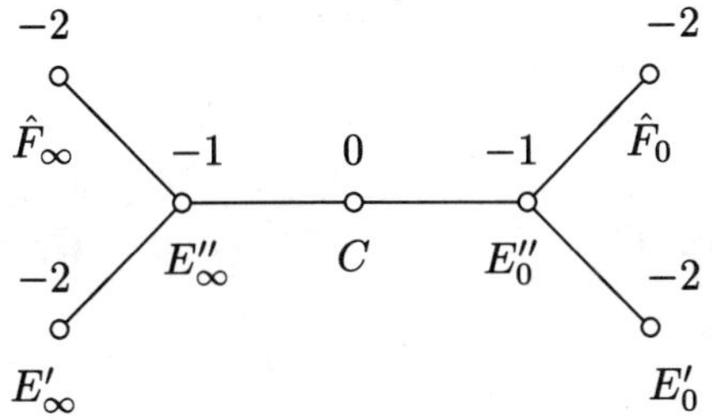


(3) *Type* B1, $s = 2, k_1 = k_2 = 2$.

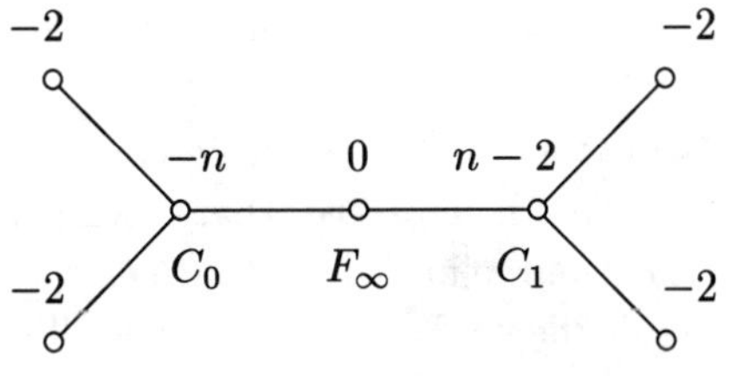


$n \geq 1$

(4) *Type* B2, $s = 2, k_1 = k_2 = 2$.

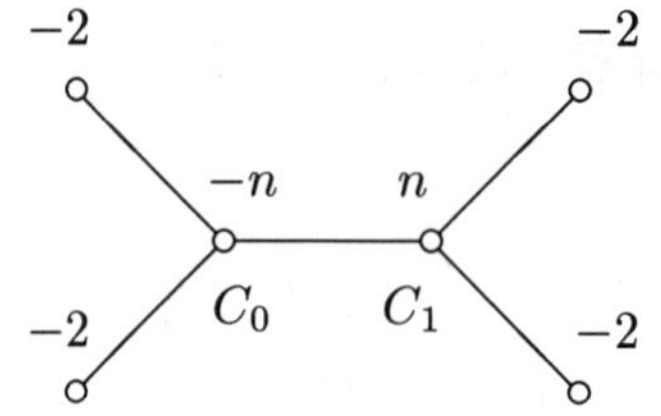


$n \geq 1$

PROOF. From (5.13) we obtain that $\bar{k}(X) = 0$ iff $\delta(V, D, p) = 0$. Hence there are only the following possibilities for $(s; k_1, \dots, k_s)$ :

A1 : (1;2)
A2 : (0)
B1 and B2 : (2;2,2).

If $k_\sigma = 2$, then the fibre $F_\sigma$ has the following dual graph:

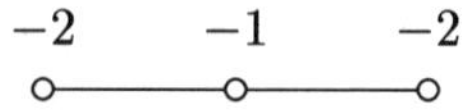


(indeed, blowing up in further intersection points would lead to a multiplicity $\geq 3$ of the $(-1)$–curve). In order to prove (1) it remains to compute $C^2$. We have on $\Sigma(n)$ that $\overline{C} = 2C_\infty + (n+1)F$, see the remark following (5.7), and so $\overline{C}^2 = 4$. Since $\overline{C}$ is blown up 3 times, we obtain $C^2 = 1$. For (2) there are 4 blowing ups of $\overline{C}$, and so $C^2 = 0$ in this case.

Now assume that we are in case (3) or (4). We can also assume that $\overline{C}_0 = C_\infty$, and that no blowing up have been done on $\overline{C}_0$ in order to obtain $V$, since otherwise we can perform elementary transformations. Hence $C_0^2 = \overline{C}_0^2 = -n$ in both cases. If we are in case B1, then $\overline{C}_1 \sim C_\infty + n\overline{F}$ because of $\overline{C}_1\overline{C}_0 = 0$, and so $\overline{C}_1^2 = n$. Since there are 2 blowing ups on $\overline{C}_1^2$, we get $C_1^2 = n - 2$. Similarly, in case B2 we have: $\overline{C}_1 \sim C_\infty + (n+1)\overline{F}$, and so $C_1^2 = n$. □

*Remark* 5.15. (1) The surface $(V, D, p)$ described in (5.14) (1) is isomorphic to the surface in (5.14) (4) if $n = 1$. This is easily seen by considering for the surface in (5.14) (4) the fibration given by $H = 2C_0 + \Gamma_1 + \Gamma_2$, where $\Gamma_1, \Gamma_2$ are the $(-2)$–curves neighbouring $C_0$. The fibration is a twisted $\mathbb{C}^*$–fibration with horizontal curve $C_1$, since $HC_1 = 2$.

(2) Similarly, if in (5.14) (3) $n = 1$, the resulting triple $(V, D, p)$ is isomorphic to the one described in (5.14) (2). Observe, that after performing elementary transformations on $V$ at the 0–curve in (5.14) (3), it is always possible to achieve $m = 1$, i.e. the affine parts of the surfaces in 2), 3) are all isomorphic.

(3) The type described in (5.14) (4) was called $H[n,-n]$ in [**13**], whereas (5.14) (2) is the type $H[-1,0,-1]$ in the classification of Fujita. We remark also that by [**13**] (8.5) any surface of type $H[-1,0,-1]$ is the one given in (5.14) (2). Moreover, if $(V,D)$ is of type $H[n,-n]$, then by [**13**] (8.25) it is of type as described in (4). In particular, it follows that there is only one surface of type $H[n,-n]$ and $H[-1,0,1]$.

(4) In [**13**] there appear only three other surfaces of logarithmic Kodaira dimension 0, which are pre–$\mathbb{Q}$–acyclic and strictly minimal, called $Y\{3,3,3\}$, $Y\{2,4,4\}$ and $Y\{2,3,6\}$, see FIGURE 16 for the dual graphs of D. For these surfaces an explicit construction, starting from the configuration of four lines in general position in $\mathbb{P}^2$, is given in [**7**] (4.13). Inspecting the arguments of Fujita, see [**13**] (8.35), (8.36) for the case $Y[3,3,3]$, (8.53), (8.54) for the case $Y\{2,4,4\}$ and (8.59) and (8.61) for the case $Y\{2,3,6\}$, it is easily seen that each of those surfaces is obtained by the construction of tom Dieck and Petrie [**7**], and so they are unique.

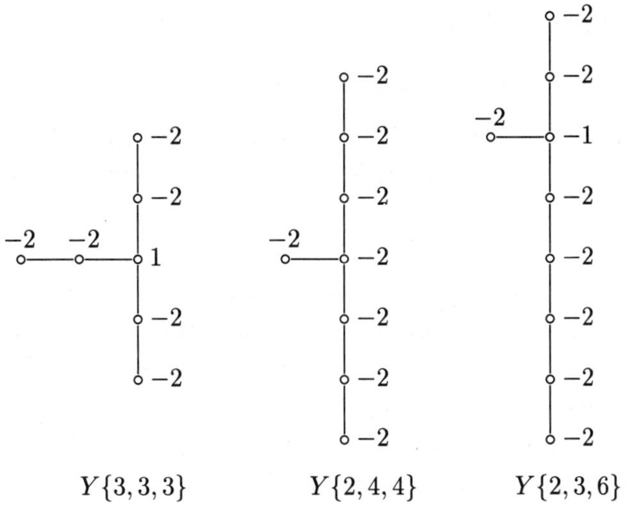


FIGURE 16.

## 6. Deformations of pre–$\mathbb{Q}$–acyclic surfaces of logarithmic Kodaira dimension 0 and 1

THEOREM 6.1. *Let $X$ be a smooth algebraic surface of logarithmic Kodaira dimension $\geq 0$. Let $(V,D)$ and $(V',D')$ be two SNC–completions which are minimal. Then*

(1) $h^i(V,\Theta_V\langle D\rangle) = h^i\left(V',\Theta_{V'}\langle D'\rangle\right)$, $i=1,2,3$, *and*

(2) *the bases of the logarithmic deformations of* $(V, D)$ *and* $(V', D')$ *are isomorphic.*

PROOF. By (4.14) the dual graph $\Gamma$ of $D$ has no outer non–admissible chain, and so $(V', D')$ can be obtained from $(V, D)$ by a sequence of inner modifications in $D$, see (A.4). Hence the conclusion follows from (1.5) (4) and (1.6). $\square$

PROPOSITION 6.2. *Let* $(V, D)$ *be an SNC–pair. Assume that* $V$ *admits a* $\mathbb{P}^1$*–fibration* $\pi : V \to C$ *over some smooth curve* $C$*, such that* $D \cdot F \le 3$ *for the generic fibre* $F$*. Then* $h^2(\Theta_V\langle D\rangle) = 0$*. In particular, the logarithmic deformations of* $(V, D)$ *are unobstructed.*

PROOF. By Serre-duality,

$$H^2(\Theta_V\langle D\rangle)^* \simeq H^0(\Omega^1_V\langle D\rangle \otimes \omega_V),$$

and $\Omega^1_V\langle D\rangle \otimes \omega_V \simeq \Theta_V\langle D\rangle \otimes \omega^2_V(D)$. It is well known that $2K + D$ is linearly equivalent to a divisor $\Delta$, which does not contain any component of $D$. We show that

$$H^0(\Theta_V\langle D\rangle(\Delta)) = 0.$$

Assume that the latter group contains a section $\delta$, which is then a vector field tangent to $D$, with pole and vanishing orders prescribed by $\Delta$. Restricting $\delta$ onto the normal bundle of the generic fibre $F$, we obtain a section in $H^0\Big(\mathcal{N}_{F/V}(\Delta)\Big)$, which is zero, since $\mathcal{N}_{F/V} = \mathcal{O}_F$ and $\Delta F \le -1$. Hence $\delta$ is tangent to the fibers of $\pi$. Now the generic fibre of $F$ meets $D$ transversally, hence $\delta$, being tangent along $F$ and $D$, must have a zero along $D^h$, where $D^h \subseteq D$ is the horizontal part of $D$ (possibly empty). Since $\Delta$ does not contain any component of $D$, we obtain that $\delta \mid F$ gives a section of $\Theta_F(-DF + \Delta F) = \Theta_F(-4) \simeq \mathcal{O}_F(-2)$, hence $\delta \mid F = 0$ and so $\delta = 0$. $\square$

COROLLARY 6.3. *Let* $(V, D)$ *be an SNC–pair and* $X = V \setminus D$*. Then*

$$H^2(\Theta_V\langle D\rangle) = 0.$$

*Thus the logarithmic deformation are unobstructed in each of the following cases:*

(1) $\overline{k}(X) = -\infty$,

(2) $(V, D)$ *is pre–*$\mathbb{Q}$*–acyclic and* $\overline{k}(X) = 1$.

PROOF. The condition that $H^2(\Theta_V\langle D\rangle) = 0$ does not change under blowing up a point of $D$ (see 1.5). Also, deleting a curve of selfintersection $\leq -1$ does not effect $H^2(\Theta_V\langle D\rangle)$, see (1.7). Hence it is sufficient to prove (1), (2) for some SNC–completion of $X$. If $\overline{k}(X) = -\infty$ and either $h^1(\mathcal{O}_X) > 0$ or $D$ is connected, then for a suitable SNC–completion $(V, D)$ of $X$ there is a $\mathbb{P}^1$–fibration $V \to C$ for some smooth curve $C$, such that $DF \leq 1$ for the generic fibre, see [**24**] I.2; [**30**]. If $D$ is not connected, then for some SNC-completion $(V, D)$ of $X$ there is a $\mathbb{P}^1$–fibration $V \to \mathbb{P}^1$ with $F{\cdot}D \leq 2$ for the generic fibre $F$ (see [**23**]). In any case, we can apply (6.2). Similarly, if $(V, D)$ is pre–$\mathbb{Q}$–acyclic, then, by the remarks at the beginning of the proof, we can assume that $(V, D)$ is strictly minimal. If $\overline{k}(X) = 1$, then by the results of section 5 there is $\mathbb{P}^1$–fibration $V \to \mathbb{P}^1$ with $F \cdot D \leq 2$ for the generic fibre. Hence, also in this case (6.2) applies. $\square$

In the proposition below we will compare deformations of strictly minimal $\mathbb{C}^\star$–triples and the image of this triple after blowing down to some Hirzebruch surface. More precisely, let

$$\begin{array}{ccc} V & \xrightarrow{\ \pi\ } & \Sigma := \Sigma(n) \\ & {\scriptstyle p}\searrow \quad \swarrow{\scriptstyle \bar{p}} & \\ & \mathbb{P}^1 & \end{array}$$

be as in (5.3), i.e. $D^h \to \pi(D^h)$ is an isomorphism, where $D^h \subseteq D$ is the horizontal part of $D$. If $(V, D, p)$ is of twisted type, we will always assume that the branching points of $D^h \xrightarrow{p} \mathbb{P}^1$ are lying over $0, \infty \in \mathbb{P}^1$. Let $\overline{D} \subseteq \Sigma$ be the union of $\pi(D^h)$ and all fibers $\overline{F}_\sigma$, such that $\overline{F}_\sigma \neq \overline{F}_0, \overline{F}_\infty$ and $D$, and $F_\sigma$ have some component in common, i.e. $\overline{D}$ is $\pi(D)$ with the fibers over 0 and $\infty$ (possibly) deleted. With these notations we have

PROPOSITION 6.4. *Assume that $(V, D, p)$ is a strictly minimal $\mathbb{C}^\star$–triple, and assume that $D$ is connected. Then $H^i(\Theta_V\langle D\rangle) \simeq H^i\Big(\Theta_\Sigma\langle\overline{D}\rangle\Big)$ for all $i$. Moreover, the logarithmic deformations of $(V, D)$ and $(\Sigma, \overline{D})$ are the same.*

PROOF. Let $F \subseteq V$ be a singular fibre of broken chain type. Let $G$ be the unique $(-1)$–curve in $F$. Adding $G$ to $D$ and blowing down $F$ successively does not affect $H^i(\Theta_V\langle D\rangle)$, nor the logarithmic deformations of the pair, see (1.6) and (1.7). If $(V, D, p)$ is of untwisted type, then by (5.5) we have only singular fibers of broken chain type and so, after blowing them, we are on the Hirzebruch surface. Thus (6.4) follows in this case. So assume from now on that $(V, D, p)$ is of twisted type. With the same arguments as above we may restrict to the case that $p$ has no singular fibers, except possibly $F_0$ and $F_\infty$. If both $F_0$ and $F_\infty$ are nonsingular, then there is nothing to prove. So assume that $F_\infty$ is singular. Blowing down $F_\infty$ to a smooth fibre, we get a surface $V^\star$ with the nonsingular fiber $F^\star_\infty$ of the induced fibering $V^\star \to \mathbb{P}^1$. Let $D^\star \subseteq V^\star$ be the divisor $\sigma_\star(D) - F^\star_\infty$, where $\sigma : V \to V^\star$ is the canonical map. Let $C \subseteq V$, $C^\star \subseteq V^\star$ be the horizontal components, and let $x \in V^\star$ denote the ramification point of $C^\star \to \mathbb{P}^1$. Let $C \cup F_\infty$ have dual graph as shown in FIGURE 17, where $\hat{F}_\infty$ is the proper transform of $F^\star_\infty$. Then $D = C + E + E' + \hat{F}_\infty + D'$

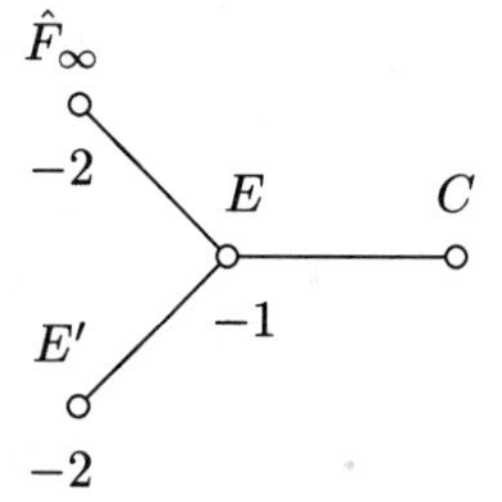


,

FIGURE 17.

for some divisor $D' \subseteq D$ that does not meet $F_\infty$. Using (1.5) (3), (4) we get

$$\mathcal{M} := \pi_\star\Big(\Theta_V\langle C + E + E' + D'\rangle\Big) = \mathfrak{m}_x\Theta_{V^\star} \cap \Theta_{V^\star}\langle C^\star + D'\rangle,$$

and $R^1\pi_\star(\Theta_V\langle C + E + E' + D'\rangle) = 0$. From the exact sequence

$$0 \to \Theta_V\langle D\rangle \to \Theta_V\langle C + E + E' + D'\rangle \to \mathcal{N}_{\hat{F}_\infty/V} \to 0$$

we obtain the exact sequence

$$0 \to \pi_\star(\Theta_V\langle D\rangle) \to \mathcal{M} \xrightarrow{\zeta} \mathcal{N}_{\hat{F}_\infty/V} \simeq \mathcal{O}_{\hat{F}_\infty}(-2) \to R^1\pi_\star(\Theta_V\langle D\rangle) \to 0.$$

By Lemma (6.5) below, $\zeta$ is surjective. Hence $R^1\pi_\star(\Theta_V\langle D\rangle) = 0$. Restricting a vector field to the normal bundle of $F^\star$ in $V^\star$, we get by (6.5) a surjective map

$$\Theta_{V^\star}\langle C^\star + D'\rangle \xrightarrow{\tilde{\rho}} \mathcal{N}_{F^\star/V^\star}(-x) \simeq \mathcal{O}_{F^\star}(-1)$$

with Ker $\rho =$ Ker $\tilde{\rho}$. Hence we have an exact sequence

$$0 \to \pi_\star(\Theta_V\langle D\rangle) \to \Theta_{V^\star}\langle D^\star\rangle \to \mathcal{O}_{F^\star}(-1) \to 0.$$

By the long exact cohomology sequence, we obtain

$$H^i(\Theta_V\langle D\rangle) \simeq H^i(\Theta_{V^\star}\langle D^\star\rangle).$$

That the logarithmic deformations of $(V, D)$ and $(V^\star, D^\star)$ are the same, follows now from the fact that both are unobstructed and that the tangent map for the bases of the semiuniversal deformations is an isomorphism. This proves (6.4). □

LEMMA 6.5. *Let $C, F \subseteq \mathbb{C}^2$ be the curves $C = \{x = 0\}$, $F = \{x - y^2 = 0\}$. Then the natural maps*

$$\Theta_{\mathbb{C}^2}\langle C\rangle \xrightarrow{\rho} \mathcal{N}_{F/\mathbb{C}^2} \quad \text{and} \quad \Theta_{\mathbb{C}^2}\langle C\rangle \cap \mathfrak{m}\Theta_{\mathbb{C}^2} \xrightarrow{\tilde{\rho}} \mathcal{N}_{F/\mathbb{C}^2}$$

*have the same kernel, and* $\mathrm{im}(\rho) = \mathcal{N}_{F/\mathbb{C}^2}(-P)$, $\mathrm{im}(\tilde{\rho}) = \mathcal{N}_{F/\mathbb{C}^2}(-2P)$, *where $P$ denotes the origin in $\mathbb{C}^2$ and $\mathfrak{m}$ is the maximal ideal of $P$.*

PROOF. $\Theta_{\mathbb{C}^2}\langle C\rangle$ has generators $x\frac{\partial}{\partial x}, \frac{\partial}{\partial y}$, and $\rho$ is given by mapping $\delta$ onto $\delta(x - y^2)$ in $\mathcal{N}_{F/\mathbb{C}^2} \simeq \mathcal{O}_F$. Since $x\frac{\partial}{\partial x}(x - y^2) = 0$ in $\mathcal{O}_F$ and $\frac{\partial}{\partial y}(x - y^2)$, restricted to $F$, has a simple zero, the image of $\rho$ is $\mathcal{N}_{F/\mathbb{C}}(-P)$. Similarly, it follows that $\mathrm{im}(\tilde{\rho}) = \mathcal{N}_{F/\mathbb{C}^2}(-2P)$. It remains to show that $\ker(\rho) = \ker(\tilde{\rho})$. Let $\delta = xf\frac{\partial}{\partial x} + g\frac{\partial}{\partial y} \in \ker \rho$. Then $xf - 2gy = 0$, when restricted to $F = \{x - y^2 = 0\}$. But this obviously implies that $g(0) = 0$, i.e. $\delta \in \Theta_{\mathbb{C}^2}\langle C\rangle \cap \mathfrak{m}\Theta_{\mathbb{C}^2}$. □

THEOREM 6.6. *Let $(V, D, p)$ be a strictly minimal $\mathbb{C}^*$-triple, and assume that $D$ is connected. Let $s$ be the number of those fibers $F = p^{-1}(x)$, such that $F$ and $D$ have a common component and $D^h \to \mathbb{P}^1$ is unramified in the points of $D^h \cap F$. Then the logarithmic deformations of $(V, D)$ are unobstructed, and the dimension of bases of the semiuniversal deformation is given by $h^1 = h^1(\Theta_V\langle D\rangle)$. Moreover, $\chi := \chi(\Theta_V\langle D\rangle)$, $h^i := h^i(\Theta_V\langle D\rangle)$ are given by* TABLE 6 *below, where $C_0, C_1 \subseteq D$ denote the horizontal curves, if $(V, D, p)$ is of untwisted type, see* (5.2).

TABLE 6

| | Type | $s$ | $\chi$ | $h^0$ | $h^1$ |
|---|---|---|---|---|---|
| (a) | twisted | 0 | 1 | 1 | 0 |
| (b) | twisted | $s \geq 1$ | $1 - s$ | 0 | $s - 1$ |
| (c) | untwisted $C_0C_1 = 1$ | 0 | 2 | 2 | 0 |
| (d) | " | 1 | 1 | 1 | 0 |
| (e) | " | $s \geq 2$ | $2 - s$ | 0 | $s - 2$ |
| (f) | untwisted $C_0C_1 = 0$ | 0 | 4 | 4 | 0 |
| (g) | " | 1 | 3 | 3 | 0 |
| (h) | " | 2 | 2 | 2 | 0 |
| (i) | " | $\geq 3$ | $4 - s$ | 1 | $s - 3$ |

*Remark.* If $(V, D, p)$ is in addition pre–$\mathbb{Q}$–acyclic, then in case A1, A2 we are in (a) and (b), whereas type B1 resp. B2 corresponds to (f) – (i) resp. (c) – (e), see (5.7) and (5.8).

Since by (6.3) $h^2 = 0$, we have $h^1 = \chi - h^0$, and so (6.6) follows from the lemmas (6.7) – (6.8) below.

LEMMA 6.7. *Let $(V, D, p)$ be as in (6.6). Then $\chi = \chi(\Theta_V\langle D\rangle)$ is given by*

$$\chi = \begin{cases} 1 - s & \text{if } (V, D, p) \text{ is of twisted type,} \\ 4 - s - 2C_0C_1 & \text{if } (V, D, p) \text{ is of untwisted type.} \end{cases}$$

PROOF. By (6.4) we may assume that $V = \Sigma(n)$. From the exact sequence

$$0 \to \Theta_V\langle D\rangle \to \Theta_V \to \bigoplus_i \mathcal{O}_{D_i}(D_i^2) \to 0$$

we get $\chi = \chi(\Theta_V) - \Sigma(D_i^2 + 1)$, where $D_i$ are the irreducible components of $D$. By (1.3), $\chi(\Theta_V) = 6$. Hence, if we are in the twisted case and $C$ is the horizontal curve, then by the remark after (5.7), $C^2 = 4$ and so $\chi = 1 - s$. If we are in the twisted case, then we can write $C_0 \sim C_1 + kF$ for a fibre $F$. Multiplying by $C_0$ and $C_1$ gives $C_0^2 = C_0C_1 + k, C_0C_1 = C_1^2 + k$, and so $C_0^2 + C_1^2 = 2C_0C_1$. □

LEMMA 6.8. *Under the assumptions of* (6.6), *the dimension $h^0 = h^0(\Theta_V\langle D\rangle)$ is as in the list in* (6.6).

PROOF. By (6.4) we may assume $V = \Sigma(n)$. We will use the description of vector fields in the affine chart $(u, v)$, given in (2.5). First we treat the twisted case. Let $C \subseteq V$ be the horizontal curve. If $F \subseteq V$ is a fibre intersecting $C$ transversally, then $H^0(\Theta_V\langle C \cup F\rangle) = 0$ because of the list in (2.7). In particular, $h^0(\Theta_V\langle C \cup F_1 \cup \ldots \cup F_s\rangle) = 0$ if $s \geq 1$. Since $\chi(\Theta_V\langle C \cup F_1 \cup \ldots \cup F_s\rangle) = 1 - s$ and $h^2(\Theta_V\langle C \cup F_1 \cup \ldots \cup F_s\rangle) = 0$ by (6.7) and (6.3), we get also that $h^1(\Theta_V\langle C \cup F_1 \cup \ldots \cup F_s\rangle) = s - 1$. If $s = 0$, then in the exact cohomology sequence

$$\to H^1(\Theta_V\langle C \cup F\rangle) \to H^1(\Theta_V\langle C\rangle) \to H^1(\mathcal{N}_{F/V})$$

we have $H^1(\mathcal{N}_{F/V}) = 0$, since $\mathcal{N}_{F/V} = \mathcal{O}_F$. Hence $H^1(\Theta_V\langle C\rangle) = 0$, and so $H^0(\Theta_V\langle C\rangle) = 1$ because of $\chi = 1$ in this case.

Now assume that we are in the untwisted type and that $C_0C_1 = 1$, where $C_0, C_1 \subseteq V = \Sigma(n)$ are the horizontal curves. If we add two different fibers $F, F'$ intersecting $C_0 \cup C_1$ transversally, then by our lists (2.4) and (2.7) $h^0(\Theta_V\langle C_0 \cup C_1 \cup F \cup F'\rangle) = 0$. In particular, $h^0(\Theta_V\langle C_0 \cup C_1 \cup F_1 \ldots \cup F_s\rangle) = 0$ for $s \geq 2$.

Now $\chi(\Theta_V\langle C_0 \cup C_1 \cup F \cup F'\rangle) = 0$ by (6.7) and so $h^1(\Theta_V\langle C_0 \cup C_1 \cup F \cup F'\rangle) = 0$. Using again the exact cohomology sequences associated to

$$0 \to \Theta_V\langle C_0 \cup C_1 \cup F \cup F'\rangle \to \Theta_V\langle C_0 \cup C_1 \cup F\rangle \to \mathcal{O}_{F'} \to 0$$

and

$$0 \to \Theta_V\langle C_0 \cup C_1 \cup F'\rangle \to \Theta_V\langle C_0 \cup C_1\rangle \to \mathcal{O}_F \to 0,$$

we get as above that $h^1(\Theta_V\langle C_0 \cup C_1 \cup F_1 \cup \ldots \cup F_s\rangle) = 0$ if $s \leq 2$. Hence $h^0(\Theta_V\langle C_0 \cup C_1 \cup F_1 \cup \ldots \cup F_s\rangle) = 2 - s$ for $s \leq 2$, as desired.

Finally consider the case, where the $\mathbb{C}^\star$-triple is untwisted, but $C_0 C_1 = 0$. We have a canonical exact sequence

$$0 \to \Theta_{V/\mathbb{P}^1} \to \Theta_V \to \pi^\star(\Theta_{\mathbb{P}^1}) \to 0.$$

Here $\Theta_{V/\mathbb{P}^1}$ is the sheaf of vertical vector fields, which is isomorphic to $\mathcal{O}(2C_\infty + nF)$, since $\omega_{\Sigma(n)} = \mathcal{O}(-2C_\infty - (n+2)F)$. Hence, if a vector field is tangent to three fibers $F_1, F_2, F_3$, then it is vertical because of $H^0(\pi^\star(\Theta_{\mathbb{P}^1})(-F_1 - F_2 - F_3)) = 0$. If a vertical vector field is tangent along $C_0 \cup C_1$, then it must be zero along $C_0 \cup C_1$, and so it is a section of $\Theta_{V/\mathbb{P}^1}(-C_0 - C_1) \simeq \mathcal{O}_V$, since $C_0 + C_1 = 2C_\infty + nF$ because of $C_0 C_1 = 0$. Hence we obtain that $h^0(\Theta_V\langle C_0 + C_1 + F_1 + \ldots + F_s\rangle) = 1$ for $s \geq 3$. Now with the same arguments as above it is easily seen, that $h^0(\Theta_V\langle C_0 + C_1 + F_1 + \ldots + F_s\rangle) = 4 - s$ for $s \leq 3$. □

*Example* 6.9. We will give an example of a 1–dimensional family $(V_t, D_t)$, such that $X_t = V_t \setminus D_t$ are topologically contractible pairwise non isomorphic surfaces of logarithmic Kodaira dimension 1. We start from the Hirzebruch surface $\Sigma := \Sigma\ (1)$. Let $\overline{F}_0, \overline{F}_1, \overline{F}_\infty$ be the fibers over $0, 1, \infty$ resp., and consider disjoint sections $\overline{C}_\infty, \overline{C}_0$, where $\overline{C}_\infty^2 = -1$. Let $P \in \overline{F}_\infty$ be a fixed point. Then we perform blowing ups in the points $P$, $\overline{C}_0 \cap \overline{F}_0$ and $\overline{C}_0 \cap \overline{F}_1$, and obtain a surface $q : \Sigma' \to \Sigma$. Blowing up again the singular points of the fibers $q^{-1}(\overline{F}_0), q^{-1}(\overline{F}_1)$, and then on the resulting surface also the intersection point of the proper transform of $\overline{F}_1$ with the unique $(-1)$–curves in the fibre, we obtain a modification $\pi : V' \to \Sigma$. In the fibers $F_0 := \pi^{-1}(\overline{F}_0)$ and $F_1 := \pi^{-1}(\overline{F}_1)$ there are unique $(-1)$–curves $G_0$ resp. $G_1$. On $V'$ we consider the SNC–divisor

$$D' := C_0 + C_1 + F_\infty + (F_{0_{red}} - G_0) + (F_{1_{red}} - G_1),$$

see FIGURE 18. Let $E$ be the $(-1)$–curve $\pi^{-1}(P)$. For every point $t \in E$, which is not on the proper transform $\hat{F}_\infty$ of $\overline{F}_\infty$, we can perform a half-point detachment in $t$ and obtain a pair $(V_t, D_t)$. By **[16]** (3.3), the affine surface $X_t = V_t \setminus D_t$ is contractible.

We claim that if $X_t$ and $X_{t'}$ are isomorphic, then necessarily $t = t'$. In fact, every isomorphism $X_t \to X_t'$ gives a birational map $(V_t, D_t) \xrightarrow{\varphi} (V_{t'}, D_{t'})$. Since the dual graph of $D_t$ (see FIGURE 19) has only admissible chains, by (A.3) $(V_t, D_t)$ is absolutely minimal, i.e. $\varphi$ is regular everywhere.

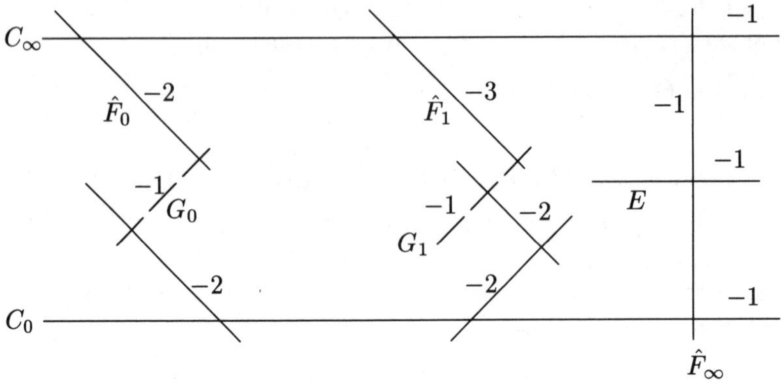


FIGURE 18.

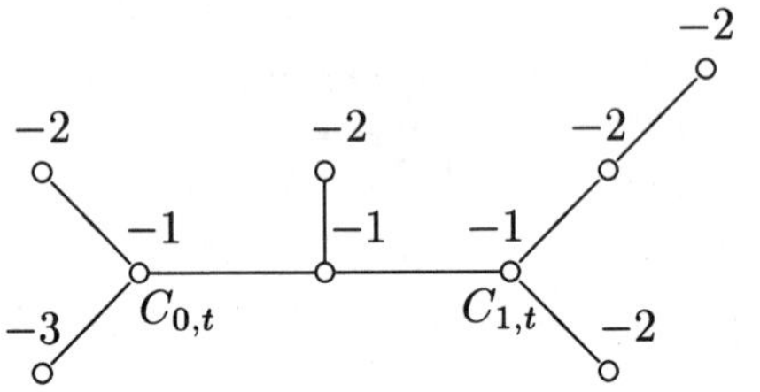


FIGURE 19.

Let $F \subseteq V_t$, $F' \subseteq V_{t'}$ be the generic fibers of the given fiberings $V_t \to \mathbb{P}^1$ resp. $V_{t'} \to \mathbb{P}^1$. Then $\varphi(F) \cdot D_i' = F \cdot D_i$ for every component $D_i'$ of $D_t'$, since $\varphi$ induces the identical isomorphism of the dual graphs of $D_t$ resp. $D_{t'}$. Since the $D_i'$ is a basis for $\operatorname{Pic}(V')$, we get $\varphi(F) \sim F'$, and so $\varphi$ is compatible with the projections to $\mathbb{P}^1$. In particular, the fibre of $V_t \to \mathbb{P}^1$ over $0, 1, \infty$ resp. is mapped onto the corresponding fibre of $V_{t'} \to \mathbb{P}^1$. It follows, that by blowing down $V_t$ and $V_{t'}$ to $\Sigma$ (1), we obtain a morphism $\overline{\varphi} : \Sigma(1) \to \Sigma(1)$ mapping the fibers over $0, 1, \infty$ onto itself and fixing the point $P$. From the description in (2.5) (1) we get that $\overline{\varphi} = \mathrm{id}_{\Sigma(1)}$ and so the induced map $\varphi' : V' \to V'$ is the identity. This forces $t = t'$. □

*Remark* 6.10. If $(V, D)$ is an SNC–pair and $X = V \setminus D$ is contractible, then by the $h$–cobordism theorem $X \times \mathbb{C}$ is diffeomorphic to $\mathbb{C}^3$, see [**26**]. Hence by the previous example and the cancellation theorem of [**18**], there is a 1–parameter family of pairwise non–isomorphic algebraic structures on $\mathbb{C}^3$, i.e. there are moduli for exotic algebraic structures on $\mathbb{C}^3$, cf. [**32**], [**33**]. Observe that replacing in (6.10) the half–point detachment by a comb–attachment with

dual graph

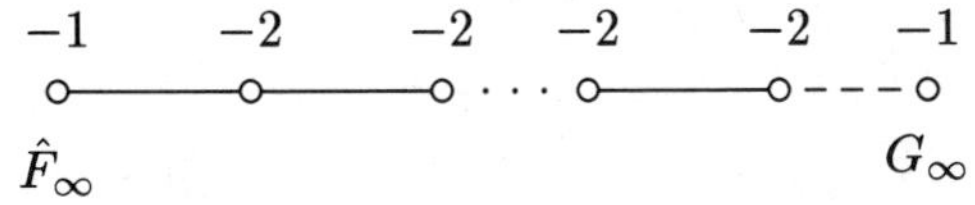


we can also obtain an arbitrary large number of moduli.

In the next proposition we describe the deformations in the case of logarithmic Kodaira dimension 0.

PROPOSITION 6.11. *Let $(V, D)$ be a strictly minimal SNC–pair of logarithmic Kodaira dimension 0, and assume that $(V, D)$ is pre–$\mathbb{Q}$–acyclic. Then $h^i(\Theta_V\langle D\rangle) = 0$ for all $i$.*

PROOF. $(V, D)$ is one of the surfaces $H[k, -k]$, $H[-1, 0, -1]$, or $Y\{3, 3, 3\}$, $Y\{2, 4, 4\}$, $Y\{2, 3, 6\}$, see (5.14), (5.15) (4). The cases $H[k, -k]$ and $H[-1, 0, -1]$ are of the type described in (5.14), and so in this case the conclusion follows from (6.6). Now assume that $(V, D)$ is one of the types $Y\{\dots\}$. By **[5]** (4.13) we obtain these types from one of the line configurations (4.2) – (4.6) in **[5]** by blowing up over certain intersection points and cutting cycles. Thus (6.11) follows from (6.12) below. □

We will add a few remarks about the case of logarithmic Kodaira dimension 2. In this case, up to now there is no full classification of $\mathbb{Q}$– or $\mathbb{Z}$–acyclic surfaces. Most of the examples, which are known **[26]**, **[16]**, **[31]**, **[32]**, **[33]**, etc., can be constructed from line configurations in the projective plane. A classification of the line configurations, which lead to $\mathbb{Z}$–acyclic surfaces, has been given in **[7]**, **[8]**, **[9]**. It is also shown there that those configurations of lines, leading to surfaces of logarithmic Kodaira dimension 2, are given by their combinatorial data, which implies that they admit no nontrivial logarithmic deformations with a basis of positive dimension. Slightly more generally, we will prove

PROPOSITION 6.12. *Let $(V, D)$ be a $\mathbb{Q}$–acyclic surface obtained from one of the line configurations* (4.2) – (4-6) *in* **[5]**. *Then $h^i(\Theta_V\langle D\rangle) = 0$ for $i = 0, 1, 2$.*

For the proof of (6.12) we need a few preparations. Let $\overline{V}$ be a compact smooth surface and $C \subseteq \overline{V}$ a reduced curve, such that in every singular point $p \in C$ the branches of $C$ are smooth and have no common tangent; in **[7]** these singularities are called *tidy*. Let $(V, D)$ be an SNC–pair such that

(1) $V$ is obtained from $\overline{V}$ by a sequence of inner blowing ups (i.e. every blowing up takes place in a singular point of the reduced preimage of $C$),
(2) $D = \pi^{-1}(C)_{red} - \Sigma E_i$, where $\pi : V \to \overline{V}$ is the canonical map and $E_i \subseteq V$ are $(-1)$–curves, which are blown down by $\pi$.

Using (1.5) (4) and (1.7) (3) it is easily seen, that the numbers $h^i(\Theta_V\langle D\rangle)$ only depend on the configuration $C \subseteq \overline{V}$. Hence we will denote this numbers by $h^i(C)$. The following observations will be useful in the proof of (6.12).

LEMMA 6.13. (1) *If $\sigma : V' \to V$ is a blowing up in a singular point of $C$ and $C' = \sigma^{-1}(C)_{red}$, then $h^i(C) = h^i C')$ for all $i$.*

(2) *Let $C = C_1 \cup \ldots \cup C_n$ be the irreducible components of $C$ and set $\tilde{C} = C_2 \cup \ldots \cup C_n$. Assume that the number of points $P$ on $C_1$, which are singular points of $\tilde{C}$, is $C_1^2 + 1$. Then $h^i(\tilde{C}) = h^i(C)$ for all $i$.*

PROOF. (1) follows easily from (1.5), (4) and (1.7), (3). For the proof of (2), let $V \to \overline{V}$ be the blowing up of $\overline{V}$ in the points, where the multiplicity of $C$ is at least three. Let $D$ resp. $\tilde{D}$ be the reduced preimage of $C$ resp. $\tilde{C}$, and denote by $\hat{C}_1$ the proper transform of $C_1$. Then $\hat{C}_1^2 = -1$ by our assumption, $h^i(C) = h^i(\Theta_V\langle D\rangle)$ and $h^i(\tilde{C}) = h^i(\Theta_V\langle\tilde{D}\rangle)$. Hence (2) follows from (1.7), (3). □

LEMMA 6.14. (1) *If $C = l_1 \cup l_2 \cup l_3 \cup \lambda_1 \cup \lambda_2 \cup \lambda_3 \subseteq \mathbb{P}^1 \times \mathbb{P}^1$ is the union of three vertical lines $l_1, l_2, l_3$ and three horizontal lines $\lambda_1, \lambda_2, \lambda_3$, then $h^i(C) = 0$ for all $i$.*

(2) *If $C^\star$ is the union of four lines in general position in $\mathbb{P}^2$, then $h^i(C^\star) = 0$ for all $i$.*

PROOF. (1) follows from the fact that $\Theta_{\mathbb{P}^1\times\mathbb{P}^1}\langle C\rangle = \mathcal{O}(-1,0) \oplus \mathcal{O}(0,-1)$ and the Künneth formula. For the proof of (2), let $P_1, P_2$ be singular points of $C^\star$, which are not connected by a line of $C^\star$. Let $\lambda$ be the line connecting $P_1$ and $P_2$. By (6.13) (2), $h^i(C^\star) = h^i(C^\star \cup \lambda)$. Blowing up $P_1, P_2$ and blowing down the proper transform of $\lambda$, we arrive at the line configuration $C \subseteq \mathbb{P}^1 \times \mathbb{P}^1$ of (1). Since $h^i(C) = h^i(C^\star \cup \lambda)$ by (6.13), we get (2). □

PROOF OF (6.12): For each of the line configurations in (4.3), (4.4), (4.6) and also the first line configuration in (4.5) of **[7]** there is a line with 2 points of multiplicity $\geq 3$. Deleting this line does not change the $h^i$ by (6.13), (2), and after a finite number of steps we arrive at the configuration (4.2), which is the configuration described in (6.14), (2). Hence in each of these cases $h^i(C) = 0$ for all $i$.

It remains to show the assertion in the case of the second line configuration in (4.5) of **[7]**, which is shown on the left hand side of FIGURE 20. Here $\omega$ denotes a 3rd root of 1, and one must add the lines $l_\infty$ at $\infty$ and $l_\omega := \{z_2 = \omega z_1\}$ , which passes through the points marked by a circle. Denote by $P_1$ resp. $P_2$ the

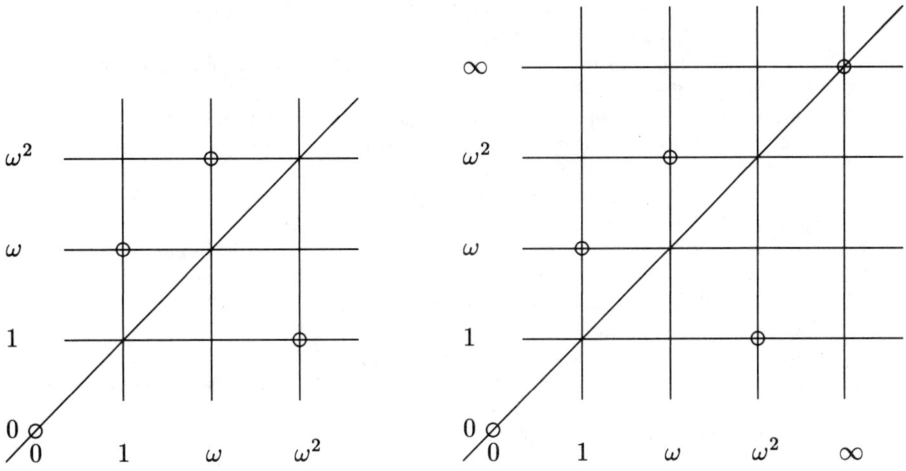


FIGURE 20

intersection points of the three vertical resp. horizontal lines with $l_\infty$. Blowing up $P_1$ and $P_2$ and blowing down the proper transform of $l_\infty$, we do not change the $h^i$ and obtain the configuration on $\mathbb{P}^1 \times \mathbb{P}^1$, which is shown on the right hand side of FIGURE 20. Using (6.13) (2), we can delete first the vertical and horizontal lines over $\infty$, and then $\{z_1 = z_2\}$ and $\{wz_1 = z_2\}$, without changing $h^i$. Now (6.14) (1) yields the vanishing of the $h^i$, as desired. □

## REFERENCES


1. W. Barth, C. Peters, A. Van de Ven, *Compact Complex Surfaces*, Springer-Verlag: Berlin–Heidelberg–New York 1984
2. A. Beauville, *Surfaces algebriques complexes*, Asterisque **54** (1978)
3. J. Bingener, *Offenheit der Versalität in der analytischen Geometrie*, Math. Z. **173** (1980), 241 – 281
4. W. Danielewski, *On a cancellation problem and automorphism group of affine algebraic varieties*, Preprint Warsawa (1989)
5. T. tom Dieck, T. Petrie, *Homology planes. An announcement and survey*, in: Topological Methods in Algebraic Transformation Groups. Progress in Math. Vol. **80**. Boston: Birkhäuser 1989, 27 – 48
6. T. tom Dieck, T. Petrie, *Contractible affine surfaces of Kodaira dimension one*, Japan J. Math. **16** – 1 (1990), 147 – 169
7. T. tom Dieck, T. Petrie, *Homology planes and algebraic curves*, I. Preprint. Math. Gott. **6** (1989), 1 – 62
8. T. tom Dieck, T. Petrie, *Arrangements of lines with tree resolutions*, Arch. Math. **56** (1991), 189-196
9. T. tom Dieck, *Linear plane divisors of homology planes*, J. Fac. Sci. Univ. Tokyo (Sec. 1A) **37** (1990), 33 – 69
10. T. tom Dieck, *Symmetric homology planes*, Math. Ann. **286** (1990), 143 – 152
11. T. tom Dieck, *Homology planes without cancellation property*, Preprint. Math. Gott. **4** (1991), 1 – 15

12. H. Flenner, *Über Deformationen holomorpher Abbildungen*, Thesis; Osnabrücker Schriften zur Mathematik: Osnabrück, 1978
13. T. Fujita, *On the topology of non–complete algebraic surfaces*, J. Fac. Sci. Univ. Tokyo (Sec. 1A) **29** (1982), 503 – 566
14. T. Fujita, *On Zariski problem*, Proc. Japan Acad. (Ser. A) **55** (1979), 106 – 110
15. M. H. Gizatullin, *Invariants of incomplete algebraic surfaces that can be obtained by means of completions*, Izvestia Akad. Nauk SSSR, Ser. Mat. **35** (1971), 485-497
16. R. V. Gurjar, M. Miyanishi, *Affine surfaces with* $k \leq 1$, In: Algebraic Geometry and Commutative Algebra in Honor of M. Nagata, 1987, 99 – 124
17. R. V. Gurjar, A. R. Shastri, *On the rationality of complex homology 2–cells, I, II*, J. Math. Soc. Japan **41** – 1 (1989), 37 – 56; **41** – 2 (1989), 175 – 212
18. Sh. Iitaka, T. Fujita, *Cancellation theorem for algebraic varieties*, J. Fac. Sci. Univ. Tokyo (sec. 1a) **24** (1977), 123 – 127
19. Y. Kawamata, *On the classification of non–complete algebraic surfaces*, in: Proc. Algebraic Geometry, Kopenhagen 1978. Lecture Notes in Math. **732**, Springer–Verlag: Berlin–Heidelberg New–York 1978, 215 – 232
20. Y. Kawamata, *On deformations of compactifiable complex manifolds*, Math. Ann. **235** (1978), 247 – 265
21. K. Kodaira, *Holomorphic mappings of polydiscs into compact complex manifolds*, J. Diff. Geometry **6** (1971), 33–46
22. M. Miyanishi, T. Sugie, *Affine surfaces containing cylinderlike open sets*, J. Math. Kyoto Univ. **20** (1980), 11 – 42
23. M. Miyanishi, S. Tsunoda, *Open algebraic surfaces with Kodaira dimension* $-\infty$, in: Symposia in Pure Mathematics **46** – 1 (1987), 435 – 450
24. M. Miyanishi, *Non–complete algebraic surfaces*, Lect. Notes in Math. **857**, Springer–Verlag: Berlin–Heidelberg–New York 1981
25. J. A. Morrow, *Minimal compactifications of* $\mathbb{C}^2$, Rice Univ. Studies **59** (1973), 97 – 112
26. C. P. Ramanujam, *A topological characterization of the affine plane as an algebraic variety*, Ann. Math. **94** (1971), 69 – 88
27. D. S. Rim, *Formal deformation theory*, in: SGA 7 II, Lecture Notes in Math. **288**. Springer–Verlag: Berlin–Heidelberg–New York (1972)
28. O.Riemenschneider, *Deformations of rational singularities and their resolutions*, Rice Univ. Studies **59** (1973), 119–130
29. P. Russell, *On affine–ruled rational surfaces*, Math. Ann. **255** (1981), 287 – 302
30. T. Sugie, *On a characterization of surfaces containing cylinderlike open sets*, Osaka J. Math. **17** (1980), 363 – 376
31. T. Sugie, *On T. Petrie's problem concerning homology planes*, J. Math. Kyoto Univ. **30** – 2 (1990), 317 – 342
32. M. G. Zaidenberg, *Ramanujam's surfaces and exotic algebraic structures on* $\mathbb{C}^n$, $n \geq 3$, Soviet Math. Dokl. **42** – 2 (1991), 636 – 640
33. M. G. Zaidenberg, *An analytic cancellation theorem and exotic algebraic structures on* $\mathbb{C}^n$, $n \geq 3$, preprint MPI / **91** – 26 (1991), 1 – 33



Mathematisches Institut der Georg–August–Universität, Bunsenstr.3–5, D–3400 Göttingen

Université de Grenoble, Institut Fourier, B.P. 74, F–38402 Saint–Martin–d'Hères

Contemporary Mathematics
Volume **162**, 1994

# ON A QUESTION OF J. KOLLÁR

A. GRASSI

This work is motivated by some recent conjectures by J. Kollár [Ko2, 3.5]. I would like to thank Kollár for showing me an earlier version of his paper and useful discussions. Thanks also to Mark Gross for some suggestions. We first present the results of this paper and their application to Kollár's conjecture [Ko2, 3.5].

**Theorem 1.5.** *Let $X$ be a minimal threefold of Kodaira dimension* 1 *and let $\pi : X \to C$ be associated Iitaka fibration. Then*

(1) *either $K_X \cdot c_2 = 0$ and the smooth fibers are abelian or bi-elliptic surfaces or*
(2) *there exists an $\epsilon > 0$, independent of $X$, such that $K_X \cdot c_2 \geq \epsilon > 0$.*

**Proposition 2.5.** *Let $X$ be a minimal threefold of Kodaira dimension* 2 *and let $\pi : X \to S$ be the associated elliptic fibration. If the $j$ invariant of the fibers of $\pi$ is not a constant function, then there exists an $\epsilon > 0$, independent of $X$, such that $K_X \cdot c_2(X) \geq \epsilon$.*

**Theorem 2.7.** *Let $X$ be an $n$ dimensional variety with terminal singularities. Assume that $|mK_X|$ is basepoint free for some integer $m$, and that $\mathrm{kod}(X) = n-1$, that is the associated morphism $\pi : X \to S$ is an elliptic fibration. (In particular $X$ is minimal and $K_X$ is nef.)*

*Then $(K_X)^{n-2} \cdot c_2(X) = 0$ if and only if there is no monodromy over a divisor on $S$. In particular the $j$ invariant of the fibers is a constant function.*

If $dim(X) \leq 3$, then the hypothesis of $X$ minimal implies $|mK_X|$ base point free, for some $m$ [Ka2], [Ko et al] and the associated morphism is the Iitaka fibration.

## §1

**Theorem 1.1 [Mo].** *Let $X$ be a minimal threefold of Kodaira dimension* 1 *and let $\pi : X \to C$ be the associated Iitaka fibration. Then there exists an integer $m_0$ independent of $X$, such that $m_0 K_X = \pi^*(mD)$, with $D$ ample line bundle on $C$.*

We start by recalling the following:

1991 *Mathematics Subject Classification.* Primary 14E30, Secondary 14J10.
Research at MSRI supported in part by NSF grant #DMS 9022140
This paper is in final form and no version of it will be submitted for publication elsewhere.

**Theorem 1.2 [Mk, Th 2.3] [Ko1, Th 2.6].** *Let $\tilde{\pi} : \tilde{X} \to \tilde{S}$ be a surjective map between smooth projective varieties. Let $\Sigma$ be a normal crossings divisor such that $\pi$ is smooth above $S - \Sigma$. Then the sheaves $R^q\pi_*(\mathcal{O}_{\tilde{X}})$ are locally free , $\forall q$ and $R^q\pi_*(\mathcal{O}_{\tilde{X}}) = 0$ for $q \geq dim\tilde{X} - dim\tilde{S}$.*

**Proposition 1.3.** *Let $\pi : X \to C$ be as in 1.1 and $\mu : \tilde{X} \to X$ a resolution. Then there exist the following short exact sequences and isomorphisms:*

$$H^0(\tilde{X}, \mu^*(rm_0K_X)) = H^0(C, rD)$$
$$0 \to H^1(C, rD) \to H^1(\tilde{X}, \mu^*(rm_0K_X)) \to H^0(C, rD \otimes R^1(\mu\pi)_*\mathcal{O}_{\tilde{X}}) \to 0$$
$$H^3(\tilde{X}, \mu^*(rm_0K_X)) = H^3(C, rD \otimes R^2(\mu\pi)_*\mathcal{O}_{\tilde{X}})$$

*where $r$ is an integral multiple of $m_0$.*

*Proof.* It follows immediately from Theorem 1.2 and the spectral sequence:

$$E_2^{p,q} = H^p(C, R^q(\pi\mu)_*(\mu^*\pi^*rD))$$
$$\Rightarrow H^{p+q}(\tilde{X}, \mu^*\pi^*(rD)) = H^{p+q}(\tilde{X}, \mu^*rm_0K_X). \quad \square$$

**Corollary 1.4.** *Let $\pi : X \to C$ be as in 1.1 and let $F$ be the general fiber of $\pi$. Then*

(1.4.1) $$\chi(\tilde{X}, r\mu^*(K_{\tilde{X}})) = \chi(C, rD) - \chi(C, rD \otimes R^1(\mu\pi)_*(\mathcal{O}_{\tilde{X}})) + \chi(C, rD \otimes R^2(\mu\pi)_*(\mathcal{O}_{\tilde{X}}))$$

(1.4.2) $$K_X \cdot c_2(X) = \chi(\mathcal{O}_F) \cdot \deg D$$

*Proof.* (1.4.1) follows from 1.3.

(1.4.2) By [Ko1, Th. 2.6] (see 1.2) the sheaves $R^i(\mu\pi)_*(\mathcal{O}_{\tilde{X}})$ are locally free and

$$c_1(rD \otimes R^i(\mu\pi)_*(\mathcal{O}_{\tilde{X}})) = rh^i(F, \mathcal{O}_F)\deg D + h^{i-1}(F, \mathcal{O}_F) \cdot c_1(R^i(\mu\pi)_*(\mathcal{O}_{\tilde{X}})).$$

On the other hand the Riemann-Roch Theorem for threefolds implies $\chi(\tilde{X}, r\mu^*(K_{\tilde{X}})) = \chi(\mathcal{O}_{\tilde{X}}) + rK_X \cdot c_2$. $\square$

The following result follows from (1.4.2) and Theorem 1.1:

**Theorem 1.5.** *Let $X$ be a minimal threefold of Kodaira dimension 1 and let $\pi : X \to C$ be associated Iitaka fibration. Then*

(1) *either $K_X \cdot c_2 = 0$ and the smooth fibers are abelian or bi-elliptic surfaces or*

(2) *there exists a $m_0 > 0$, independent of $X$, such that $K_X \cdot c_2 \geq \frac{1}{m_0} > 0$.*

## §2

We start by recalling some definitions and properties regarding elliptic fibrations.

In the following, $\tilde{X}$ is a $n$-dimensional variety and $\tilde{\pi} : \tilde{X} \to \tilde{S}$ an elliptic fibration, with $\tilde{S}$ smooth. Without loss of generality we can assume that the ramification locus is a divisor with simple normal crossings and that the $j$ invariant of the fibers extends to a morphism $j : \tilde{S} \to \mathbb{P}^1$.

**Definition-Theorem 2.1 [Kd],[Ka1], [Mk].**

(1) Set:

$$\begin{cases} \mathcal{O}_{\tilde{S}}(\tilde{\Delta}) = \tilde{\pi}_*(K_{\tilde{X}/\tilde{S}}) \\ \tilde{\Lambda} = \tilde{\Delta} + \sum \frac{m_i - 1}{m_i} Y_i \end{cases}$$

where the fiber over the general point of $Y_i$ is a multiple fiber of multiplicity $m_i$. Then

$$12\tilde{\pi}_*(K_{\tilde{X}/\tilde{S}}) = 12\mathcal{O}_S(\tilde{\Delta}) = \mathcal{O}_S(\sum 12a_k D_k) + j^{-1}(c)),$$

where the numbers $a_k \in \frac{1}{12}\mathbb{N}$ depend on the type of the singular fiber (with monodromy) over the general point of $D_k$ and are described explicitly in [Kd]; $c$ is a general point in $\mathbb{P}^1$. If there is no monodromy $\tilde{\Delta} = 0$. $\tilde{\Lambda}$ is a $\mathbb{Q}$-divisor with no reduced part.

(2) If $\pi : X \to S$ is *any* elliptic fibration (with no assumption on $X$, $S$ or the ramification locus), we consider a birationally equivalent fibration

$$\begin{array}{ccc} \tilde{X} & \xrightarrow{\mu} & X \\ {\scriptstyle\tilde{\pi}}\downarrow & & \downarrow{\scriptstyle\pi} \\ \tilde{S} & \xrightarrow{\phi} & S \end{array}$$

with $\tilde{\pi} : \tilde{X} \to \tilde{S}$ as above and define $\Lambda = \phi_*(\tilde{\Lambda})$, $\Delta = \phi_*(\tilde{\Delta})$.

$(S, \Lambda)$ is a log surface (see for example, [Ko et al]).

If $K_X = \pi^*(K_S + \Lambda)$, and $X$ has canonical singularities then $(S, \Lambda)$ has log terminal singularities [Na, 0.4].

**Proposition 2.2.** *Let $\pi : X \to S$ be an elliptic fibration, with $\dim(X) = n$ and $\tilde{\pi} : \tilde{X} \to \tilde{S}$ a birationally equivalent fibration as in 2.1. Assume that $mK_X = \pi^* m(K_S + \Lambda)$, for some integer $m$ and let $r$ any integral multiple of $m$. Then*

$$E_2^{p,q} = H^p(\tilde{S}, (R^q\tilde{\pi}_*)(r\tilde{\pi}^*\phi^*(K_S + \Lambda))) \Rightarrow H^{p+q}(\tilde{X}, \tilde{\pi}^*(r\phi^*(K_S + \Lambda)))$$

*and there exists the following long exact sequence: (Gysin)*

$$\cdots \to H^\ell(\tilde{S}, r\phi^*(K_S + \Lambda)) \to H^\ell(\tilde{X}, r\mu^* K_X) \to H^{\ell-1}(\tilde{S}, r\phi^*(K_S + \Lambda) - \tilde{\Delta}) \to$$

$$H^{\ell+1}(\tilde{S}, r\phi^*(K_S + \Lambda)) \to H^{\ell+1}(\tilde{X}, r\mu^* K_X) \to H^\ell(\tilde{S}, r\phi^*(K_S + \Lambda) - \tilde{\Delta}) \to \cdots$$

*( $0 < \ell \leq n - 2$). Furthermore the following isomorphisms hold :*

$$H^n(\tilde{X}, r\mu^* K_X) \sim H^{n-1}(\tilde{S}, r\phi^*(K_S + \Lambda) - \tilde{\Delta})$$

and $H^0(\tilde{X}, r\mu^* K_X) \sim H^0(\tilde{S}, r\phi^*(K_S + \Lambda))$.

*Proof.* Consider the spectral sequence for $\tilde{\pi}^*(r\phi^*(K_S+\Lambda))$: by the projection formula, ( 2.1) and ( 1.2) we have

$$(R^i\tilde{\pi}_*)\tilde{\pi}^*\phi^*(r(K_S+\Lambda)) = r\phi^*(K_S+\Lambda) + R^i\tilde{\pi}_*\mathcal{O}_{\tilde{X}} = \begin{cases} r\phi^*(K_S+\Lambda) \text{ if } i=0 \\ r\phi^*(K_S+\Lambda) - \tilde{\Delta} \text{ if } i=1 \\ 0 \text{ if } i \geq 2. \end{cases}$$

Consequently $d_3 = 0$.
At the $E_2$ level we have as non zero terms:

$$\begin{matrix} E_2^{0,1} & E_2^{1,1} & E_2^{2,1}\cdots \\ E_2^{0,0} & E_2^{1,0} & E_2^{2,0}\cdots \end{matrix}$$

We have the following exact sequence:

$$\cdots \to E^{p,q} \to H^p \to E_2^{p-1,1} \to E^{p+1,0} \to H^{p+1} \to E^{p,1} \to \cdots$$

Now $\tilde{\pi}^*(r\phi^*(K_S+\Lambda)) = \mu^*\pi^*r(K_S+\Lambda) = \mu^*(rK_X)$ by commutativity of the diagram and by hypothesis. □

J. Kollár pointed out that one can get a faster proof by writing the exact sequence for $K_X = \pi^*(K_S+\Lambda)$ and using the vanishing theorem for higher direct images of the dualizing sheaf [Kol]. The argument given above can be extended to any Kodaira dimension.

**Corollary 2.3.**

(2.3.1) $\chi(\tilde{X}, r\mu^*(K_{\tilde{X}})) = \chi(\tilde{S}, r\phi^*(K_S+\Lambda)) - \chi(\tilde{S}, r\phi^*(K_S+\Lambda) - \Delta)$.

(2.3.2) $$(K_X)^{n-2}\cdot c_2(X) = 12(K_S+\Lambda)^{n-2}\cdot\Delta$$

*Proof.* (2.3.1) follows immediately from (2.2). Apply the Riemann-Roch Theorem to both sides of ( 2.3.1) and compare the coefficients of $r^{n-2}$. On the left hand side we have

$$\frac{1}{12}({K_{\tilde{X}}}^2 + c_2)\cdot\frac{n-1}{(n-2)!}\mu^*(K_X)^{n-2} = \frac{n-1}{12(n-2)!}(K_{\tilde{X}})^{n-2}\cdot c_2,$$

in fact $K_{\tilde{X}}^n = 0$ by hypothesis. The right hand side gives $\frac{n-1}{(n-2)!}(K_S+\Lambda)^{n-2}\cdot\Delta$.
( 2.3.2) follows from the two formulas combined. □

If $X$ is minimal $K_S+\Lambda$ is nef; since $12\Delta$ is effective (see 2.1), we have:

$$(K_X)^{n-2}\cdot c_2(X) = 12(K_S+\Lambda)^{n-2}\cdot\Delta \geq 0;$$

this is what expected [Mi].

**Corollary 2.4.** *If $\pi$ has no monodromy over any divisor of $S$, then $(K_X)^{n-2} \cdot c_2(X) = 0$.*

*Proof.* If $\pi$ has no monodromy over any divisor of $S$, then $\Delta = 0$ [Kd], (see part (1) in 2.1). □

**Proposition 2.5.** *Let $X$ be a minimal threefold of Kodaira dimension 2 and let $\pi : X \to S$ be the associated elliptic fibration. If the $j$ invariant of the fibers of $\pi$ is not a constant function, then there exists an $\epsilon > 0$, independent of $X$, such that $K_X \cdot c_2(X) \geq \epsilon$.*

*Proof.* If $j$ is non constant then $Supp(\Lambda)$ contains the reduced components $j^{-1}(c)$ ( 2.1, part 2); furthermore $K_S + \Lambda$ is ample. Now [Ko2, 6. 2.1 and 6.5] apply and we can find a positive number $\epsilon$, independent of $(S, \Lambda)$ such that $(K_S + \Lambda) \cdot j^{-1}(c) \geq \epsilon$. We deduce then that

$$K_X \cdot c_2(X) = 12(K_S + \Lambda) \cdot \Delta \geq 12(K_S + \Lambda) \cdot \frac{1}{12} j^{-1}(c) \geq \epsilon. \quad \square$$

**Remark 2.6.** *There exist many threefolds of Kodaira dimension 2 with $j$ constant and*
*$K_X \cdot c_2 >> 0$.*

*For example, let $\alpha$ be the local equation of a smooth curve $C$ of degree 5 in $\mathbb{P}^2$. Then $y^2 = x^3 + \alpha^3$ and $y^2 = X^3 + \alpha^2 x$ are the Weierstrass equations of two elliptic threefolds over $\mathbb{P}^2$, $W_0$ and $W_1$. By construction $W_0$ and $W_1$ have Kodaira dimension 2 and the $j$ invariant of the fibers is constant and equal to 0 and 1 respectively. $W_0$ and $W_1$ have singular fibers (of type $I_0^*$) only over $C \in \mathbb{P}^2$. Using ( 2.3.2) we find $K_X \cdot c_2 = 48$.*

The hypothesis on $j$ non constant comes from the nature of Kollár's original statement in [Ko2, 6.4, 6.5]. It might be possible to complete the proof of Kollár's conjecture by analyzing separately the particular cases of $j$ constant.

**Theorem 2.7.** *Let $X$ be an $n$ dimensional variety with terminal singularities. Assume that $|mK_X|$ is basepoint free for some integer $m$, and that $\operatorname{kod}(X) = n-1$, that is the associated morphism $\pi : X \to S$ is an elliptic fibration. (In particular $X$ is minimal and $K_X$ is nef.)*

*Then $(K_X)^{n-2} \cdot c_2(X) = 0$ if and only if there is no monodromy over a divisor on $S$, that is the rank of the relative Picard group $\rho(X/S) = 1$. In particular the $j$ invariant of the fibers is a constant function.*

*Proof.* Let $\pi : X \to S$ be the given elliptic fibration; then $K_X \equiv \pi^*(K_S + \Lambda)$, where $K_S + \Lambda$ is an ample divisor. By [Na, 0.4] $\Lambda$ is as in 2.1. and $(S, \Lambda)$ has log terminal singularities. Thus (2.3.2) applies and $K_X^{n-2} \cdot c_2(X) = 0$ if and only if $\Delta = \emptyset$. □

## References

[Ka1] Y. Kawamata, *Kodaira dimension of certain algebraic fiber spaces*, J. Fac. Sci. Tokyo Univ. IA **30** (1983), 1 –24.

[Ka2] Y. Kawamata, *Abundance Theorem for Threefolds*, Preprint (1991).

[Kd] K. Kodaira, *On compact analytic surfaces II, III*, Annals of Math **77,78** (1963), 563–626, 1–40.

[Ko1] J. Kollár, *Higher direct images of dualizing sheaves II*, Annals of Math **124** (1986), 171–202.

[Ko2] J. Kollár, *Log surfaces of general type: Some conjectures*, Preprint (1992).

[Ko et al] J. Kollár et al., *Log Flips and Abundance*, Preprint (1992).

[Mi] Y. Miyaoka, *The Chern classes and Kodaira dimension of a minimal variety*, Proc. Sym. Alg. Geom. Sendai 1985, Adv. Stud. Pure Math. **10**, Kinokuniya, Tokyo, 1987, pp. 449–476.

[Mk] A. Moriwaki, *Torsion freeness of higher direct images of canonical bundle*, Math. Ann. **276** (1987), 385–398.

[Mo] S. Mori, *A canonical bundle formula*, Preprint (1992).

[Na] N. Nakayama, *On Weierstrass models*, "Algebraic Geometry and Commutative Algebra", in honor of M.Nagata, Kinokuniya (1987), 405–431.



Mathematical Sciences Research Institute, 1000 Centennial Dr, Berkeley, CA 94720

*E-mail address*: grassi@msri.org

Contemporary Mathematics
Volume **162**, 1994

# BOUNDEDNESS FOR SOME SPECIAL FAMILIES OF EMBEDDED MANIFOLDS

PALTIN IONESCU AND MATEI TOMA

ABSTRACT. We are studying boundedness and classification problems for some special $r$-folds embedded in $\mathbb{P}^{2r}$ and $\mathbb{P}^{2r-1}$ (e.g. Fano fibrations over curves, scrolls over a surface).

It is conjectured that the Hilbert scheme of non-general type $r$-dimensional manifolds embedded in $\mathbb{P}^N$ for $N \leqslant 2r$ has only a finite number of components, or equivalently, that the degree of such manifolds is bounded. This was proved by Ellingsrud and Peskine for $r = 2$, $N = 4$ (cf. [4]), and subsequently in [3] for $r = 3$, $N = 5$. The proofs seem to depend on the fact that in these cases the codimension equals two. In [9] Schneider proves the conjecture for $N \leqslant 2r - 2$ by using the positivity of the Schur polynomials for the normal bundle and the fact that in this situation the canonical bundle is induced.

In this paper, we are using Schneider's method in order to prove boundedness for some special classes of embedded manifolds, e.g. Fano manifolds if $N = 2r$ (Proposition 1) and Fano fibrations over curves if $N = 2r - 1$ (Proposition 2). A different argument shows boundedness for scrolls over a surface, in case $N = 2r - 1$ (Proposition 4). We also give some existence results and a partial classification for the two special classes mentioned above when $N = 2r - 1$ (Propositions 3, 4 and 5).

ACKNOWLEDGEMENT. Most of this work was done while both authors were visiting the University of Bayreuth in the framework of the research program "Komplexe Mannigfaltigkeiten". We are grateful to all algebraic geometers working at the department for stimulating discussions and warm hospitality. Special thanks are due to Prof. Dr. Michael Schneider for his kind invitation.

Let $X$ be an $r$-dimensional manifold embedded in $\mathbb{P}^N(\mathbb{C})$ with degree $d$.

1991 *Mathematics Subject Classification.* Primary 14M07, 14J40; Secondary 14J10, 14E25.
This paper is in final form and no version of it will be submitted for publication elsewhere.

If $K_X$, $H_X$ denote the canonical and the hyperplane class on $X$ respectively, and $a$ is a nonnegative integer we shall say that $X$ *is in class* $C_a$ if

$$K_X \cdot H_X^{r-1} \leqslant a \cdot d,$$

(or equivalently, if $2g-2 \leqslant (a+r-1)d$, where $g = \frac{1}{2}(K \cdot H^{r-1} + (r-1)H^r) + 1$ is the sectional genus of $X$).

PROPOSITION 1. *If $X$ is in $C_a$, $X \subset \mathbb{P}^{2r}$, then*

$$d < (a + 2r + 1)^r.$$

*In particular if $X$ is an $r$-dimensional Fano manifold (i.e. $-K_X$ is ample) embedded in $\mathbb{P}^{2r}$ or if $K_X$ is numerically trivial and $X \subset \mathbb{P}^{2r}$ then $d$ is bounded by $(2r+1)^r$.*

*Proof.* Let $N_{X/\mathbb{P}^N}$ be the normal bundle of $X$ in $\mathbb{P}^N$. By the self-intersection formula (cf. [7]) we have

$$c_{N-r}(N_{X/\mathbb{P}^N}) = d \cdot H_X^{N-r}.$$

In our case

$$c_r(N_{X/\mathbb{P}^{2r}}) = d^2.$$

Since $N_{X/\mathbb{P}^N}$ is ample the Schur polynomials $(c_{k-1} \cdot c_1 - c_k)(N_{X/\mathbb{P}^N})$ are positive by the Fulton-Lazarsfeld theorem (cf. e.g. [5]) hence one gets

$$\begin{aligned} d^2 &= c_r(N_{X/\mathbb{P}^{2r}}) < c_1(N_{X/\mathbb{P}^{2r}})^r \\ &= (K_X + (2r+1)H_X)^r. \end{aligned}$$

We now apply the generalized Hodge index theorem ([2]) which asserts that for $D$ nef and $H$ ample on $X$ and $j = 1, \dots, r-1$ one has

$$(H^{j+1} \cdot D^{r-j-1})(H^{j-1} \cdot D^{r-j+1}) \leqslant (H^j \cdot D^{r-j})^2.$$

An easy consequence of it is that

$$D^r \leqslant \frac{(H^{r-1} \cdot D)^r}{(H^r)^{r-1}},$$

(cf. [2]).

Taking in our case $D = K_X + (2r+1)H_X$ and $H = H_X$ we get

$$\begin{aligned} (K_X + (2r+1)H_X)^r &\leqslant \frac{[H_X^{r-1} \cdot (K_X + (2r+1)H_X)]^r}{(H_X^r)^{r-1}} \\ &\leqslant \frac{(a+2r+1)^r d^r}{d^{r-1}} = (a+2r+1)^r d \end{aligned}$$

and the conclusion follows. □

REMARKS. 1. It is not difficult to give examples of $r$-dimensional Fano manifolds in $\mathbb{P}^{2r}$, e.g. complete intersections of type $(2, 2, \dots, 2)$.

2. Let $C$ denote the curve obtained by intersecting $X \subset \mathbb{P}^{2r}$ by an $(r+1)$-dimensional generic linear variety. If one assumes $h^1(\mathcal{O}_C(H)) = 0$ it follows $h^0(\mathcal{O}_C(H)) = d + 1 - g > 0$ so by the above proposition the degree of $X$ is bounded.

The next manifolds we investigate are $r$-dimensional fibrations $X$ over smooth curves, embedded in $\mathbb{P}^{2r-1}$. By Barth's theorem, [1], $X$ is regular hence the base curve must be $\mathbb{P}^1$.

PROPOSITION 2. *If $X$ is an $r$-dimensional manifold embedded in $\mathbb{P}^{2r-1}$ with degree $d$, admitting a fibering over $\mathbb{P}^1$ with a smooth fiber in class $C_a$, then*

$$d < (a+2r)^{r-1}.$$

*In particular if $X \subset \mathbb{P}^{2r-1}$ is an $r$-dimensional Fano fibration over $\mathbb{P}^1$ then*

$$d < (2r)^{r-1}.$$

*Proof.* Let $F$ be a smooth fiber in class $C_a$ of our fibration $X \longrightarrow \mathbb{P}^1$. The normal bundle sequence

$$0 \longrightarrow N_{F/X} \longrightarrow N_{F/\mathbb{P}^{2r-1}} \longrightarrow N_{X/\mathbb{P}^{2r-1}}|_F \longrightarrow 0$$

implies

$$c_i(N_{F/\mathbb{P}^{2r-1}}) = c_i(N_{X/\mathbb{P}^{2r-1}}|_F), \quad \text{for } 0 < i < r.$$

Making again use of the self-intersection formula for $X$, the positivity of the Schur polynomials and the generalized Hodge index theorem on $F$ we obtain

$$\begin{aligned} d \cdot \deg F = d \cdot H_F^{r-1} &= c_{r-1}(N_{X/\mathbb{P}^{2r-1}}|_F) \\ &< c_1^{r-1}(N_{X/\mathbb{P}^{2r-1}}|_F) = c_1^{r-1}(N_{F/\mathbb{P}^{2r-1}}) \\ = (K_F + 2rH_F)^{r-1} &\leqslant \frac{[H_F^{r-2} \cdot (K_F + 2rH_F)]}{(H_F^{r-1})^{r-2}} \\ &\leqslant \frac{[(a+2r)\deg F]^{r-1}}{(\deg F)^{r-2}} = (a+2r)^{r-1} \deg F \end{aligned}$$

and the conclusion follows. □

We now wish to illustrate the preceding result by a class of examples.

An $r$-dimensional manifold embedded in $\mathbb{P}^N$ is said to be a *(degree $f$)-hyper surface fibration over* $\mathbb{P}^1$ if it admits a map $p : X \to \mathbb{P}^1$ such that any of its fibers is a hypersurface of degree $f$ in some $r$-plane of $\mathbb{P}^N$.

We shall show the existence of such fibrations for $N = 2r - 1$. Remark that any smooth fiber is in class $C_f$, but in this case one easily gets a precise relation connecting $d$, $f$ and $r$. Indeed, letting $F$ be a smooth fiber of $p$ we have as in the proof of Proposition 2

$$\begin{aligned} df = dH_F^{r-1} &= c_{r-1}(N_{X/\mathbb{P}^{2r-1}}|_F) \\ = c_{r-1}(N_{F/\mathbb{P}^{2r-1}}) &= c_{r-1}(\mathcal{O}_F(f) \oplus \mathcal{O}_F(1)^{\oplus(r-1)}) = f\,[(r-1)f + 1], \end{aligned}$$

hence $d = (r-1)f + 1$.

PROPOSITION 3. *For any integers $r \geqslant 2$, $f \geqslant 1$, there exist $r$-dimensional (degree $f$)-hypersurface fibrations $X$ over $\mathbb{P}^1$ embedded in $\mathbb{P}^{2r-1}$.*

*Proof.* We may assume $f \geqslant 2$ since for $f = 1$ a solution is given by the Segre embedding $\mathbb{P}^{r-1} \times \mathbb{P}^1 \hookrightarrow \mathbb{P}^{2r-1}$.

Let $E = \mathcal{O}_{\mathbb{P}^1} \oplus \mathcal{O}_{\mathbb{P}^1} \oplus \mathcal{O}_{\mathbb{P}^1}(1)^{\oplus(r-1)}$ (we make no distinction between a vector bundle and its associated sheaf of sections), $Y = \mathbb{P}_{\mathbb{P}^1}(E)$ the projectivized bundle over $\mathbb{P}^1$, $\pi : Y \longrightarrow \mathbb{P}^1$ the projection, $L = \mathcal{O}_{\mathbb{P}(E)}(1)$ and $A = \pi^*\mathcal{O}_{\mathbb{P}^1}(1)$.

Our $X$ will be a general member of $|fL + A|$. First notice that since $(fL + A)^{r+1} = f^{r+1}(r-1) + (r+1)f^r > 0$ it follows by Bertini's theorem that $X$ is smooth and connected.

Let $\varphi_L : Y \to \mathbb{P}^{2r-1}$ be the morphism induced by $|L|$ ($L$ is generated by sections). Since $\varphi_L$ embeds the fibres of $\pi$ as linear subspaces in $\mathbb{P}^{2r-1}$, $p := \pi|_X : X \longrightarrow \mathbb{P}^1$ will be a (degree $f$)-hypersurface fibration if we prove that $\varphi_L|_X : X \longrightarrow \mathbb{P}^{2r-1}$ is an embedding.

Let $E' = \mathcal{O}_{\mathbb{P}^1}(1)^{\oplus(r-1)}$, $E'' = \mathcal{O}_{\mathbb{P}^1} \oplus \mathcal{O}_{\mathbb{P}^1}$ and $S = \mathbb{P}_{\mathbb{P}^1}(E'')$. Then $S$ is naturally embedded in $Y$ and $\varphi_L(S)$ is a line $\ell$ in $\mathbb{P}^{2r-1}$. In fact $p_1 := \pi|_S$ and $p_2 := \varphi_L|_S$ are the two projections of $S \cong \mathbb{P}^1 \times \mathbb{P}^1$. Moreover, by general facts on projective bundles, $\varphi_L(Y)$ is a cone with vertex $\ell$ over $\varphi_L(\mathbb{P}_{\mathbb{P}^1}(E'))$ and $\varphi_L|_{Y\setminus S} : Y \setminus S \longrightarrow \mathbb{P}^{2r-1} \setminus \ell$ is an embedding.

So the fact that $\varphi_L|_X$ is an embedding remains to be checked only around $S$.

Let on $S$, $F_i := p_i^*(\mathcal{O}_{\mathbb{P}^1}(1))$, $i = 1, 2$. Then $(fL + A)|_S = fF_2 + F_1$. Since $(fF_2 + F_1)^2 = 2f > 0$ and $(fF_2 + F_1)F_2 = 1$ we may choose a smooth $X$ in $|fL+A|$ such that its intersection with $S$ is smooth again and $\varphi_L|_{X\cap S} : X \cap S \to \ell$ is an isomorphism.

It follows that $\varphi_L|_X$ is injective. In order to see that it is also immersive we shall verify that the only tangent vectors to $Y$ killed by $d\varphi_L$ are tangent to the fibers of $S \to \ell$. (Such vectors cannot be tangent to $X$ too).

Consider an open subset $U$ of $\mathbb{P}^1$ over which $E$ is trivial and coordinates $(t, (x_0 : \cdots : x_r))$ on $Y_U \cong U \times \mathbb{P}^r$. Let $S_U \cong U \times \mathbb{P}_1 = \{x_2 = \cdots = x_r = 0\} \subset Y_U$. With respect to the chosen coordinates $\varphi_L$ is given by $(t, (x_0 : \cdots : x_r)) \mapsto (x_0 : \cdots : x_r : tx_2 : tx_3 : \cdots : tx_r)$. Since we are interested to compute everything around $S$, we may further assume $x_0 = 1$ thus getting the map $(t, x_1, \ldots, x_r) \mapsto (x_1, \ldots, x_r, tx_2, \cdots, tx_r)$. A direct computation of its Jacobian proves now our claim. $\square$

The simplest class of Fano fibrations is the class of scrolls. $X \subset \mathbb{P}^N$ is called a *scroll* over an $s$-dimensional manifold $B$ if there exists a surjective morphism $\pi : X \longrightarrow B$ whose fibres are $(r-s)$-dimensional linear subspaces of $\mathbb{P}^N$. Letting $E = \pi_*\mathcal{O}_X(H)$ we may identify $X$ to $\mathbb{P}_B(E)$ and $\mathcal{O}_X(H)$ to the tautological line bundle. It is an easy exercise to see that an $r$-dimensional scroll in $\mathbb{P}^{2r-1}$ over a curve is necessarily isomorphic to the Segre embedding of $\mathbb{P}^1 \times \mathbb{P}^{r-1}$. Now we would like to investigate $r$-dimensional scrolls in $\mathbb{P}^{2r-1}$ over a two-dimensional base.

When $r = 3$ Ottaviani ([8]) gave a complete classification of such objects. The following proposition partly extends his results to all dimensions.

PROPOSITION 4. *If $X = \mathbb{P}_B(E) \subset \mathbb{P}^{2r-1}$ is an $r$-dimensional scroll of degree $d$*

*over the smooth surface* $B$, *then:*

$$d < \frac{r^6}{2}$$

*and* $X$ *belongs to one of the following five classes:*

i) $r = 3$, $B$ *is a cubic in* $\mathbb{P}^3$, $d = 7$ (cf. [8]);

ii) $r \geqslant 3$, $X$ *is the isomorphic projection to* $\mathbb{P}^{2r-1}$ *of the Segre embedding* $\mathbb{P}^2 \times \mathbb{P}^{r-2} \subset \mathbb{P}^{3r-4}$, *having degree* $d = \frac{r(r-1)}{2}$;

iii) $r \geqslant 3$, $B \cong \mathbb{P}^2$, $d = \frac{r(r+1)}{2}$, $X$ *is the locus where a suitable morphism* $u : \mathcal{O}^3_{\mathbb{P}^{2r-1}} \longrightarrow \mathcal{O}^{\oplus(r+1)}_{\mathbb{P}^{2r-1}}(1)$ *drops rank;*

iv) $r \geqslant 3$, $B$ *is a* $K3$ *surface,* $d = r^2$;

v) $r \geqslant 7$, $B$ *is a regular surface with ample effective canonical divisor.*

REMARK. Actually we suspect that the last class never occurs. When $r = 3$, Ottaviani, [8], excludes class v) using codimension two arguments. When $r = 4$, R. Braun, (work in progress), rules it out by a divisibility argument, which inspired the proof below for the boundedness. The cases $r = 5$ and $r = 6$ may also be excluded by divisibility manipulations.

*Proof.* Keeping the above notations, let us consider the standard exact sequences:

$$0 \longrightarrow \mathcal{O}_{\mathbb{P}^{2r-1}}(-H) \longrightarrow \mathcal{O}^{2r}_{\mathbb{P}^{2r-1}} \longrightarrow T_{\mathbb{P}^{2r-1}}(-H) \longrightarrow 0$$
$$0 \longrightarrow T_X(-H) \longrightarrow T_{\mathbb{P}^{2r-1}}(-H)|_X \longrightarrow N_{X/\mathbb{P}^{2r-1}}(-H) \longrightarrow 0$$
$$0 \longrightarrow \mathcal{O}_X(-H) \longrightarrow \pi^* E^\vee \longrightarrow T_X(-H) \longrightarrow \pi^* T_B(-H) \longrightarrow 0.$$

One infers:

$$c(N')c(\pi^* E^\vee)c(\pi^* T_B(-H)) = 1,$$

where $N' := N_{X/\mathbb{P}^{2r-1}}(-H)$.

Hence $c(N') = c(\pi^* E^\vee)^{-1} c(\pi^* T_B(-H))^{-1}$.

Letting $c_1(\pi^* E^\vee) =: y_1$, $(c_1^2 - c_2)(\pi^* E^\vee) =: y_2$ we can immediately write down:

$$c(\pi^* E^\vee)^{-1} = 1 - y_1 + y_2.$$

We consider now the Segre classes of $\pi^* T_B(-H)$,

$$s(\pi^* T_B(-H)) := c(\pi^* T_B(-H))^{-1}.$$

Then

$$s_t(\pi^* T_B(-H))$$
$$= \binom{t+1}{1} H^t + \binom{t+1}{2} s_1(B) H^{t-1} + \binom{t+1}{3} s_2(B) H^{t-2},$$

cf. [5], p. 49. Denoting $c_1(B) = -s_1(B) =: x_1$, $(c_1^2 - c_2)(B) = s_2(B) =: x_2$, we can express

$$c_t(N') = \binom{t+1}{1} H^t - \left(\binom{t}{1} y_1 + \binom{t+1}{2} x_1\right) H^{t-1} + \left(\binom{t-1}{1} y_2 + \binom{t}{2} x_1 y_1 + \binom{t+1}{3} x_2\right) H^{t-2}.$$

We have a first equation

$$c_r(N') = 0. \tag{1}$$

Using the Hirsch-Leray formula

$$H^{r-1} + c_1(E^\vee) H^{r-2} + c_2(E^\vee) H^{r-3} = 0$$

we get

$$(*) \qquad y_2 H^{r-2} = d, \quad x_1 y_1 H^{r-2} = -x_1 H^{r-1}, \quad y_1^2 H^{r-2} = -y_1 H^{r-1}.$$

Introducing all these in (1) one has

$$2d - (r x_1 + y_1) H^{r-1} + \frac{r^2 - 1}{6} x_2 H^{r-2} = 0. \tag{1'}$$

Next one notices that

$$c_{r-1}(N_{X/\mathbb{P}^{2r-1}}) = \sum_{t=0}^{r-1} c_t(N') H^{r-1-t},$$

hence

$$c_{r-1}(N_{X/\mathbb{P}^{2r-1}}) = \binom{r+1}{2} H^{r-1} - \left(\binom{r}{2} y_1 + \binom{r+1}{3} x_1\right) H^{r-2} + \left(\binom{r-1}{2} y_2 + \binom{r}{3} x_1 y_1 + (r+1) x_2\right) H^{r-3},$$

by applying the formula $\sum_{t=0}^{s} \binom{t+k}{k} = \binom{s+k+1}{k+1}$.

Multiplying the self-intersection formula,

$$c_{r-1}(N_{X/\mathbb{P}^{2r-1}}) = d \cdot H^{r-1},$$

by $H$ and replacing the entities from $(*)$ we get a second equation

$$d^2 = (r^2 - r + 1)d - \binom{r}{2} y_1 H^{r-1} - \frac{r(r-1)(2r-1)}{6} x_1 H^{r-1} + \binom{r+1}{4} x_2 H^{r-2} \tag{2}$$

The third needed equation is obtained by multiplying the self-intersection formula by $y_1$:

$$(3) \qquad \left[(d-r^2)y_1 - \binom{r+1}{3}x_1\right]\cdot H^{r-1} = 0.$$

If we let $L := \det E$ and $K := K_B$, it follows from (1′), (2), (3):

$$L^2 = \frac{2(r^2-1)d(2d-r^2-2)}{2(r^2+2)d - r^2(r^2+5)},$$

$$K^2 = \left[\frac{d-r^2}{\binom{r+1}{3}}\right]^2 \cdot L^2,$$

$$K\cdot L = \frac{d-r^2}{\binom{r+1}{3}}\cdot L^2,$$

$$\chi(\mathcal{O}_B) = \frac{1}{12}(c_1^2 + c_2)(B)$$

$$= 1 + \frac{[2d-r(r+1)][2d-r(r-1)][6d^2-6(r^2+1)d+r^2(r^2+5)]}{r^2(r^2-1)[2(r^2+2)d - r^2(r^2+5)]}.$$

We now may write $L^2 = \frac{P(d)}{Q(d)}$, where

$$P := (r^2-1)d(2d-r^2-2),$$

$$Q := (r^2+2)d - \frac{1}{2}r^2(r^2+5)$$

seen as polynomials in $d$. Since $L^2$ is an integer, $Q(d)$ divides $\frac{1}{2}r^2(r^2-4)(r^2-1)(r^2+5)$ which is the remainder of the division of $(r^2+2)^2P$ by $Q$. Assuming $Q(d) > 0$ we have

$$Q(d) \leqslant \frac{1}{2}r^2(r^2-4)(r^2-1)(r^2+5)$$

which gives the desired bound

$$d < \frac{r^6}{2}.$$

If $Q(d) \leqslant 0$ one has an even better bound.

Now let us remark that

$$K^2 \cdot L^2 = (KL)^2,$$

by the preceding formulae.

The Hodge index theorem implies that $K$ and $L$ are numerically proportional. Since $L$ is the determinant of an ample vector bundle, it is itself ample so we distinguish the following three cases:

a) $-K$ is ample, i.e. $B$ is a Del Pezzo surface,

b) $K$ is numerically trivial,

c) $K$ is ample.

Next we show that case a) leads to one of the classes i), ii), iii), b) leads to class iv) and c) to v). Remark that, since $X$ lies in $\mathbb{P}^{2r-1}$ it must be regular by Barth's theorem so $B$ is also regular.

Assume now that $-K$ is ample. By the well-known classification of Del Pezzo surfaces we must have $1 \leqslant K^2 \leqslant 9$ and $K^2 = 9$ holds only if $B \cong \mathbb{P}^2$. Knowing that $\chi(\mathcal{O}_B) = 1$ we get from the above expression of $\chi(\mathcal{O}_B)$ the following equation:

$$(4) \qquad [2d - r(r+1)]\,[2d - r(r-1)]\,\left[6d^2 - 6(r^2+1)d + r^2(r^2+5)\right] = 0.$$

Two of its solutions are $d = \frac{r(r+1)}{2}$ and $d = \frac{r(r-1)}{2}$. For these values of $d$, $K^2 = 9$ so $B \cong \mathbb{P}^2$.

For $d = \frac{r(r+1)}{2}$ we are in case iii) of the proposition; we postpone the argument that $X$ is a determinantal variety until the next proposition.

When $d = \frac{r(r-1)}{2}$ one verifies that $\deg E = \operatorname{rank} E = r - 1$. Since the restriction of $E$ to any line of $\mathbb{P}^2$ is ample, this must be $\mathcal{O}_{\mathbb{P}^1}(1)^{\oplus(r-1)}$. By [10], $E \cong \mathcal{O}_{\mathbb{P}^2}(1)^{\oplus(r-1)}$ and we are in case ii).

Still assuming that we are in case a) we may suppose that $d$ satisfies the remaining degree two equation from (4). Putting $\lambda := K^2$, we consider the above expression of $K^2$ as an equation of degree four in $d$ with parameter $\lambda$. The remainder of its division by $6d^2 - 6(r^2+1)d + r^2(r^2+5)$ is

$$-2(r^2-1)[(6-\lambda)r^2 + 2(12-\lambda)]d + (r^2-1)r^2(r^2+5)(8-\lambda),$$

hence $d = \frac{r^2(r^2+5)(8-\lambda)}{2[(6-\lambda)r^2+2(12-\lambda)]}$. Since this value satisfies $6d^2 - 6(r^2+1)d + r^2(r^2+5) = 0$, we get a quadratic equation in $r^2$. Giving to the parameter $\lambda$ values from 1 to 9 we see that the only solutions are: $r = \lambda = 3$, $d = 7$, which is case i) of the proposition, and $\lambda = 9$, $r = 3$, $d = 3$ which is a special case of ii).

Let now $K$ be numerically trivial. Since $L^2 > 0$ and $K^2 = 0$ we get from our formulas $d = r^2$ and $\chi(\mathcal{O}_B) = 2$. By surface classification $B$ must be $K3$ and we are in case iv).

We come to the last case when $K$ is ample. Since now $KL > 0$ we get $d > r^2$ and $\chi(\mathcal{O}_B) > 2$. Taking into account the remark following the proposition, the proof is complete. □

When $r = 3$ the existence of classes i) – iv) was proved in [8].

PROPOSITION 5. *The class* ii) *from Proposition 4 occurs for all* $r \geqslant 3$. *Any scroll in class* iii) *is the degeneracy locus of a morphism* $u\colon \mathcal{O}^3_{\mathbb{P}^{2r-1}} \longrightarrow \mathcal{O}^{\oplus(r+1)}_{\mathbb{P}^{2r-1}}(1)$ *and moreover, by taking* $u$ *to be generic their existence follows for all* $r \geqslant 3$.

*Proof.* For ii) it is enough to see that the secant variety of the Segre embedding $\mathbb{P}^2 \times \mathbb{P}^{r-2} \subset \mathbb{P}^{3r-4}$ is $(2r-1)$ -dimensional. We give the following (certainly classical) argument for the reader's convenience. The secant variety of the restricted scroll over any line in $\mathbb{P}^2$ is a $(2r-3)$-plane in $\mathbb{P}^{3r-4}$. But any secant to

$\mathbb{P}^2 \times \mathbb{P}^{r-2}$ is a secant to such a restricted scroll over a line. Hence by counting parameters, we get the assertion.

Let us prove now the existence for class iii). Consider a generic morphism

$$u : \mathcal{O}^3_{\mathbb{P}^{2r-1}} \longrightarrow \mathcal{O}^{\oplus(r+1)}_{\mathbb{P}^{2r-1}}(1).$$

By [6] the locus where $u$ has rank $\leqslant 1$ (resp. $\leqslant 2$ ) is empty (resp. smooth connected of dimension $r$). Let $X$ be the locus where $u$ drops rank and $\mathcal{K} = \operatorname{Ker} u|_X$, which is a rank 1 subbundle of $\mathcal{O}^3_X$. Taking the associated projective bundles (this time lines instead of hyperplanes!) we get

$$X = \mathbb{P}(\mathcal{K}) \hookrightarrow \mathbb{P}(\mathcal{O}^3_X) = X \times \mathbb{P}^2$$

which gives a natural map $\pi : X \longrightarrow \mathbb{P}^2$ by composing with the second projection. By construction there are $r+1$ linear forms vanishing on each fiber of $\pi$. The genericity assumption on $u$ implies that for all fibers of $\pi$ the corresponding $r+1$ linear forms are independent. So $\pi : X \longrightarrow \mathbb{P}^2$ is a scroll and an easy application of Porteous' formula, (cf. [5]), gives $\deg X = \frac{r(r+1)}{2}$.

Conversely, consider $\pi : X = \mathbb{P}(E) \longrightarrow \mathbb{P}^2$ an $r$-dimensional scroll of degree $\frac{r(r+1)}{2}$ embedded in $\mathbb{P}^{2r-1}$. Letting $\ell \subset \mathbb{P}^2$ be any line and $D = \pi^{-1}(\ell)$, we first show that $D \subset \mathbb{P}^{2r-1}$ is nondegenerate and linearly normal. If $D$ were contained in a hyperplane, the self-intersection formula would give

$$(\deg D - (r-1))(\deg D - r) = 0.$$

But in our case $\deg D = \deg E_{|\ell} = r+1$, by the formulas given in the proof of Proposition 4. So $D$ is nondegenerate. Next, from the exact sequence

$$0 \longrightarrow \mathcal{O}_X(H-D) \longrightarrow \mathcal{O}_X(H) \longrightarrow \mathcal{O}_D(H) \longrightarrow 0$$

it follows

$$2r \leqslant h^0(\mathcal{O}_X(H)) \leqslant h^0(\mathcal{O}_D(H)) = h^0(E_{|\ell}) = 2r.$$

So both $D$ and $X$ ar linearly normal.

Let us recall a classical determinantal description of $D$ as a rational normal scroll. Fixing a point $P_0 \in \ell$, any hyperplane containing $\pi^{-1}(P_0)$ lies in an unique pencil of hyperplanes sweeping out $D$ fiberwise; this corresponds to the natural map

$$H^0(\ell, \mathcal{O}(1)) \otimes H^0(\ell, E_{|\ell}(-1)) \to H^0(\ell, E_{|\ell}).$$

Thus, letting $F_0 := \pi^{-1}(P_0)$, $F_1 := \pi^{-1}(P_1)$ be two fibers and $s_{0,1}, \dots, s_{0,r+1}$ a basis of $H^0(D, \mathcal{O}(H - F_0)) \subset H^0(D, \mathcal{O}(H)) \cong H^0(\mathbb{P}^{2r-1}, \mathcal{O}(H))$, one obtains a basis $s_{1,1}, \dots, s_{1,r+1}$ for $H^0(D, \mathcal{O}(H - F_1)) \subset H^0(\mathbb{P}^{2r-1}, \mathcal{O}(H))$ by multiplying with a linear form vanishing at $P_1$. The classical fact is that $D$ is precisely the degeneracy locus of the matrix of linear forms $(s_{ij})_{\substack{i=0,1;\\ j=1,\dots,r+1}}$.

This construction may be adapted to $X$ as follows. Let $P_0, P_1, P_2$ be non-colinear points in $\mathbb{P}^2$, $s_{0,1}, \dots, s_{0,r+1}$ a basis for $H^0(\mathbb{P}^{2r-1}, \mathcal{I}_{F_0}(H)) \subset H^0(\mathbb{P}^{2r-1}, \mathcal{O}(H))$. We transport this basis to a basis $(s_{1,j})_{j=1,\dots,r+1}$ of $H^0(\mathbb{P}^{2r-1}, \mathcal{I}_{F_1}(H))$, (resp. $(s_{2,j})_{j=1,\dots,r+1}$ of $H^0(\mathbb{P}^{2r-1}, \mathcal{I}_{F_2}(H))$) along the line $P_0P_1$ (resp. $P_0P_2$) by the above procedure. We claim that $X$ is the degeneracy locus of the matrix of linear forms $s := (s_{ij})_{\substack{i=0,1,2;\\ j=1,\dots,r+1}}$.

The main point is that for any choice of points $P_0$,$P_1$, $P_2$ by transporting our basis successively along $P_0P_1$, $P_1P_2$, $P_2P_0$ we get the same basis up to a multiplicative factor. In order to see this, consider the natural morphism

$$\alpha : \mathbb{P}^2 \times \mathbb{P}^2 \longrightarrow \mathrm{Aut}(\mathbb{P}(H^0(\mathbb{P}^{2r-1}, \mathcal{I}_{F_0}(H))))$$

obtained by fixing $P_0$ and associating to the pair $(P_1, P_2)$ the automorphism induced by our successive transport. $\alpha$ has to be constant since it maps to an affine variety.

This remark enables one to see that any fiber of $\pi : X \longrightarrow \mathbb{P}^2$ appears as the common zero locus of the entries of a suitable linear combination of the three lines of our matrix $s$. In particular any $3 \times 3$ minor of $s$ vanishes on $X$. If we let $\Delta_i(s) := \{x \in \mathbb{P}^{2r-1} \mid \mathrm{rank}\, s(x) \leqslant i\}$ we have shown that $X$ lies in $\Delta_2(s)$.

Next we notice that $\Delta_1(s)$ is empty. Indeed, throwing out one line of $s$, say the first, $\Delta_1(s)$ is contained in the degeneracy locus of the remaining matrix which is a rational normal scroll over the line $P_1P_2$. Since the lines $P_0P_1$, $P_1P_2$, $P_2P_0$ have no common point, $\Delta_1(s)$ is empty. As in the existence proof above this ensures that $\Delta_2(s)$ is an $r$-dimensional scroll over $\mathbb{P}^2$ and so $X = \Delta_2(s)$. □

REMARKS. 1. The degeneracy locus of a generic morphism

$$\mathcal{O}_{\mathbb{P}^N}^{t+1} \longrightarrow \mathcal{O}_{\mathbb{P}^N}^{\oplus(N-r+t)}(1)$$

is an $r$-dimensional scroll over $\mathbb{P}^t$ of degree $\binom{N-r+t}{t}$ if $N \geqslant 2r-1$ and $r \geqslant 2t-1$. The proof is the same as in the case $t = 2$, $N = 2r-1$.

2. Assuming that $X = \mathbb{P}(E) \subset \mathbb{P}^N$ is a scroll over $\mathbb{P}^t$ such that for any line $\ell$ in $\mathbb{P}^t$ the restricted scroll over $\ell$ is linearly normal and nondegenerate in $\mathbb{P}^N$, the preceding argument applies to give that $X$ is the degeneracy locus of a morphism as in 1.

## REFERENCES

1. Barth, W., Transplanting cohomology classes in complex-projective space, Amer. J. Math. **92** (1970), 951–967.
2. Beltrametti, M., Biancofiore, A., Sommese, A. J., Projective $N$-folds of log-general type I, Trans. AMS **314** (1989), 825–849.
3. Braun, R., Ottaviani, G., Schneider, M., Schreyer, F. O., Boundedness for non-general type 3-folds in $\mathbb{P}^5$, preprint, 1991.
4. Ellingsrud, G., Peskine, C., Sur les surfaces lisses de $\mathbb{P}^4$, Invent. Math. **95** (1989), 1–11.
5. Fulton, W., *Intersection theory*, Springer-Verlag, 1984.
6. Kleiman, S. L., Geometry on Grassmannians and applications to splitting bundles and smoothing cycles, IHES Publ. Math **36** (1969), 281–297.
7. Lascu, A. T., Mumford, D., Scott, D. B., The self-intersection formula and the "formule-clef", Proc. Camb. Phil. Soc. **78** (1975), 117–123.
8. Ottaviani, G., On 3-folds in $\mathbb{P}^5$ which are scrolls, preprint, 1991.
9. Schneider, M., Boundedness of low-codimensional submanifolds of projective space, preprint, 1991.

10. Van de Ven, A., On uniform vector bundles, Math. Ann. **195** (1972), 245–248.

PALTIN IONESCU, UNIVERSITY OF BUCHAREST, ACADEMIEI 14, 70109 BUCHAREST, ROMANIA.
*E-mail address*: pionescu@roearn.bitnet

MATEI TOMA, INSTITUTE OF MATHEMATICS OF THE ROMANIAN ACADEMY, P.O.BOX 1–764, 70700 BUCHAREST, ROMANIA.
*E-mail address*: mtoma@roearn.bitnet

Contemporary Mathematics
Volume **162**, 1994

# Generalized Euler Sequence and Toric Varieties

Krzysztof Jaczewski

## 1 Introduction

A line bundle generated by global sections determines a mapping into projective space. In this paper we shall give a generalization of this correspondence: a mapping that leads to some torus embedding determined by some reducible divisor. Our starting point is the observation that many properties of the sheaf $\mathcal{O}_{P^n}(1)$ follows immediately from the fact that the direct sum of $n+1$ copies of this sheaf is the nontrivial extension of the tangent sheaf by the structural one. Dualizing the corresponding short exact sequence we obtain a sequence of the form:

$$0 \to \Omega_{P^n} \to \oplus\mathcal{O}_{P^n}(-1) \to \mathcal{O}_{P^n} \to 0 ,$$

where $\Omega_{P^n}$ denotes the cotangent sheaf, so we may idetify the fibre of the trivial bundle with the space $H^1(\Omega_{P^n})$ . On any nonsingular variety one may find a short exact sequence of the similar form that has functorial properties (see def. 2.1). It turns out that torus embeddings are the only varieties for which the sheaves obtained as such extensions are the direct sums of the invertible sheaves which correspond to nonsingular divisors intersecting transversally (thm. 3.1). In this context it is natural to ask which divisors, on arbitrary nonsingular variety $Y$ , are pullbacks of the distinguished divisors on torus embeddings. For a vast divisor (def. 4.3) there exists a mapping leading to a torus embedding and such that this divisor is the pullback of the distinguished one on the torus embedding. The paper is organized as follows. In this section we recall briefly the basic objects and constructions concerning the toric varieties. In the second section we give the definition of the Euler sequence and the potential sheaf, and we discuss their properties. We prove theorem 2.5 which describes the link between the potential sheaf and some sheaves of logarithmic differentials. Using this result we find the potential sheaf on a hypersurface in a projective space. We also notice that the class of the exterior power of the Euler sequence in the appropriate Ext group is equal to the element determined by the intersection index of divisors (proposition 2.3). The third section contains the proof of the characterization

1991 Mathematics Subject Classification. Primary 14M25; Secondary 14C05
This paper is in final form and no version of it will be submitted for publication elsewhere.

of the torus embeddings in terms of the potential sheaf and the discussion of corollaries.

In the fourth section we present the construction of a regular morphism leading from a given variety into a torus embbeding which is determined by a vast divisor.

The author would like to thank Andrzej Białynicki-Birula, Jerzy Jurkiewicz and Michał Szurek for numerous talks and remarks.

## 1.1 Notation and terminology

As many our constructions refer to the theory of toric varieties, we shall briefly present the basic notions and constructions from this theory (see [2], [3], [7], [8], [9], [11], [12]).

Let $N$ be a free abelian group and $M$ the dual group. We shall use the following notation :

$$N_R = N \otimes_Z R\ , \quad M_R = M \otimes_Z R\ ,$$

and $< , > : N_R \otimes M_R \longrightarrow R$ for the evaluation of the linear forms on vectors ($R$ denotes the real numbers). A convex polyhedral cone $\sigma$ in the space $N_R$ is by definition the intersection of a *finite* number of linear halfspaces. Such a cone is called rational if it is given by a finite number of the inequalities of the form:

$$x \in \sigma \ \ if \ \ < x, m_i > \geq 0 \ \ for\ i = 1 \ldots r,\ m_i \in M\ .$$

From now on, we shall use the word *cone* instead of the expression *finitely generated rational convex polyhedral cone*, since we shall not deal with any other cones. A complex of cones $\Sigma$ is a set of cones that fulfills the following conditions:
*(i) if* $\sigma \in \Sigma$ *and* $\tau$ *is a face of* $\sigma$ , *then* $\tau \in \Sigma$ ,
*(ii) if* $\sigma_1 \in \Sigma$ *and* $\sigma_2 \in \Sigma$ , *then* $\sigma_1 \cap \sigma_2$ *is a face of each of them.*
The one dimensional cones will be sometimes called the edges of the complex, and will be denoted by the same letters as the primitve vectors from $N$ that generate them.

Let us attach to a cone in the space $N_R$ finitely generated semigroup $\sigma^\vee \cap M$ ([12] chap.1) and the affine variety $X_\sigma = Spec\ IC[\sigma^\vee \cap M]$ . (Our cones are finitely generated and rational). The inclusion of semigroups $\sigma^\vee \cap M \subseteq M$ induces the embedding:

$$S = Spec\ IC[M] \rightarrow Spec\ IC[\sigma^\vee \cap M] = X_\sigma$$

of the algebraic torus $S$ into the variety $X_\sigma$ . Moreover, the action of the torus $S$ on itself by the translations may be extended to the action of $S$ on the variety $X_\sigma$ . Then, the group $M$ may be identified with the group of characters and $N$ with the group of one-parameter subgroups in $S$ . We get the following correspondence (comp.[12] chap 1,2).
*(i) The cones with vertices in* $N_R$ *are in one-to-one correspondence with the normal affine varieties equipped with the action of the torus* $S$ *and containing* $S$ *as a dense subset.*
*(ii) This correspondence may be extended to the one between the complexes of the*

*cones (with vertices) in the space $N_R$, and normal varieties (now not necessarily affine) equipped with the action of the torus $S$ and containing $S$ as a dense subset.*
*(iii) For a morphism of the lattices $N' \to N$ which induces a morphism of complexes $\Sigma'$ to $\Sigma$, there exists an equivariant mapping $X_{\Sigma'} \to X_\Sigma$.*

The variety corresponding to the complex of cones $\Sigma$ is called a torus embedding or a toric variety and is denoted by $X_\Sigma$.

Our notation concerning homological algebra will not be too rigorous. We hope that this will not lead to misunderstanding but will make this paper easier to read. For example the term 'kernel' may refer to the object as well as to the pair consisting of the object and monomorphism. Usually by a diagram we shall mean a commutative diagram with exact rows and columns. We shall not distinguish between locally free sheaves and the corresponding vector bundles. In particular, the stalk of the sheaf will be denoted by the same symbol as the fibre of the corresponding bundle. For a vector space $V$ and a variety $X$, by $V_X$ we denote the trivial bundle $V \otimes \mathcal{O}_X$ with fibre $V$. The symbol $^\vee$ indicates the dual object : i.e. dual bundle, homomorphism, vector space etc.. $1_X$ denotes the constant function equal to 1 for any $x \in X$. Usually we shall omit the superscript 1 in the symbol of the sheaf of differentials, so $\Omega_X$ means $\Omega^1_X$. By $P(V)$ we mean the projective space parametrizing the lines in a vector space $V$.

All our varieties will be defined over $IC$.

# 2 Generalized Euler Sequence

For a complete algebraic variety $X$, let $H$ denote the space $H^1(X, \Omega_X)$:

$$H = H^1(X, \Omega_X) ,$$

and $H_X$ the sheaf $H \otimes \mathcal{O}_X$. The extensions of the sheaf $\Omega_X$ by the trivial one $H_X$ are classified by the following space:

$$Ext^1(H_X, \Omega_X) = H^\vee \otimes Ext^1(\mathcal{O}_X, \Omega_X) = H^\vee \otimes H = Hom(H, H) .$$

**Definition 2.1** *The short exact sequence of sheaves*

$$0 \to \Omega_X \to L_X^\vee \to H_X \to 0 ,$$

*which corresponds to $id_H \in Hom(H,H) = Ext^1(H_X, \Omega_X)$ is called the Euler sequence on the variety $X$. The sheaf $L_X$ will be called the sheaf of potentials or the potential sheaf.*

For a vector space $I$, the extensions of the sheaf $\Omega_X$ by the sheaf $I_X$ are classified by the space:

$$Ext^1(I_X, \Omega_X) = I^\vee \otimes Ext^1(\mathcal{O}_X, \Omega_X) = I^\vee \otimes H = Hom(I, H) .$$

The class of such an extension

$$0 \to \Omega_X \to A \to I_X \to 0 ,$$

is represented by the connecting homomorphism $\delta : I \to H$. Hence each such extension is induced from the Euler sequence by the morphism $\delta \otimes 1_X$ . and, by virtue of the five homomorphisms lemma, $\delta$ is an isomorphism if and only if the sheaf $A$ is isomorphic to the potential sheaf.

**Example 2.2** *Let $V$ be a vector space and let $X$ be the corresponding projective space of lines in $V$ . Then the Euler sequence (in the sense of our definition) has the following form*

$$0 \to \Omega_X \to V_X^\vee(-1) \to \mathcal{O}_X \to 0$$

*and it is dual to the well-known Euler sequence on projective space.*

*(The homomorphism $\delta$ is an isomorphism, because $h^0(V_X^\vee(-1))$ and $h^1(V_X^\vee(-1))$ are zero.)*

The Euler sequence is functorial with respect to morphisms between varieties in the following sense. For any regular map $f : X' \to X''$ between complete varieties there exists a morphism $L(f)^\vee : f^* L_{X''}^\vee \to L_{X'}^\vee$ , such that the following diagram is commutative:

$$\begin{array}{ccccccccc}
0 & \longrightarrow & f^*\Omega_{X''} & \longrightarrow & f^*L_{X''}^\vee & \longrightarrow & H''_{X''} & \longrightarrow & 0 \\
 & & \downarrow & & \downarrow & & \downarrow & & \\
0 & \longrightarrow & \Omega_{X'} & \longrightarrow & L_{X'}^\vee & \longrightarrow & H'_{X'} & \longrightarrow & 0
\end{array}$$

where $H' = H^1(\Omega_{X'})$, $H'' = H^1(\Omega_{X''})$, and the morphisms between the sheaves of differentials and between the trivial bundles $H'_X$ and $H''_X$ are induced by the pullback of differential forms.

For example: let $P$ denote the projectivization of a vector bundle $E$ on a complete variety $X$ , and let $f$ denote the projection $f : P \to X$ . Then the map $L^\vee(f)$ is a monomorphism and the following sequence is exact:

$$0 \to f^* L_X^\vee \xrightarrow{L^\vee(f)} L_P^\vee \longrightarrow f^* E^\vee(-1) \to 0 ,$$

where $(-1)$ means tensoring by the tautological sheaf $\mathcal{O}_P(-1)$.

Let us point out that the connected component of neutral element in the automorphism group $G_0$ acts on the sheaf $L_X^\vee$ (the sheaf $L_X^\vee$ has natural $G_0$-linearization which makes the maps in the Euler sequence invariant). The extension of the $G_0$-action from the variety $X$ to the sheaf $\Omega_X$ is given by the maps dual to the differentials of the automorphisms, and the action of the group $G_0$ on the trivial bundle $H_X$ is a product of the trivial action on the space $H$ with the action on $X$ .

Let $\iota_p$ be the element in the space

$$Ext^p(Sym^p H \, , \Omega_X^p) = Sym^p H^\vee \otimes H^p(\Omega_X^p) \, .$$

corresponding to the sequence

$$0 \to \Omega^p_X \to \Lambda^p L^\vee_X \to \Lambda^{p-1} L^\vee_X \otimes H \to \ldots \to Sym^p H \to 0$$

obtained from the Euler sequence by taking the p-th exterior power. Let $\mu_p$ be the restriction of the multiplication in the cohomology ring to the space $H$ (the intersection of divisors map). We regard $\mu_p$ as the element of $Sym^p H^\vee \otimes H^p(\Omega^p_X)$.

**Proposition 2.3** *For* $p \leq dim\, X$

$$\iota_p = \mu_p \quad \textit{in the space} \quad Ext^p(Sym^p H\,,\, \Omega^p_X) = Sym^p H^\vee \otimes H^p(\Omega^p_X)\,.$$

*Moreover all the connecting homomorphisms in the short exact sequences obtained by breaking up the above long exact sequence are induced from exterior multiplication of differential forms.*

PROOF. We take the Dolbeault resolvents of the elements of the Euler sequence and its tensor powers to note the second claim. Hence both $\iota_p$ and $\mu_p$ are induced by the exterior multiplication of differential forms. ☐

**Remark 2.4** *By breaking up the above long exact sequence into the short ones we see that the following conditions are equivalent*
*(i) The Hodge numbers* $h^{p,q}$ *are zero for* $p \neq q$ .
*(ii) The spaces* $H^q(\Lambda^p L^\vee_X)$ *are trivial for* $q > p$ .
*Moreover,* $H^p(\Lambda^p L^\vee_X) = 0$ *iff the products of divisors generate the space* $H^p(\Omega^p_X)$ .

Now, we shall describe a link between the sheaves of logarithmic differentials and the sheaf of potentials. Let $D$ be an effective divisor. For a morphism of sheaves

$$m \,:\, \mathcal{O}_X(-D) \to L^\vee_X\,,$$

let $m'$ denote the corresponding global section from the space

$$H^0(L^\vee_X(D)) = Hom(\mathcal{O}_X(-D), L^\vee_X)\,.$$

**Theorem 2.5**
*Let* $X$ *be a nonsingular and complete algebraic manifold. Then*

*(i) For any global section* $s$ *of the sheaf* $\mathcal{O}_X(D)$ *which vanishes on a nonsingular connected and reduced divisor* $D$ , *there exists a monomorphism* $m$

$$m \,:\, \mathcal{O}_X(-D) \to L^\vee_X\,,$$

*with a locally free cokernel. The monomorphism* $m$ *extends the embedding*

$$j \,:\, \mathcal{O}_D(-D) \to \Omega_X|D$$

*of the conormal sheaf for submanifold* $D$ *into the sheaf of differentials* $\Omega_X$ *restricted to* $D$ .

(ii) *Let $h = \dim H$ and let $D = \sum D_e$ be a normal crossing divisor with exactly $h$ irreducible components such that the first Chern classes of the sheaves $\mathcal{O}_X(D_e)$ generate the space $H$. Then the following sequence is exact with natural morphisms:*

$$0 \to \oplus \mathcal{O}_X(-D_e) \to L_X^\vee \to \Omega_X(\log D) \to 0 ,$$

*and the monomorphism in this sequence is the sum of the $m$'s from (i).*

PROOF. First assume that the divisor $D$ is nonsingular connected and reduced.

We write down the diagram that comes from the Euler sequence written together with the extension induced from it by the monomorphism $\Omega_X \to \Omega_X(\log D)$ (we have used the snake lemma to obtain the bottom row):

$$\begin{array}{ccccccccc}
 & & 0 & & 0 & & 0 & & \\
 & & \downarrow & & \downarrow & & \downarrow & & \\
0 & \longrightarrow & \Omega_X & \longrightarrow & L_X^\vee & \longrightarrow & H_X & \longrightarrow & 0 \\
 & & \downarrow & & \downarrow & & \downarrow & & \\
0 & \longrightarrow & \Omega_X(\log D) & \longrightarrow & Ind & \longrightarrow & H_X & \longrightarrow & 0 \\
 & & \downarrow & & \downarrow \kappa & & \downarrow & & \\
 & & \mathcal{O}_D & \longrightarrow & \mathcal{O}_D & \longrightarrow & 0 & & \\
 & & \downarrow & & \downarrow & & & & \\
 & & 0 & & 0 & & & &
\end{array} \tag{1}$$

In the cohomology sequence attached to the first column of this diagram the section $1_D$ is mapped onto the Chern class $c_1(\mathcal{O}_X(D))$ in the space $H^1(\Omega_X)$. The first row is the Euler sequence, so the connecting homomorphism $\delta$ is an isomorphism between the spaces $H^0(H_X)$ and $H^1(\Omega_X)$. Since the diagram is commutative, the connecting homomorphism in the cohomology sequence attached to the middle column sends the section $1_D$ to zero in the space $H^1(L_X^\vee)$. Thus there exists a section $t$ of the sheaf $Ind$ which goes onto $1_D$. This section $t$ vanishes nowhere, because the morphism $\iota$ sends it onto nowhere-vanishing section $c_1(\mathcal{O}_X(D)) \otimes 1_X$ of the trivial bundle $H_X$. We have proved that the sheaf $Ind$ contains $\mathcal{O}_X t = \mathcal{O}_X$ as a subbundle, and the morphism $\kappa$ restricted to this bundle sends the section $t$ to $1_D$. So the kernel of $\kappa$ is equal to the sheaf $\mathcal{O}_X(-D)$, and therefore we get the monomorphism $m$ from (i).

We construct the next diagram by applying the evaluation morphisms for the sections $t$ and $1_D$ to the middle column of the previous one:

$$
\begin{array}{ccccccccc}
 & & & & 0 & \longrightarrow & \mathcal{O}_X(-D) & & \\
 & & & & \downarrow & & \downarrow & & \\
 & & 0 & \longrightarrow & \mathcal{O}_X & \longrightarrow & \mathcal{O}_X & \longrightarrow & 0 \\
 & & \downarrow & & \downarrow & & \downarrow & & \\
0 & \longrightarrow & L_X^\vee & \longrightarrow & Ind & \longrightarrow & \mathcal{O}_D & \longrightarrow & 0 \\
 & & \downarrow & & \downarrow & & \downarrow & & \\
 & & L_X^\vee & \longrightarrow & Ind_0 & \longrightarrow & 0 & & \\
 & & \downarrow & & \downarrow & & & & \\
 & & 0 & & 0 & & & &
\end{array}
\tag{2}
$$

$Ind_0$ denotes the quotient. The monomorphism $m$ appears here as the one from the snake lemma.

Now, we can prove that $m$ extends the embedding $j : \mathcal{O}_D(-D) \to \Omega_X|D$. Consider the following diagram:

$$
\begin{array}{ccccccccc}
 & & & & 0 & & 0 & & \\
 & & & & \downarrow & & \downarrow & & \\
 & & 0 & \longrightarrow & \mathcal{O}_X(-D) & \longrightarrow & \mathcal{O}_X & & \\
 & & \downarrow & & \downarrow & & \downarrow & & \\
0 & \longrightarrow & \Omega_X & \longrightarrow & L_X^\vee & \longrightarrow & H_X & \longrightarrow & 0 \\
 & & \downarrow & & \downarrow & & \downarrow & & \\
0 & \longrightarrow & \Omega_X(log D) & \longrightarrow & Ind_0 & \longrightarrow & H'_X & \longrightarrow & 0 \\
 & & \downarrow & & \downarrow & & & & \\
 & & \mathcal{O}_D & \longrightarrow & 0 & & & &
\end{array}
\tag{3}
$$

$H'$ denotes here the quotient of the space $H$ by the subspace generated by $c_1(\mathcal{O}_X(D))$ . Since $Tor_1^{\mathcal{O}_X}(\mathcal{O}_D,\mathcal{O}_D) = \mathcal{O}_D(-D)$ , tensoring by the sheaf $\mathcal{O}_D$

we get the following diagram:

$$\begin{array}{ccccccccc}
 & & 0 & & 0 & & 0 & & \\
 & & \downarrow & & \downarrow & & \downarrow & & \\
0 & \longrightarrow & \mathcal{O}_D(-D) & \rightarrow & \mathcal{O}_D(-D) & \longrightarrow & \mathcal{O}_D & & \\
 & & \downarrow & & \downarrow & & \downarrow & & \\
0 & \longrightarrow & \Omega_X|D & \longrightarrow & L_X^{\vee}|D & \longrightarrow & H_D & \longrightarrow & 0 \\
 & & \downarrow & & \downarrow & & \downarrow & & \\
0 & \rightarrow & \Omega_X(log D)|D & \rightarrow & Ind_0|D & \longrightarrow & H'_D & \longrightarrow & 0 \\
 & & \downarrow & & \downarrow & & & & \\
 & & \mathcal{O}_D & \longrightarrow & 0 & & & &
\end{array} \tag{4}$$

The morphism $\Omega_{X|D} \to \Omega_X(log D)|D$ factors through $\Omega_D$ , and the embedding of the conormal sheaf is its kernel. So the commutativity of the "upper-left" square means that the monomorphism $\mathcal{O}_X(-D) \to L_X^{\vee}$ extends the embedding of the conormal sheaf.

Now, we are going to prove the second part of the theorem. Let $D = \sum_e D_e$ be a divisor whose irreducible components are nonsingular, reduced and intersect transversally each the others. Assume that the Chern classes $c_1(\mathcal{O}_X(D_e))$ generate the space $H$ .

As before, at the begining we consider the diagram:

$$\begin{array}{ccccccccc}
 & & 0 & & & & 0 & & \\
 & & \downarrow & & \downarrow & & \downarrow & & \\
0 & \longrightarrow & \Omega_X & \longrightarrow & L_X^{\vee} & \longrightarrow & H_X & \longrightarrow & 0 \\
 & & \downarrow & & \downarrow & & \downarrow & & \\
0 & \rightarrow & \Omega_X(log D) & \rightarrow & Ind & \longrightarrow & H_X & \longrightarrow & 0 \\
 & & \downarrow & & \downarrow & & \downarrow & & \\
 & & \bigoplus_{e\in E} \mathcal{O}_{D_e} & \longrightarrow & \bigoplus_{e\in E} \mathcal{O}_{D_e} & \longrightarrow & 0 & & \\
 & & \downarrow & & \downarrow & & & & \\
 & & 0 & & 0 & & & &
\end{array} \tag{5}$$

For any $e$, there is a section $t_e$ of the sheaf $Ind$ which does not vanish anywhere and which is sent onto $1_{D_e}$ in the space $H^0(\oplus \mathcal{O}_{D_e})$ . Next, look at the middle row of the diagram. The subspace of the space $H^0(Ind)$ generated by the sections $t_e$ , is mapped isomorphically onto the space $H$ . The composition of the inverse of this isomorphism with the evaluation of the sections $t_e$ gives the splitting of the middle row. The trivial subbundle of $Ind$ is mapped onto the sheaf $\oplus \mathcal{O}_{D_e}$ and the kernel of this map is contained in $L_X^\vee$ so we get the sequence from (ii) of the theorem.

For a fixed index $e_0$ we restrict the constructed monomorphism $\oplus \mathcal{O}_X(-D_e) \to L_X^\vee$ firstly to the direct summand determined by $e_0$ , then to the subvariety $D_{e_0}$ . Again repeating the arguments from the proof of the item (i), we see that this monomorphism extends the embedding

$$\mathcal{O}_D(-D_{e_0}) \to \Omega_X|D_{e_0} .$$

□

When $X = P(V)$ both cases of theorem 2.5 refer to a nonsingular hypersurface $D$. Let $k$ be its degree and let $i$ denote the embedding of $D$ into the projective space. By the theorem we have the following exact sequence:

$$0 \to \mathcal{O}_X(-k) \to V_X^\vee(-1) \to \Omega_X(log D) \to 0$$

If $k = 1$ , i.e. $D$ is a hyperplane, this sequence comes from the map of vector spaces:

$$IC \to V^\vee$$

that sends 1 to the linear form corresponding to $D$ . The sequence splits up and we get the equality:

$$\Omega_X(log D) = \oplus \mathcal{O}_X(-1) .$$

If $k = 2$, then $D$ is a nonsingular quadric, and the equation of $D$ determines the nondegenerate bilinear form on the space $V$ . We may identify the space $V$ and $V^\vee$ using this form. The sequence becomes isomorphic to the dual to the Euler one tensored by $\mathcal{O}_X(-1)$ . So

$$\Omega_X(log D) = T_X(-2) ,$$

where $T_X$ denotes the tangent bundle.

**Remark 2.6** *For $k > 1$ the exact sequence from the proposition may be regarded as the pullback of Euler sequence on the dual projective space under the finite map defined by the partial derivatives of the equation of the divisor with respect to some coordinate system in $V$ .*

**Corollary 2.7** *Let $D$ be a nonsingular connected hypersurface of degree $k$ ( $k > 1$ ) in the projective space $X = P(V)$ of the the dimension bigger than 3 , and therefore $h^1(\Omega_D) = 1$ .*

*Then the sheaf dual to the potential one on the manifold $D$ is isomorphic to the restriction of the sheaf $\Omega_X(log D)$:*

$$L_D^\vee = \Omega_P^n(log D)|D .$$

PROOF. Restrict the considered sequence to $D$. □

# 3 Vector Bundle Characterization of Torus Embeddings

**Theorem 3.1**
*A nonsingular complete and connected variety $X$ is isomorphic to a torus embedding if and only if there exists a divisor $D$ with normal crossing such that the following formula holds for the potential sheaf:*

$$L_X = \oplus_{e \in E} \mathcal{O}_X(D_e)$$

*where $D_e$ are the irreducible components of the divisor $D$.*

*The divisors $D_e$ are then the closures of codimension one orbits of the torus.*

PROOF. First, we shall prove the implication: if there exists a divisor $D$ with the given properties then the variety $X$ has the structure of a torus embedding.

**Lemma 3.2** *Under our assumptions we have:*

$$h^0(\Omega_X) = 0 \ , \qquad h^1(\mathcal{O}_X) = 0 \ ,$$
$$h^0(L_X^\vee) = 0 \ , \qquad h^1(L_X^\vee) = 0 \ .$$

PROOF. We look at the cohomology sequence attached to the Euler sequence. By the assumption the sheaf $L_X^\vee$ is a direct sum of the ideal sheaves $\mathcal{O}_X(-D_e)$ , so $h^0(L_X^\vee)$ is zero. Therefore also $h^0(\Omega_X)$ is zero. By the symmetry of the Hodge numbers we obtain that $h^1(\mathcal{O}_X)$ is zero, too and $h^1(H_X) = 0$ . The connecting homomorphism $\delta$ is an isomorphism between the spaces $H$ and $H^1(\Omega_X)$ , so $H^1(L_X^\vee) = 0$ . □

We denote by $Ind$ the sheaf that appears in the extension of the sheaf $H_X$ by $\Omega_X(log\ D)$ induced from the Euler sequence by the monomorphism $\Omega_X \to \Omega_X(log D)$ . Using the snake lemma we obtain the following diagram:

$$\begin{array}{ccccccccc}
 & & 0 & & 0 & & & & \\
 & & \downarrow & & \downarrow & & & & \\
0 & \longrightarrow & \Omega_X & \longrightarrow & L_X^\vee & \longrightarrow & H_X & \longrightarrow & 0 \\
 & & \downarrow & & \downarrow & & \downarrow & & \\
0 & \longrightarrow & \Omega_X(log D) & \longrightarrow & Ind & \longrightarrow & H_X & \longrightarrow & 0 \\
 & & \downarrow & & \downarrow & & \downarrow & & \\
 & & \bigoplus_{e \in E} \mathcal{O}_{D_e} & \longrightarrow & \bigoplus_{e \in E} \mathcal{O}_{D_e} & \longrightarrow & 0 & & \\
 & & \downarrow & & \downarrow & & & & \\
 & & 0 & & 0 & & & &
\end{array} \qquad (6)$$

**Lemma 3.3** *The bundle Ind is trivial and the middle column of the diagram is a direct sum of the short exact sequences of the form*

$$0 \to \mathcal{O}_X(-D_e) \to \mathcal{O}_X \to \mathcal{O}_{D_e} \to 0 .$$

PROOF. By the previous lemma we know that $h^0(L_X^\vee)$ and $h^1(L_X^\vee)$ are zero, hence in the cohomology sequence corresponding to the middle column we have the isomorphism

$$H^0(Ind) = H^0(\oplus_e \mathcal{O}_{D_e}) .$$

We denote by $1_e$ the section of the sheaf $\mathcal{O}_{D_e}$ which is equal 1 in the stalks over the points from the variety $D_e$ , and by $i_e$ the corresponding section of the sheaf $Ind$ . The section $1_e$ is sent onto the first Chern class of the sheaf $\mathcal{O}_X(D_e)$ by the connecting homomorphism

$$\delta : H^0(\oplus_e \mathcal{O}_{D_e}) \to H^1(\Omega_X) .$$

The section $i_e$ vanishes nowhere, because its image – $c_1(\mathcal{O}_X(D_e)) \otimes 1_X$ is such.

We shall prove that the evaluation of the global sections is an isomorphism between the sheaves $H^0(Ind)_X$ and $Ind$ .

First, we notice that the both bundles have the same rank, equal to the rank of the bundle $L_X^\vee$ . Next, we notice that the sections $\{i_e\}_{e\in E}$ span the fibres of the bundle $Ind$ at least at the points of $D$ , because their images $1_e$ are linearly independent. By the Nakayama lemma there exists an open neighbourhood of the subvariety $D$ in $X$ , such that the sections $\{i_e\}_{e\in E}$ span the fibres of $Ind$ over this set. Therefore the kernel of the evaluation morphism is a sheaf of rank zero.

On the other hand this kernel is a subsheaf of the torsion free sheaf $\oplus_e \mathcal{O}_X(-D_e)$ , that itself is a kernel of the evaluation morphism for the sheaf $\oplus_e \mathcal{O}_{D_e}$ . Therefore the considered sheaf is zero, as a torsion free sheaf of rank zero.

Now, we compare the middle column of the diagram (6) with the short exact sequence which describes the evaluation morphism for the sheaf $\oplus_e \mathcal{O}_{D_e}$ . The support of the cokernel of the evaluation morphism for the bundle $Ind$ should have been contained in the variety $D$ , but we have already checked that the global sections $\{i_e\}_{e\in E}$ span this bundle over $D$ ; so the support is empty and the evaluation of global sections map is an isomorphism. □

**Corollary 3.4** *The bundle* $\Omega_X(log D)$ *is trivial.*

PROOF. Since $h^1(\mathcal{O}_X)$ is zero (lemma 3.1) and the bundle $Ind$ is trivial, by counting dimensions we see that the sequence of spaces of global sections:

$$0 \to H^0(\Omega_X(log D)) \to H^0(Ind) \to H \to 0$$

is exact. The corresponding sequence of the trivial bundles is mapped to

$$0 \to \Omega_X(log D) \to Ind \to H_X \to 0$$

by the evaluation morphisms. By the five homomorphisms lemma we get that the bundle $\Omega_X(log D)$ is trivial. □

**Remark 3.5** *As a by-product of the above we have proved that the first Chern classes of the sheaves $\mathcal{O}_X(D_e)$ generate the space $H$ .*

Now, we are going to prove that the group $G$ of the automorphisms of the manifold $X$ of dimension $n$ contains the $n$-dimensional torus $S$. The proof will be carried out in several steps.

Let $G_0$ denote the connected component of identity in the group $G$ , and let $K$ be the subgroup of $G_0$ composed of all automorphisms preserving the subvarieties $D_e$ .

For any automorphism $g \in G$ and a meromorphic 1-form $\eta$ on $X_{hol}$ that has at least logarithmic poles in $D$ and is holomorphic outside $D$, the 1-form $g^*\eta$ has at least logarithmic poles in $g^{-1}(D)$ . In particular, if $g \in K$ , then $g^*\eta$ is a section of the sheaf $\Omega_{X_{hol}}(log D)$ . So there is an action of the group $K$ on the bundle $\Omega_X(log D)$ ($K$-linearization of the bundle) such that the monomorphism

$$\Omega_X \to \Omega_X(log D)$$

is $K$-invariant.(comp.[8])

The potential sheaf has a natural $G_0$-linearization and hence a $K$-linearization making the homorphisms in the Euler sequence invariant, so all arrows in the diagram (6) are $K$-invariant. We are going to consider the diagram dual to (6), but first let us simplify the notation. Let $V$ denote the $n$-dimensional space dual to the fibre of the trivial bundle $\Omega_X(log D)$ (corollary 3.4), and let $U$ be the dual space to $H^0(Ind)$ . We have already considered the basis of this space, namely $\{i_e\}_{e \in E}$ , and let $\{s_e\}_{e \in E}$ denote the dual basis. Applying the functor $Hom(\ ,\mathcal{O}_X)$ to the middle column of the diagram (6) we get the following exact sequence (lemma 3.3):

$$0 \to U_X \to L_X \to \oplus_e \mathcal{O}_{D_e}(D_e) \to 0 .$$

With this notation we write down the diagram dual to (6):

$$\begin{array}{ccccccccc}
 & & 0 & & 0 & & 0 & & \\
 & & \downarrow & & \downarrow & & \downarrow & & \\
0 & \longrightarrow & H_X^\vee & \longrightarrow & U_X & \longrightarrow & V_X & \longrightarrow & 0 \\
 & & \downarrow & & \downarrow{\scriptstyle s} & & \downarrow{\scriptstyle ev} & & \\
0 & \longrightarrow & H_X^\vee & \longrightarrow & L_X & \longrightarrow & T_X & \longrightarrow & 0 \\
 & & \downarrow & & \downarrow & & \downarrow & & \\
 & & 0 & \longrightarrow & \bigoplus_{e \in E} \mathcal{O}_{D_e}(D_e) & \longrightarrow & \bigoplus_{e \in E} \mathcal{O}_{D_e}(D_e) & \longrightarrow & 0 \\
 & & & & \downarrow & & \downarrow & & \\
 & & & & 0 & & 0 & &
\end{array}$$

(7)

We have put $s = \oplus s_e$ and $ev$ denotes the evaluation map. All the morphisms in this diagram are $K$-invariant, the rows and columns are exact.

**Lemma 3.6** *We consider the space $H^0(T_X)$ with Poisson bracket of vector fields as the Lie algebra of group $G_0$. Then*

$$Lie\, K = V\ .$$

PROOF. First, we notice that a vector field $w$ belongs to the space $Lie\, K$ if and only if it is tangent to the varieties $D_e$. In fact, let $g$ denote an automorphism of the form $\exp(t\, ad_w)$ for $t \in IC^*$ and $w \in Lie\, K$.

The automorphisms of such form are dense in the component of the identity of the group $K$. From the well-known theorem on existence and uniqueness of solutions of ordinary differential equations applied to the variety $D_e$, it follows that the field $w$ is tangent to $D_e$ if and only if the corresponding automorphisms fulfill the condition $g(D_e) = D_e$.

Next, we consider the last column of the diagram (11). The cokernel of the morphism $ev$ is a direct sum of the normal sheaves to the subvarieties $D_e$. Therefore if a global vector field is tangent to all subvarieties $D_e$, it comes from the space $V$. Hence the inclusion:

$$Lie\, K \subset V\ .$$

In the cohomology sequence corresponding to the middle column, the homomorphism between the spaces $H^0(L_X)$ and $H^0(\oplus \mathcal{O}_{D_e}(D_e))$, sends a section to its restriction to the subvariety $D$. Since $h^1(\mathcal{O}_X)$ is zero, the space $H^0(L_X)$ is mapped onto the space $H^0(T_X)$. So the homomorphism from the space $H^0(T_X)$ to $H^0(\oplus \mathcal{O}_{D_e}(D_e))$ is also the restriction of sections to $D$. Therefore a vector field coming from the space $V$ is tangent to all subvarieties $D_e$. □

**Lemma 3.7** *The group $K$ acts trivially on the spaces $U$ and $V$.*

PROOF. We shall describe the action on the space $U$. The group $K$ preserves the subvarieties $D_e = zero\ s_e$. So for any $g \in K$ we have $g\, s_e = a_e(g)\, s_e$ $a_e \in IC^*$. On the other hand, the dual space $U^\vee$ is mapped invatriantly onto $H$, where the group $K$ acts trivially. The images of the elements of the dual basis $\{i_e\}_{e\in E}$ are nonzero, so $a_e = 1$. The space $V$ is the image of the trivial $K$-space $U$ therefore $K$ acts trivially also on $V$. □

Let $S$ denotes the connected component of unity in the group $K$.

**Corollary 3.8** *$S$ is a commutative group of dimension $n$ (equal to $\dim X$).*

PROOF. Apply (to the group $S$) the theorem of Borel ([1] chap.1 sec.3.8), which says that for a connected algebraic group $G$ defined over a field of characteristic zero, $ker\ Ad\, G$ is equal to the centre of $G$. □

We shall prove now that the group $S$ is an algebraic torus. The variety $X\setminus D$ is an orbit of $S$, since the vector fields from the space $V$ span the tangent spaces

$T_{X,x}$ for $x \in X \backslash D$ (comp. the last column of the diagram (11)). The isotropy group of this orbit is finite, because

$$dim\, S = dim\, X \backslash D \ .$$

It sufficies to prove that the quotient $S'$ of $S$ by this finite group is a torus. We have

$$S' = X \backslash D \ .$$

The group $S'$ does not contain abelian variety, because such a subgroup should act transitively on the manifold $X$ ; and we know that the Lie algebra of the group $S$ is generated by the vector fields vanishing on the divisors $D_e$ (lemma 3.6 and diagram (11)).

Now we apply the theorem of Deligne ([4] sec.3.5), which asserts that there exists a spectral sequence starting from the array

$$E^{p,q} = H^q(\Omega^p_X(logD))$$

and abutting to $H^{p+q}(X\backslash D; IC)$. By the lemma 3.2 we know that $H^1(\mathcal{O}_X) = 0$, so

$$H^1(X\backslash D; IC) = H^0(\Omega_X(logD)) \ .$$

According to the corollary 3.8, this space is $n$-dimensional. Thus $S'$ must be an algebraic torus - the only $n$-dimensional commutative linear group which has the space of the first cohomology of dimension $n$. Then $X$ is a torus embedding as a variety equipped with the effective action of a torus of the same dimension.([12] chap. 1,2). We have then proved the first implication.

Now, we shall describe the Euler sequence on a torus embedding $X_\Sigma$ , where $\Sigma$ is a complete nonsingular complex of cones (see sec.1). The set of one dimensional cones we shall denote by $E$. The elements of this set corresponds one-to-one to divisors $D_e$ , which are the closures of codimension one orbits of the torus. Let $D$ denote their sum $D = \sum_e D_e$ . This divisor has normal crossings, because we assume that $X_\Sigma$ is nonsingular (comp. [12]). Again we look at diagram (6). The first column is the sequence used by Danilov to calculate the Chern classes of toric varieties (see [2],[8]). Recall the arguments showing that the bundle $\Omega_X(logD)$ is trivial. Choose a cone $\sigma$ of maximal dimension and the corresponding basis $x_1, ..., x_n$ of the group of characters $M$ . This means that the elements of this basis span the dual cone $\sigma^\vee$ ). Over the affine variety $X_\sigma$ (isomorphic to the space $IC^n$ ) the forms $\frac{dx_i}{x_i}$ for $i = 1, \ldots n$ are the nonvanishing sections of the sheaf $\Omega_X(logD)|_{X_\sigma}$ . For two such cones $\sigma$ and $\sigma'$ , over the set $X_\sigma \cap X_{\sigma'}$ both systems of sections are linearly dependent by the matrix with integer elements. Thus the forms $\frac{dx_i}{x_i}$ can be extended to the global nonvanishing sections of the sheaf $\Omega_X(logD)$ and these sections determine its trivialization.

Since torus embedding is a rational variety, the space $H^1(\mathcal{O}_X)$ is zero. So the middle row of the diagram (6) splits and the bundle $Ind$ has to be trivial. Since $H^0(\Omega_X) = 0$ , from the cohomology sequence corresponding to the Euler sequence we get

$$H^0(L_X^\vee) = 0 \quad and \quad H^1(L_X^\vee) = 0 \ .$$

So the middle column of the diagram must be the sequence determined by the evaluation of global sections of the sheaf $\oplus_{e\in E}\mathcal{O}_{D_e}$ , and $L_X^\vee = \oplus_{e\in E}\mathcal{O}_X(-D_e)$.

□

**Corollary 3.9** *Let $X = X_\Sigma$ be a complete nonsingular torus embedding, and E be the set of one dimensional cones. Then*

$$H^p(\oplus_{e\in E}\mathcal{O}_X(D_e)) = H^p(T_X) \quad for\ p > 0$$

$$H^p(\oplus_{e\in E}\mathcal{O}_X(-D_e)) = 0 \quad for\ p = 0, 1 \ldots.$$

*where $D_e$ are the closures of codimension one orbits.*

*The following sequence is exact :*

$$0 \to H^\vee \to H^0(\oplus_{e\in E}\mathcal{O}_X(D_e)) \to H^0(T_X) \to 0 ,$$

*and the images of the sections $s_e$ vanishing on the subvarieties $D_e$ , generate the Lie algebra of the torus (comp[8]).* □

For the completé nonsingular torus embeddings Danilov proved that

$$h^p(\Omega_X^q) = 0 \quad for \ \ p \neq q .$$

Moreover, for these varieties the Chow ring is equal to the cohomoloy one and is generated by the classes of divisors ([2],[6],[8]). So, from the theorem 3.1 and the proposition 2.3 it follows that:

**Corollary 3.10** *On the nonsingular complete torus embedding*

$$H^0(\Lambda^p L_X^\vee) = 0 \quad for\ any\ p ,$$

$$H^q(\Lambda^p L_X^\vee) = 0 \quad for \ \ q \geq p .$$

It is possible to prove this result directly, using the methods of calculation of the cohomology groups of line bundles on toric varities, hence one may deduce that

$$h^p(\Omega_X^q) = 0 \quad for \ \ p \neq q .$$

Also the structure of the Chow ring can be re-discovered in this way.

The next corollary follows from our description of potential sheaf for toric varieties and theorem 2.5.

**Corollary 3.11** *For a cone $\sigma$ of maximal dimension we put $D_\sigma = \sum_{e\notin\sigma} D_e$ . Then*

$$\Omega_X(log D_\sigma) = \oplus_{e\in\sigma}\ \mathcal{O}_X(-D_e) .$$

**Corollary 3.12** *Let $Y$ be a m-dimensional nonsingular subvariety in an n-dimensional torus embedding $X$ and suppose that the divisor $B = D \cap Y$ has normal crossings and the varieties $B_e = D_e \cap Y$ are connected. Then the following sequence is exact:*

$$0 \to N_{X/Y}^\vee \to V_Y \to \Omega_Y(log B) \to 0 ,$$

*where $N_{X/Y}^\vee$ is the conormal sheaf, and $V$ is a fiber of the trivial bundle $\Omega_X(log D)$. Therefore the bundle $\Omega_Y(log B)$ determines a morphism of the variety $Y$ to the Grassman manifold $G(n, m)$.*

# 4 Vast Divisors

In this section we shall discuss the question when a given divisor $B$ with normal crossings on a complete algebraic variety $Y$ is a pullback of the distinguished divisor $D$ on some torus embedding, that means when there exists a mapping

$$f : Y \to X_\Sigma ,$$

where $\Sigma$ is a nonsingular and complete complex of cones, such that

$$\mathcal{O}_Y(B) = f^*\mathcal{O}_{X_\Sigma}(D)$$

(see [9],[11]). We shall use the description 3.9 of the Euler sequence on torus embeddings.

We shall need the following notation. For an effective Weil divisor $B = \sum_{e\in E} B_e$, where $E$ denotes the set of irreducible components, let $P(B)$ denote the free abelian group generated by the components and $\delta$ be the map from this group to the cohomology group $H^2(Y,\mathbb{Z})$ that associates a cohomology class to an algebraic cycle. We shall consider the following groups:

$$M(B) = ker\ \delta\ , \quad N(B) = Hom(M(B), Z) \quad H(B) = im\ \delta \otimes \mathbb{C}\ .$$

When the variety $Y$ is nonsingular and the divisor $B$ has normal crossings, the map $\delta$ can be interpreted as the connecting homomorphism in the cohomology sequence associated to the sequence

$$0 \to \Omega_Y \to \Omega_Y(log B) \to \oplus_{e\in E}\mathcal{O}_{B_e} \to 0$$

The group $P(B)$ has the distiguished set of generators parametrized by $E$ . The dual system generates the group $Hom(P(B), Z)$, which is mapped onto the group $N(B)$. So we have the distinguished elements $n_e$ in $N(B)$. Now let us look at the space $H(B)$. It is generated by the Chern classes of the line bundles so, according to the theorem on differential forms of the type (1,1) ([4] chapt.1), we may regard it as a subspace in the space $H = H^1(\Omega_Y)$ . Let $P(B)_Y$ denote the trivial bundle $P(B)\otimes\mathcal{O}_Y$. Attaching to each divisor $B_e$ the section $1_{B_e} \in H^0(\mathcal{O}_{B_e})$ , we obtain an isomorphism between the space $P(B) \otimes \mathbb{C}$ and the one of global sections of the sheaf $\oplus_{e\in E}\mathcal{O}_{B_e}$ . This leads to the following exact sequences:

$$0 \to \oplus_{e\in E}\ \mathcal{O}_Y(-B_e) \to P(B)_Y \to \oplus_{e\in E}\mathcal{O}_{B_e} \to 0\ ,$$

and, applying the functor $Hom(\ ,\mathcal{O}_Y)$ :

$$0 \to P(B)_Y^\vee \to \oplus_{e\in E}\mathcal{O}_Y(B_e) \to \oplus_{e\in E}\mathcal{O}_{B_e}(B_e) \to 0\ .$$

So the space $(P(B)\otimes \mathbb{C})^\vee$ may be regarded as a subspace of $H^0(\oplus_{e\in E}\mathcal{O}_Y(B_e))$ and it is generated by the sections $s_e \in H^0(\mathcal{O}_Y(B_e))$ which satisfy the conditions

$$zero\ s_e = B_e\ .$$

The composition of the monomorphism $H(B)_Y^\vee \to P(B)_Y^\vee$ with the evaluation of sections $s_e$ (the monomorphism in the exact sequence above) gives the monomorphism of sheaves $H(B)_Y^\vee \to \oplus_{e\in E}\mathcal{O}_Y(B_e)$ . Let $T(B)$ denotes its cokernel. We are asking when (in terms of divisor $B$), there exists a mapping from the variety $Y$ into some torus embbeding for which the sequence just constructed

$$0 \to H(B)_Y^\vee \to \oplus_{e\in E}\mathcal{O}_Y(B_e) \to T(B) \to 0$$

is a pullback of the sequence dual to the Euler one. In particular we want to know when the sheaf $T(B)$ is locally free.

From now on, we assume that the divisor $B$ has normal crossings. According to the theorem 2.5 for a section $s_e$ of the sheaf $\mathcal{O}_Y(B_e)$ there exists a monomorphism $\mathcal{O}_Y(-B_e) \to L_Y^\vee$ . Let us choose such a morphism for each $e$ , and let $L(B)^\vee$ be their sum:

$$L(B)^\vee : \oplus_{e\in E}\mathcal{O}_Y(-B_e) \to L_Y^\vee .$$

We get the following commutative diagram associated with the divisor $B$ :

$$\begin{array}{ccccccccc} 0 & \longrightarrow & H_Y^\vee & \longrightarrow & L_Y & \longrightarrow & T_Y & \longrightarrow & 0 \\ & & \downarrow & & \downarrow & & \downarrow & & \\ 0 & \longrightarrow & H(B)_Y^\vee & \longrightarrow & \oplus\mathcal{O}_Y(B_e) & \longrightarrow & T(B) & \longrightarrow & 0 \end{array} \tag{8}$$

Let $\sigma$ be a cone spanned by a basis of the space $N(B)\otimes R$ composed of the elements $n_e$ . We put $Y_\sigma = Y \setminus \cup_{n_e \notin \sigma} B_e$ .

**Lemma 4.1** *The Chern classes of the sheaves $\mathcal{O}_Y(B_e)$ for $n_e \notin \sigma$ form a basis of the space $H(B)$ .*

PROOF. This is a lemma from linear algebra. Let $\varphi$ denote the isomorphism of the group $P(B)$ with its dual $P(B)^\vee$ that sends the distinguished system of generators onto the dual one. Let $i$ be the monomorphism $M(B) \to P(B)$ and let $\delta'$ be the compositon of the isomorphism $\varphi$ with the epimorphism $i^\vee$ $\delta' = i^\vee \circ \varphi$ . With these notations, we have to prove that

$$ker\,\delta \cap ker\,\delta' = 0 .$$

Assume that there exists an element $p$ in both these spaces. So let $p \in M$ , such that $\varphi(p)$ be a functional that vanishes on the elements of $M$ . In particular

$$\varphi(p)(p) = 0 .$$

Writing down $p = \sum_{e\in E} p_e B_e$, we obtain that $\sum_{e\in E} p_e^2 = 0$ so $p = 0$. $\square$

**Lemma 4.2** *On the manifold $Y_\sigma$ we have*

$$T(B)_{Y_\sigma} = \oplus_{n_e \in \sigma} \mathcal{O}_{Y_\sigma}(B_e) .$$

PROOF. Since the system $\{n_e\}_{n_e \in \sigma}$ generates the group $N(B)$ , the Chern classes of the sheaves $\mathcal{O}_Y(B_e)$ $n_e \notin \sigma$ form a basis of the space $H(B)$ . We consider the following diagram:

$$\begin{array}{ccccccccc}
 & & & & 0 & & 0 & & \\
 & & & & \downarrow & & \downarrow & & \\
 & & 0 & \to & \oplus_{e\in\sigma}\mathcal{O}_Y(B_e) & \to & T(B) & & \\
 & & \downarrow & & \downarrow & & \downarrow & & \\
0 & \to & H_Y^\vee & \to & \oplus_e \mathcal{O}_Y(B_e) & \to & T(B) & \to & 0 \\
 & & \downarrow & & \downarrow & & \downarrow & & \\
0 & \to & \oplus_{e\notin\sigma}\mathcal{O}_Y(B_e) & = & \oplus_{e\notin\sigma}\mathcal{O}_Y(B_e) & \to & 0 & & \\
 & & \downarrow & & \downarrow & & & & \\
 & & \oplus_{e\notin\sigma}\mathcal{O}_{B_e}(B_e) & \to & 0 & & & &
\end{array} \tag{9}$$

The first column is a sum of the sequences

$$0 \to \mathcal{O}_Y \to \mathcal{O}_Y(B_e) \to \mathcal{O}_{B_e}(B_e) \to 0 .$$

From the snake lemma we obtain the exact sequence

$$0 \to \oplus_{n_e\in\sigma}\mathcal{O}_Y(B_e) \to T(B) \to \oplus_{n_e\notin\sigma} \mathcal{O}_{B_e}(B_e) \to 0 .$$

So the support of the sheaf $\oplus_{n_e\notin\sigma} \mathcal{O}_{B_e}(B_e)$ lies off of $Y_\sigma$ . □

**Definition 4.3** *A normal crossing divisor $B = \sum_e B_e$ on a complete variety $Y$ will be called vast if the following conditions are satisfied:*

*(i) there exists a complete and nonsingular complex of cones $\Sigma(B)$ in the space $N(B)_R$, with the one dimensional cones spanned by the distinguished elements $n_e$ (in particular $n_e$ should be primitive);*

*(ii) for any cone $\sigma$ of maximal dimension we denote by $Y_\sigma$ the variety $Y \setminus \cup_{e\notin\sigma} B_e$ and require the family $\{Y_\sigma\}_\sigma$ to be a covering of $Y$ .*

*(iii) for a linear combination of the divisors $B_e$ the homological equivalence implies the linear one.*

**Example 4.4** *Assume that all the components $B_e$ divisor $B = \sum_e B_e$ are linearly equivalent. Let $n+1$ denote the number of them. We fix one of them, say $e_0$, and let $s_e$ be the sections of the sheaf $\mathcal{O}_Y(B_{e_0})$, such that zero $s_e = B_e$. In the group $N(B)$ there is only one relation between the distinguished elements, namely*

$$\sum_e n_e = 0 ,$$

*so every $n$ elements among $n_e$ generate this group. We take all such systems for the generators of the $n$-dimensional cones and obtain the complex corresponding to the projective space. The conditions (i) and (iii) in the definition 4.3 are satisfied automatically, and the condition (ii) means that the system of divisors $B_e$ has no base points, so the rational map sending a point $y$ onto the point with projective coordinates $\{s_1(y) : \ldots : s_n(y)\}$ is everywhere defined.*

**Theorem 4.5**
*For a vast divisor $B$ there exists a map $f_B$ of the variety $Y$ into the torus embedding $X_{\Sigma(B)}$ such that the sequence*

$$0 \to H(B)^\vee_Y \to \oplus_{e\in E}\mathcal{O}_Y(B_e) \to T(B) \to 0$$

*is the pullback of the sequence dual to the Euler one on the manifold $X_{\Sigma(B)}$.*

*Moreover, if the variety $Y$ is nonsingular then the following conditions are equivalent:*

*(i) the sheaf that is the kernel of the homomorphism $L(B)$ has the rank equal to the codimension of the space $H(B)$ in the space $H$;*

*(ii) the map $f_B$ induces embeddings on the tangent spaces.*

*Any of these conditions implies that the sheaf described in (i) is a locally free $\mathcal{O}_Y$ − module.*

PROOF. First, we shall construct the mapping $f_B$. The space $N(B) \otimes \mathbb{C}$ may be regarded as a subspace in the space of sections $H^0(T(B))$. Let $w_e$ denote the sections of the sheaf $T(B)$ corresponding to the elements $n_e$. For any cone $\sigma$ of maximal dimension, the set $\{w_e\}_{e\in\sigma}$ makes up a basis of this subspace. For any such cone we put $w_\sigma = \sum_{e\in\sigma} w_e$. By the condition (ii) in the definition of a vast divisor, we know that the family $\{Y_\sigma\}_\sigma$ covers $Y$. Thus, by the lemma 4.2, the sheaf $T(B)$ is locally free. By the condition (iii) and the description of the group $N(B)$ as the group of relationships between the Chern classes, for any $n_{e_0} \in \sigma$ the divisor $B_{e_0}$ is linearly equivalent to some linear combination of the divisors $B_e$ for $e \notin \sigma$. Hence $\mathcal{O}_{Y_\sigma}(B_{e_0}) = \mathcal{O}_{Y_\sigma}$. We have got the trivialization (comp.lemma 4.2): $T(B)_{Y_\sigma} \to \oplus_{n_e\in\sigma}\mathcal{O}_{Y_\sigma}$. This isomorphism is a direct sum of isomorphisms of the invertible sheaves with the structural sheaf which has a distinguished section, namely $1_{Y_\sigma}$, so we have a distinguished basis $\{b_e\}_{n_e\in\sigma}$ in every fibre (in a vector bundle sense) $T(B)_y$ for any $y \in Y_\sigma$. Hence we may write $w_e(y) = k_e(y)\, b_e$ for $y \in Y_\sigma$, and $w_\sigma(y) = \sum_{e\in\sigma} k_e(y)\, b_e$ for $y \in Y_\sigma$, where $k_e(y)$ is a regular function in some affine neighbourhood of the point $y$ vanishing only at the points from the subvarieties $B_e$ . The condition (i) from the definiton of vast divisor implies $ord_{B_e}(k_e) = 1$ , since the divisor $B_e$ is irreducible and reduced.

The tangent space to the affine torus embedding $X_\sigma$ at the fixed point with respect to the action of torus is isomorphic as an algebraic variety equipped with the action of torus to the variety $X_\sigma$ itself in the following way. To each vector $n_e$ that generates an edge of the cone $\sigma$ we attach the derivation $\delta_e$ of the algebra of regular functions on $X_\sigma$ defined by: $\delta_e = \frac{\delta}{\delta x_e}$ , where $x_e \in IC[\sigma^\vee \cap M(B)]$ is the character of the torus that corresponds to the element $n_e^\vee \in M(B)$ via the isomorphism between the groups $N(B)$ and $M(B)$ determined by the choise of the basis $\{n_e\}_{e\in\sigma}$ . We get the isomorphisms of the algebras:

$$Sym(T^\vee_{X_\sigma,a}) = IC[\{x_e\}_{e\in\sigma}] = IC[\sigma^\vee \cap M(B)] \ ,$$

which gives the isomorphism of affine varieties. We attach to $b_e$ the derivation $\delta_e$ . Again we obtain an isomorphism, now between the space $T(B)_y$ and the tangent space to the affine torus embedding $X_\sigma$ at the fixed point. Let $J_\sigma(y)$ denote the composition of these isomorphisms $J_\sigma(y) : T(B)_y \to X_\sigma$ . We define the map $f_B(\sigma) : Y_\sigma \to X_\sigma$ as the composition $J_\sigma(y) \circ w_\sigma(y)$ . So locally $f_B$ can be expressed by the formula:

$$f_B(\sigma)(y) = (\ k_{e_1}(y)^{m_{1,1}} \cdots k_{e_n}(y)^{m_{1,n}}, \dots, k_{e_1}(y)^{m_{n,1}} \cdots k_{e_n}(y)^{m_{n,n}}\ ) \ ,$$

where the set $\{(m_{1,1},\dots,m_{1,n}),\dots,(m_{n,1},\dots,m_{n,n})\}$ is a basis of the dual cone $\sigma^\vee$ . For two cones of maximal dimension $\sigma$ and $\sigma'$ the two systems of sections $\{w_e\}_{e\in\sigma}$ and $\{w'_e\}_{e'\in\sigma'}$ are related by the same linear transformation (from the group $SL(N(B))$ ) as the corresponding systems of vector fields on the torus embedding $X_{\Sigma(B)}$ . Therefore the maps $f_B(\sigma)$ can be glued together and define a morphism

$$f_B : Y \to X_{\Sigma(B)} \ .$$

As the functions $k_e$ vanish only on the subvarieties $B_e$ and $ord_{B_e}(k_e) = 1$ we see that $f_B^*(\mathcal{O}_{X_{\Sigma(B)}}(D_e)) = \mathcal{O}_Y(B_e)$ . The composition of the pullback of the homomorphism $\mathcal{O}_{X_{\Sigma(B)}}(-D_e) \to L^\vee_{X_{\Sigma(B)}}$ (thm. 2.5) with the mapping $f_B^* L^\vee_{X_{\Sigma(B)}} \to L^\vee_Y$ gives a morphism $\mathcal{O}_Y(-B_e) \to L^\vee_Y$ , which extends the embedding of the conormal sheaf into the sheaf of differentials:

$$\mathcal{O}_{B_e}(-B_e) \to \Omega_{1B_e} \ .$$

Hence, the sequence

$$0 \to H(B)^\vee_Y \to \oplus_{e\in E}\mathcal{O}_Y(B_e) \to T(B) \to 0$$

is a pullback of the dual sequence to the Euler one over the manifold $X_{\Sigma(B)}$ . From the snake lemma applied to the diagram (9) we obtain the description of the kernel of the homomorphism $L(B)$ and the equivalence of the conditions (i) and (ii). □

**Corollary 4.6** *If the kernel and cokernel of the homomomorphism $L(B)$ are locally free, then the manifolds $Y_\sigma$ and $Y \setminus B$ are affine.*

**Remark 4.7** *In general the divisor B determines in a unique way only the one dimensional skeleton of the complex* $\Sigma$ *, and there exist finitely many complexes spanned by this skeleton for which the condition (ii) from definition 4.3 may be fulfilled. If* $2\ dimY > dim\Sigma$ *, divisor B is vast, and the mapping* $f_B$ *induces the monomorphism of the middle cohomology group of the torus embedding* $X_\Sigma$ *, then the complex of cones* $\Sigma$ *is the only one possible and can be described in the following way. Let n denote the rank of the group* $N(B)$ *and let* $k = [n/2]$ *. First we define the k dimensional cones. The system* $\{n_{i_1}, ..., n_{i_k}\}$ *does not span a cone, if the intersection of the divisors* $(B_{i_1}, ..., B_{i_k})$ *is zero. This condition is fulfilled for the torus embeddings (as it follows from the description of the Chow ring see [6],[2],[8]) and determines the* $k-1$ *dimensional skeleton of the simplicial cone that arises as the intersection of the complex of cones with the sphere* $S^{n-1}$ *in the space* $N_R$ *. Since* $k \geq (n-1)/2$ *, this skeleton determines the triangulation.*

# 5 References


[1] *A.Borel*, Linear Algebraic Groups (W.A.Benjamin 1969)
[2] *W.I.Danilov*, Geometrija toriczeskich mnogoobrazijej (Usp.M.N., vol.32, 1978)
[3] *M.Demazure*, Sous-groupes algebriques de rang maximum ... (Ann. Sc. Ec. Norm. Sup. (4)1970,p. 259-268)
[4] *P.Griffiths, J.Harris*, Principles of Algebraic Geometry ( John Wiley and Sons 1978)
[5] *R.Hartshorne*, Algebraic Geometry (Grad.Texts in Math. 52 Springer-Verlag 1977)
[6] *J.Jurkiewicz*, Chow Ring of Projective Nonsingular Torus Embedding (Coll. Mathem. vol.XLIII, 1980, fasc 2)
[7] *T.Oda, K.Miyake*, Lectures on Torus Embeddings and Applications (Tata Inst. 58, 1978)
[8] *T.Oda*, Convex Bodies and Algebraic Geometry (Ergeb,Math. Grenzgeb., vol.15, 1988)
[9] *T.Oda*, Geometry of Toric Varieties (preprint)
[10] *R.Remmert, A.van de Ven*, Zur Funktionentheorie Homogener Komplexer Mannigfaltigkeiten (Topology, vol.2,p. 137-157)
[11] *A.Ash, D.Mumford, M.Rapoport, Y.Tai*, Smooth Compactification of Locally Symmetric Varieties (Math.Sci.Press, Brooklyne, Mass,1975)
[12] *G.Kempf, F.Knudsen, D.Mumford, B.Saint-Donat*, Toroidal Embeddings I (Lect. Notes in Math.339)



Institute of Mathematics, Warsaw University, Banache 2-02-097
Warsaw, Poland

Contemporary Mathematics
Volume **162**, 1994

# A GENERALIZED TEISSIER–PLÜCKER FORMULA

STEVEN L. KLEIMAN

ABSTRACT. We generalize Teissier's Plücker formula for hypersurfaces with isolated singularities to arbitrary varieties with isolated singularities, replacing the ordinary multiplicity of the Jacobian ideal with the Buchsbaum–Rim multiplicity of the Jacobian matrix. We use two strikingly different methods: first we intersect with a complementary polar variety, but this method works only for complete intersections; second we use a recently proved intersection-theoretic expression for the multiplicity, and we obtain the general formula.

## 1. INTRODUCTION

In 1756, Goudin and du Séjour noted [**12**, p. 554] that a real plane curve of degree $d$ has this property: the number $d'$ of tangents with a given direction is at most $d(d-1)$. In 1820, Poncelet showed [**1**, fn. 92), p. 343] that, for a complex plane curve, $d'$ is equal to $d(d-1)$ if the curve is smooth, but that each node decreases the number $d(d-1)$ by 2 and each cusp by at least 3. In 1834, Plücker noted [**12**, p. 857] that the decrease is exactly 3; thus he obtained the celebrated formula,

$$d' = d(d-1) - 2\delta - 3\kappa$$

where $\delta$ is the number of nodes and $\kappa$ is the number of cusps. Subsequently, many geometers have generalized Plücker's formula. In 1975, Teissier [**17**, App. II] proved the following generalization to a hypersurface of dimension $r$ with isolated singularities:

$$d' = d(d-1)^r - \sum\nolimits_{x \in \mathrm{Sing}(X)} e(x) \qquad (1\text{-}1)$$

where $e(x)$ is the ordinary multiplicity of the Jacobian ideal at $x$.

In this article, we generalize Teissier's formula (1-1) to an arbitrary variety with isolated singularities $x$, correcting for the presence of the $x$ with the

---

1991 *Mathematics Subject Classification.* Primary 14N10; Secondary 14C17, 14B05.
The author was supported in part by NSF grant 9106444-DMS.
This paper is in final form and no version of it will be submitted for publication elsewhere.

Buchsbaum–Rim multiplicity $e(x)$ of the Jacobian matrix. Indeed, the article may serve as an introduction to the use of this new invariant of singularities in algebraic geometry. In Section 2, we generalize (1-1) to complete intersections by using a complementary polar variety and some recent results of Gaffney and Rees about the Buchsbaum–Rim multiplicity. In Section 3, we generalize (1-1) to arbitrary varieties, recovering the case of complete intersections, via an entirely different method; it is based on an intersection-theoretic expression for the multiplicity, which was proved recently and for local complete intersections by Henry and Merle [**6**, 4.2.3] and independently and for arbitrary varieties by Thorup and the author [**11**, (5.7)]. We conclude the article with some remarks and examples. In the rest of this section, we discuss the general setup, including the definitions and basic properties of the class $d'$ and the Buchsbaum–Rim multiplicity $e(x)$.

For the rest of the article, fix a closed subvariety (that is, reduced subscheme) $X$ of the projective $n$-space $\mathbf{P}^n$ over an algebraically closed ground field of arbitrary characteristic. Say $X$ has degree $d$, dimension $r$, and codimension $\rho$; so

$$r + \rho = n.$$

Assume $1 \leq r < n$. Denote the set of singular points by $\mathrm{Sing}(X)$, and denote its complement, the set where $X$ is smooth, by $\mathrm{Sm}(X)$. Denote the dual projective space by $\check{\mathbf{P}}^n$. In $X \times \check{\mathbf{P}}^n$ form the *conormal variety* $CX$; by definition, $CX$ is the (reduced) closure of the set of point-hyperplane pairs $(x, H)$ such that $x \in \mathrm{Sm}(X)$ and $H \supseteq T_x$ where $T_x$ denotes the tangent space to $X$ at $x$. Clearly $CX$ has dimension $n - 1$. Let

$$p\colon CX \to X \text{ and } p'\colon CX \to \check{\mathbf{P}}^n$$

denote the restrictions of the projections. Finally, set

$$X' := p'CX.$$

So $X'$ is, by definition, the *dual variety* of $X$.

Suppose that $X$ is a hypersurface. Then clearly $p\colon CX \to X$ restricts to an isomorphism over $\mathrm{Sm}(X)$. Denote the induced map by

$$g\colon \mathrm{Sm}(X) \longrightarrow \check{\mathbf{P}}^n.$$

Obviously, $g(x) := T_x$. So $g$ is, by definition, the *Gauss* map of $X$.

When $X$ is arbitrary, define its *class* $d'$ formally as an intersection number:

$$d' := \int \ell'^{\,n-1}[CX] \text{ where } \ell' = c_1(\mathcal{O}_{\check{\mathbf{P}}^n}(1)).$$

To interpret $d'$, let $A$ be a linear space of codimension 2 in general position in $\mathbf{P}^n$, and consider the dual variety $A'$. Obviously, $A'$ is equal to the line in $\check{\mathbf{P}}^n$ parameterizing the hyperplanes $H$ of $\mathbf{P}^n$ through $A$. Since $A$ is in general position, so is $A'$. Hence $A'$ intersects $X'$ in only finitely many points, and they all lie outside any given closed set of dimension at most $n-2$. So they lie outside $p'p^{-1}\operatorname{Sing}(X)$. Moreover, $p'^{-1}A'$ is finite. Thus

$$p'^{-1}A' = \{(x,H) | x \in \operatorname{Sm}(X) \text{ and } H \supseteq T_x, A\}.$$

Since $p^{-1}\operatorname{Sm}(X)$ is smooth, it follows that

$$d' = \sum\nolimits_{(x,H)\in p'^{-1}A'} \operatorname{length}(\mathcal{O}_{p'^{-1}A',(x,H)}). \tag{1-2}$$

Since $A'$ is in general position, it does not meet $\operatorname{Sing}(X)$. Moreover, at each point $x'$ of $A' \cap X'$, the intersection is transversal by Bertini's theorem. Furthermore, each fiber $p'^{-1}x'$ consists of $\operatorname{s.deg} p'$ distinct pairs $(x,H)$, and each pair appears with multiplicity $\operatorname{i.deg} p'$, where $\operatorname{s.deg} p'$ and $\operatorname{i.deg} p'$ denote the separable and inseparable degrees of $p'$. Of course, $\operatorname{i.deg} p' = 1$ if the characteristic is 0; moreover, whenever $\operatorname{i.deg} p' = 1$, then $\operatorname{s.deg} p' = 1$ by the Monge–Segre–Wallace criterion [**10**, (4), p.169]. Now, a simple count of constants shows that, at each pair $(x,H)$, the span of $T_x$ and $A$ is equal to $H$, again because $A$ is in general position. Thus $d'$ is equal to the weighted number of points $x$ in $\operatorname{Sm}(X)$ such that $T_x$ and $A$ span a hyperplane; moreover, the weighting factor is $\operatorname{i.deg} p'$.

From now on, assume that $X$ has isolated singularities; in other words, $\operatorname{Sing}(X)$ is finite. Then, associated to each $x \in \operatorname{Sing}(X)$, there are two different natural "tangential" multiplicities: the ordinary multiplicity of the Jacobian ideal and the Buchsbaum–Rim multiplicity of the Jacobian matrix. By definition, the *Jacobian ideal* $\mathcal{J}$ is the $r$th Fitting ideal of the sheaf of differentials $\Omega^1_X$. By Samuel's definition [**15**, top p. 186], the multiplicity of $\mathcal{J}_x$ is the coefficient of $m^r/r!$ in the following polynomial of degree $r$:

$$\lambda(m) := \operatorname{length}(\mathcal{O}_{X,x}/\mathcal{J}_x^m).$$

Clearly, this multiplicity is an intrinsic local invariant of $X$ at $x$; in fact, it depends only on the completion $\widehat{\mathcal{O}}_{X,x}$.

The Jacobian ideal $\mathcal{J}_x$ at $x$ can be computed as follows. Let $\mathcal{I}$ be the sheaf of ideals of $X$ in $\mathbf{P}^n$, and form the usual right exact sequence,

$$\mathcal{I}/\mathcal{I}^2 \xrightarrow{\delta} \Omega^1_{\mathbf{P}^n}|X \longrightarrow \Omega^1_X \longrightarrow 0. \tag{1-3}$$

Although $(\mathcal{I}/\mathcal{I}^2)_x$ need not be free, nevertheless $\mathcal{J}_x$ is the ideal in $\mathcal{O}_{X,x}$ of $\rho$ by $\rho$ minors of any matrix representing $\delta$ at $x$. Since locally $\delta$ carries a function $f$ vanishing on $X$ to its differential $df$, the transpose $\delta^*$ is the usual *Jacobian map* and any matrix representing $\delta^*$ at $x$ is a *Jacobian matrix* $J_x$. Thus the Jacobian ideal $\mathcal{J}_x$ is the ideal of $\rho$ by $\rho$ minors of any Jacobian matrix $J_x$.

Suppose that $X$ is a complete intersection locally at $x$. Then $\mathcal{I}/\mathcal{I}^2$ is free locally at $x$. Hence, so is the *normal sheaf*,

$$\mathcal{N} := (\mathcal{I}/\mathcal{I}^2)^*.$$

Moreover, since $\delta$ is injective on the smooth locus $\mathrm{Sm}(X)$ and since $X$ is reduced, $\delta$ is therefore injective at $x$. Hence (1-3) yields the right exact sequence,

$$(\Omega^1_{\mathbf{P}^n}|X)^*_x \xrightarrow{\delta^*_x} \mathcal{N}_x \longrightarrow \mathrm{Ext}^1(\Omega^1_{X,x}, \mathcal{O}_{X,x}) \longrightarrow 0.$$

In other words, the cokernel $\mathrm{Cok}(\delta^*_x)$ is equal to the indicated $\mathrm{Ext}^1$, which is obviously an intrinsic local invariant; in fact, it depends only on the completion $\widehat{\mathcal{O}}_{X,x}$. Moreover, as is well known, it is equal to the module $T^1(\mathcal{O}_{X,x})$ of deformation theory.

To any map $u\colon R^p \to R^q$ of free modules over a Noetherian local ring $R$, such that the cokernel $\mathrm{Cok}(u)$ has finite length, Buchsbaum and Rim [**2**, pp. 213–14] associated a multiplicity $e(u)$ as follows. They proved that the cokernel of the $m$th symmetric power $\mathrm{Cok}(\mathrm{Sym}^m(u))$ has finite length, which is eventually given by a polynomial in $m$ of degree precisely $s$ where $s := q - 1 + \dim(R)$, and they defined $e(u)$ to be the coefficient of $m^s/s!$. In particular, when $q = 1$, they recovered Samuel's definition of the ordinary multiplicity of the ideal $\mathrm{Im}(u)$. In general, $u$ splits into the direct sum of a minimal presentation of $\mathrm{Cok}(u)$ and of a surjection of free modules; it follows easily [**2**, 3.3, pp. 211] that $e(u)$ depends only on $\mathrm{Cok}(u)$.

When $X$ is a complete intersection locally at $x$ with an isolated singularity at $x$, then the Buchsbaum–Rim theory yields an intrinsic local invariant $e(x)$; namely, $e(x)$ is the multiplicity of the Jacobian map $\delta^*_x$, or of any representing Jacobian matrix $J_x$. Clearly, $e(x)$ depends only on the analytic type of $X$ at $x$; that is, two pairs $(X, x)$ have the same $e(x)$ if the two completions $\widehat{\mathcal{O}}_{X,x}$ are isomorphic. Over the complex numbers, $e(x)$ is, in fact, a topological invariant. Indeed, Gaffney [**4**, par. after Def. 1.2] proved that $e(x)$ is equal to the sum of the Milnor number of $X$ at $x$ and of that of a general hyperplane section through $x$. Furthermore, Gaffney has generalized much of Teissier's work on complex hypersurface germs with isolated singularities to complete intersection germs with isolated singularities. Notably, Gaffney proved this theorem [**4**, Thm. 1] (compare [**3**, 5.11, p. 44]): if $e(x)$ and all the polar multiplicities remain constant as $(X, x)$ varies analytically, then the family is Whitney equisingular, and conversely.

If $X$ is a hypersurface, then $(\mathcal{I}/\mathcal{I}^2)_x$ is isomorphic to $\mathcal{O}_{X,x}$, and the isomorphism identifies the image of the Jacobian map with the Jacobian ideal; hence, $e(x)$ is equal to the ordinary multiplicity of the Jacobian ideal in this case. In general, the two numbers differ. Here is an example (Gaffney, private comm., Dec. 2, 1990): $X$ is the complete intersection surface in $\mathbf{C}^4$, which sits as the member $X(0)$ in the following family parameterized by $t$:

$$X(t) : X_1^2 + X_2^2 + X_3^2 + tX_4 = 0, X_1^4 + X_2^4 + X_3^4 + X_4^2 = 0.$$

Henry and Merle [**5**, pp. 578-79] introduced this remarkable example in 1981. They proved that it has these three properties at the origin $x$: (1) it is Whitney equisingular; (2) if $t \neq 0$, then $X(t)$ is analytically isomorphic to a hypersurface; and (3) at $t = 0$, there is a change in the dimension of the fiber over $x$ of the relative Nash modification of the total space. Because of (1), the multiplicity of the Jacobian matrix remains constant by virtue of Gaffney's theorem above. Because each $X(t)$ is a complete intersection, the relative Nash modification is equal to the blowup of the relative Jacobian ideal. Hence, because of (3), the multiplicity at $X$ of the Jacobian ideal of $X(t)$ changes at $t = 0$. Finally, because of (2), the two multiplicities are equal if $t \neq 0$. Therefore, they differ at $t = 0$.

When $X$ is arbitrary, it is natural to extend the definition of $e(x)$ as follows. First, note that the sequence (1-3) yields this short exact sequence:

$$0 \longrightarrow \mathrm{Im}(\delta_x^*) \longrightarrow (\mathrm{Im}\,\delta_x)^* \longrightarrow \mathrm{Ext}^1(\Omega^1_{X,x}, \mathcal{O}_{X,x}) \longrightarrow 0. \tag{1-4}$$

Next, form the following two graded $\mathcal{O}_X$-algebra $\mathcal{G}$ and $\mathcal{G}'$ (the twist by $\mathcal{O}_X(-1)$ is unnecessary here, but is introduced for use in Section 3):

$$\mathcal{G}' := \mathrm{Sym}(\mathrm{Im}(\delta^*)(-1))/X\text{-torsion};\ \mathcal{G} := \mathrm{Sym}((\mathrm{Im}\,\delta)^*(-1))/X\text{-torsion}. \tag{1-5}$$

The two algebras are equal on $\mathrm{Sm}(X)$; hence, $\mathcal{G}'$ is a graded subalgebra of $\mathcal{G}$, and, at $x$, the quotient $\mathcal{G}_m/\mathcal{G}'_m$ of the $m$th graded pieces has finite length, $\lambda(m)$ say. It is a theorem of Kirby's [**7**, Thm. 6, p. 20] that $\lambda(m)$ is eventually given by a polynomial in $m$ of degree at most $n-1$ (for different proofs of more general results, see [**8**, §4] and [**11**, (5.10)]). Finally, define $e(x)$ to be the coefficient of $m^{n-1}/(n-1)!$. Then $e(x)$ is an intrinsic local analytic invariant. In fact, it can be computed using, in place of (1-3), any presentation of $\Omega^1_{X,x}$ as a quotient of a free $\mathcal{O}_{X,x}$-module. Indeed, any such presentation splits as a direct sum of the minimal presentation and of a presentation of the zero module as a quotient of a free module $\mathcal{E}$. Hence, in the counterpart of (1-4), the second map splits off $1_{\mathcal{E}^*}$ as a direct summand, and it is easy to check the equality of the coefficients in question as in the original case of Buchsbaum and Rim.

## 2. Complete intersections

In this section, we generalize Teissier's Plücker formula (1-1) when $X$ is a complete intersection by following the traditional method of intersecting $X$ with a polar variety of complementary dimension; the generalized formula is given in Theorem 1 below. We build up in stages: first $X$ is a plane curve, then a hypersurface, and finally arbitrary. Teissier proved (1-1) essentially as done below, although he worked over the complex numbers. More recently, van Gastel [**18**, Cor. 3.6, p. 273] gave a more "dynamic" version of the proof. For an arbitrary complete intersection, the proof below retains the original flavor, but requires some new ingredients.

First, suppose that $X$ is a plane curve; that is, $n = 2$ and $r = 1$. Choose coordinates $X_0, X_1, X_2$ for $\mathbf{P}^2$, and say

$$X : F(X_0, X_1, X_2) = 0$$

where $F$ is a homogeneous polynomial of degree $d$. Then, at a point $x$ of the smooth locus $\mathrm{Sm}(X)$, the tangent line $T_x$ is given as follows:

$$T_x : \sum_{i=0}^{2} F_i(x)X_i = 0 \text{ where } F_i := \partial F/\partial X_i.$$

Let $A$ be a point of $\mathbf{P}^2$; say $A = (a_0, a_1, a_2)$. Then $A \in T_x$ if and only if

$$\sum_{i=0}^{2} F_i(x)a_i = 0.$$

We can reinterpret this equation as saying that $x \in X \cap \Pi$ where $\Pi$ is the *polar curve* of $X$ with respect to $A$, namely, the curve defined as follows:

$$\Pi : \sum_{i=0}^{2} F_i(X_0, X_1, X_2)a_i = 0.$$

Clearly, $\Pi$ has degree $d-1$. So, by Bezout's theorem, the global intersection number $\int X \cdot \Pi$ is equal to $d(d-1)$. At any $x$ in $X \cap \Pi$, the local intersection number is equal to

$$\operatorname{length}(\mathcal{O}_{X,x}/(\textstyle\sum F_i a_i)) \tag{2-1}$$

where $(\sum F_i a_i)$ denotes the ideal in the local ring $\mathcal{O}_{X,x}$ generated by the homogeneous function $\sum F_i(X_0, X_1, X_2)a_i$.

Fix $x \in X \cap \Pi$. Suppose $x \in \operatorname{Sm}(X)$. Then clearly $\mathcal{O}_{X,x}/(\sum F_i a_i)$ is equal to the local ring at $x$ of the scheme $g^{-1}A'$ where $g\colon \operatorname{Sm}(X) \to \check{\mathbf{P}}^2$ is the Gauss map,

$$g(x) = (F_0(x), F_1(x), F_2(x)),$$

where $A'$ is the line in $\check{\mathbf{P}}^2$ dual to $A$. If $A$ is in general position, then the sum,

$$\sum_{x \in \operatorname{Sm}(X)} \operatorname{length}(\mathcal{O}_{X,x}/(\textstyle\sum F_i a_i)),$$

is equal to the class $d'$ of $X$ by Equation (1-2) because $p\colon CX \to X$ is an isomorphism over $\operatorname{Sm}(X)$.

Fix $x \in \operatorname{Sing}(X)$. Obviously $x \in X \cap \Pi$. Suppose $A$ is in general position. Then the length (2-1) is equal to the multiplicity of the ideal of $\mathcal{O}_{X,x}$ arising from the three homogeneous partial derivatives $F_i$, namely, the Jacobian ideal $\mathcal{J}_x$ of $X$ at $x$; indeed, $\mathcal{J}_x$ and $(\sum F_i a_i)$ generate the same ideal in the local ring of any point over $x$ in the normalization of $X$. Thus, this length is equal to $e(x)$. Hence Teissier's Plücker formula (1-1) holds in the case $r = 1$.

Next, suppose $X$ is a hypersurface of arbitrary dimension $r$. Choose coordinates $X_0, \ldots, X_n$ for $\mathbf{P}^n$, and say $X : F = 0$ where $F$ is a homogeneous polynomial of degree $d$. Let $A$ be an arbitrary linear space of dimension 2 in $\mathbf{P}^n$. Represent $A$ as the column space of an $n+1$ by $n-1$ matrix $(a_{ij})$; in other words, $A$ is the span of the $n-1$ points represented by the columns. Then, obviously, at a point $x$ in $\operatorname{Sm}(X)$, the tangent hyperplane $T_x$ to $X$ passes through $A$ if and only if

$$\sum_{i=0}^{n} F_i(x)a_{ij} = 0 \text{ for } j = 1, \ldots, n-1 \text{ where } F_i := \partial F/\partial X_i.$$

We can reinterpret these equations as saying that $x \in X \cap \Pi$ where $\Pi$ is the *polar locus* of $X$ with respect to $A$, namely, the following subscheme of $\mathbf{P}^n$:

$$\Pi : \sum_{i=0}^{n} F_i(X_0, \ldots, X_n)a_{ij} = 0 \text{ for } j = 1, \ldots, n-1.$$

Suppose that $A$ is in general position. Then $\Pi$ is a complete intersection of dimension 1, because it is defined by $n-1$ equations and its intersection with $X$ is finite; indeed, $X \cap \Pi$ is equal to the union of $\operatorname{Sing}(X)$ and $g^{-1}A'$, where $g\colon \operatorname{Sm}(X) \to \check{\mathbf{P}}^n$ is the Gauss map and where $A'$ is the line in $\check{\mathbf{P}}^n$ dual to $A$. Hence, by Bezout's theorem, the global intersection number $\int X \cdot \Pi$ is equal to $d(d-1)^r$. The local intersection number at any $x$ in $X \cap \Pi$ is equal to the length of the residue ring $\mathcal{O}_{X,x}/\mathcal{J}'_x$ where $\mathcal{J}'_x$ denotes the ideal arising from the $n-1$ homogeneous polynomials $\sum_i F_i a_{ij}$. If $x \in \operatorname{Sm}(X)$, then clearly that ring is equal to the local ring at $x$ of the scheme $g^{-1}A'$. Since $A$ is in general position, the sum

$$\sum_{x \in \operatorname{Sm}(X)} \operatorname{length}(\mathcal{O}_{X,x}/\mathcal{J}'_x)$$

is equal to the class $d'$ of $X$ by Equation (1-2) because $p\colon CX \to X$ is an isomorphism over $\operatorname{Sm}(X)$. If $x \in \operatorname{Sing}(X)$, then, by a well-known result of ordinary multiplicity theory (see, for example, [**15**, proof of Thm. 5, p. 186]), the length of $\mathcal{O}_{X,x}/\mathcal{J}'_x$ is equal to the multiplicity of the ideal arising from the $n+1$ partial derivatives $F_i$, namely, the Jacobian ideal $\mathcal{J}_x$ of $X$ at $x$; thus, this length is equal to $e(x)$. Hence, Teissier's formula (1-1) is proved.

Finally, suppose that $X$ is an arbitrary complete intersection. Say

$$X : F^1 = 0, \dots, F^\rho = 0$$

where $F^k$ is a homogeneous polynomial in $X_0, \dots, X_n$ of degree $d_k$. By Bezout's theorem, the product $d_1 \cdots d_\rho$ is equal to the degree $d$ of $X$. Let $\pi$ be the $r$th complete symmetric function in the numbers $d_k - 1$; by definition, $\pi$ is the sum of all monomials of degree $r$ in the $d_k - 1$,

$$\pi := \sum_{i_1 + \cdots + i_\rho = r} (d_1 - 1)^{i_1} \dots (d_\rho - 1)^{i_\rho}.$$

**Theorem 1.** *Under the conditions specified in the preceding paragraph,*

$$d' = d\pi - \sum_{x \in Sing(X)} e(x).$$

Indeed, consider a pair $(x, H)$ where $x$ is a point in $\operatorname{Sm}(X)$ and $H$ is a hyperplane. Then $H$ contains the tangent $r$-plane $T_x$ to $X$ at $x$ if and only if there exist $\rho$ constants $b_k$ such that $H$ is given as follows:

$$H : \sum_{k,i} b_k F_i^k(x) X_i = 0 \text{ where } F_i^k := \partial F / \partial X_i.$$

So $H$ contains a given linear space $A$ of codimension 2, represented as the column space of an $n+1$ by $n-1$ matrix $(a_{ij})$, if and only if the following system of $n-1$ equations is satisfied:

$$\sum_{k,i} b_k F_i^k(x) a_{ij} = 0 \text{ for } j = 1, \dots, n-1. \tag{2-2}$$

Now, given $x \in \operatorname{Sm}(X)$, there exist $\rho$ constants $b_k$ satisfying the system of equations (2-2) if and only if the maximal minors $M_l$ of the $\rho$ by $n-1$ matrix

$$J' := \left[\sum_i F_i^k a_{ij}\right]$$

all vanish. In other words, there exists a hyperplane $H$ containing both $T_x$ and $A$ if and only if $x$ lies in the intersection $X \cap \Pi$ where $\Pi$ is the *polar locus* of $X$ with respect to $A$, namely, the scheme $\Pi : M_l = 0$.

Suppose that $A$ is in general position. Then $\Pi$ has dimension at most $\rho$ because it intersects $X$ in a finite set of points $x$, namely, the union of $\mathrm{Sing}(X)$ and $pp'^{-1}A'$ where $A'$ is the dual of $A$. Hence $\Pi$ is Cohen–Macaulay of dimension $\rho$ by a celebrated theorem, which was proved in the generality needed here by Cohen in 1949 [**2**, Remark, p. 207]. Moreover, $\Pi$ is of degree $\pi$ by Porteous's formula applied to the induced map $\mathcal{O}_{\mathbf{P}^n}^{n-1} \to \mathcal{O}_{\mathbf{P}^n}^{\rho}(d-1)$ associated to $J'$. So, by Bezout's theorem, the global intersection number $\int X \cdot \Pi$ is equal to $d\pi$. Moreover, since $X$ is a complete intersection and $\Pi$ is Cohen-Macaulay, the local intersection number at any $x$ in $X \cap \Pi$ is equal to

$$\mathrm{length}(\mathcal{O}_{X,x}/(M_l)) \tag{2-3}$$

where $(M_l)$ denotes the ideal generated by the $M_l$.

Fix $x \in X \cap \Pi$. Suppose $x \in \mathrm{Sm}(X)$. Then Equations (2-2) say that, on a neighborhood of $x$ in $X$, the induced map $p'^{-1}A' \to X$ is a bundle of linear spaces; more precisely, as a scheme, $p'^{-1}A'$ is equal, on a neighborhood of $x$ to $\mathbf{P}(\mathcal{K})$ where $\mathcal{K}$ is the cokernel of the map of free $\mathcal{O}_X$-modules represented by the matrix $J'$. Now, $p'^{-1}A'$ is finite. Hence, the fiber $\mathcal{K}(x)$ is a vector space of dimension 1. So, by Nakayama's lemma, $\mathcal{K}$ is cyclic at $x$. Therefore, over a neighborhood of $x$, the structure map embeds $\mathbf{P}(\mathcal{K})$ into $X$ as a closed subscheme supported at $x$, namely, the subscheme defined by the annihilator of $\mathcal{K}$. Again because $\mathcal{K}$ is cyclic, its annihilator is equal to its zeroth Fitting ideal, which is the ideal of $\mathcal{O}_{X,x}$ generated by the minors $M_l$. However, the latter is the ideal of $X \cap \Pi$ in $X$. Now, the degree of the zero scheme $p'^{-1}A'$ is equal to the class $d'$ of $X$ by Equation (1-2). Therefore, as $x$ varies over $\mathrm{Sm}(X) \cap \Pi$, the sum of the lengths (2-3) is equal to $d'$.

Finally, let $x \in \mathrm{Sing}(X)$, and follow, more or less, Gaffney's line of reasoning in [**4**, 2.6], just as van Gastel did in [**18**, p. 273]. Let $C'$ be the column space of $J'$. Then

$$\mathrm{length}(\mathcal{O}_{X,x}/(M_l)) = \mathrm{length}(\mathcal{O}_{X,x}^r/C') = e(J') = e(x).$$

Indeed, because $\mathcal{O}_{X,x}$ is Cohen–Macaulay, the first two equations hold by virtue of some theorems of Buchsbaum and Rim [**2**, 2.4, 4.3, 4.5]. The last equation was conjectured by Gaffney, and proved by Rees on the basis of some of Kirby's work, see [**4**, Prop. 2.5 iii)]; proofs are also given in [**8**, 1.5, 8.1], in [**6**, 4.2.4], and in [**11**, (6.2) and (6.8)]. Thus Theorem 1 is proved.

## 3. Arbitrary varieties

In this section, we generalize Theorem 1 of Section 2 when $X$ is arbitrary. The only change necessary to the formula is to replace the term $d\pi$ by the number $s_r$ defined as follows: let $\mathcal{G}$ be the graded $\mathcal{O}_X$-algebra of (1-5), and set

$$P := \mathrm{Proj}(\mathcal{G}),\ \ell_P := c_1\mathcal{O}_P(1),\ \text{and } s_r := \int \ell_P^{n-1}[P].$$

**Theorem 2.** *Under the conditions specified in the preceding paragraph,*

$$d' = s_r - \sum\nolimits_{x \in Sing(X)} e(x).$$

Consider the case where $X$ is a local complete intersection. Then the normal sheaf $\mathcal{N}$ is locally free of rank $n-r$, and $\mathcal{G}$ is equal to the twisted symmetric algebra $\mathrm{Sym}(\mathcal{N}(-1))$. Hence $P = \mathbf{P}(\mathcal{N}(-1))$. Therefore, $P$ is a variety of dimension $n-1$, and $s_r$ is the top Segre class of the negatively twisted normal sheaf,

$$s_r = s_r(\mathcal{N}(-1)).$$

If $X$ is the global complete intersection of hypersurfaces of degrees $d_k$, then $\mathcal{N}$ is equal to $\bigoplus_k \mathcal{O}_X(d_k)$; hence, $s_r$ is equal to $d\pi$, and we recover Theorem 1.

In general, $s_r$ is not divisible by $d$, even if $X$ is smooth. Hence, there is no complementary global polar variety $\Pi$, and the argument in Section 2 cannot be generalized. One well-known example is the twisted cubic $X$ in $\mathbf{P}^3$. Indeed, $\mathcal{O}_x(1) = \mathcal{O}_X(3[x])$ where $[x]$ is the cycle of a point. Now, $X$ lies on a smooth quadric surface, and the corresponding exact sequence of conormal sheaves is

$$0 \longrightarrow \mathcal{O}_X(-6[x]) \longrightarrow \mathcal{I}/\mathcal{I}^2 \longrightarrow \mathcal{O}_X(-4[x]) \longrightarrow 0.$$

Hence $s_1(\mathcal{N}(-1)) = 3+1 = 4$, but 4 is not divisible by 3.

The proof of Theorem 2 is entirely different from that of Theorem 1 in Section 2. It is remarkably simpler and less delicate. In spirit, it is a natural generalization of Piene's proof of Teissier's Plücker formula [**13**, Cor. 3.7, p. 266] (see also [**9**, pp. 356–7]). Its setup can also be compared with that in Sections 4–6 of Speiser's article [**16**, pp. 136–43]. (However, contrary to the impression left by that article, Formula (6.3) of [**16**, p. 142] is not always true because $B_i$ is not always true because $B_i$ is not equal to $J_i \times L$.)

To prove Theorem 2, consider the global Jacobian map,

$$J: \mathcal{O}_X^{n+1} \longrightarrow \mathcal{N}(-1),$$

which arises from the first map $\delta$ in (1-3) and the natural inclusion map of $\Omega^1_{\mathbf{P}^n}$ into $\mathcal{O}_X^{n+1}(-1)$. So $\mathrm{Im}(J) = \mathrm{Im}(\delta^*(-1))$. Hence the graded $\mathcal{O}_X$-algebra $\mathcal{G}'$ of (1-5) is a natural quotient of the symmetric algebra $\mathrm{Sym}(\mathcal{O}_X^{n+1})$. Correspondingly, the conormal variety $CX$ is equal to the scheme $\mathrm{Proj}(\mathcal{G}')$ (Gaffney, private comm., May 1990). Indeed, $J$ is surjective on the smooth locus $\mathrm{Sm}(X)$; so the two schemes are equal over $\mathrm{Sm}(X)$. On the other hand, $\mathrm{Proj}(\mathcal{G}')$ is reduced and every component lies over a component of $X$ because $\mathcal{G}'$ has no $\mathcal{O}_X$-torsion. Hence, $CX = \mathrm{Proj}(\mathcal{G}')$.

Let $Z$ be the subscheme of $P$ associated to the ideal of $\mathcal{G}$ generated by $\mathrm{Im}(J)$. Let $B$ be the blowup of $P$ along $Z$, and $E$ the exceptional divisor. Then $B$ is equal to the join of $P$ and $CX$ because $Z$ is the center of the central projection associated to the inclusion of $\mathcal{G}'$ into $\mathcal{G}$; see [**11**, (2.1)]. Over $\mathrm{Sm}(X)$, the three schemes $P$, $B$ and $CX$ are obviously equal. Moreover, $P$ is reduced and every component lies over a component of $X$ because $\mathcal{G}$ has no $\mathcal{O}_X$-torsion. Hence $P$,

$B$ and $CX$ are birationally equivalent varieties of dimension $n-1$. In particular, $s_r$ is well defined.

By a fundamental theorem of the modern theory of the Buchsbaum–Rim multiplicity ([**6**, 4.2.3] and [**11**, (5.7)]),

$$\sum\nolimits_{x\in Sing(X)} e(x) = \sum\nolimits_{i=0}^{n-2} \int \ell_P^i \ell'^{n-2-i}[E]. \tag{3-1}$$

Now, $[E] = \ell_P - \ell'$ by the general theory of central projections [**11**, (2.2)]. Hence

$$\sum\nolimits_{x\in Sing(X)} e(x) = \int \ell_P^{n-1}[B] - \int \ell'^{n-1}[B].$$

Therefore, by the projection formula applied to the birational maps $B \to CX$ and $B \to P$, the right hand side is equal to $s_r - d'$. Thus Theorem 2 is proved.

The fundamental theorem generalizes a well-known geometric interpretation of the ordinary multiplicity of an ideal, often credited to Ramanujam [**14**, p. 119]; namely, the multiplicity is equal to the degree of the projectivized tangent cone of the subscheme defined by the ideal (and the cone is just the exceptional divisor of the blowup along the subscheme). This interpretation is obvious in view of Samuel's definition of the multiplicity, because the degree of a projective variety is given by the leading coefficient of the Hilbert polynomial of its homogeneous coordinate ring, and the ring of the cone is the associated graded ring of the ideal.

Fix $x \in \mathrm{Sing}(X)$. If $X$ is a complete intersection locally at $x$, then $e(x) \neq 0$ because the generalized Hilbert–Samuel function that defines it has degree precisely $n-1$ by the Buchsbaum–Rim theory [**2**, Thm. 3.4, pp. 212]. In general, however, $e(x)$ can vanish. It vanishes whenever $\mathcal{G}_x$ is a finitely generated $\mathcal{G}'_x$-module, because then $\mathcal{G}_x/\mathcal{G}'_x$ is a finitely generated $\mathcal{G}'_x$-module whose support on $CX$ has dimension at most $n-2$; see [**11**, (6.3)]. By [**11**, (10.5)], therefore $e(x)$ vanishes whenever the fiber of $CX$ over $x$ has dimension at most $n-3$, that is, codimension at least 2.

Here are some examples. First, suppose that $X$ is a curve in $\mathbf{P}^3$ which is analytically equivalent at $x$ to the union of three noncoplanar lines, say the $u$-axis, the $v$-axis, and the $w$-axis. Work with the completion $R := \widehat{\mathcal{O}}_{X,x}$. Then clearly

$$R = k[\![u,v,w]\!]/(vw, uw, uv) = \{a + uf(u) + vg(v) + wh(w)\}$$

where $k$ is the ground field, where $a$ is an arbitrary element of $k$, and where $f$, $g$ and $h$ are arbitrary power series. Using the images of $u$, $v$ and $w$ in $(\mathcal{I}/\mathcal{I}^2)_x$ as generators, represent $\widehat{\delta}_x$ by the matrix,

$$(\widehat{\delta}_x) = \begin{pmatrix} 0 & w & v \\ w & 0 & u \\ v & u & 0 \end{pmatrix}.$$

So, in terms of spanning vectors,

$$\operatorname{Ker} \widehat{\delta}_x = \left\langle \begin{pmatrix} u \\ 0 \\ 0 \end{pmatrix}, \begin{pmatrix} 0 \\ v \\ 0 \end{pmatrix}, \begin{pmatrix} 0 \\ 0 \\ w \end{pmatrix} \right\rangle.$$

Now, $(\operatorname{Im}\widehat{\delta}_x)^*$ is equal to the orthogonal complement in $R^3$ of $\operatorname{Ker}\widehat{\delta}_x$; whence, we easily find

$$(\operatorname{Im}\widehat{\delta}_x)^* = \left\langle \begin{pmatrix} v \\ 0 \\ 0 \end{pmatrix}, \begin{pmatrix} w \\ 0 \\ 0 \end{pmatrix}, \begin{pmatrix} 0 \\ u \\ 0 \end{pmatrix}, \begin{pmatrix} 0 \\ w \\ 0 \end{pmatrix}, \begin{pmatrix} 0 \\ 0 \\ u \end{pmatrix}, \begin{pmatrix} 0 \\ 0 \\ v \end{pmatrix} \right\rangle.$$

Write $\operatorname{Sym}(R^3)$ as $R[U, V, W]$. Then $\operatorname{Im}(\widehat{\delta}_x^*)$ and $(\operatorname{Im}\widehat{\delta}_x)^*$ generate the subalgebras,

$$R[wV + vW,\ wU + uW,\ vU + uV] \text{ and } R[vU, wU, uV, wV, uW, vW].$$

The latter is integral over the former because

$$(vU)^2 - (vU + uV)vU = 0,\ (wU)^2 - (wU + uW)wU = 0, \text{ and so forth.}$$

Now, the two algebras are equal to $\mathcal{G}'_x \otimes R$ and $\mathcal{G}_x \otimes R$. Hence $e(x) = 0$.

It is interesting to look at the corresponding computation when $X$ is analytically isomorphic to the union of the $u$-axis and the $v$-axis in the plane. Then

$$R = k[\![u, v]\!]/(uv) = \{a + uf(u) + vg(v)\}.$$

Moreover, $\widehat{\delta}_x$ is represented by $\binom{v}{u}$. So $\operatorname{Ker}\widehat{\delta}_x$ vanishes, as it must. Write $\operatorname{Sym}(R)$ as $R[T]$. Then the subalgebras in question are $R[uT, vT]$ and $R[T]$ itself. Hence $\lambda(m) = 2m - 1$ and so $e(x) = 2$. If $X$ is, in fact, the union of the two axes, then $CX$ is the union of the two corresponding conormal varieties; hence, $d' = 0$, and the Plücker formula reduces to the equation $0 = 2(2 - 1) - 2$.

For the last example, suppose that $X$ is a surface in $\mathbf{P}^4$ which is analytically equivalent at $x$ to the union of two transverse planes, say the $uv$-plane and the $yz$-plane. Then

$$R = k[\![u, v, y, z]\!]/(uy, vy, uz, vz) = \{a + f(u, v) + g(y, z) | f(0, 0) = 0,\ g(0, 0) = 0\},$$

$$\widehat{\delta}_x = \begin{pmatrix} y & 0 & z & 0 \\ 0 & y & 0 & z \\ u & v & 0 & 0 \\ 0 & 0 & u & v \end{pmatrix} \text{ and } \operatorname{Ker}\widehat{\delta}_x = \left\langle \begin{pmatrix} -v \\ u \\ 0 \\ 0 \end{pmatrix}, \begin{pmatrix} 0 \\ 0 \\ -v \\ u \end{pmatrix}, \begin{pmatrix} -z \\ 0 \\ y \\ 0 \end{pmatrix}, \begin{pmatrix} 0 \\ -z \\ 0 \\ y \end{pmatrix} \right\rangle.$$

Hence $(\operatorname{Im}\widehat{\delta}_x)^*$ is equal to $\operatorname{Im}(\widehat{\delta}_x^*)$. Therefore, $e(x) = 0$. Alternatively, we can see that $e(x) = 0$ by a geometric analysis; indeed, it suffices to observe that, clearly, the fiber of $CX$ over $x$ is the disjoint union of two projective lines, so it has codimension 2.

In closing, note that, if $X$ is a local complete intersection and it is varied in a connected flat family, then the nearby varieties are also local complete intersections, and the formation of the normal sheaf of the total space commutes with passage to these nearby varieties; hence, $s_r$ remains constant near $X$. It seems plausible that $s_r$ remains constant in general, when $X$ is arbitrary and varied in a connected flat family. Suppose so. Then Theorem 2 gives a formula that expresses the class $d'$ of $X$ as the difference between a continuous invariant $s_r$ and a sum of intrinsic local invariants $e(x)$. The continuous invariant may be viewed as the *virtual* class of $X$, and the sum as a term that corrects for the acquisition by $X$ of special singularities. In this sense, the formula is a generalized Plücker formula.

## REFERENCES


1. L. Berzolari, *Allgemeine Theorie der höheren ebenen algebraischen Kurven*, Enzyklopädie der Math. Wissenschaften [III C 4, $2^1$], Teubner, Leipzig, 1906, pp. 313–455.
2. D. A. Buchsbaum and D. S. Rim, *A generalized Koszul complex. II. Depth and multiplicity*, Trans. Amer. Math. Soc. **111** (1964), 197–224.
3. T. Gaffney, *Integral closure of modules and Whitney equisingularity*, Invent. Math. **107** (1992), 301–22.
4. T. Gaffney, *Multiplicities and equisingularity of ICIS germs*, manuscript, spring 1990.
5. J.-P. Henry and M. Merle, *Limites de normales*, Singularities, Proc. Symposia pure math. vol. **40**, part 1, Amer. Math. Soc., 1983, pp. 575–84.
6. J.-P. Henry and M. Merle, *Conormal space and Jacobian module. A short dictionary*, preprint, spring 1991.
7. D. Kirby, *Graded multiplicity theory and Hilbert functions*, J. London Math. Soc. (2) **36** (1987), 16–22.
8. D. Kirby and D. Rees, *Hilbert functions of multigraded modules and the Buchsbaum-Rim multiplicity*, manuscript dated May 25, 1991.
9. S. Kleiman, *The enumerative theory of singularities*, Real and complex singularities. Proceedings, Oslo 1976 (P. Holm, eds.), Sitjhoff & Noorhoof, 1977, pp. 297–396.
10. S. Kleiman, *Tangency and duality*, Proc. 1984 Vancouver Conf. in Algebraic Geometry (J. Carrell, A. V. Geramita, P. Russell, eds.), CMS Conf. Proc. **6**, Amer. Math. Soc., 1986, pp. 163–226.
11. S. Kleiman and A. Thorup, *A geometric theory of the Buchsbaum–Rim multiplicity* (to appear).
12. M. Kline, *Mathematical thought from ancient to modern times*, Oxford Univ. Press, 1972.
13. R. Piene, *Polar classes of singular varieties*, Ann. Sci. École Norm. Sup. **11** (1978), 247–76.
14. C. P. Ramanujam, *On a geometric interpretation of multiplicity*, Invent. Math. **22** (1973), 63–67.
15. P. Samuel, *La notion de multiplicité en algèbra et en géométrie algébrique*, J. Math. Pures Appl. **30** (1951), 159–274.
16. R. Speiser, *Limits of conormal schemes*, Algebraic Geometry: Sundance 1988, vol. 116, 1991, pp. 129–46.
17. B. Teissier, *Sur diverse conditions numériques d'équisingularité des familles de courbes (et un principe de specialisation de la dépendance intégrale)*, preprint, Centre de Math., École Polytechnique (1975).
18. L. J. van Gastel, *Degenerations of conormal varieties*, Enumerative Algebraic Geometry, Proceedings of the 1989 Zeuthen Symposium, vol. 123, 1991, pp. 267–78.


2-278 MIT, CAMBRIDGE MA 02139-4307, USA
*E-mail address*: Kleiman@math.MIT.edu

Contemporary Mathematics
Volume **162**, 1994

# Log Surfaces of General Type; Some Conjectures

JÁNOS KOLLÁR

Below I outline a series of conjectures for pairs $(S,\Delta)$ where $S$ is a proper surface and $\Delta$ is a $\mathbb{Q}$-divisor. My main interest here is in the case $K_S+\Delta$ ample. Although I formulate everything for surfaces, all the conjectures can be posed for higher dimensional varieties. However the surface case is the only one where some positive results are known.

The case where $-(K_S+\Delta)$ is nef was investigated in [Alexeev89,92; Shokurov-88,91; Kollár et al.92]. Most of the conjectures posed here have their counterparts in one of these articles. Mori pointed out to me that the results of [Xiao91,92] can be interpreted in terms of log surfaces. This gave the impetus to consider the case $K_S+\Delta$ ample in more detail.

## 1. The Main Conjecture

1.1 DEFINITION. (1.1.1) Let $\mathcal{C}\subset[0,1]$ be a subset. For us the most important cases are:

$$\mathcal{C}_1=\{1\},\quad \mathcal{C}_2=\{1-1/n|n\in\mathbb{N}\cup\{\infty\}\}\quad\text{and}\quad \mathcal{C}_3=\mathcal{C}_2\cup\{1/3,1/4,1/6,1/12\}.$$

(1.1.2) Let $\mathcal{S}(\mathcal{C})$ be the set of all pairs $(S,\Delta)$ satisfying the following conditions:

(1.1.2.1) $S$ is a proper surface with quotient singularities (in higher dimensions $S$ should be proper with log terminal singularities);

(1.1.2.2) $\Delta=\sum d_iD_i$ is an $\mathbb{R}$-divisor on $S$ such that $d_i\in\mathcal{C}$;

(1.1.2.3) $(S,\Delta)$ is log canonical.

(1.1.3) Let $K^2(\mathcal{C})\subset\mathbb{R}$ be the set of all $(K_S+\Delta)^2$ as $(S,\Delta)$ runs through those pairs in $\mathcal{S}(\mathcal{C})$ for which $K_S+\Delta$ is ample. (In higher dimensions $K_S+\Delta$ nef and

1991 *Mathematics Subject Classification.* 14E30, 14E35, 14J50.
Partial financial support was provided by the NSF under grant numbers DMS-8707320 and DMS-8946082 and by an A. P. Sloan Research Fellowship.
This paper is in final form and no version of it will be submitted for publication elsewhere.

big might be a better condition, since then we can assume that the singularities are $\mathbb{Q}$-factorial.)

1.2 REMARK. The class $\mathcal{S}(\mathcal{C}_2)$ is very natural and is closed under some important operations:

(1.2.1) Let $(S, \Delta) \in \mathcal{S}(\mathcal{C}_2)$ and let $G \subset \mathrm{Aut}(S, \Delta)$ be a finite subgroup. (By $\mathrm{Aut}(S, \Delta)$ I mean those automorphisms of $S$ which map $\Delta$ into itself and if $\Delta_i \subset \Delta$ has coefficient $d_i$ then so does the image of $\Delta_i$.)

Consider the surface $X = S/G$ and let $f : S \to X$ be the quotient morphism. We define a divisor $\Theta$ on $X$ as follows. Let $C \subset X$ be an ireducible curve. Let $e(C)$ be the ramification index of $f$ over the generic point of $C$ and let $1 - 1/c(C)$ be the coefficient of $\operatorname{red} f^{-1}(C)$ in $\Delta$ (these are well defined since $f$ is Galois). Set

$$\Theta = \sum \left(1 - \frac{1}{e(C)c(C)}\right)[C].$$

It is easy to see that $K_S + \Delta = f^*(K_X + \Theta)$. Thus by [Kollár et al.92, 20.3], $(K_X + \Theta)$ is log canonical, hence $(X, \Theta) \in \mathcal{S}(\mathcal{C}_2)$. We will call $(X, \Theta)$ the quotient of $(S, \Delta)$ by $G$.

(1.2.2) Let $(S, \Delta) \in \mathcal{S}(\mathcal{C}_2)$ and assume that $m(K + \Delta) \sim 0$ (which is always satisfied locally). This defines a cyclic cover given by

$$Y = \mathrm{Spec}_S \sum_{i=0}^{m-1} \mathcal{O}_S(iK + \llcorner i\Delta \lrcorner) \xrightarrow{p} S.$$

Let $\Theta = p^{-1}(\llcorner \Delta \lrcorner)$. Let $C \subset \Delta$ be an irreducible curve with coefficient $1 - 1/c(C)$. The ramification index of $p$ over $C$ is $c(C)$. This implies that $K_Y + \Theta = p^*(K_S + \Delta)$. Thus $\mathcal{S}(\mathcal{C}_2)$ is the smallest class containing $\mathcal{S}(\mathcal{C}_1)$ which is closed under quotients by finite (or even cyclic) groups.

(1.2.3) It is easy to see that if $(S, \Delta) \in \mathcal{S}(\mathcal{C}_2)$ and $Z \subset \llcorner \Delta \lrcorner$ is an irreducible component then $(Z, \mathrm{Diff}_Z(\Delta - Z)) \in \mathcal{S}(\mathcal{C}_2)$. (See [Kollár et al.92,Ch. 16] for the definition of Diff.) This is a codimension two question, thus it is sufficient to consider the case when $S$ is a surface. Pick a point $p \in Z$ and let

$$\Delta' = Z + \sum (1 - 1/m_i) C_i$$

be the components of $\Delta$ passing through $p$. By assumption $\Delta'$ is log canonical at $p$. Since $p \in S$ is a quotient singularity, we can pull everything back to a smooth point. Easy computation shows that one of the following conditions is satisfied:

(1.2.3.1) $\Delta' = Z$ and $\Delta'$ is log terminal;

(1.2.3.2) $\Delta' = Z + (1 - 1/m_1)C_1$ and $Z + C_1$ is log canonical at $p$;

(1.2.3.3) all the $m_i = 2$ and $\Delta'$ is not log terminal at $p$.

Let $a(P)$ be the coefficient of $P$ in $\mathrm{Diff}_Z(\Delta - Z)$ and let $g(P)$ be the order of the local fundamental group of $p \in S$. In the above cases $a(P)$ is computed as follows (cf. [Kollár et al.92, 16.6]):

(1.2.3.1) $a(P) = 1 - 1/g(P)$;

(1.2.3.2) $a(P) = 1 - 1/m_1 g(P)$;

(1.2.3.3) $a(P) = 1$. □

The following is the most general conjecture, and I do not have much evidence to support it. It however fits very nicely within the framework of the results and conjectures developped in [Shokurov91; Xiao91,92; Kollár et al.92, Ch.18].

1.3 CONJECTURE. *Assume that $\mathcal{C}$ satisfies the descending chain condition. Then*

*(1.3.1) $K^2(\mathcal{C})$ also satisfies the descending chain condition.*

*(1.3.2) There are only finitely many deformation types of pairs $(S, \Delta) \in \mathcal{S}(\mathcal{C})$ with fixed $(K_S + \Delta)^2$ and $K_S + \Delta$ ample.*

Next we discuss some weaker versions of (1.3), in some cases giving explicit values for some bounds whose existence is implied by (1.3).

## 2. Application to Automorphism Groups of Surfaces

These were inspired by [Xiao91,92]. It is quite likely that his methods will prove (2.1).

2.1 CONJECTURE. *Notation as above. Let $(S, \Delta) \in \mathcal{S}(\mathcal{C}_2)$. Then*

$$(K_S + \Delta)^2 \geq \frac{1}{3261636} = \left(\frac{1}{2 \cdot 3 \cdot 7 \cdot 43}\right)^2,$$

*with equality holding only for*

$$S = \mathbb{P}^2 \quad \text{and} \quad \Delta = \frac{1}{2}L_1 + \frac{2}{3}L_2 + \frac{6}{7}L_3 + \frac{42}{43}L_4,$$

*where $L_i \subset \mathbb{P}^2$ are lines, no three of which intersect at a point.*

2.2 COROLLARY - CONJECTURE. *Let $(S, \Delta) \in \mathcal{S}(\mathcal{C}_2)$ be a surface such that $K_S + \Delta$ is ample. Then*

$$\# \operatorname{Aut}(S, \Delta) \leq 3261636(K_S + \Delta)^2.$$

PROOF. As in (1.2.1) consider the surface $(X, \Theta) = (S, \Delta)/\operatorname{Aut}(S, \Delta)$ and let $f : S \to X$ be the quotient morphism. Then

$$(K_S + \Delta)^2 = \deg f \cdot (K_X + \Theta)^2 = \# \operatorname{Aut}(S, \Delta) \cdot (K_X + \Theta)^2.$$

Thus (2.1) implies (2.2). □

2.3 REMARK. Assuming (2.1), the bound in (2.2) is sharp only for the example given in (2.1). Unfortunately, in the $\Delta = 0$ case it is considerably above the optimal bound $1764K_S^2$ [Xiao92] which is achieved by the square of a curve with maximal number of automorphisms.

## 3. Application to Plurigenera of Threefolds

3.1 CONJECTURE. *(3.1.1) Assume that $\mathcal{C}$ satisfies the descending chain condition. Then there is an $m(\mathcal{C}) > 0$ such that if $(S, \Delta) \in \mathcal{S}(\mathcal{C})$ and $K_S + \Delta$ is ample then*

$$h^0(S, \mathcal{O}(mK_S + \llcorner m\Delta \lrcorner)) > 0 \quad \text{for some} \quad 0 < m \leq m(\mathcal{C}).$$

*(3.1.2)* $m(\mathcal{C}_2) = 42$.
*(3.1.3)* $m(\mathcal{C}_3) = 24492$.

3.2 EXAMPLE. The example of (2.1) shows that 42 in (3.1.2) cannot be improved. The following example shows that 24492 in (3.1.3) is again the best possible:

$$S = \mathbb{P}^2 \quad \text{and} \quad \Delta = \frac{1}{12}L_1 + \frac{12}{13}L_2 + \frac{156}{157}L_3 + \frac{24492}{24493}L_4.$$

In connection with the next conjecture it is worth while to remark that the latter example can not arise from an elliptic threefold. Thus the value 24492 in the next conjecture may not be optimal.

3.3 COROLLARY - CONJECTURE. *Let $X$ be a smooth threefold with $\kappa(X) = 2$. Then $h^0(X, \mathcal{O}(mK_X)) > 0$ for some $m \leq 24492$.*

PROOF. Choose a model for the elliptic fibration $f : X \to S$ such that $X$ and $S$ are smooth and $f$ is smooth outside a divisor with normal crossings $D = \sum D_i \subset S$. We may assume furthermore that the $j$ invariant of the fibers gives a morphism $j : S \to \mathbb{P}^1$. Choose coefficients $d_i \in \mathcal{C}_3$ as in [Fujita86, 2.6] to obtain a divisor $\Delta = \sum d_i D_i + (1/12)[j^{-1}(c)]$ where $c \in \mathbb{P}^1$ is a general point. (Thus $[j^{-1}(c)] = 0$ iff $j$ is constant.) By [Fujita86, 2.15] there is a natural isomorphism

$$H^0(S, \mathcal{O}(mK_S + \llcorner m\Delta \lrcorner)) \cong H^0(X, \mathcal{O}(mK_X)).$$

We can replace $S$ by its canonical model to get $(S', \Delta')$ such that $K_{S'} + \Delta'$ is ample and

$$H^0(S', \mathcal{O}(mK_{S'} + \llcorner m\Delta' \lrcorner)) \cong H^0(X, \mathcal{O}(mK_X)).$$

Thus (3.1) implies (3.3). □

3.4 REMARK. Similarly, one should be able to find a number $M$ such that if $X$ is a threefold with $\kappa = 2$ then the $m$-canonical map gives the Iitaka fibration for every $m \geq M$.

A related problem is the following:

3.5 CONJECTURE. *There is an $\epsilon > 0$ with the following property:*
*Let $X$ be a minimal threefold with $\kappa(X) = 2$. Then either $K_X \cdot c_2(X) = 0$, or $K_X \cdot c_2(X) \geq \epsilon$.*

3.6 REMARK. One could ask the same about any minimal threefold $X$. By [Miyaoka87] we know that $K_X \cdot c_2(X) \geq 0$. Hopefully one can give a structure theory for those threefolds with $K_X \cdot c_2(X) = 0$ and find a universal lower bound $K_X \cdot c_2(X) \geq \epsilon > 0$ for the rest.

3.7 NOTE. A large part of (3.5) is settled in [Grassi93]. She also pointed out that my original (more precise) formulation of (3.5) was incorrect.

## 4. Application to Degenerations of Surfaces

A candidate for a geometrically meaningful compactification of the moduli of surfaces of general type was proposed in [KSB88]. The following is the only open problem left to settle:

4.1 CONJECTURE. *[KSB88, 5.4.1] Let $f : X \to C$ be a stable degeneration of surfaces of general type. Let $S$ be the general fiber. Then the index of $X$ can be bounded by a function of $K_S^2$.*

4.2 PROPOSITION. *Notation as above. Then (1.3) for $\mathcal{C}_1$ implies (4.1).*

PROOF. Let $X_0$ be a singular fiber with normalization $\bar{X}_0 = \cup S_i$. Let $\Delta_i \subset S_i$ be the preimage of the double curves of $X_0$. Then

$$K_S^2 = K_{X_0}^2 = \sum (K_{S_i} + \Delta_i)^2. \tag{4.2.1}$$

It is easy to see that by (1.3.1) there are only finitely many possible values for $(K_{S_i} + \Delta_i)^2$ once $K_S^2$ is fixed. Thus by (1.3.2) we can bound the index of $S_i$. The index of $K_X$ (near $X_0$) is the l.c.m. of the indices of the $S_i$. □

## 5. 1/42

The aim of this section is to compute some numbers related to the previous definitions. These bounds will be used in later chapters to get partial results concerning some of the conjectures. Familiarity with [Kollár et al.92, Chs. 16-17] will be helpful for some of the proofs.

The abbreviations klt (resp. lc, lt) stand for Kawamata log terminal (resp. log canonical, log terminal). See e.g. [Kollár et al.92, 2.13] for the definitions.

5.1 DEFINITION. Fix a set $\mathcal{C} \subset [0, 1]$.

(5.1.1) Let $\delta(\mathcal{C})$ be the maximal $t$ such that if $(S, \Delta = \sum d_i D_i) \in \mathcal{S}(\mathcal{C})$, $d_0 \in \mathcal{C}$ and $K_S + \Delta + (d_0 - c_0)D_0$ is lc and numerically trivial for some $d_0 \geq c_0 > 0$ then $c_0 \geq t$.

(5.1.2) Let $\delta_1(\mathcal{C})$ be the maximal $t$ such that if $(S, \Delta = \sum d_i D_i) \in \mathcal{S}(\mathcal{C})$ and $K_S + \Delta + (1 - c_0)D_0$ is lc and numerically trivial for some $c_0 > 0$ then $c_0 \geq t$. Clearly $\delta_1(\mathcal{C}) \geq \delta(\mathcal{C})$.

(5.1.3) Let $\epsilon_1(\mathcal{C})$ be the maximal $t$ such that if $(S, \Delta = \sum d_i D_i) \in \mathcal{S}(\mathcal{C} \cup \{1\})$, $d_1 = 1$ and $D_1 \cdot (K_S + \Delta) > 0$ then $D_1 \cdot (K_S + \Delta) \geq t$.

5.1.4 REMARK. One could also define $\epsilon(\mathcal{C})$ as above but without the assumption $d_1 = 1$. Unfortunately $\epsilon(\mathcal{C}_2) = 0$ as shown by the following example. Let $D \subset S$ be a smooth rational curve with $D^2 = -2n$. Then

$$\left(K + \left(1 - \frac{1}{n-k}\right) D\right) \cdot D = \frac{2k}{n-k}.$$

Another example is the following. Let $f : S \to \mathbb{P}^1$ be a smooth ruled surface. Fix $0 < d_C < 1$. Let $C \subset S$ be a smooth section of $f$ with $C^2 = n$. Let $F_i$ be fibers of $f$ and consider $\Delta = d_C C + \sum (1 - m_i^{-1}) F_i$. It is easy to see that by choosing suitable $n$ and $m_i$ we can make $C \cdot (K_S + \Delta)$ arbitrary small (positive).

5.1.5 REMARK. It is clear from the definition that $\delta(\mathcal{C}) \leq \inf(\mathcal{C})$. (Let for instance $S$ be a K3 surface and $\Delta = D_0$ be a smooth and ample curve.)

It is not hard to se that $\epsilon_1(\mathcal{C}) \leq 1/42$ holds for every $\mathcal{C}$. To see this let $S$ be a surface containing a curve $C$ with the following properties:

$C \cong \mathbb{P}^1$, $S$ has three singular points $P_2, P_3, P_7$ along $C$ and locally at $P_i$ the equations of $C \subset S$ are $(x = z = 0) \subset (xy - z^i = 0)$. Then $C \cdot (K_S + C) = 1/42$.

5.2 THEOREM. *[Alexeev92] Assume that $\mathcal{C}$ satisfies the descending chain condition. Then $\delta(\mathcal{C}) > 0$.*

PROOF. Assume that $K_S + \sum d_i D_i + (d_0 - c_0) D_0$ is lc and numerically trivial. If $d_0 - c_0 = 0$ then $c_0 \geq \inf \mathcal{C}$. Otherwise take the

$$K_S + \sum_{i>0} d_i D_i + (d_0 - c_0 - \epsilon) D_0 \text{ minimal model program}$$

for $0 < \epsilon \ll 1$. We obtain a series of birational contractions and these do not contract $D_0$. When the program stops we are reduced to the case when either $S$ is a Del Pezzo surface (with quotient singularities) or we are reduced to the similar one dimensional problem.

In the Del Pezzo case $(d_0 - c_0) + \sum d_i \leq 3$ by [ibid, 5.1], thus the number of the irreducible components of $\sum D_i$ is bounded from below by $3/\inf(\mathcal{C})$. Assume that we have an infinite sequence of examples as in (5.1.1)

$$K_{S^j} + \sum d_i^j D_i^j + (d_0^j - c_0^j) D_0^j, \quad j = 1, \dots \quad \text{and } c_0^j \to 0.$$

By passing to a subsequence we may assume that $\Delta^j$ has the same number of irreducible components for very $j$. Furthermore, we may also assume that $d_i^1 \leq d_i^2 \leq \dots$ holds for every $i$. Since $c_0^i \to 0$, by passing to a further subsequence we may assume that $d_0^1 - c_0^1 < d_0^2 - c_0^2 < \dots$ also holds. By [ibid, 5.3] this is impossible. $\square$

5.2.1 REMARK. The above agrument, and also the method of [Alexeev92] study the accumulation points of the possible values of the numbers $d_i$, thus they give no information about the value of $\delta(\mathcal{C})$.

5.3 THEOREM. $\delta_1(\mathcal{C}_2) = 1/42$.

The proof of this equality is rather long. The main reason for presenting it is to show the variety of techniques that can be used in similar problems.

(5.3.1) First of all we see that equality is achieved for the example

$$S = \mathbb{P}^2 \quad \text{and} \quad \Delta = \frac{1}{2} L_1 + \frac{2}{3} L_2 + \frac{6}{7} L_3 + L_4,$$

where $L_i \subset \mathbb{P}^2$ are lines, no three of which intersect at a point.

(5.3.2) Assume that $(S, \Delta = \sum d_i D_i) \in \mathcal{S}(\mathcal{C})$ and $K_S + \Delta + (1 - c) D_0$ is numerically trivial. As in [Kollár et al.92, 18.19.4] we can apply the $(K_S + \Delta + (1 - c - \epsilon) D_0)$-MMP to reduce the problem to the special case when the Picard number of $S$ is 1. Thus $S$ is a $\mathbb{Q}$-Del Pezzo surface.

If there exists a $D_1 \in \llcorner\Delta\lrcorner$ then consider $\operatorname{Diff}_{D_1}(\Delta + (1-c)D_0)$. This gives an equation $2 = \sum(1 - 1/n_i) + c/n_0$, which implies that $c \geq 1/42$. Thus we may assume that $\llcorner\Delta\lrcorner = 0$. This will be important for technical reasons.

Next we reduce the computation of $\delta_1(\mathcal{C}_2)$ to a three dimensional local problem:

5.3.3 DEFINITION. Let $\delta'(n, \mathcal{C})$ be the infimum of $(1-c)$ such that there exists an $n$-dimensional $\mathbb{Q}$-factorial log terminal pair $(X, \Delta = \sum d_i D_i)$, $d_i \in \mathcal{C}$ and an integral divisor $B$ such that $K + \Delta + B$ is not lc but $K + \Delta + cB$ is.

5.3.4 CLAIM. $\delta_1(\mathcal{C}_2) \geq \delta'(3, \mathcal{C}_1)$. *(We will see that in fact equality holds.)*

PROOF. Consider $(S, \Delta)$ as in (5.3.2). Let $L$ be a Cartier divisor on $S$ such that $L + \Delta$ is ample and set

$$X = \operatorname{Spec}_S \sum_{i=0}^{\infty} \mathcal{O}_S(iL + \llcorner i\Delta\lrcorner) \quad \text{with projection } p : X \to S.$$

The zero section defines an embedding $S \subset X$. $X$ has cyclic quotient singularities along the components of $\operatorname{Supp}\Delta$. Let $D_X = p^* D_0$. By [Kollár et al.92, 16.7] $\operatorname{Diff}_S(0) = \Delta$, thus by [Kollár et al.92, 17.7] $(X, S)$ is klt. $S \subset X$ is contractible to a point $f : (E \subset X) \to (0 \in Y)$. By construction $K_X + S|S \equiv -(1-c)D_0$, which implies that $(Y, \emptyset)$ is log terminal and $(Y, cf(D_X))$ is lc. □

5.4 THEOREM. $\delta'(1, \mathcal{C}_1) = 1/2$, $\delta'(2, \mathcal{C}_1) = 1/6$, $\delta'(3, \mathcal{C}_1) = 1/42$.

5.4.1 REMARK. The conjectures [Kollár et al.92, 18.16] imply that $\delta'(n, \mathcal{C}_1) > 0$ for every $n$. Unfortunately, $\delta'(n, \mathcal{C}_1)$ is rather small for $n$ large:

Choose $a_1 = 2$, $a_{k+1} = a_1 \cdots a_k + 1$ and consider

$$\left(\mathbb{P}^n, (x_0 = 0) + \sum_{i \geq 1} \left(1 - \frac{1}{a_i}\right)(x_i = 0)\right).$$

As in the proof of (5.3.4) this shows that

$$\delta'(n+1, \mathcal{C}_1) \leq 1 - \sum_1^n \frac{1}{a_i} = \frac{1}{a_{n+1} - 1} \leq 2^{-2^{n-1}}.$$

PROOF. The first two values are essentially in [Reid83]. $\delta'(1, \mathcal{C}_1) = 1/2$ is clear. The two dimensional case can be reduced to the case when $X$ is smooth (similarly to the following 3 dimensional argument) and the rest is an easy computation analogous to (5.5.7).

In dimension three we start with a reduction step (which, at least conjecturally, works in all dimensions):

Let $(X, \Delta = \sum D_i)$ be a 3-dimensional $\mathbb{Q}$-factorial log terminal pair, and $B$ an integral divisor such that $K + \Delta + B$ is not lc but $K + \Delta + cB$ is. If $g : X' \to X$ is a finite cover, étale in codimension one, then by [Kollár et al.92, 20.3] $(X, \Delta + cB)$ is lc iff $(X', g^*\Delta + cg^*B)$ is lc. Thus by repeated cyclic covers we are reduced to the case when $K_X, B$ and $\Delta$ are all Cartier. In particular $X$ is canonical of index 1.

Let $h : X'' \to X$ be a factorial crepant terminalisation.

$$K_{X''} + h^*\Delta + ch^*B = h^*(K_X + \Delta + cB),$$

thus again $(X'', h^*\Delta + ch^*B)$ is lc iff $(X, \Delta + cB)$ is.

Thus (5.4) is implied by the following:

5.5 PROPOSITION. *Let $X$ be a 3-dimensional factorial terminal singularity and let $B, D \subset X$ be Cartier divisors. Then one of the following holds:*

*(5.5.1) $(X, D + B)$ is lc; or*

*(5.5.2) $(X, D + cB)$ is not lc for every $c > 41/42$.*

PROOF. The guiding hope is that the leading terms of the equations pretty much decide everything. Of course this can only be expected if we look at the leading terms with suitable weights.

Before starting the proof we collect some easy technical lemmas.

5.5.3 LEMMA. *(cf. [Reid80, 4.3]) Let $D_i = (f_i(x_1, \dots, x_n) = 0)$. Assume that the pair $(\mathbb{A}^n, \sum c_i D_i)$ is log canonical. Let $\alpha(x_i) \in \mathbb{R}^+$ be weights. Then $\sum c_i \alpha(f_i) \le \sum \alpha(x_j)$.* □

5.5.4 LEMMA. *Notation as above. Let $f^1$ be the lowest weight part of $f$ (with respect to some weight $\alpha$). Then $(f = 0)$ is a flat deformation of $(f^1 = 0)$.* □

5.5.5 LEMMA. *Let $D_t$ be a flat deformation of a log canonical surface singularity $D_0$. Then $D_t$ is log canonical for small $t$.* □

5.5.6 LEMMA. *Let $D$ be a normal surface singularity. If the projectivised tangent cone $C$ of $D$ is a reduced curve with at worst nodes such that $\omega_C \cong \mathcal{O}_C$ then $D$ is lc.* □

The following is the hardest part of the proof:

5.5.7 LEMMA. *Let $(f = 0) \subset \mathbb{A}^3$ be a surface singularity. Then one of the following holds:*

*(5.5.7.1) $(f = 0)$ is lc; or*

*(5.5.7.2) there are weights $\alpha(x), \alpha(y), \alpha(z)$ such that $\frac{41}{42}\alpha(f) \ge \alpha(x) + \alpha(y) + \alpha(z)$.*

PROOF. We distinguish several cases:

(1) $\operatorname{ord}(f) \ge 4$. Set $\alpha(x) = 1/4; \alpha(y) = 1/4; \alpha(z) = 1/4$.

(2.1) $\operatorname{ord}(f) = 3$ and $f_3$ has at worst nodes. Then $f_3 = 0$ is lc and so is $f = 0$.

(2.2) $\operatorname{ord}(f) = 3$ and $f_3$ has a tacnode, cusp or worse singularity. In suitable coordinates $f = azx^2 + bxy^2 + cy^3 + (\deg \ge 4)$. Set $\alpha(x) = 3/8; \alpha(y) = 1/3; \alpha(z) = 1/4$.

(3) $\operatorname{ord} f = 2$. Write $f = x^2 + g(y, z)$.

(3.1) $\operatorname{ord} g \ge 4$ and $g_4$ has a triple root. Thus $g = ay^3z + by^4 + (\deg \ge 5)$. Set $\alpha(x) = 1/2; \alpha(y) = 4/15; \alpha(z) = 1/5$.

(3.2) $\operatorname{ord} g \ge 4$ and $g_4$ has only simple or double roots. Then $x^2 + f_4 = 0$ is lc and so is $f = 0$.

(3.3) $\operatorname{ord} g = 3$. Write $g = y^3 + yh_4(z) + h_6(z)$.

(3.3.1) Assume that setting $\alpha(x) = 1/2; \alpha(y) = 1/3; \alpha(z) = 1/7$ we get $\alpha(f) \geq 1$. $41/42$ is sharp for $f = x^2 + y^3 + z^7$.

(3.3.2) Otherwise $\operatorname{ord} h_4 \leq 4$ or $\operatorname{ord} h_6 \leq 6$. If one of these is a strict inequality then $(f = 0)$ is a $DV$ point. Otherwise it is lc.

(3.4) $\operatorname{ord} g = 2$. Then $f = x^2 + y^2 + z^m$ defines a $DV$ point.

(4) $\operatorname{ord} f = 1$. Then $(f = 0)$ is smooth. □

(5.6) PROOF OF (5.5).

(5.6.1). Assume that $D \neq 0$. By [Kollár et al.92, 17.7], $(D, \operatorname{Diff}_D(cB))$ is lc iff $(X, D + cB)$ is. Thus we are reduced to the easier two dimensional case and we could even use $c \geq 5/6$ instead of $c \geq 41/42$. For the rest of the proof assume that $D = 0$.

(5.6.2). Let $(g = 0)$ be the defining equation of $X \subset \mathbb{A}^4$ and let $(f = 0)$ be a defining equation of $B$. By [Kollár et al.92, 17.7] $(X, c(f = 0))$ is lc iff $(\mathbb{A}^4, (g = 0) + c(f = 0))$ is lc.

(5.6.3) $\operatorname{ord} g = 1$ OR $\operatorname{ord} f = 1$. In suitable coordinates one of them is $x_4 = 0$ and the other equation has the form $h_1(x_1, x_2, x_3) + x_4 h_2 = 0$. If $h(x_1, x_2, x_3) = 0$ is lc then $(\mathbb{A}^4, (x_4 = 0) + (h_1 + x_4 h_2 = 0))$ is lc by [Kollár et al.92, 17.7] (applied twice). Otherwise choose weights $\alpha(x_i)$ $(i = 1, 2, 3)$ as in (5.5.7.2). We can normalise the choice such that $\alpha(h_1) = 1$. Set $\alpha(x_4) = 1$. Then $\alpha(x_4) = \alpha(h_1 + x_4 h_2) = 1$ and $\sum \alpha(x_i) \leq 1 + 41/42$. Thus if $\alpha(g) + c\alpha(f) \leq 1 + 41/42$ then $c \leq 41/42$.

(5.6.4) $\operatorname{ord}(g) \geq 2$ AND $\operatorname{ord}(f) \geq 2$. Let $Q_1$ and $Q_2$ be the quadratic parts of $f$ and $g$. We distinguish two cases:

(5.6.4.1) THE PENCIL $\langle Q_1, Q_2 \rangle$ DOES NOT CONTAIN ANY RANK 4 QUADRICS. Then in suitable coordinates $Q_i$ can be written as $Q_1 = Q_1(x_1, x_2, x_3)$ and $Q_2 = Q_2(x_1, x_2, x_3)$. Set weigths $\alpha(x_1) = \alpha(x_2) = \alpha(x_3) = 1/2$ and $\alpha(x_4) = 1/3$. Then

$$1 + c \leq \alpha(g) + c\alpha(f) \leq 3\frac{1}{2} + \frac{1}{3}$$

implies that $c \leq 5/6$.

(5.6.4.2) THE PENCIL $\langle Q_1, Q_2 \rangle$ DOES CONTAIN A RANK 4 QUADRIC. Fix a value of $c$ such that $(\mathbb{A}^4, (g = 0) + c(f = 0))$ is lc. As in the proof of [Kollár et al.92, 18.22] it is easy to see that $(\mathbb{A}^4, (g + tf = 0) + c(f = 0))$ is also lc for general $t$. Note that $(\mathbb{A}^4, (g + tf = 0) + (f = 0))$ is still not lc since we have not changed the intersection of the two hypersurfaces. Let $Y = (g + tf = 0)$. By [Kollár et al.92, 17.7] $(Y, Y \cap (f = 0))$ is not lc but $(Y, c(Y \cap (f = 0)))$ is lc. $Y$ is analytically isomorphic to $(xy - uv = 0)$, thus it has a small resolution $h : Y' \to Y$. Let $B' = h^*(Y \cap (f = 0))$. Then $(Y', B')$ is not lc but $(Y', cB')$ is. $Y'$ is smooth thus by (5.5.7) $c \leq 41/42$. □

5.7 PROPOSITION. *(5.7.1) Assume that $\mathcal{C}$ satisfies the descending chain condition. Then $\epsilon_1(\mathcal{C}) > 0$.*

*(5.7.2) $\epsilon_1(\mathcal{C}_2) = 1/42$.*

*(5.7.3) Assume that $\mathcal{C}$ satisfies the descending chain condition and $1 \in \mathcal{C}$. Let $(S, \Delta) \in \mathcal{S}(\mathcal{C})$ and let $|C|$ be a positive dimensional linear system without fixed components on $S$. Assume that $(K+\Delta)\cdot C > 0$. Then $(K+\Delta+C)\cdot C \geq \epsilon_1(\mathcal{C})$.*

PROOF. Let $C$ be an irreducible component of $\llcorner\Delta\lrcorner$ and let $C' \to C$ be the normalisation. Then $C \cdot (K_S + \Delta) = \deg K_{C'} + \mathrm{Diff}_{C'}(\Delta - C)$. Set

$$\mathrm{Diff}(\mathcal{C}) \stackrel{def}{=\!=} \left\{1 - \frac{1-\sum k_i c_i}{m} | m \in \mathbb{N} \cup \{\infty\},\ k_i \in \mathbb{N},\ c_i \in \mathcal{C}\right\} \cap [0,1].$$

By [Kollár et al.92, 16.6] $\mathrm{Diff}_{C'}(\Delta - C)$ is of the form $\sum d(P)[P]$ for some points $P \in C'$ and $d(P) \in \mathrm{Diff}(\mathcal{C})$. Let furthermore $\sum \mathrm{Diff}(\mathcal{C}) \subset [0, \infty)$ be the set of those numbers that can be written as a sum of elements of $\mathrm{Diff}(\mathcal{C})$. It is easy to see that if $\mathcal{C}$ satisfies the descending chain condition then so does $\sum \mathrm{Diff}(\mathcal{C})$. In particular, there is $\epsilon_1(\mathcal{C}) > 0$ such that if $c \in \sum \mathrm{Diff}(\mathcal{C})$ and $c > 0$ (resp. $c > 2$) then $c \geq \epsilon_1(\mathcal{C})$ (resp. $c \geq 2 + \epsilon_1(\mathcal{C})$).

We assumed that

$$\deg \mathrm{Diff}_{C'}(\Delta - C) = 2g(C') - 2 + \sum d(P) > 0.$$

thus $\deg \mathrm{Diff}_{C'}(\Delta - C) \geq \epsilon_1(\mathcal{C}) > 0$.

By (1.2.3) $\mathrm{Diff}(\mathcal{C}_2) = \mathcal{C}_2$ from which it is easy to see that $\epsilon_1(\mathcal{C}_2) = 1/42$ with equality only for $C' = \mathbb{P}^1$ and $\mathrm{Diff}_{C'}(\Delta - C) = (1/2, 2/3, 6/7)$.

Let $C_g$ be a general member of $|C|$ and let $f : S' \to S$ be the log canonical model of $(S, \Delta + C_g)$. The proper transform of a curve on $S'$ will be denoted by . Let $E_i \subset S'$ be the $f$-exceptional divisors. Then

$$K + \Delta' + C_g' + \sum E_i \equiv f^*(K + \Delta + C_g) - \sum e_i E_i$$

for some $e_i \geq 0$ (see e.g. [Kollár et al.92, 2.21]). Thus

$$\begin{aligned}(K + \Delta + C_g)\cdot C_g &= f^*(K + \Delta + C_g)\cdot C_g' \\ &\geq (K + \Delta' + C_g' + \sum E_i)\cdot C_g'.\end{aligned}$$

Furthermore,

$$K + \Delta' + C_g' + \sum E_i = f^*(K + \Delta) + C_g' + \sum a_\ell(E_i, \Delta) E_i,$$

and $(C_g')^2 \geq 0$. (Here the $a_\ell(E_i, \Delta) \geq 0$ are the unique coefficients that make the above equality valid; see e.g. [Kollár et al.92, 2.5.4].) Hence

$$(K + \Delta' + C_g' + \sum E_i)\cdot C_g' > 0.$$

Therefore by the first part

$$(K + \Delta + C_g)\cdot C_g \geq \epsilon_1(\mathcal{C}). \quad \square$$

In the applications we will need a slightly stronger form of (5.7.3):

5.7.4 COMPLEMENT. *Assume that $\mathcal{C}$ satisfies the descending chain condition and $1 \in \mathcal{C}$. Let $(S, \Delta) \in \mathcal{S}(\mathcal{C})$ and let $\{C_t\}$ be a covering family of curves on $S$. Assume that $(K + \Delta) \cdot C_t > 0$. Then $(K + \Delta + C_t) \cdot C_t \geq \epsilon_1(\mathcal{C})$.*

PROOF. Assume first that $K+\Delta+C_t$ is lc at the singular points of $(S, \Delta)$ for general $t$. Let $C' \to C$ be the normalisation of $C$ at the singular points of $(S, \Delta)$. Then the proof of (5.7.1) goes through to give that $(K + \Delta + C_t) \cdot C_t \geq \epsilon_1(\mathcal{C})$.

For arbitrary $\{C_t\}$ we work as in the proof of (5.7.3). We choose $f : S' \to S$ such that $f^{-1}$ is an isomorphism outside the singular points of $(S, \Delta)$. The rest of the proof applies without changes. □

## 6. Lower Bounds for $K^2(\mathcal{C})$

In this section we prove some lower bounds for $(K_S + \Delta)^2$ for certain pairs $(S, \Delta)$. Applications of these bounds will be presented in this and in the next chapter.

6.1 THEOREM. *Let $(S, \Delta) \in \mathcal{S}(\mathcal{C})$. Assume that $K_S + \Delta$ is ample and let $C \subset \operatorname{Supp} \Delta$ be an irreducible curve. Then*

$$(K_S + \Delta)^2 \geq \delta(\mathcal{C}) C \cdot (K_S + \Delta).$$

PROOF. If $K_S + \Delta - \delta(\mathcal{C})C$ is the sum of a nef and of an effective divisor, then

$$(K_S + \Delta)^2 \geq \delta(\mathcal{C}) C \cdot (K_S + \Delta).$$

Let $d_C$ be the coefficient of $C$ in $\Delta$. Consider divisors of the form $K_S+\Delta-tC$, $0 \leq t \leq d_C$. There is a smallest value $t_1$ such that $K_S + \Delta - t_1 C$ is nef but not ample. If $t_1 \geq \delta(\mathcal{C})$ then we are done. Otherwise $K_S + \Delta - t_1 C$ induces a morphism

$$f_1 : S \to S_2 \quad \text{such that} \quad K + \Delta - t_1 C \equiv f_1^*(K + \Delta_2 - t_1 C_2),$$

where $\Delta_2 = f_1(\Delta)$ and $C_2 = f_1(C)$. If $f_1$ is birational then $C_2$ is a curve, i.e. $C$ is not contracted by $f_1$. Assume the contrary. Then $C^2 < 0$ and

$$0 = (K + \Delta - t_1 C) \cdot C = (K + \Delta) \cdot C - t_1 C^2 > -t_1 C^2 \geq 0.$$

Next we increase $t_1$ to $t_2 > t_1$ and obtain a morphism

$$f_2 : S_2 \to S_3 \quad \text{such that} \quad K + \Delta_2 - t_2 C_2 \equiv f_2^*(K + \Delta_3 - t_2 C_3).$$

Thus

$$K + \Delta - t_2 C \equiv (f_2 f_1)^*(K + \Delta_3 - t_2 C_3) + (t_2 - t_1)\left(f_1^*(C_2) - C\right).$$

We can continue this procedure until we get $t_{n-1} \geq \delta(\mathcal{C})$ or $\dim S_n < 2$. Thus we obtain $g : S \to S_n$ such that

$$K + \Delta - t_{n-1} C \equiv f_1^*(K + \Delta_n - t_{n-1} C_n) + (\text{effective divisor}).$$

Consider first the case when $S_n = $ point. Then $K + (\Delta_{n-1} - d_C C_{n-1})$ is log terminal and

$$K + \Delta_{n-1} - t_{n-1} C_{n-1} = K + (\Delta_{n-1} - d_C C_{n-1}) + (d_C - t_{n-1}) C_{n-1}$$

is numerically trivial. By definition, $t_{n-1} \geq \delta(\mathcal{C})$. If $S_n$ is a curve, we apply the same argument to the general fiber of $f_{n-1}$. □

6.2 COROLLARY. *Assume that $\mathcal{C}$ satisfies the descending chain condition. Let $(S, \Delta) \in \mathcal{S}(\mathcal{C})$. If $\llcorner\Delta\lrcorner \neq 0$ and $K_S + \Delta$ is ample then*

$$(K_S + \Delta)^2 \geq \delta_1(\mathcal{C})\epsilon_1(\mathcal{C}).$$

PROOF. Apply (6.1) for $C \subset \llcorner\Delta\lrcorner$. □

6.2.1 REMARK. By (6.2), if $(S, \Delta) \in \mathcal{S}(\mathcal{C}_2)$, $\llcorner\Delta\lrcorner \neq 0$ and $K_S + \Delta$ is ample then $(K_S + \Delta)^2 \geq 1/1764$. This bound is in fact sharp for the example given in (5.3.1).

By the methods of section 2, with the above assumptions

$$\#\operatorname{Aut}(S, \Delta) \leq 1764(K_S + \Delta)^2.$$

6.3 COROLLARY. *Let $f : X \to C$ be a stable degeneration of surfaces of general type. Let $S$ be the general fiber and let $S_0$ be a special fiber. Then the number of irreducible components of $S_0$ is bounded by $1764K_S^2$.* □

6.4 THEOREM. *Assume that $\mathcal{C}$ satisfies the descending chain condition. Let $(S, \Delta) \in \mathcal{S}(\mathcal{C})$. Assume that $K_S + \Delta$ is ample and let $(K + \Delta)^2 = d^2$. Let $\{C_t\}$ be a covering family of curves. Then*

$$(K_S + \Delta) \cdot C_t \geq \min\left\{d\sqrt{\epsilon_1(\mathcal{C})/4}, \epsilon_1(\mathcal{C})/4\right\} \geq d\epsilon_1(\mathcal{C})/4.$$

PROOF. Let $(K_S + \Delta) \cdot C_t = c$. We need to exclude the possibility that $c$ is very small. Fix $m > 2c/d^2$ and let $k \gg 1$ such that $km(K + \Delta)$ is Cartier. Then

$$h^0(\mathcal{O}(km(K + \Delta))) \sim \frac{m^2 d^2}{2} k^2.$$

It is $kmc + 1$ conditions for a section of $\mathcal{O}(km(K + \Delta))$ to vanish along $C_t$. Thus we can find a section $s \in H^0(\mathcal{O}(km(K + \Delta)))$ which vanishes along $k$ different curves $C_{t_1}, \dots, C_{t_k}$, since

$$\frac{m^2 d^2}{2} k^2 > k(kmc + 1).$$

Thus

$$m(K + \Delta) \equiv C_t + (\text{effective divisor}).$$

By (5.4.4) we know that $(K + \Delta + C_t) \cdot C_t \geq \epsilon_1(\mathcal{C})$, which implies that

$$(m + 1)(K + \Delta) \cdot C_t \geq (K + \Delta + C_t) \cdot C_t \geq \epsilon_1(\mathcal{C}).$$

This yields the inequality

$$\left(2c/d^2 + 2\right) c \geq \epsilon_1(\mathcal{C}),$$

which can be rearranged to

$$c^2 + d^2 c \geq \frac{d^2}{2}\epsilon_1(\mathcal{C}).$$

One of the summands on the left hand side is at least $d^2\epsilon_1(\mathcal{C})/4$. □

6.5 THEOREM. *Assume that $\mathcal{C}$ satisfies the descending chain condition. Let $(S, \Delta) \in \mathcal{S}(\mathcal{C})$. Assume that $K_S + \Delta$ is ample. Let $\{C_t\}$ be a covering family of curves and assume that $C_s \subset \operatorname{Supp} \Delta$ for some value $s$ of the parameter $t$. Then*

$$(K_S + \Delta)^2 \geq \frac{1}{16}\delta(\mathcal{C})^2\epsilon_1(\mathcal{C})^2.$$

PROOF. Applying (6.1) with $C = C_s$ and then (6.4) we obtain:

$$(K_S + \Delta)^2 \geq \delta(\mathcal{C})(K_S + \Delta) \cdot C_t \geq \delta(\mathcal{C})\sqrt{(K_S + \Delta)^2\epsilon_1(\mathcal{C})/4}. \quad \square$$

## 7. Plurigenera of Elliptic Threefolds

7.1 DEFINITION. Let $X$ be a variety. A subvariety $\mathbf{Z} \subset T \times X$ is called a covering family of $X$ if $T$ is irreducible, the second projection $p_X : \mathbf{Z} \to X$ is dominant and the first projection $p_T : \mathbf{Z} \to T$ is proper and flat with irreducible and reduced fibers.

We will frequently denote a covering family by $\{Z_t\}$ where $\{Z_t : t \in T\}$ supposed to run through all fibers of $p_T : \mathbf{Z} \to T$. While this is somewhat ambiguous, in the present situation this will not cause any problems.

By the countability of Hilbert scheme we know that there are countably many divisors $D_i \subset X$ such that if $Z \subset X$ is an irreducible and reduced subvariety such that $Z \not\subset \cup D_i$ then $Z$ occurs as a fiber in a covering family.

A point $x \in X \setminus \cup D_i$ will be refered to as a very general point of $X$.

7.2 DEFINITION. Let $Y$ be a normal projective variety of dimension $n$ and let $D$ be a nef and big $\mathbb{Q}$-Cartier $\mathbb{Q}$-divisor on $Y$. We say that $D$ satisfies the condition (7.2.$m_0$) if for every covering family $\{Z_t\}$

$$D^{\dim Z_t} \cdot Z_t \geq \left(\frac{n^2 + n + 1}{2m_0}\right)^{\dim Z_t}.$$

(Here $\dim Z_t$ denotes the dimension of the cycle $Z_t$ and not the dimension of the family.)

It is clear that $D$ satisfies the condition (7.2.$m_0$) for $m_0 \gg 1$. Also, if $D$ is Cartier then it satisfies condition (7.2.$m_0$) for $m_0 \geq (n^2 + n + 1)/2$.

7.3 PROPOSITION. *Let $Y$ be a normal projective variety of dimension $n$ and let $D$ be a nef and big $\mathbb{Q}$-Cartier $\mathbb{Q}$-divisor on $Y$. Assume that $D$ satisfies condition* (7.2.$m_0$).

*Then $h^0(X, \omega_X(\ulcorner mD \urcorner)) \neq 0$ for every $m \geq m_0$.*

PROOF. We apply [Kollár92, 3.2] by setting $S = Y$, $g : X \to S$ a log resolution of singularities, $L = 2m/(n^2 + n + 1)D$, $M = 0$, $\Delta = \ulcorner g^*mD \urcorner - g^*mD$ and $N = K_X + \ulcorner g^*mD \urcorner \equiv K_X + g^*mD + \Delta$. $\square$

Frequently we need to understand $h^0(\omega_X(\llcorner mD \lrcorner))$ instead of $h^0(\omega_X(\ulcorner mD \urcorner))$. One can transfrom (7.3) into a result of the desired form if $\llcorner (m+k)D \lrcorner \geq \ulcorner mD \urcorner$ for some $k$.

7.4 COROLLARY. *Let $Y$ be a normal projective variety of dimension $n$ and let $D = \sum d_i D_i$ be a nef and big $\mathbb{Q}$-Cartier $\mathbb{Q}$-divisor on $Y$. Assume that $D$ satisfies condition (7.2.$m_0$). Assume furthermore that $kd_i \geq 1$ for every $i$ (in particular, $D$ is effective).*

*Then $h^0(X, \omega_X(\llcorner mD \lrcorner)) \neq 0$ for every $m \geq m_0 + k$.*

PROOF. This follows from (7.3) since $H^0(X, \omega_X(\llcorner mD \lrcorner)) \supset H^0(X, \omega_X(\ulcorner (m-k)D \urcorner))$. □

The following result seems more useful for log surfaces and elliptic threefolds.

7.5 THEOREM. *Let $Y$ be a normal projective variety of dimension $n$ and let $D = \sum d_i D_i$ be a $\mathbb{Q}$-divisor on $Y$ such that $K + D$ is $\mathbb{Q}$-Cartier, nef and big. Assume that $K + D$ satisfies condition (7.2.$m_0$). Assume furthermore that for some $k \in \mathbb{N}$*

$$\{d_1, d_2, \dots\} \subset \{1 - 1/n | n \in \mathbb{N}\} \cup \{j/k | j = 1, \dots, k\} \cup \{x \in \mathbb{Q} | x \geq 1\}.$$

*Then $h^0(X, \mathcal{O}_X(mK_X + \llcorner mD \lrcorner)) \neq 0$ if either $m \geq m_0 + 1$ and $m \equiv 0, 1 \mod k$ or $k \geq 3$ and $m \geq (k-1)(m_0 + k)$.*

PROOF. The condition on the $d_i$ implies that $\llcorner m d_i \lrcorner \geq \ulcorner (m-1) d_i \urcorner$ if $m \equiv 0, 1 \mod k$. Thus $H^0(X, \mathcal{O}_X(mK_X + \llcorner mD \lrcorner)) \supset H^0(X, \omega_X((m-1)K_X + \ulcorner (m-1)D \urcorner))$ if $m \equiv 0, 1 \mod k$. Hence (7.3) implies that first part.

Pick $m_0 + 1 \leq s \leq m_0 + k$ such that $s \equiv 1 \mod k$. If $m \geq (k-1)(m_0 + k)$ and $m \equiv i \mod k$ $(1 < i < k)$ then write $m = (i-1)s + [m - (i-1)s]$. By the first part, $h^0(X, \mathcal{O}_X(kK_X + \llcorner kD \lrcorner)) \neq 0$ for $k = s$ and $k = [m - (i-1)s]$. Thus $h^0(X, \mathcal{O}_X(mK_X + \llcorner mD \lrcorner)) \neq 0$. □

7.6 COMMENT. Fix $\mathcal{C}$ such that $\mathcal{C} \subset \{1 - 1/n | n \in \mathbb{N}\} \cup \{j/k | j = 1, \dots, k\}$ for some $k$. Assume that $K^2(\mathcal{C})$ is bounded from below by a constant $\epsilon' > 0$. Then (6.4) and (7.5) imply (3.1.1) with a value $m(\mathcal{C})$ depending only on $k$ and on $\epsilon'$.

7.7 COROLLARY. *Let $X$ be a smooth threefold with $\kappa(X) = 2$. Assume that the $j$-invariant of the elliptic fibration is not constant. Then there is a constant $M$, independent of $X$, such that*

$$h^0(X, \mathcal{O}(mK_X)) > 0 \quad \text{for } m \geq M.$$

PROOF. Let $(S, \Delta)$ be the corresponding surface in $\mathcal{S}(\mathcal{C}_3)$. The level set of the $j$ function gives a component $C \subset \operatorname{Supp} \Delta$ which is a member of a covering family of curves. By (6.5) $(K + \Delta)^2$ is bounded from below and (6.4) provides a lower bound for $(K_S + \Delta) \cdot C_t$ where $\{C_t\}$ is any covering family. Hence (7.5) gives the result. □

7.7.1 REMARK. The above proof does not give an actual bound on $M$ since the value of $\delta(\mathcal{C}_3)$ is not known. An actual bound for $\epsilon_1(\mathcal{C}_3)$ is also not known; however it should be possible to obtain one with the present methods.

For lower values of $\kappa$ the following results are known:

7.8 THEOREM. *(7.8.1) [Kawamata86] There is a number* $m_0 > 0$ *such that if* $X$ *is a smooth threefold with* $\kappa(X) = 0$ *then* $h^0(X, \mathcal{O}(m_0K_X)) > 0$.

*(7.8.2) [Mori92] There is a number* $m_1 > 0$ *such that if* $X$ *is a smooth threefold with* $\kappa(X) = 1$ *then* $h^0(X, \mathcal{O}(m_1K_X)) > 0$.

ACKNOWLEDGEMENT. I am greatful to V. Alexeev and A. Grassi for calling my attention to several errors in earlier versions.

## REFERENCES

V. A. Alexeev, *Fractional indices of log Del Pezzo surfaces*, Math. USSR Izv. **8** (1989), 613-629.

V. A. Alexeev, *Two two-dimensional Terminations*, Duke Math. J. **69** (1992).

T. Fujita, *Zariski decomposition and canonical rings of elliptic threefolds*, J. Math. Soc. Japan **38** (1986), 19-37.

A. Grassi, *On a question of J. Kollár*, (in this volume).

Y. Kawamata, *On the plurigenera of minimal algebraic threefolds with* $K \approx 0$, Math. Ann. **275** (1986), 539-546.

J. Kollár, *Effective Base Point Freeness*, (to appear), Math. Annalen.

J. Kollár et al., *Flips and Abundance for Algebraic Threefolds*, (to appear), Astérisque.

J. Kollár - N. Shepherd-Barron, *Threefolds and deformations of surface singularities*, Inv. Math. **91** (1988), 299-338.

Y. Miyaoka, *The Chern Classes and Kodaira Dimension of a Minimal Variety*, Advanced Studies in Pure Math. 10, Algebraic Geometry Sendai 85, Kinokuniya-North Holland, 1987, pp. 449-476.

S. Mori, *A canonical bundle formula*, (in preparation).

M. Reid, *Canonical Threefolds*, Géométrie Algébrique Angers, A. Beauville ed., Sijthoff & Noordhoff, 1980, pp. 273-310.

V. V. Shokurov, *Problems About Fano varieties*, Birational Geometry of Algebraic Varieties, Open Problems (preprint), 1988, pp. 30-32.

V. Shokurov, *3-fold log flips*, Izv. Russ. A. N. Ser. Mat. **56** (1992), 105-203.

G. Xiao, *Bound of automorphisms of surfaces of general type, I-II*, (to appear), Ann. Math..

University of Utah, Salt Lake City, UT 84112
e-mail: kollar@math.utah.edu

Contemporary Mathematics
Volume **162**, 1994

# On Triple Covers of $\mathbf{P}^n$ as Very Ample Divisors

A. LANTERI, M. PALLESCHI, A. J. SOMMESE

ABSTRACT. A contribution to the classification of complex projective manifolds admitting a smooth triple cover of the projective space among their hyperplane sections is given.

## Introduction

This paper is aimed to classifying the projective manifolds containing a triple cover of a projective space among their hyperplane sections. More precisely we want to describe the pairs $(X, L)$ where

(*) $X$ is a smooth complex projective $(n+1)$-fold and $L$ is a very ample line bundle on it whose complete linear system $|L|$ contains a smooth element $A$ endowed with a finite morphism of degree 3, $\pi : A \to \mathbf{P}^n$.

Of course for $n = 1$ this problem is related to the study of surfaces having a trigonal curve among their hyperplane sections (e. g. see [Fa2] and [Se]). Here we assume that $n \geq 2$.

In any dimension $n \geq 2$ there are obvious examples fitting the above situation:

($a_n$) $(\mathbf{P}^{n+1}, \mathcal{O}_{\mathbf{P}^{n+1}}(3))$;

($b_n$) a cubic hypersurface of $\mathbf{P}^{n+2}$ polarized by its hyperplane bundle.

A nonobvious example is given by

($c_n$) $(X, 3\mathcal{L})$ where $(X, \mathcal{L})$ is a del Pezzo manifold (i. e. $X$ has an ample anti-canonical bundle $-K_X = n\,\mathcal{L}$) with $\mathcal{L}^{n+1} = 1$.

To show that ($c_n$) satisfies (*) the main point is to prove that the line bundle $3\mathcal{L}$ is actually very ample. Sec. 1 is devoted to proving this fact (Theorem 1.2). Example ($c_2$) is also related to standard conjectures on the spannedness of the adjoint bundle $K_X+L$ to a very ample line bundle $L$ on a smooth 3-fold $X$. Moreover it provides an

1991 Mathematics Subject Classification. Primary 14C20, 14J30.

Supported by the MURST of the Italian Government and by the NSF (DMS 89-21702).
This paper is in final form and no version of it will be submitted for publication elsewhere.

interesting example in connection with a recent Reider-type result for smooth 3-folds obtained by Ein and Lazarsfeld [EL].

In sec. 2 we get our classification result in high dimension: it says that the above examples ($a_n$), ($b_n$), ($c_n$) are the only pairs occurring for $n > 3$ (Theorem 2.5). This is also the case for $n = 3$ provided that $\pi : A \to \mathbf{P}^n$ is a cyclic cover. To prove this we extend to the present case arguments worked out in [LPS2] to study the same problem for double solids.

As one can expect the situation is far more complicated for $n = 2$ and becomes richer even if we impose further restrictions on $A$, as the following one:

(**) $$K_A = \pi^* \mathcal{O}_{\mathbf{P}^2}(k) \quad \text{for some integer } k.$$

Note that condition (**) is satisfied in cases ($a_2$), ($b_2$), ($c_2$) and also by any cyclic triple plane. For pairs verifying (**) we prove that either $k = -1$ or $k > 0$ (Lemma 3.1). In case $k = -1$, which is discussed in sec. 3, $A$ is a del Pezzo surface of degree 3 and so our problem is a special case of a more general setting considered in [LPS3]. Here we simply state the result (Theorem 3.3) adding some indication for the proof.

Pairs ($a_2$), ($b_2$) having been found in the above case, our aim is to characterize ($c_2$) among those pairs verifying (**) with $k > 0$. To do this we consider the general element $C$ in the linear system $|\pi^* \mathcal{O}_{\mathbf{P}^2}(1)|$ and its degree $\delta$ in the projective embedding given by $|L_C|$. We prove that $8 \leq \delta \leq 9$ (Theorem 4.3) and we show that ($c_2$) is the only pair occurring for $\delta = 9$ (Theorem 4.4). The main technical tool for this purpose is an adjunction theoretic result for 3-folds whose surface sections have ample canonical bundle, which we prove in sec. 0 (Theorem 0.1). By standard inductive arguments it can be easily extended to higher dimensions.

## 0. Notation and background

We work over the complex number field. A projective $n$-fold is an irreducible smooth projective scheme of dimension $n$. Vector bundles are holomorphic vector bundles. We use standard notation from algebraic geometry. We also adopt some current abuses. Everywhere we do not distinguish between line bundles and invertible sheaves. We adopt the additive notation for line bundles, while multiplicative notation omitting "." is reserved for the intersection product in the Chow rings.

Let $V$ be a projective $n$-fold and let $\mathcal{L}$ be a line bundle on $V$. We let $\mathcal{L}^r = c_1(\mathcal{L})^r$. $\mathcal{L}_W$ will denote the restriction of $\mathcal{L}$ to a subvariety $W$ of $V$. $K_V$ will stand for the canonical bundle of $V$.

A line bundle $\mathcal{L}$ on $V$ is said to be nef (i.e. numerically effective) if $\mathcal{L}C \geq 0$ for every curve $C$ on $V$. In addition $\mathcal{L}$ is said to be big if $\mathcal{L}^n > 0$. We say that $\mathcal{L}$ is spanned if it is spanned by $\Gamma(V, \mathcal{L})$.

For an ample line bundle $\mathcal{L}$ on $V$, the sectional genus $g(V, \mathcal{L})$ is defined by the genus formula $2g(V, \mathcal{L}) - 2 = (K_V + (n-1)\mathcal{L})\mathcal{L}^{n-1}$. If $\mathcal{L}$ is also spanned, then $g(V, \mathcal{L})$ is simply the geometric genus $g(C)$ of the smooth curve $C$ obtained by intersecting $n$-1 general elements of the complete linear system $|\mathcal{L}|$. We also set $d(V, \mathcal{L}) = \mathcal{L}^n$. The $\Delta$-genus of $(V, \mathcal{L})$ is defined as $\Delta(V, \mathcal{L}) = d(V, \mathcal{L}) + n - h^0(\mathcal{L})$.

We denote by $\mathbf{Q}^n$ a smooth quadric hypersurface of $\mathbf{P}^{n+1}$. We also recall the terminology concerning some special varieties frequently occurring in the sequel. Let $V$ be a projective $n$-fold and let $\mathcal{L}$ be an ample line bundle on $V$. The pair $(V, \mathcal{L})$ is said to be a Del Pezzo manifold if $-K_V = (n-1)\mathcal{L}$. According to the tradition, for $n = 2$, we call Del Pezzo surface any smooth surface $S$ with $-K_S$ ample. We say that $(V, \mathcal{L})$ is a scroll (respectively a quadric fibration) over a normal variety $W$ of dimension $h$, if there exists a surjective morphism with connected fibres $p:V \to W$ and an ample line bundle $H$ on $W$ such that $K_V + (n-h+1)\mathcal{L} = p^*H$ (respectively $K_V + (n-h)\mathcal{L} = p^*H$). In particular, if $(V, \mathcal{L})$ is a scroll over a curve or a surface $W$ and $n-h > 0$, then $W$ is smooth, $V$ is a $\mathbf{P}^{n-h}$-bundle over $W$ and $\mathcal{L}_f = \mathcal{O}_{\mathbf{P}^{n-h}}(1)$ for every fibre $f$ of $p$.

For the properties of reductions and all the results of adjunction theory we need in the sequel for pairs $(V, \mathcal{L})$ with $\mathcal{L}$ very ample, we refer to [SV], [S2] and [BFS]. Here we prove the following general result we need in section 4.

(0.1) THEOREM. *Let $(X, \mathcal{L})$ be the reduction of a pair $(X', \mathcal{L}')$ consisting of a smooth projective threefold $X'$ polarized by a very ample line bundle $\mathcal{L}'$. Assume that $K_X+\mathcal{L}$ is nef and big and set $d_i = \mathcal{L}^{3-i}(K_X+\mathcal{L})^i$. Let $A$ be a smooth element of $|\mathcal{L}|$ and furthermore assume that $K_A$ is ample. If $d_2^2 < (d_1+1)d_3$ then $K_X+\mathcal{L}$ is ample; in particular if $d_2^2 = d_1d_3$, then $K_X+\mathcal{L}$ is ample and $d_2(K_X+\mathcal{L}) \equiv d_3\mathcal{L}$ (numerical equivalence).*

*Proof.* This needs a second reduction argument (see [BFS, sec. 0]). Since $K_X+\mathcal{L}$ is nef and big, it follows from the Kawamata-Shokurov base-point-free theorem [KMM, § 3] that $\Gamma(m(K_X+\mathcal{L}))$ gives rise to a morphism $\varphi: X \to Y$ for some $m > 0$, and for $m$ large enough we can assume that $\varphi$ has connected fibres and normal image. Thus there is an ample line bundle $\mathcal{K}$ on $Y$ such that

$$K_X + \mathcal{L} = \varphi^*\mathcal{K}. \tag{0.1.1}$$

The map $\varphi$ is known as the second reduction morphism. It is an isomorphism outside of an algebraic subset $Z$ of dimension $\leq 1$ of $Y$ and its positive dimensional fibres are completely classified [S1, (1.0.1)] and [Fa1, (1.0)] (see also [BFS, (0.2.1)]). However, due to the assumption that $K_A$ is ample, $\varphi$ cannot have 2-dimensional fibres. Otherwise, were $F$ such a fibre, then there would be an effective divisor $E = F \cap A$ on $A$, $A$ being ample on $X$, satisfying the condition $E K_A = 0$ since $m(K_X+\mathcal{L}) F = 0$; a contradiction. It thus follows from [BFS, (0.2.1)]) that $\varphi: X \to Y$ is the blow-up of $Y$ along $Z$, $Z$ consisting of some smooth curves contained in $Y \setminus \mathrm{Sing}(Y)$. In particular we have that $Y$ is smooth since so is $X$, and by (0.1.1) we have $\mathcal{L} = \varphi^*\mathcal{M} - \varphi^{-1}(Z)$, for some line bundle $\mathcal{M}$ on $Y$. Note that

$$d_1 = \mathcal{L}^2(K_X+\mathcal{L}) = \varphi^*\mathcal{K}(\varphi^*\mathcal{M} - \varphi^{-1}(Z))^2 = \mathcal{K}\mathcal{M}^2 + (\varphi^*\mathcal{K})_{\varphi^{-1}(Z)}(\varphi^{-1}(Z))_{\varphi^{-1}(Z)}$$

$$= \mathcal{K}\mathcal{M}^2 - \mathcal{K}Z,$$

$$d_2 = \mathcal{L}(K_X+\mathcal{L})^2 = \mathcal{K}^2\mathcal{M} \qquad \text{and} \qquad d_3 = (K_X+\mathcal{L})^3 = \mathcal{K}^3.$$

On the other hand, since $\mathcal{K}$ is ample, by the generalized Hodge index theorem [BBS,

(0.15.1) letting $M = \mathcal{K}, N = \mathcal{M}$ and $j = 2$] we have the inequality

$$(0.1.2) \qquad (\mathcal{K}^3)(\mathcal{K}\mathcal{M}^2) \le (\mathcal{K}^2\mathcal{M})^2.$$

Note that the nefness assumption on $\mathcal{M}$, needed for the use of [BBS, (0.15.1)], is not required in our case. Actually $|m\mathcal{K}|$ contains a smooth surface $S$ for $m$ large; so (0.1.2) follows directly from the inequality $\mathcal{K}_S{}^2 \mathcal{M}_S{}^2 \le (\mathcal{K}_S \mathcal{M}_S)^2$ given by the Hodge index theorem on $S$. Note also that (0.1.2) can be rewritten as $d_3(d_1 + \mathcal{K}Z) \le d_2{}^2$. By combining it with the inequality in our assumption we get $\mathcal{K}Z = 0$, which implies $Z = 0$, $\mathcal{K}$ being ample. It thus follows that $\varphi$ is a biholomorphism, and then $K_X + \mathcal{L} = \mathcal{K}$ is ample. Finally assume that $d_2{}^2 = d_1 d_3$; then $K_X + \mathcal{L}$ is ample in view of the above conclusion, since $d_3 > 0$. The last assertion follows from [BBS, (0.15.2)], since in this case (0.1.2) is an equality.

Note that by an inductive argument one can prove the following higher dimensional version of (0.1).

(0.2) THEOREM. *Let $(X, \mathcal{L})$ be the reduction of a pair $(X', \mathcal{L}')$ consisting of a smooth projective n-fold $X'$ polarized by a very ample line bundle $\mathcal{L}'$. Assume that $K_X+(n-2)\mathcal{L}$ is nef and big and set $d_i = \mathcal{L}^{n-i}(K_X+(n-2)\mathcal{L})^i$. Let $A$ be the 2-dimensional element of a smooth ladder of $\mathcal{L}$ and furthermore assume that $K_A$ is ample. If $d_2{}^2 < (d_1+1)d_3$ then $K_X+(n-2)\mathcal{L}$ is ample; in particular if $d_2{}^2 = d_1 d_3$, then $K_X+(n-2)\mathcal{L}$ is ample and $d_2(K_X+(n-2)\mathcal{L}) \equiv d_3\mathcal{L}$ (numerical equivalence).*

## 1. A special example

(1.1) Let $(Y, \mathcal{L})$ be a Del Pezzo manifold of dimension $m \ge 3$, with $\mathcal{L}^m = 1$. Then $\mathrm{Pic}(Y) \cong \mathbf{Z}$ and $K_Y = -(m-1)\mathcal{L}$. Moreover, we know from [F1, § 14] (see also [Is, (4.2), (6.10)]), that

(1.1.1) $\mathcal{L}$ is spanned off a single point, say $x$, and $h^0(\mathcal{L}) = m$.

(1.1.2) $2\mathcal{L}$ is spanned and the map associated with $\Gamma(2\mathcal{L})$ is a morphism of degree 2 onto a cone over the Veronese manifold $(\mathbf{P}^{m-1}, \mathcal{O}_{\mathbf{P}^{m-1}}(2))$.

The aim of this section is to prove the following

(1.2) THEOREM. *Let $(Y, \mathcal{L})$ be as in* (1.1). *Then $3\mathcal{L}$ is very ample.*

*Proof.* As a first thing, note that

$$(1.2.1) \qquad 3\mathcal{L} = K_Y + (m+2)\mathcal{L}.$$

Of course [LPS1, (1.2)] does not apply, since $\mathcal{L}$ is not spanned everywhere. We have to prove the following facts: a) $3\mathcal{L}$ is spanned; b) the morphism $\varphi$ associated with $\Gamma(3\mathcal{L})$ is an immersion at every point of $Y$; c) the morphism $\varphi$ separates every couple of points of $Y$.

a) In view of (1.1.1) and (1.1.2) we only need to show that $3\mathcal{L}$ is spanned at $x$, the unique base point of $|\mathcal{L}|$. This is equivalent to showing that

$$(1.2.2) \qquad H^1((K_Y + (m+2)\mathcal{L}) \otimes \mathcal{J}_x) = 0,$$

where $\mathcal{J}_x$ stands for the ideal sheaf of $x$. Let $\sigma: Y' \to Y$ be the blowing-up of $Y$ at $x$ and let $E = \sigma^{-1}(x)$. Then (1.2.2) is equivalent to $H^1(\sigma^*(K_Y + (m+2)\mathcal{L}) - E) = 0$, and recalling that $K_{Y'} = \sigma^* K_Y + (m-1)E$, this is in turn equivalent to proving that

$$(1.2.3) \qquad H^1(K_{Y'} + (m+2)\,\sigma^*\mathcal{L} - mE) = 0.$$

Note that $\sigma^*\mathcal{L} - E$ is nef, since $x$ is the unique base point of $|\mathcal{L}|$ [LPS1, (0.5.1)]. Hence $(m+2)\,\sigma^*\mathcal{L} - mE = 2\,\sigma^*\mathcal{L} + m\,(\sigma^*\mathcal{L} - E)$ is nef and big (and in fact even ample). Therefore (1.2.3) follows from the Kawamata-Viehweg vanishing theorem. By the way note also that

$$(1.2.4) \qquad (m+2)\,\sigma^*\mathcal{L} - (m+1)E = \sigma^*\mathcal{L} + (m+1)\,(\sigma^*\mathcal{L} - E) \text{ is nef and big.}$$

b) In view of (1.2.4), the same argument as in a) shows that $\varphi$ is an immersion at $x$. Now consider a point $y \in Y \setminus \{x\}$. Assume for a moment that

(1.2.5) the general element of $|2\mathcal{L} - y|$ is singular at $y$.

Let $\sigma: Y' \to Y$ be the blowing-up of $Y$ at $y$ and let $E = \sigma^{-1}(y)$. By reasoning as in a), to prove that $H^1((K_Y + (m+2)\mathcal{L}) \otimes \mathcal{J}_y^2) = 0$ we show that

$$(1.2.6) \qquad H^1(K_{Y'} + (m+2)\sigma^*\mathcal{L} - (m+1)E) = 0.$$

Note that $|2\mathcal{L} - y| = |2\mathcal{L} - 2y|$, by (1.2.5), hence the line bundle $2\,\sigma^*\mathcal{L} - 2E$ is nef [LPS1, (0.5.1)], and then $\sigma^*\mathcal{L} - E$ is nef as well. Therefore $(m+2)\,\sigma^*\mathcal{L} - (m+1)E$ is nef and big, which implies (1.2.6), by the Kawamata-Viehweg vanishing theorem. So we can assume that

(1.2.7) the general element of $|2\mathcal{L} - y|$ is smooth at $y$.

By contradiction assume that there exists a tangent vector $\tau_y$ to $Y$ at $y$ such that $d\varphi_y(\tau_y) = 0$. As a first thing assume that $\tau_y$ is not tangent to the general $D \in |2\mathcal{L} - y|$. Let $s_D \in \Gamma(2\mathcal{L})$ be a section defining $D$ and let $\alpha \in \Gamma(\mathcal{L})$ be a section such that $\alpha(y) \neq 0$. Then the two sections $\alpha s_D$, $\alpha^3 \in \Gamma(3\mathcal{L})$ define a morphism in a neighbourhood of $y$, whose differential at $y$ does not vanish on $\tau_y$; henceforth $d\varphi_y(\tau_y) \neq 0$, a contradiction. So we can assume that $\tau_y$ is tangent to the general $D \in |2\mathcal{L} - y|$. This implies, in view of (1.1.2), that $y$ is a branch point of the morphism associated with $\Gamma(2\mathcal{L})$; in particular we have

$$(1.2.8) \qquad \mathrm{Bs}\,|2\mathcal{L} - y| = \{y\}.$$

It thus follows from (1.2.7), (1.2.8) and the Bertini theorem that there is a smooth element $D \in |2\mathcal{L} - y|$, and we can also assume in view of (1.2.8) that $D$ does not contain $x$. Then $\mathcal{L}_D$ is spanned and by [LPS1, (1.2)] we have that the line bundle $K_D + m\mathcal{L}_D$ is very ample. Then, from the exact cohomology sequence of

$$(1.2.9) \qquad 0 \to K_Y + m\mathcal{L} \to K_Y + (m+2)\mathcal{L} \to K_D + m\mathcal{L}_D \to 0,$$

we conclude that $\varphi$ is an immersion at $y$.

c) Let $y, z \in Y \setminus \{x\}, y \neq z$. Note that if $|2\mathcal{L} - y| \neq |2\mathcal{L} - z|$, then obviously $\varphi(y) \neq \varphi(z)$. So we can assume that $|2\mathcal{L} - y| = |2\mathcal{L} - z|$ and then $\mathrm{Bs}|2\mathcal{L} - y| = \{y, z\}$, in view of (1.1.2). Let $D$ be the general element of $|2\mathcal{L} - y|$; then $D$ is smooth. Otherwise, by Bertini's theorem $D$ would be singular at some point of $\{y, z\}$ and then

$$2 = D\,\mathcal{L}^{m-1} \geq \mathrm{mult}_y(D) + \mathrm{mult}_z(D) \geq 3,$$

contradiction. Moreover $D$ does not contain $x$ and then $\mathcal{L}_D$ is spanned. Thus, by [LPS1, (1.2)] again we have that the line bundle $K_D + m\mathcal{L}_D$ is very ample and, as before, from the exact cohomology sequence of (1.2.9), we conclude that $\varphi$ separates $y$ and $z$. As a last thing we have to show that $\varphi(x) \neq \varphi(y)$, for every $y \in Y \setminus \{x\}$. To do this, let $\sigma \in \Gamma(\mathcal{L})$ be a section such that $\sigma(y) \neq 0$ and let $t \in \Gamma(2\mathcal{L})$ be such that $t(x) \neq 0$, $t(y) \neq 0$. Then the section $s = \sigma t \in \Gamma(3\mathcal{L})$ satisfies $s(x) = 0$ and $s(y) \neq 0$. This concludes the proof.

(1.3) REMARK. Let $(Y, \mathcal{L})$ be as in (1.1) with $m = 3$. Then the pair $(Y, L{=}3\mathcal{L})$ provides an interesting example in connection with [S2, Conjecture 2.3 and (2.2.3)]. Actually, $Y$ is a threefold of negative Kodaira dimension, $L$ is very ample by (1.3), and $(Y, L)$ coincides with its reduction since $L \in 3\,\mathrm{Pic}(Y)$. Note that $\mathcal{L} = K_Y + L = K_Y + (m\text{-}2)L$ is not spanned; note also that $2\mathcal{L} = 2(K_Y + L)$ is spanned. The pair $(Y, L)$ also provides an interesting example in connection with a recent result of Ein and Lazarsfeld [EL, Thm. 1*]. Actually since $L^3 = 27$, it shows that condition $B^3 > 27$ cannot be relaxed in order to get the spannedness of $K_X + B$ for a nef and big line bundle $B$ on a smooth complex projective 3-fold $X$.

## 2. High dimension result

We recall the general set-up of this paper.

(2.0) Let $A$ be a smooth complex projective $n$-fold and let $\pi: A \to \mathbf{P}^n$ be a finite morphism of degree 3. Let $X$ be a smooth projective $(n+1)$-fold and assume that $A \subset X$ is a very ample divisor. We let $L = [A]$.

(2.1) EXAMPLES. Some obvious pairs $(X, L)$ as in (2.0) are the following:

1) Let $(X, L) = (\mathbf{P}^{n+1}, \mathcal{O}_{\mathbf{P}^{n+1}}(3))$; then $A$ is a smooth hypersurface of degree 3 and $\pi: A \to \mathbf{P}^n$ is the triple cover obtained by projecting $A$ from a point of $\mathbf{P}^{n+1} \setminus A$ onto a hyperplane.

2) Let $(X, L) = (V_{(3)}, \mathcal{O}_{\mathbf{P}^{n+2}}(1)_V)$ be a smooth hypersurface of degree 3 of $\mathbf{P}^{n+2}$ polarized by its hyperplane bundle; $A$ is again a smooth hypersurface of degree 3 and

$\pi: A \to \mathbf{P}^n$ is the triple cover obtained by restricting to $A$ the projection of $X$ from a point of $\mathbf{P}^{n+2} \setminus X$ onto a hyperplane.

A less obvious example is

3) Let $(X, L) = (Y, 3\mathcal{L})$, where $(Y, \mathcal{L})$ is the De Pezzo manifold of dimension $n+1$ as in (1.1), i.e. $-K_Y = n\,\mathcal{L}$ with $\mathcal{L}^{n+1} = 1$. Then $3\mathcal{L}$ is very ample by (1.2). Let $x$ be the unique base point of $|\mathcal{L}|$ and let $A \in |3\mathcal{L}|$ be a smooth hypersurface not passing through $x$. Then, from the exact cohomology sequence of

$$0 \to -2\mathcal{L} \to \mathcal{L} \to \mathcal{L}_A \to 0,$$

by the Kodaira vanishing theorem we get

(*) $$h^0(\mathcal{L}_A) = h^0(\mathcal{L}) = n+1.$$

Moreover $\mathcal{L}_A$ is spanned, since $x \notin A$ and then the map $\pi: A \to \mathbf{P}^n$ associated with $|\mathcal{L}_A|$ is a finite morphism of degree 3.

(2.2) Let $(X, L)$ be as in (2.0). If $n > 3$, then by [La, (3.2)] we get $H_2(A, \mathbf{Z}) = \mathbf{Z}$ and $h^1(\mathcal{O}_A) = 0$. As a consequence, we have that $\mathrm{Pic}(A) \cong \mathbf{Z}$, generated by $\pi^* \mathcal{O}_{\mathbf{P}^n}(1)$. By the Lefschetz theorem we have also $\mathrm{Pic}(X) \cong \mathbf{Z}$, generated by an ample line bundle which we call $\mathcal{H}$, extending $\pi^* \mathcal{O}_{\mathbf{P}^n}(1)$. We thus have $L = a\,\mathcal{H}$, $a \geq 1$; furthermore $K_X = r\,\mathcal{H}$.

By [LPS2, Lemma 1.3] we have the following fact.

(2.3) REMARK. Assume that $\pi: A \to \mathbf{P}^n$ is associated with a proper sublinear system of $|\mathcal{H}_A|$, i.e. $h^0(\mathcal{H}_A) > n+1$. Then the morphism $q: A \to \mathbf{P}^{n+t}$ ($t \geq 1$) associated with $|\mathcal{H}_A|$ is birational and its image $q(A)$ is a variety of degree 3 in $\mathbf{P}^{n+t}$. In this case we have of course $t \leq 2$, since $3 \geq \mathrm{codim}(q(A), \mathbf{P}^{n+t}) + 1 = t+1$.

(2.4) REMARK. If there is a morphism $q: A \to \mathbf{P}^{n+t}$ ($t = 1, 2$) as in (2.3), by slicing down $\mathbf{P}^n$ with general hyperplanes up to getting a line we obtain a ladder as follows:

$$\begin{array}{ccccc}
\mathbf{P}^{1+t} & \mathbf{P}^{2+t} & & & \mathbf{P}^{n+t} \\
\uparrow & \uparrow & & & \uparrow \\
C \subset & S & \subset \dots \subset & & A \\
\downarrow & \downarrow & & & \downarrow \\
\mathbf{P}^1 & \mathbf{P}^2 & & & \mathbf{P}^n ,
\end{array}$$

where $C$ is a smooth curve and $S$ a smooth surface. By (2.2) the morphism $q_C$ is birational and since it is associated with a divisor of $|\mathcal{H}_C|$, which has degree 3, we conclude that $q(C)$ is a cubic curve.

Note that $t = 1$ in the above examples (2.1;1), (2.1;2), while $t = 0$ in case (2.1;3) as (*) shows. For $n > 3$ we have the following result.

(2.5) THEOREM. *Let things be as in* (2.0) *and assume that* $n > 3$. *Then* $(X, L)$ *is one of the pairs in* (2.1).

*Proof.* Assume that $h^0(\mathcal{H}_A) = n+1+t$, $t > 0$. By slicing down the triple cover as in (2.4) we get a smooth curve $C$ such that $q(C)$ is an irreducible curve of degree 3, so that $g(C)$ is either 0 or 1.

Let $g(C) = 0$. Then looking at the surface $S$ in the ladder in (2.4) we see that $(S, \mathcal{H}_S)$ is either: i) $(\mathbf{P}^2, \mathcal{O}_{\mathbf{P}^2}(e))$, $e = 1, 2$ or ii) a rational scroll. Case i) cannot occur since otherwise we would have $3 = \mathcal{H}_S{}^2 = e^2$, contradiction. In case ii), going backwords over the ladder in (2.4) we would conclude that $(A, \mathcal{H}_A)$ is either $(\mathbf{Q}^n, \mathcal{O}_{\mathbf{Q}^n}(1))$ or a rational scroll [F2, (12.1) and (5.10)]. On the other hand in the former case we would have the contradiction $3 = \mathcal{H}_A{}^n = 2$, while in the latter one it would be $\mathrm{Pic}(A) \cong \mathbf{Z}^2$, contradicting (2.2).

Let $g(C) = 1$. Then $(S, \mathcal{H}_S)$ is either: i) a Del Pezzo surface polarized by the anticanonical bundle, or ii) an elliptic scroll. The scroll case can be ruled out as before looking at $H_2(A, \mathbf{Z})$ and recalling (2.2). In case i), going backwards over the ladder in (2.4), we see that $(A, \mathcal{H}_A)$ is a Del Pezzo manifold and since $\mathcal{H}_A{}^2 = 3$, $A$ is a cubic hypersurface of $\mathbf{P}^{n+1}$ [F2, (12.3) and (6.7)]. Thus $(X, L)$ is either $(\mathbf{P}^{n+1}, \mathcal{O}_{\mathbf{P}^{n+1}}(3))$, or a cubic hypersurface of $\mathbf{P}^{n+2}$ polarized by its hyperplane bundle. This gives the pairs in (2.1;1), (2.1;2).

To complete the proof assume that

(2.5.1) $$h^0(\mathcal{H}_A) = n+1$$

and recall that $L = a\,\mathcal{H}$, $a \geq 1$. As a first thing we show that

(2.5.2) $$a \geq 2.$$

By contradiction, let $a = 1$. Then $\mathcal{H} = L$, so that $\mathcal{H}_A = L_A$ is very ample. But this gives a contradiction since $\pi : A \to \mathbf{P}^n$, which in view of (2.5.1) is the map associated to $|\mathcal{H}_A|$, would have degree 1. Now look at the exact sequence

$$0 \to (1-a)\mathcal{H} \to \mathcal{H} \to \mathcal{H}_A \to 0.$$

Since $a \geq 2$, by the Kodaira vanishing theorem we get $h^0(\mathcal{H}) = h^0(\mathcal{H}_A)$, so that $h^0(\mathcal{H}) = n+1$, by (2.5.1). Hence $\pi$ extends to a meromorphic map $\Pi : X \dashrightarrow \mathbf{P}^n$, whose indeterminacy locus must be a finite set, $A$ being very ample. Let $\Gamma$ be a general fibre of the projection onto $\mathbf{P}^n$ of the graph of $\Pi$. By identifying $\Gamma$ with its image in $X$, $\Gamma$ is the smooth intersection of $n$ general elements of $|\mathcal{H}|$. Therefore

(2.5.3) $$3 = \Gamma A = a\,\Gamma\,\mathcal{H},$$

and so, by (2.5.2), we get $a = 3$, $\Gamma\,\mathcal{H} = 1$. Recalling (2.2) we have $K_X = r\,\mathcal{H}$, for some integer $r$; so by adjunction we get

(2.5.4) $$2g(\Gamma) - 2 = \deg K_\Gamma = (K_X + n\,\mathcal{H})\,\Gamma = r+n.$$

Now, since $L$ is very ample, (2.5.3) implies that $g(\Gamma) \leq 1$. For $g(\Gamma) = 0$ (2.5.4) gives $r = -(n+2) = -(\dim X + 1)$, hence $(X, L) = (\mathbf{P}^{n+1}, \mathcal{O}_{\mathbf{P}^{n+1}}(3))$, contradicting (2.5.1). For $g(\Gamma) = 1$ (2.5.4) gives $r = -n = -(\dim X - 1)$, hence $(X, \mathcal{H})$ is a Del Pezzo manifold and $L = 3\mathcal{H}$; moreover $\mathcal{H}^{n+1} = 1$, by (2.5.3). So $(X, L)$ is the pair in (2.1;3) and this concludes the proof.

(2.6) REMARK. By a result of Cornalba [C] we still have $H_2(A, \mathbf{Z}) = \mathbf{Z}$ and $h^1(\mathcal{O}_A) = 0$, in case $n = 3$, provided that $\pi$ is a cyclic cover. So all the above arguments apply in this case too allowing us to get the same conclusion as in (2.5).

## 3. Special triple covers of $\mathbf{P}^2$.

Let $\pi: A \to \mathbf{P}^2$ be a triple cover; then $K_A = \pi^* K_{\mathbf{P}^2} + R$, where $R$ is the ramification divisor. We will restrict our attention to triple covers satisfying the following assumption:

(3.0) $$K_A = \pi^* \mathcal{O}_{\mathbf{P}^2}(k), \text{ for some } k \in \mathbf{Z}.$$

(3.1) REMARKS. i) Let $(Y, \mathcal{L})$ be a Del Pezzo threefold as in (1.1) and let $A \in |3\mathcal{L}|$ be a general element. Then $A$ is a smooth surface by (1.2) and the triple cover $\pi: A \to \mathbf{P}^2$ defined by $|\mathcal{L}_A|$ satisfies (3.0). Actually, by adjunction, we have

$$K_A = (K_Y + 3\mathcal{L})_A = \mathcal{L}_A = \pi^* \mathcal{O}_{\mathbf{P}^2}(1).$$

By the way note that $K_A$ is ample, so that $A$ is a minimal surface of general type. Moreover from (*) in (2.1;3) we get $3 = h^0(\mathcal{L}) = h^0(\mathcal{L}_A) = p_g(A)$ and so $\pi$ is simply the canonical map of $A$. Furthermore $K_A{}^2 = 3\mathcal{L} \cdot \mathcal{L}^2 = 3\mathcal{L}^3 = 3$. Finally $q(A) = h^{1,0}(Y) = 0$, by the Lefschetz theorem.

ii) Any cyclic triple cover $\pi: A \to \mathbf{P}^2$ satisfies (3.0). In fact in this case we have $\pi_* \mathcal{O}_A = \mathcal{O}_{\mathbf{P}^2} \oplus \mathcal{O}_{\mathbf{P}^2}(-b) \oplus \mathcal{O}_{\mathbf{P}^2}(-2b)$ for some $b \geq 1$ and

$$K_A = \pi^*(K_{\mathbf{P}^2} \otimes \mathcal{O}_{\mathbf{P}^2}(2b)) = \pi^* \mathcal{O}_{\mathbf{P}^2}(2b-3).$$

(3.2) LEMMA. *Assume that* (3.0) *holds and let* $\mathcal{D} = \pi^* \mathcal{O}_{\mathbf{P}^2}(1)$. *Then either* $k > 0$, *or* $k = -1$, *in which case* $(A, \mathcal{D})$ *is a cubic surface of* $\mathbf{P}^3$ *polarized by its hyperplane section.*

*Proof.* First of all note that case $k = 0$ cannot occur. Otherwise $K_A$ would be trivial by (3.0) and then $\mathcal{D}^2 + \mathcal{D}K_A = \mathcal{D}^2 = 3$ would be odd, contradicting genus formula. Now let $k < 0$. It thus follows from (3.0) that $A$ is a Del Pezzo surface. Since $-K_A = -k\,\mathcal{D}$, it can only be $k = -3, -2$ or $-1$. But from $K_A{}^2 = k^2 \mathcal{D}^2 = 3k^2$ and the inequality $K_A{}^2 \leq 9$, we immediately get $k = -1$. So $A$ is a Del Pezzo surface of degree 3, which proves the assertion.

Now come back to our original problem, $\pi: A \to \mathbf{P}^2$ being a triple cover satisfying (3.0). Case $k > 0$ will be studied in the next section. If $k = -1$ the problem of determining which pairs $(X, L)$ can occur fits into a more general problem we are dealing with in a separate paper [LPS3]. Here we simply state the result with a few comments, referring to [LPS3] for the proof.

(3.3) THEOREM. *Let $(X, L)$ be as in (2.0), $\pi: A \to \mathbf{P}^2$ being a triple cover such that $K_A = \pi^* \mathcal{O}_{\mathbf{P}^2}(-1)$. Then $(X, L)$ is one of the following pairs:*
(3.3.1) *A quadric fibration over $\mathbf{P}^1$. More precisely $X$ embeds fibrewise in a projective bundle $\mathbf{P}(\mathcal{F})$, with $\mathcal{F} = \oplus_{i=0,\dots,3} \mathcal{O}_{\mathbf{P}^1}(a_i)$, $a_i \geq 0$; we have $X \in |2\xi - bF|$, $L = \xi_X$, where $\xi$ and $F$ denote respectively the tautological bundle of $\mathcal{F}$ and a fibre of $\mathbf{P}(\mathcal{F})$ and $b \in \mathbf{Z}$; the "a priori" possible values of the invariants $b$, $\delta = c_1(\mathcal{F})$, $d = d(X, L)$, $g = g(X, L)$ are listed in the table below:*

| $b$ | $\delta$ | $d$ | $g$ |
|---|---|---|---|
| 1 | 4 | 7 | 2 |
| 3 | 7 | 11 | 3 |
| 5 | 10 | 15 | 4 . |

(3.3.2) $X = \mathbf{P}(\mathcal{E})$, *where $\mathcal{E}$ is a rank-2 vector bundle on a smooth surface $S$, $L$ is the tautological bundle of $\mathcal{E}$ and $S$, $\mathcal{E}$ and the invariants $d$, $g$ are listed in the table below:*

| $S$ | $\mathcal{E}$ | $d$ | $g$ |
|---|---|---|---|
| $\mathbf{P}^2$ | *given by* $0 \to \mathcal{O}_{\mathbf{P}^2}^{\oplus 2} \to \mathcal{O}_{\mathbf{P}^2}(1)^{\oplus 4} \to \mathcal{E} \to 0$ | 10 | 3 |
| $\mathbf{P}^1 \times \mathbf{P}^1$ | $\mathcal{O}_{\mathbf{P}^1 \times \mathbf{P}^1}(1,2) \oplus \mathcal{O}_{\mathbf{P}^1 \times \mathbf{P}^1}(2,1)$ | 13 | 4 . |

(3.3.3) *A cubic hypersurface of $\mathbf{P}^4$ polarized by its hyperplane section.*
(3.3.4) $(\mathbf{P}^3, \mathcal{O}_{\mathbf{P}^3}(3))$.
(3.3.5) $(X, L)$ *admits* $(\mathbf{Q}^3, \mathcal{O}_{\mathbf{Q}^3}(2))$ *as reduction, the reduction morphism $\rho: X \to \mathbf{Q}^3$ being the blow-up at a single point $x \in \mathbf{Q}^3$.*
(3.3.6) *$X$ is $\mathbf{P}^3$ blown-up along a line $\ell$ and $L = \sigma^*[\mathcal{O}_{\mathbf{P}^3}(3) - \ell]$, where $\sigma: X \to \mathbf{P}^3$ is the blowing-up.*

(3.4) REMARKS. i) The first set of invariants in case (3.3.1) does really occur and $(X, L)$ is the same as the pair described in [Io, Th. 3.4, case iii)], i. e. $a_i = 1$ for all $i$ 's. So $\mathbf{P}(\mathcal{F}) = \mathbf{P}^1 \times \mathbf{P}^3$ and $X \in |\mathcal{O}_{\mathbf{P}^1 \times \mathbf{P}^3}(1,2)|$, $L = \mathcal{O}_{\mathbf{P}^1 \times \mathbf{P}^3}(1,1)_X$. Of course such an $L$ is in fact very ample; $|L|$ embeds $X$ in $\mathbf{P}^7$ and its general element is isomorphic to $\mathbf{P}^2$ blown-up at six points in general position.

ii) In case (3.3.2) $(X, L)$ is a scroll over $S$, so $K_X + 2L = q^*H$, where $q: X \to S$ is the scroll projection and $H$ is an ample line bundle on $S$. For such pairs, by adjunction, (3.0) gives $-K_X = L + \mathcal{H}$, where, as in [LPS2, Lemma 1.5, step 1], one can show that $\mathcal{H}$ is nef. Therefore $X$ is Fano and so pairs $(X, L)$ as above can be found in the list of [SzW, main theorem]. Taking into account that $X = \mathbf{P}(\mathcal{E})$, with $\mathcal{E} = q_* L$, we have that $H = K_S + \det \mathcal{E}$ must be ample, while $\mathcal{E}$ has to be very ample, since so is its tautological bundle $L$; so a close inspection of the list in [SzW] shows that only four pairs satisfy this condition. They are the two pairs in (3.3.2) and $(\mathbf{P}^2, \mathcal{O}_{\mathbf{P}^2}(3)\oplus\mathcal{O}_{\mathbf{P}^2}(2))$, $(\mathbf{F}_1,[2s+3f]\oplus[s+2f])$, where $\mathbf{F}_1$ denotes the first Segre-Hirzebruch surface and $s, f$ stand for the section of minimal self-intersection and a fibre of it. Note that for all these pairs $L$ is in fact very ample since so is $\mathcal{E}$, and the general element of $|L|$ is the image of a meromorphic (non-holomorphic) section of $q$, containing exactly $c_2(\mathcal{E})$ fibres of $q$. However in the last two cases $-K_A$ is not very ample. Actually, since in both cases $\mathcal{E} = [-K_S] \oplus \mathcal{L}$, one immediately sees that $-K_A$ restricts trivially to the section of $q$ corresponding to the surjection $\mathcal{E} \to \mathcal{L}$.

Finally, to compute the invariants $d$ and $g$ in the two remaining cases, it is enough to recall that $(S, \det \mathcal{E})$ is simply the reduction of the pair $(A, L_A)$, hence $d(X,L) = d(A, L_A) = d(S, \det \mathcal{E}) - c_2(\mathcal{E})$, while $g(X,L) = g(A, L_A) = g(S, \det \mathcal{E})$.

Note that a pair $(X, L)$ as in the first case is given by the section of the Segre product $\mathbf{P}^2 \times \mathbf{P}^3 \subset \mathbf{P}^{11}$ with two general hyperplanes.

iii) Cases (3.3.5), (3.3.6) come from the study of Veronese bundles over $\mathbf{P}^1$. They are the analogues of cases (3.2.4), (3.2.6) in [LPS2] and in both cases $L$ is in fact very ample. Note that in case (3.3.5) an element $A \in |L|$ is isomorphic to a smooth cubic surface provided that the point $x$ does not lie on a line of the complete intersection of two quadrics $\rho(A)$.

## 4. Recovering the special example in dimension 3.

According to the program stated in sec. 3, in this section we study pairs $(X, L)$ when $\pi: A \to \mathbf{P}^2$ is a triple cover such that

(4.0) $$K_A = \pi^* \mathcal{O}_{\mathbf{P}^2}(k), \quad k > 0.$$

By Bertini's theorem the general element $C \in |\pi^* \mathcal{O}_{\mathbf{P}^2}(1)|$ is a smooth curve and by (4.0) the genus formula gives

(4.0.1) $$2g(C) - 2 = (k+1)\,(\pi^* \mathcal{O}_{\mathbf{P}^2}(1))^2 = 3(k+1).$$

Therefore:

(4.0.2) $k$ is odd,

(4.0.3) $g(C) \geq 4$.

(4.1) LEMMA. *There exists a nef line bundle* $\mathcal{H} \in \mathrm{Pic}(X)$ *such that* $\mathcal{H}_A = \pi^* \mathcal{O}_{\mathbf{P}^2}(1)$.

*Proof.* By adjunction $(K_X + L)_A = K_A = \pi^* \mathcal{O}_{\mathbf{P}^2}(k)$. So since the restriction homomorphism $\mathrm{Pic}(X) \to \mathrm{Pic}(A)$ is injective with torsion free cokernel, there exists $\mathcal{H} \in \mathrm{Pic}(X)$ such that $\mathcal{H}_A = \pi^* \mathcal{O}_{\mathbf{P}^2}(1)$. Note that since $K_A$ is ample, $A$ cannot contain (-1)-curves; then the same argument used in [LPS2, Thm. 2.2] proves that $K_X + L$ is nef. Since $K_X + L = k\,\mathcal{H}$, this shows that $\mathcal{H}$ itself is nef.

(4.2) LEMMA. *$\mathcal{H}$ is big, i.e.* $\mathcal{H}^3 \geq 1$.

*Proof.* By (4.1) we have $\mathcal{H}^3 \geq 0$. Assume that equality holds; then $(X, L)$ is a conic bundle over a surface $S$, the conic bundle map $r: X \to S$ being given by $\Gamma(n(K_X + L))$, for some $n > 0$, via Stein factorization. Moreover the surface $S$ is smooth by [Be]. In general $r_A$, the conic bundle map restricted to $A$, is generically finite of degree 2, but in our case it is finite. Actually, let $D$ be a curve in $A$ contracted by $r$; then $(K_X + L)D = 0$ and so $K_A D = 0$, contradicting the ampleness of $K_A$. So $K_A{}^2$ is even, as $A$ is a double cover of $S$, and this implies that $k$ is even, since $K_A{}^2 = 3k^2$, which contradicts (4.0.2).

So $\mathcal{H}$ is nef and big. Set $\delta = \mathcal{H}L^2$ and note that $\delta = \deg L_C$, for $C \in |\mathcal{H}_A|$ a general element; then $\delta \geq 6$ due to (4.0.3) and the very ampleness of $L$. In fact we have

(4.3) THEOREM. $\mathcal{H}^3 = 1$ *and* $8 \leq \delta \leq 9$.

*Proof.* In view of (4.1), the generalized Hodge index theorem [BBS, (0.15.1) letting $M = L$, $N = \mathcal{H}$ and $j = 1$] provides the inequality

(4.3.1) $$\mathcal{H}^3\, \delta \leq (L\, \mathcal{H}^2)^2 = 9.$$

As already observed $\delta \geq 6$. So by (4.2), (4.3.1) we get both $\mathcal{H}^3 = 1$ and $\delta \leq 9$. It remains to show that it cannot be $\delta = 6$ or $7$. To do this we will argue by contradiction and first of all we show that

(4.3.2) $$k = 1 \quad \text{and} \quad g(C) = 4.$$

To see this, note that $|\mathcal{H}_C|$ endows $C$ with a $g^1{}_3$. So $C$ embedded by $|L_C|$ cannot be a plane curve. Hence $h^0(L_C) \geq 4$, but then by the Castelnuovo inequality and (4.0.1) we get $g(C) \leq 4$, hence $g(C) = 4$, by (4.0.3). Then (4.0.1) gives $k = 1$. Next we show that

(4.3.3) $$h^0(K_X) = h^1(K_X) = 0.$$

By (4.3.2) we have $K_X + L = \mathcal{H}$. So $K_X\, \mathcal{H}^2 = \mathcal{H}^3 - L\, \mathcal{H}^2 = 1 - 3 < 0$ and we conclude that $h^0(K_X) = 0$. Note that $K_X + L$ is nef by (4.1); hence by [S2, main theorem] there exists a divisor $D \in |\mathcal{H}|$. Note also that $(X, L)$ coincides with its reduction since $K_X + 2L = L + K_X + L$ is ample. Now consider the second reduction morphism $\varphi: X \to X''$ [BFS, (0.2)]. Since $K_A$ is ample by (4.0), the first part of the proof of (0.1) shows that $X''$ is smooth and $\varphi$ is simply the blow-up of $X''$ along a smooth 1-dimensional cycle $Z$. Then $L = \varphi^*L'' - \varphi^{-1}(Z)$, for some $L'' \in \mathrm{Pic}(X'')$; moreover $\mathcal{H}'' = K_{X''} + L''$ is ample. Let $D'' \in |\mathcal{H}''|$ be the image of $D$ on $X''$. Since $1 = \mathcal{H}^3 = \mathcal{H}''^3$ we conclude that $D''$ is irreducible and reduced. Now consider the exact cohomology sequence of

$$0 \to K_{X''} \to K_{X''} + \mathcal{H}'' \to K_{D''} \to 0.$$

Note that $K_{D''}\mathcal{H}''_{D''} = (K_{X''} + \mathcal{H}'')\mathcal{H}''^2 = (2\mathcal{H}'' - L'')\mathcal{H}''^2 = 2 - 3 < 0$, implying that $h^0(K_{D''}) = 0$. We also have $h^1(K_{X''} + \mathcal{H}'') = 0$, by the Kodaira vanishing theorem. This gives $h^1(K_{X''}) = 0$ and then $h^1(K_X) = 0$, in view of the birational invariance of $h^i(K_X)$. Now we prove that

(4.3.4) $|\mathcal{H}|$ contains a smooth surface $D$ and $h^0(K_D) = 0$.

Note that $K_A = \mathcal{H}_A$, by (4.3.2). By the exact cohomology sequence of

$$0 \to K_X \to \mathcal{H} \to K_A \to 0$$

and (4.3.3) we see that $\Gamma(\mathcal{H}) \to \Gamma(\mathcal{H}_A)$ is an isomorphism, hence $h^0(\mathcal{H}) \geq 3$. Suppose that $\mathcal{H}$ is not spanned.

(4.3.5) *Claim.* The base locus $\mathrm{Bs}\,|\mathcal{H}|$ of $|\mathcal{H}|$ consists of a single point $x$.

Note that $\mathrm{Bs}\,|\mathcal{H}|$ cannot contain a curve Z; otherwise $Z \cap A$ would be contained in the base locus $\mathrm{Bs}\,|\mathcal{H}_A|$, which is empty, since $\mathcal{H}_A$ is spanned, a contradiction. Therefore $\mathrm{Bs}\,|\mathcal{H}|$ is a finite set and then it consists of a single point $x$, since $\mathcal{H}^3 = 1$.

Now let $D, D_1, D_2$ be general elements of $|\mathcal{H}|$ and set $\Gamma = D_1 \cap D_2$. Then by Claim (4.3.5) and Bertini's theorem, $D$ is smooth off $x$; on the other hand $\Gamma D = \mathcal{H}^3 = 1$, so that $D$ is smooth also at $x$. Finally, by adjunction we have $K_D L_D = (2\mathcal{H} - L)\mathcal{H}L = 6 - \delta \leq 0$, which implies $h^0(K_D) = 0$ in view of the very ampleness of $L_D$. This proves (4.3.4).

We will get a contradiction by proving that $D$ is birational to a smooth quartic surface of $\mathbf{P}^3$. To see this first note that $C \in |L_D|$. Then, by the exact cohomology sequence of

$$0 \to K_D \to K_D + L_D \to K_C \to 0,$$

recalling (4.3.4) and (4.3.2), we get $h^0(K_D + L_D) = h^0(K_C) - h^1(K_D) \leq g(C) = 4$. Moreover $(K_D + L_D)^2 = 4\mathcal{H}^3 = 4$, so the adjunction mapping of $(D, L_D)$ has 2-dimensional image. Let $(D', L'_{D'})$ be the reduction of $(D, L_D)$. Since $L_D^2 = \delta \leq 7$, by [SV, (2.1)] we see that either $K_{D'} + L'_{D'}$ is very ample or $D'$ is a Del Pezzo surface with $K_{D'}^2 = 2$ and $L'_{D'} = -2K_{D'}$. In the latter case we would get $g(L_D) = g(L'_{D'}) = 3$, contradicting (4.3.2). In the former case we have that $|K_{D'} + L'_{D'}|$ embeds $D'$ in $\mathbf{P}^3$ as a surface of degree $(K_{D'} + L'_{D'})^2 = (K_D + L_D)^2 = 4$. This concludes the proof.

When $\delta = 8$ one can show that $k = 1$ or 3. The same argument used to rule out cases $\delta = 6,7$ in (4.3) works also for $(\delta, k) = (8,1)$. Though we cannot exclude the case $(\delta, k) = (8, 3)$, we can provide several strong restrictions on the invariants $d$ and $g$, but the arguments are too long to be reproduced here. We suspect that $\delta = 9$. The remainder of this section is devoted to showing that in case $\delta = 9$ the pair $(X, \mathcal{H})$ is exactly the Del Pezzo 3-fold considered in sec. 1. By applying (0.1) to our pair $(X, L)$ in case $\delta = 9$ we are able to prove the following

(4.4) THEOREM. *If* $\delta = 9$, *then* $K_X = -2\mathcal{H}, L = 3\mathcal{H}$, *with* $\mathcal{H}$ *ample, i.e.* $(X, \mathcal{H})$ *is the special Del Pezzo 3-fold of sec. 1.*

*Proof.* Of course our pair $(X, L)$ is the reduction of itself, since $K_X + 2L = L + K_X + L = L + k\mathcal{H}$ is ample. Moreover $K_X + L$ is nef and big by (4.1), (4.2) and $K_A$ is ample by (4.0). Letting $L = \mathcal{L}$ and recalling that $K_X + L = k\mathcal{H}$, the numerical invariants introduced in (0.1) become

$$d_1 = k\,\delta, \qquad d_2 = 3k^2, \qquad d_3 = k^3.$$

So, for $\delta = 9$ we get $d_2^2 = d_1 d_3$ and then (0.1) gives that $\mathcal{H}$ is ample and $3k^3\mathcal{H} = d_2(K_X + L) \equiv d_3 L = k^3 L$, i.e.

(4.4.1) $$L \equiv 3\mathcal{H}, \quad \mathcal{H} \text{ ample.}$$

Of course $\mathcal{H}$ cannot be spanned. Otherwise, being also ample and satisfying $\mathcal{H}^3 = 1$, we would get $(X, \mathcal{H}) = (\mathbf{P}^3, \mathcal{O}_{\mathbf{P}^3}(1))$ (e.g. see [LPS1, (0.6.1)]), which would imply $A \in |\mathcal{O}_{\mathbf{P}^3}(3)|$, contradicting (4.0). This also shows that $h^0(\mathcal{H}) \leq 3$. Otherwise it would be $\Delta(X, \mathcal{H}) = 4 - h^0(\mathcal{H}) \leq 0$, producing again the absurdity $(X, \mathcal{H}) = (\mathbf{P}^3, \mathcal{O}_{\mathbf{P}^3}(1))$ [F2, (5.10) and (4.12)]. Now look at the exact sequence

$$0 \to \mathcal{H} - L \to \mathcal{H} \to \mathcal{H}_A \to 0$$

and note that $h^i(\mathcal{H} - L) = h^{3-i}(K_X + L - \mathcal{H}) = 0$ for $i = 1, 2$, by the Kodaira vanishing theorem, since $L - \mathcal{H} \equiv 2\mathcal{H}$ is ample by (4.4.1). Since $h^0(\mathcal{H}_A) \geq 3$ we thus conclude that $h^0(\mathcal{H}) = h^0(\mathcal{H}_A) = 3$. But then Claim (4.3.5) applies and we conclude that $|\mathcal{H}|$ has a single base point $x$. Thus two general elements of $|\mathcal{H}|$ meet along a curve $\Gamma$, which is irreducible and smooth outside $x$, by Bertini's theorem; but since $\Gamma\mathcal{H} = \mathcal{H}^3 = 1$ we see that $\Gamma$ is in fact smooth. Furthermore, since $L$ is very ample and

$$\deg L_\Gamma = L\mathcal{H}^2 = 3,$$

we conclude that $g(\Gamma) = 0$ or $1$. On the other hand, recalling (4.4.1), we have

$$2g(\Gamma) - 2 = 2g(X, \mathcal{H}) - 2 = (K_X + 2\mathcal{H})\mathcal{H}^2 = (K_X + L - \mathcal{H})\mathcal{H}^2 = (k-1)\mathcal{H}^3 = k-1\,,$$

hence $k = 1$ and $g(\Gamma) = 1$, as $k > 0$. This implies that $K_X + L = \mathcal{H}$, and so $K_X \equiv -2\mathcal{H}$, by (4.4.1) again. But this means that $X$ is Fano, hence $K_X = -2\mathcal{H}$, as $\mathrm{Pic}(X) \cong H^2(X, \mathbf{Z})$. Moreover $L = 3\mathcal{H}$, and this concludes the proof.

## REFERENCES

BBS. Beltrametti, M., Biancofiore, A., Sommese, A.J., Projective n-folds of log-general type, I. Trans. Amer. Math. Soc., **314** (1989), 825-849.

BFS. Beltrametti, M., Fania, M.L., Sommese, A.J., On the adjunction theoretic classification of projective varieties. Math. Ann., **290** (1991), 31-62.

Be. Besana, G.M., On the geometry of quadric fibrations arising in adjunction theory, I. Preprint, 1991.

C. Cornalba, M., Una osservazione sulla topologia dei rivestimenti ciclici di varietà algebriche. Boll. Un.Mat.Ital., (3) **18**-A (1981), 323-328.

EL. Ein, L., Lazarsfeld, R., Global generation of pluricanonical and adjoint linear series on smooth projective threefolds. Preprint, 1992.

Fa1. Fania, M.L., Configuration of -2 rational curves on sectional surfaces of n-folds. Math. Ann., **275** (1986), 317-325.

Fa2. Fania, M.L., Trigonal hyperplane sections of projective surfaces. Manuscripta Math., **68** (1990), 17-34.

F1. Fujita, T., On the structure of polarized manifolds with total deficiency one, part III. J. Math. Soc. Japan, **32** (1984), 75-89.

F2. Fujita, T., *Classification Theories of Polarized Varieties*. London Math. Soc. Lect. Note Series, **155**, Cambridge Univ. Press, Cambridge, 1990.

KMM. Kawamata, Y., Matsuda, M., Matsuki, K., Introduction to the minimal model problem. Algebraic Geometry, Sendai, 1985, pp. 283-360. Advanced Studies in Pure Math., **10**, Kinokuniya, Tokyo, 1987.

Io. Ionescu, P., Embedded projective varieties of small invariants. Algebraic Geometry, Proc. Bucharest 1982, pp. 142-186. Lect. Notes Math. **1056**, Springer-Verlag, Berlin Heidelberg New York, 1984.

Is. Iskovskih, V.A., Fano 3-folds, I. Math. USSR Izvestija, **11** (1977), 485-527.

LPS1. Lanteri, A., Palleschi, M., Sommese, A.J., On the very ampleness of $K_X \otimes \mathcal{L}^{\dim X}$ for ample and spanned line bundles $\mathcal{L}$. Osaka J. Math., **26** (1989), 647-664.

LPS2. Lanteri, A., Palleschi, M., Sommese, A.J., Double covers of $\mathbf{P}^n$ as very ample divisors. Preprint, 1992.

LPS3. Lanteri, A., Palleschi, M., Sommese, A.J., Del Pezzo surfaces as hyperplane sections. In preparation.

La. Lazarsfeld, R., Some applications of the theory of positive vector bundles. Complete Intersections, Proc. Acireale 1983, pp. 29-61. Lect. Notes Math. **1092**, Springer-Verlag, Berlin Heidelberg New York, 1981.

Se. Serrano, F., Surfaces having a hyperplane section with a special pencil. Preprint, 1986.

S1. Sommese, A.J., Configuration of -2 rational curves on hyperplane sections of projective threefolds. Classification of Algebraic and Analytic Manifolds, pp. 465-497. Progress in Math., **39**, Birkhauser, Basel Boston, 1983.

S2. Sommese, A.J., On the nonemptiness of the adjoint linear system of a hyperplane section of a threefold. J. reine angew. Math., **402** (1989), 221-220.

SV. Sommese, A.J., Van de Ven, A., On the adjunction mapping. Math. Ann., **278** (1987), 593-603.

SzW. Szurek, M., Wisniewski, J.A., Fano bundles of rank 2 on surfaces. Compositio Math., **76** (1990), 295-305.

Antonio Lanteri & Marino Palleschi
Dipartimento di Matematica "F. Enriques", Università, Via C. Saldini 50, I-20133 Milano, Italy
*E-mail addresses*: lanteri@vmimat.mat.unimi.it
apelles@vmimat.mat.unimi.it

Andrew J. Sommese
Department of Mathematics, University of Notre Dame, Notre Dame, Indiana 46556, U.S.A.
*E-mail address*: sommese@nd.edu

Contemporary Mathematics
Volume **162**, 1994

# Component Numbers for Torsion Sections of Semistable Elliptic Surfaces

RICK MIRANDA

## Introduction

In this article we would like to address the following situation. Let $f : X \to C$ be an elliptic surface with section $S_0$. We further assume that $X$ is relatively minimal and smooth, and that all singular fibers are semistable, that is, are cycles of $\mathbb{P}^1$'s (type "$I_m$" in Kodaira's notation, see [**K**]). Given a section $S$ of $f$, it will meet one and only one component in every singular fiber. The fundamental aim of this paper is to try to say something specific about which components are being hit, in the case that $S$ is a torsion section (i.e., of finite order in the Mordell-Weil group of sections of $f$).

Most work on the Mordell-Weil group for elliptic surfaces focuses on properties of the group itself, (for example its rank, etc.) and not on properties of the elements of the group. See for example [**Shd1**] and [**C-Z**] for general properties, and papers such as [**Sch**] and [**Stl**] for computations in specific cases. Although the methods used in this article are elementary, they are suprisingly powerful, enabling us to obtain rather detailed information about the possibilities for component intersection for a torsion section.

To be specific, suppose that the $j^{th}$ singular fiber is a cycle of length $m_j$, and that the components are numbered cyclically around the cycle from 0 to $m_j - 1$, with the component numbered 0 being the one meeting the zero section $S_0$. (Note that there are two ways to do this, depending on the two orderings around the cycle.) Suppose that in this $j^{th}$ fiber, a section $S$ of order exactly $n$ hits the $k_j^{th}$ component. The main restrictions on these "component" numbers $k_j$ are the following:

---

1991 *Mathematics Subject Classification.* 14J27 secondary 14H52.
*Key words and phrases.* elliptic surfaces, Mordell-Weil, sections, torsion.
Research supported in part by the NSF under grant DMS-9104058
This paper is in final form and no version of it will be submitted elsewhere for publicaton.

**The Quadratic Relation** (Proposition 3.1):

$$\sum_j k_j(1 - k_j/m_j) = 2\chi(\mathcal{O}_X).$$

**The Component Number Sum Relation** (Corollaries 4.1 and 4.4):
If the component numbers for $S$ satisfy $k_j \leq m_j/2$, (which can always be arranged by the suitable "reorienting" of the cycle), then

$$\sum_{j=1}^{s} k_j(S) = \begin{cases} 4\chi(\mathcal{O}_X) & \text{if } n = 2 \\ 3\chi(\mathcal{O}_X) & \text{if } n \geq 3. \end{cases}$$

The Quadratic Relation as stated above follows directly (as a special case) from Shioda's construction of a height pairing on the Mordell-Weil group, see [**Shd2**, Theorem 8.6 and Table 8.16] and [**Shd3**, Proposition 3.5(iv)]. We give a direct proof which is not different from Shioda's in any essential way, but which is arranged to arrive at this one formula.

These relations hold for the general torsion section, but actually we have more to say in the case of a torsion section $S$ of prime order $p$. In this case we are able to calculate exactly the distribution of the "distances" between the components hit by $S$ and the components hit by the zero-section $S_0$, in the following sense. If a fiber has $m$ components, indexed from 0 to $m-1$ cyclically, then $S$, having order $p$, meets either the component indexed by 0 (which we take to be the component meeting $S_0$) or one of the components indexed by $im/p$ for some $i$. In other words, a section of order $p$ must meet either $1/p$ of the way around the cycle, or $2/p$ of the way around, etc. Denote by $M_{i,p}(S)$ the fraction of the total Euler number $e = \sum_j m_j$ contributed by those fibers in which $S$ meets $i/p$ of the way around the cycle, in either direction. Our main result is a computation of these numbers, which turn out to be independent of $S$, and surprisingly, independent of $i$ (as long as $i \neq 0$):

**The Distribution Numbers** (Theorem 6.3):
Let $S$ be a torsion section of odd prime order $p$. Then

$$M_{i,p}(S) = 2p/(p^2 - 1) \text{ if } 1 \leq i \leq (p-1)/2,$$

and

$$M_{0,p}(S) = 1/(p+1).$$

Thus we obtain an "equidistribution" of these distances: the total sum of those $m_j$'s contributed from fibers of type $I_{m_j}$ in which $S$ meets $i/p$ of the way around the cycle is constant, independent of $i$.

Thus a torsion section of prime order, as it travels across the elliptic surface, is forced to be completely "fair" in its choice of which components it may hit, in the above sense.

The above results generalize various statements concerning the component numbers $\{k_j\}$ for a torsion section which appeared in [**M-P1**], [**M-P2**], [**M-P3**], and [**M**].

The results in this paper could be proved by appealing to the appropriate elliptic modular surface (for those primes where such exists) and its universal properties. The equidistribution property is invariant under base change, so if it is true for the (universal) modular surface, it will be true in general. This is a computation involving the modular group, which could be made once and for all. However it is my intention to show that these properties are somehow more elementary than the theory of modular surfaces, and follow from straightforward considerations involving only basic facts concerning the Mordell-Weil group and a little intersection theory of surfaces.

In a further paper we hope to discuss the case of torsion of non-prime order more fully.

I would like to thank Igor Dolgachev, Jeanne Duflot, Ulf Persson, Miles Reid, Tetsuji Shioda, and Peter Stiller for some useful conversations. I would also like to thank the organizers of the Conference in L'Aquila for their kind invitation to allow me to speak on this subject. I would especially like to thank Prof. E. Laura Livorni for her wonderful efforts to make the conference such a success, and Prof. Ciro Ciliberto for his excellent hospitality in Rome while this work was completed.

## 1. Basic Facts

We work over the field of complex numbers. Let $C$ be a smooth curve, and let $f : X \to C$ be a smooth semistable elliptic surface over $C$. We suppose that $f$ has a chosen section $S_0$ which we treat as the zero for the group law on the sections of $f$. Let $g$ be the genus of $C$, and let $\chi = \chi(\mathcal{O}_X)$ be the holomorphic Euler characteristic of $X$. If $e$ is the topological Euler number of $X$, then $e = 12\chi$.

We suppose that the map $f$ has $s$ singular fibers $F_1, \dots, F_s$, each semistable, i.e., each a cycle of rational curves (of type "$I_m$" in Kodaira's notation [**K**]). Indeed, let us say that the fiber $F_j$ is of type $I_{m_j}$. Therefore

$$e = 12\chi = \sum_{j=1}^{s} m_j.$$

Choose an "orientation" of each fiber $F_j$ and write the $m_j$ components of $F_j$ as

$$C_0^{(j)}, C_1^{(j)}, \dots, C_{m_j-1}^{(j)},$$

where the zero section $S_0$ meets only $C_0^{(j)}$ and for each $k$, $C_k^{(j)}$ meets only $C_{k\pm1}^{(j)}$ mod $m_j$. If $m_j = 1$, then $F_j = C_0^{(j)}$ is a nodal rational curve, of self-intersection 0. If $m_j \geq 2$, then each $C_k^{(j)}$ is a smooth rational curve with self-intersection $-2$.

If we denote by $\mathrm{NS}(X)$ the Neron-Severi group of the elliptic surface $X$, we can consider the sublattice $L \subseteq \mathrm{NS}(X)$ generated by the zero-section $S_0$ and the components $C_k^{(j)}$ of the singular fibers. The class of the fiber $F$ is of course in this sublattice $L$; moreover, the only relations among these classes are that

$$(1.1) \qquad F \equiv \sum_{k=0}^{m_j-1} C_k^{(j)} \text{ for each } j = 1, \ldots, s.$$

If we let $U$ denote the sublattice of $L$ spanned by $F$ and $S_0$, then $U$ is a unimodular lattice of rank 2, isomorphic to a hyperbolic plane, and so splits off $L$. Its orthogonal complement $R$ is freely generated by the components $C_k^{(j)}$ for $j = 1 \ldots s$ and $k \neq 0$. $R$ is therefore a direct sum of $s$ root lattices of type $A$, with the $j^{th}$ summand isomorphic to $A_{m_j-1}$, generated by $C_k^{(j)}$ for $k \neq 0$.

The Shioda-Tate formula for the Mordell-Weil group $\mathrm{MW}(X)$ of sections of $X$ is derived from the exact sequence (see [**Shd1**])

$$0 \to L \to \mathrm{NS}(X) \to \mathrm{MW}(X) \to 0$$

where the first map is the inclusion of the sublattice $L$ into $\mathrm{NS}(X)$. The second map is the fiber-by-fiber summation map, sending a divisor class $D \in \mathrm{NS}(X)$ to the closure of the sum of the points of $D$ on the generic fiber of $X$. We therefore obtain information about both the rank and the torsion of the Mordell-Weil group $\mathrm{MW}(X)$. If we denote by $\rho$ the Picard number of $X$, which is the rank of the Neron-Severi group, we see that since $L$ has rank equal to $2 + \sum_{j=1}^{s}(m_j - 1)$,

$$\operatorname{rank} \mathrm{MW}(X) = \rho - 2 - \sum_{j=1}^{s}(m_j - 1) = s + \rho - 2 - e.$$

The torsion $\mathrm{MW}_{tor}(X)$ in the Mordell-Weil group corresponds to those classes in $\mathrm{NS}(X)$ for which some multiple lies in the sublattice $L$. These classes form an intermediate lattice $L^{sat}$ containing $L$ as a sublattice of finite index, and we see that

$$\mathrm{MW}_{tor}(X) \cong L^{sat}/L.$$

For any lattice $N$, denote by $N^{\#}$ the dual lattice $\mathrm{Hom}_{\mathbb{Z}}(N, \mathbb{Z})$. The inclusion of $L$ into $L^{sat}$ gives the sequence of inclusions

$$L \subseteq L^{sat} \subseteq (L^{sat})^{\#} \subseteq L^{\#}$$

which shows that the quotient group $L^{sat}/L$ is isomorphic to a subgroup of the discriminant-form group $G_L = L^{\#}/L$. Note that $L^{\#} = U^{\#} \oplus R^{\#}$ since $L = U \oplus R$ and since $U^{\#} = U$ ($U$ is unimodular), we have $G_L \cong R^{\#}/R = \bigoplus_{j=1}^{s} A_{m_j-1}^{\#}/A_{m_j-1}$. A computation shows that for the root lattices of type $A$, $A_m^{\#}/A_m \cong \mathbb{Z}/m\mathbb{Z}$; therefore we have that the torsion part $\mathrm{MW}_{tor}(X)$ of the Mordell-Weil group of sections of $X$ is isomorphic to a subgroup of $\bigoplus_{j=1}^{s} \mathbb{Z}/m_j\mathbb{Z}$. (See for example [**M-P1**] or [**M**].)

A choice of orientation for each singular fiber $F_j$ gives an identification of the cyclic group of components $C_k^{(j)}$ for $k = 0, \ldots, m_j - 1$ with $\mathbb{Z}/m_j\mathbb{Z}$. For any section $S$ of $f$, and any singular fiber $F_j$, denote by $k_j(S)$ the index of the component of $F_j$ which $S$ meets. Thus

(1.2) $$S \cdot C_l^{(j)} = \delta_{lk_j(S)}$$

where $\delta$ above is the Kronecker delta. Note that with our notation above, $k_j(S_0) = 0$ for every $j$. The numbers $\{k_j(S)\}$ will be called the *component numbers* of the section $S$. The upshot of the remarks above is that, for torsion sections, this assignment of integers $\{k_j(S)\}$ to $S$ is 1-1. Thus a torsion section is determined by its component numbers (but not all sets of components numbers can occur).

It is the purpose of this article to study the properties of the component numbers of torsion sections for semistable elliptic surfaces. Although it is clear from the above discussion that the component numbers $k_j(S)$ are well-defined modulo $m_j$, it is more useful for our purposes to take them to be integers in the range $0, \ldots, m_j - 1$.

Two facts are necessary for what follows; they can be found in [**M**] as well as the other basic references.

LEMMA 1.3.
*(1.3.1) If $S_1$ and $S_2$ are two different sections in* MW$(X)$ *with $S_1 - S_2$ torsion, then $S_1 \cdot S_2 = 0$, i.e., they are disjoint.*
*(1.3.2) If $S$ is any section in* MW$(X)$*, then $S \cdot S = -\chi$.*

## 2. The Divisor Class of a Torsion Section

Let $S$ be a torsion section of the semistable elliptic surface $f : X \to C$. In this section we wish to write down the class of $S$ in NS$(X)$.

Since the class of $S$ lies in $L^{sat}$, we have that $S$ is a $\mathbb{Q}$-linear combination of the zero section $S_0$, and the fiber components $C_k^{(j)}$. Because (1.1) is the only relation among these classes, the classes $S_0$, $F$, and $C_k^{(j)}$ for $j = 1, \ldots, s$ and $k \neq 0$ form a basis for the module spanned by all these classes.

For fixed $j = 1, \ldots, s$ and $k = 1, \ldots, m_j - 1$, set

$$\begin{aligned} D_k^{(j)} &= (m_j - k)\sum_{i=1}^{k} iC_i^{(j)} + k \sum_{i=k+1}^{m_j-1} (m_j - i)C_i^{(j)} \\ &= (m_j - k)[C_1^{(j)} + 2C_2^{(j)} + \cdots + kC_k^{(j)}] + \\ &\quad k[(m_j - k - 1)C_{k+1}^{(j)} + \cdots + 2C_{m_j-2}^{(j)} + C_{m_j-1}^{(j)}]. \end{aligned}$$

We set $D_0^{(j)} = 0$.

Note that

(2.1) $$D_k^{(j)} \cdot S_0 = 0$$

for every $k$.

LEMMA 2.2. *Fix $j = 1, \ldots, s$, and indices $k$ and $l$ with $0 \le k, l \le m_j - 1$. Then*

$$D_k^{(j)} \cdot C_l^{(j)} = \begin{cases} m_j & \text{if } l = 0 \text{ and } k \neq 0 \\ -m_j & \text{if } l = k \text{ and } k \neq 0 \\ 0 & \text{otherwise.} \end{cases}$$

PROOF. This is a straightforward check; let us first notice that if $k = 0$, then $D_k^{(j)} = 0$, so certainly the intersection product is 0. Assume then that $k \neq 0$. If $l = 0$, then since $C_0^{(j)}$ meets only $C_1^{(j)}$ and $C_{m_j-1}^{(j)}$, each once, we have

$$D_k^{(j)} \cdot C_0^{(j)} = (m_j - k)[1] + (k)[1] = m_j,$$

which proves the first statement. If $l = k$, then $C_k^{(j)}$ meets only $C_{k\pm1}^{(j)}$ (each once) and itself (-2 times), so that

$$D_k^{(j)} \cdot C_k^{(j)} = (m_j - k)[(k-1)(1) + (k)(-2)] + (k)[(m_j - k - 1)(1)] = -m_j,$$

proving this case. We leave to the reader to check that all other intersection products are 0, as stated. $\square$

It is useful to note that the above lemma can be re-expressed using the Kronecker delta function as

$$D_k^{(j_1)} \cdot C_l^{(j_2)} = m_j \delta_{j_1 j_2} (1 - \delta_{k0})(\delta_{l0} - \delta_{lk}). \tag{2.3}$$

THEOREM 2.4. *With the above notation, if $S$ is a torsion section, then*

$$S \equiv S_0 + (S - S_0 \cdot S_0)F - \sum_{j=1}^{s} \frac{1}{m_j} D_{k_j(S)}^{(j)}.$$

PROOF. Since the intersection form on $L$ is nondegenerate, we need only check that both sides of the equation intersect the generators $S_0$, $F$, and the $C_l^{(j)}$ in the same number. Let us begin with $S_0$. Using (2.1) and the equations

$$S \cdot F = S_0 \cdot F = 1,$$

we have that the right hand side intersects $S_0$ to $S \cdot S_0$, agreeing with the left hand side. Similarly, $F$ meets only the $S_0$ term on the right hand side, so the intersections with $F$ are also equal.

Now fix indices $j = 1, \ldots, s$ and $l = 0, \ldots, m_j - 1$, and let us check the intersection with $C_l^{(j)}$; Lemma 2.2 is the critical part of the computation. Intersecting with the right hand side gives $S_0 \cdot C_l^{(j)} - \frac{1}{m_j} D_{k_j(S)}^{(j)} \cdot C_l^{(j)}$, which reduces to $\delta_{l0} - (1 - \delta_{k_j(S)0})(\delta_{l0} - \delta_{lk_j(S)})$ using (2.3). This is after some simplification equal to $\delta_{lk_j(S)}$, which is also $S \cdot C_l^{(j)}$ by (1.2). $\square$

The reader familiar with [**C-Z**] will recognize the above expression as the "correction" term in their pairing on the Mordell-Weil group. They were more interested in questions of rank in that paper, and less so in the torsion sections.

The formula for the linear equivalence class of $S$ given above can be simplified somewhat in the case that $S$ is not the zero section $S_0$. Then $(S \cdot S_0) = 0$ and $(S_0 \cdot S_0) = -\chi$ by Lemma 1.3, so we have

$$S \equiv S_0 + \chi F - \sum_{j=1}^{s} \frac{1}{m_j} D^{(j)}_{k_j(S)}. \tag{2.5}$$

## 3. The Quadratic Relation for the Component Numbers

Let us now take the above formula for the torsion section $S$ and intersect it with $S$ itself. If $S = S_0$, then of course all $k_j(S) = 0$, both sides of the equation are $S_0$, so we recover no information. However if $S \neq S_0$, the formula (2.5) is nontrivial. Dotting $S$ with the right-hand side gives

$$\chi - \sum_{j=1}^{s} \frac{1}{m_j} D^{(j)}_{k_j(S)} \cdot S.$$

Since $S$ meets only the curve $C^{(j)}_{k_j(S)}$ in the $j^{th}$ singular fiber, and it meets it exactly once, we have

$$D^{(j)}_{k_j(S)} \cdot S = (m_j - k_j(S))k_j(S).$$

Therefore the above reduces to

$$\chi - \sum_{j=1}^{s} k_j(S)(1 - \frac{k_j(S)}{m_j}).$$

Finally, using Lemma 1.3.2, dotting the left-hand side with $S$ gives $-\chi$. Hence we obtain the following formula, which we call the *quadratic relation* for the component numbers.

PROPOSITION 3.1. *Let $S$ be a torsion section of $f$, not equal to the zero section $S_0$. Then*

$$\sum_{j=1}^{s} k_j(S)(1 - \frac{k_j(S)}{m_j}) = 2\chi(\mathcal{O}_X).$$

Note that the quadratic relation is independent of the choice of orientation of each singular fiber, as it should be. (If one reverses the orientation of $F_j$, then $k_j$ is replaced by $m_j - k_j$ if $k_j \neq 0$, so that the two terms of each summand are simply switched.)

As we noted in the Introduction, the quadratic relation is also a consequence of Shioda's general theory of the height pairing on the Mordell-Weil group. Miles Reid has pointed out to me that the quadratic relation given above can be derived directly via the use of the Riemann-Roch theorem for Weil divisors, as expounded in [**R**, section 9]. One should contract all components of fibers except $C_0^{(j)}$, obtaining a surface with only $A_{m_j-1}$ singularities, and consider on that surface the Weil divisor $S$. The Riemann-Roch formula for $S$ has as corrections to the usual Riemann-Roch the terms $k_j(S)(1-\frac{k_j(S)}{m_j})$. (See also [**G**, section 2].) This approach is cleaner and perhaps conceptually simpler; however in the spirit of keeping things as elementary as possible, I have decided to present this "low road" version. The interested reader will have no problem making the computation as Reid suggests and re-deriving the quadratic relation in this manner.

## 4. The Component Number Sums

In this section we will develop formulas for the component number sums $\sum_j k_j(S)$ for a torsion section $S$. Such formulas follow rather directly from the quadratic relation for the component numbers. Suppose first that $S$ has order 2. Then for each $j$ with $k_j(S) \neq 0$, we must have $m_j$ even and $k_j(S) = m_j/2$. Therefore the quadratic relation reduces to the following.

COROLLARY 4.1. *Let $S$ be an order 2 section of $f$. Then*

$$\sum_{j=1}^{s} k_j(S) = 4\chi(\mathcal{O}_X).$$

Note that in the order 2 case, each $k_j$ is either 0 or $m_j/2$, and hence the component numbers for a torsion section of order 2 are independent of the orientation of the fibers. This is clearly not true for other values of the component numbers. In general, reversing the orientation of the fiber $F_j$ changes the component number from $k_j$ to $m_j - k_j$. Hence the analogue of Corollary (4.1) for torsion sections of order at least 3 must take this into account.

To this end define a function $d_m : \{0, \dots, m-1\} \to \{0, \dots, [m/2]\}$ by setting

$$d_m(k) = \min\{k, m-k\}.$$

Note that for any given torsion section $S$, it is possible to choose the orientations of the fibers so that the component numbers $k_j(S)$ are minimal, that is, $k_j(S) = d_{m_j}(k_j(S))$ for each $j$. If this condition holds, we say that the section $S$ has *minimal component numbers.*

Next suppose that a torsion section $S$ has order $n \geq 3$. Then both $S$ and $2S$ are nonzero torsion sections of $X$. If for each index $j$, we choose the orientation of the components so that $0 \leq k_j(S) \leq m_j/2$, then we have the formulas

$$k_j(2S) = \begin{cases} 0 \text{ if } k_j(S) = m_j/2, \text{ and} \\ 2k_j(S) \text{ if } k_j(S) < m_j/2 \end{cases} \tag{4.2}$$

for every $j$. Therefore applying Proposition 3.1 to $2S$, after dividing by 2 we obtain

$$\sum_{\substack{j \text{ with} \\ k_j(S)<m_j/2}} k_j(S)(1-2\frac{k_j(S)}{m_j}) = \chi(\mathcal{O}_X).$$

Multiplying the formula of Proposition 3.1 by 2 gives

$$\sum_{\substack{j \text{ with} \\ k_j(S)=m_j/2}} k_j(S) + \sum_{\substack{j \text{ with} \\ k_j(S)<m_j/2}} k_j(S)(2-2\frac{k_j(S)}{m_j}) = 4\chi(\mathcal{O}_X),$$

and subtracting the above two equations yields the following.

COROLLARY 4.3. *Let $S$ be a section of order $n \geq 3$ with minimal component numbers $\{k_j(S)\}$. Then*

$$\sum_{j=1}^{s} k_j(S) = 3\chi(\mathcal{O}_X).$$

In case the orientations of the fibers are not chosen so as to give $S$ minimal component numbers, a similar formula holds, expressed with the $d_m$ function.

COROLLARY 4.4. *Let $S$ be a section of order $n \geq 3$. Then*

$$\sum_{j=1}^{s} d_{m_j}(k_j(S)) = 3\chi(\mathcal{O}_X).$$

Using these component number sum relations, one can re-express the quadratic relation in a simpler form.

COROLLARY 4.5. *Let $S$ be a section of order $n \geq 2$, with minimal component numbers $\{k_j = k_j(S)\}$. Then*

$$\sum_{j=1}^{s} k_j^2/m_j = \begin{cases} \chi(\mathcal{O}_X) & \text{if } n \geq 3 \\ 2\chi(\mathcal{O}_X) & \text{if } n = 2 \end{cases}$$

PROOF. The quadratic relation can be written as

$$\sum_j k_j^2/m_j = \sum_j k_j - 2\chi(\mathcal{O}_X),$$

and so the result follows from Corollaries 4.1 and 4.4 □

These results generalize a certain divisibility statement concerning the component numbers obtained in [**M-P1**, Proposition 4.6] (see also [**M-P2**, Lemma 3(a)] and [**M**, X.3.1]). Although the above cited works deal with elliptic K3 surfaces, the statement holds in general. It is:

$$\sum_{j=1}^{s} \frac{m_j-1}{2m_j} k_j^2 \quad \text{is an integer.}$$

This was proved using lattice-theoretic arguments involving the discriminant-form group $L^{\#}/L$.

We can recover this result as follows. First note that in the above sum, if any $k_j$ is replaced by $m_j - k_j$, the sum changes by an integer; hence it suffices for us to derive the divisibility result using minimal component numbers.

Now the above sum in question can be rewritten as

$$\sum_j k_j^2/2 - \sum_j k_j^2/2m_j.$$

Note that $\sum_j k_j^2$ has the same parity as $\sum_j k_j$, so that the first term $\sum_j k_j^2/2$ is equal to $\sum_j k_j/2$ modulo $\mathbb{Z}$. Now modulo $\mathbb{Z}$, we have

$$\begin{aligned} \sum_j k_j^2/2 - \sum_j k_j^2/2m_j &= \sum_j k_j/2 - \sum_j k_j^2/2m_j \\ &= \begin{cases} 4\chi/2 - 2\chi/2 & \text{if } n = 2 \\ 3\chi/2 - \chi/2 & \text{if } n \geq 3 \end{cases} \\ &= \chi \end{aligned}$$

which is an integer.

## 5. The Distribution Numbers for a Torsion Section

In this section we will apply the relations developed in the previous sections for the component numbers of a non-zero torsion section $S$ to its multiples $\alpha S$. To this end let us express the quadratic relation by using the following notation. For each singular fiber $F_j$, denote by $f_j(S)$ the fraction $k_j(S)/m_j$. Let $P(x) = x(1-x)$ for real numbers $x$. Then the quadratic relation can be expressed as

$$\sum_{j=1}^{s} P(f_j(S))\, m_j = 2\chi(\mathcal{O}_X). \tag{5.1}$$

Note again that this formula is independent of the choice of orientation of the fibers, since changing the orientation of the components of $F_j$ simply switches the two factors $f_j$ and $1 - f_j$ in each $P$ term of the above sum.

Now note that if one fixes orientations for all of the singular fibers, and if a torsion section $S$ has component numbers $\{k_j(S)\}$ using these orientations, then for any $\alpha$, the component numbers for the multiple $\alpha S$ of $S$ in these orientations will be

$$k_j(\alpha S) \equiv \alpha k_j(S) \mod m_j.$$

Let $\langle x \rangle$ denote the smallest non-negative real number in the residue class modulo $\mathbb{Z}$ of a real number $x$; we have $\langle x \rangle \in [0, 1)$ always. The above formula can be expressed using the fractions $f_j$ by

$$f_j(\alpha S) = \langle \alpha f_j(S) \rangle.$$

Hence the quadratic relation (5.1) applied to the section $\alpha S$ is

$$\sum_{j=1}^{s} P(\langle \alpha f_j(S) \rangle) m_j = 2\chi(\mathcal{O}_X). \tag{5.2}$$

We want to reorganize this sum to collect terms with the same $f_j$. Let $S$ be any torsion section whose order divides $n$. Then $nS$ is the zero-section $S_0$, all of whose component numbers are zero; hence for each $j$, $nk_j(S)$ is divisible by $m_j$. In terms of the fractions $f_j$, this is equivalent to having $f_j(S)$ being a multiple of $1/n$ for every $j$.

Define rational numbers $M'_{i,n}(S)$ for $0 \leq i \leq n-1$ by the formula

$$M'_{i,n}(S) = \Big( \sum_{\substack{j \text{ with} \\ f_j(S)=i/n}} m_j \Big) / 12\chi.$$

Roughly speaking, $M'_{i,n}(S)$ is the fraction of the total sum $\sum_j m_j = 12\chi$ contributed by fibers where $S$ meets the component which is "distance $i$" from the component meeting the zero-section $S_0$. (This distance is measured in units of $m_j/n$, and is in the direction of the orientation of the fiber.) We will call these fractions $M'_{i,n}(S)$ the *oriented distribution numbers* for the section $S$. Obviously

$$\sum_{i=0}^{n-1} M'_{i,n}(S) = 1.$$

Reorganizing the quadratic relation (5.2) for $\alpha S$ to sum over the $i$'s gives

$$\sum_{i=0}^{n-1} \sum_{\substack{j \text{ with} \\ f_j(S)=i/n}} P(\langle \alpha i/n \rangle) m_j = 2\chi(\mathcal{O}_X),$$

which, after dividing through by $12\chi(\mathcal{O}_X)$, we rewrite using the oriented distribution numbers as

$$\sum_{i=0}^{n-1} P(\langle \alpha i/n \rangle) M'_{i,n}(S) = 1/6.$$

If we disregard orientation, we are led to defining rational numbers $M_{i,n}(S)$ for $0 \leq i \leq [n/2]$ by

$$M_{i,n}(S) = \begin{cases} M'_{0,n}(S) & \text{if } i = 0, \\ M'_{i,n}(S) + M'_{n-i,n}(S) & \text{if } 1 \leq i < n/2, \text{ and} \\ M'_{n/2,n}(S) & \text{if } i = n/2. \end{cases}$$

Note that $M_{i,n}(S)$ is the fraction of the total sum $\sum_j m_j = 12\chi$ contributed by fibers where $S$ meets the component which is "distance $i$" from the component meeting the zero-section $S_0$, in either of the two orientations. We therefore refer to the $M_{i,n}(S)$ as the *unoriented distribution numbers* for the section $S$. As with the oriented distribution numbers, we have

$$(5.3) \qquad \sum_{i=0}^{[n/2]} M_{i,n}(S) = 1.$$

Since $\langle -x \rangle = 1 - \langle x \rangle$, and $\langle \ell + x \rangle = \langle x \rangle$ for integers $\ell$, we see that $\langle \alpha(n-i)/n \rangle = 1 - \langle \alpha i/n \rangle$. Therefore since the polynomial $P(x)$ has the same value at $x$ as at $1-x$, the quadratic relation written in terms of the oriented distribution numbers has the coefficient of $M'_{i,n}(S)$ equal to the coefficient of $M'_{n-i,n}(S)$. Hence we may combine these terms, obtaining the following relation for the unoriented distribution numbers. (Note that the $i = 0$ term has been dropped, since it is zero.)

LEMMA 5.4. *If $S$ is a torsion section of order exactly $n$, and $\alpha$ is not divisible by $n$, then*

$$\sum_{i=1}^{[n/2]} P(\langle \alpha i/n \rangle) M_{i,n}(S) = 1/6.$$

Note that one gets the same equation relating the unoriented distribution numbers $M_{i,n}(S)$ using either $\alpha$ or $n - \alpha$. Hence the above set of equations are, a priori, only $[n/2]$ different equations.

One can express the above set of equations in matrix form. Let $\mathbf{P}_n$ be the square matrix of size $[n/2]$ whose $\alpha i^{th}$ entry is $P(\langle \alpha i/n \rangle)$. Let $\mathbf{1}_n$ be a vector of length $[n/2]$, which has every coordinate equal to 1. Finally let $\mathbf{M}_n(S)$ be the column vector of $M_{i,n}(S)$'s, for $1 \leq i \leq [n/2]$. The equations of 5.4 can then be expressed in matrix form as follows.

LEMMA 5.5. *If $S$ is a nonzero torsion section of order exactly $n$, then*

$$\mathbf{P}_n \mathbf{M}_n(S) = (1/6)\mathbf{1}_n.$$

For example, suppose that $n = 2$, so that we are discussing a 2-torsion section $S$ of $X$. The linear system above reduces to the single equation $(1/2)M_{1,2}(S) = 1/6$, so that $M_{1,2}(S) = 1/3$ and therefore $M_{0,2}(S) = 2/3$ by (5.3).

In the case of $n = 3$, the linear system again reduces to the single equation $(2/9)M_{1,3}(S) = 1/6$, so that $M_{1,3}(S) = 3/4$ and therefore $M_{0,3}(S) = 1/4$ by (5.3).

Note that these distribution numbers are independent of $S$! This is our result in general for $p$-torsion sections, which will be discussed in the next section.

As a last example, suppose that $n = 5$. Then the linear system above is the two equations $(4/25)M_{1,5}(S) + (6/25)M_{2,5} = 1/6$ and $(6/25)M_{1,5}(S) + (4/25)M_{2,5} = 1/6$ which leads to $M_{1,5}(S) = M_{2,5}(S) = 5/12$, and $M_{0,5}(S) = 1/6$. Note again that the distribution numbers are independent of $S$, as we have seen in the previous examples. However here we see more: the distribution numbers $M_{i,5}$

for $i \neq 0$ are *equal*. This is the "equidistribution" property mentioned in the Introduction, and is our main result for general primes $p$.

## 6. Equidistribution for Torsion Sections of Prime Order

In this section we will use the linear system describing the unoriented distribution numbers $M_{i,n}(S)$ to compute these numbers for a torsion section $S$ of prime order $p$.

Our first task is to show that the matrix $\mathbf{P}_p$ is invertible, which therefore shows that the distribution numbers are determined by the linear system of Lemma 5.5.

PROPOSITION 6.1. *Fix an odd prime $p$. Let $\mathbf{P}_p$ be the square matrix of size $(p-1)/2$ whose $\alpha\, i$ entry is $P(\langle \alpha i/p\rangle)$. Then $\mathbf{P}_p$ is invertible.*

PROOF. Let $V$ be $\mathbb{C}^p$, with coordinates indexed by residue classes modulo $p$. A vector $v \in V$ will have as its $k^{th}$ coordinate the number $v_k$. The group of units $G = (\mathbb{Z}/p)^\times$ acts on $V$, by the formula

$$(a \cdot v)_k = v_{ak \mod p}$$

for $a \in G$.

Define vectors $S_\alpha \in V$ by setting $(S_\alpha)_k = P(\langle \alpha k/p\rangle)$. We obviously have $S_\alpha = S_{p-\alpha}$. Also note that $(S_\alpha)_0 = 0$ for every $\alpha$, and $(S_\alpha)_{p-k} = (S_\alpha)_k$ for every $\alpha$ and $k$.

Therefore $S_\alpha$ lies in the subspace $V^+$ defined by

$$V^+ = \{v \in V \mid v_0 = 0 \text{ and } v_{p-k} = v_k \text{ for all } k\}.$$

The dimension of $V^+$ is $(p-1)/2$. Moreover $V^+$ is stable under the $G$-action, so that we can consider $G$ to be acting on $V^+$ also, with the same formula. Further note that $a \cdot S_1 = S_a$ for every $a \in G$.

Note that the $\alpha^{th}$ row of $\mathbf{P}_p$ consists of the coordinates $(S_\alpha)_k$ for $k = 1, \dots, (p-1)/2$; hence to prove that $\mathbf{P}_p$ is invertible, it suffices to show that the vectors $S_\alpha$, for $\alpha = 1, \dots, (p-1)/2$, are independent. Equivalently, we can show that the vectors $\{S_\alpha\}$ span $V^+$.

Let $\chi : G \to \mathbb{C}^\times$, be a character of $G$. For $v \in V^+$, define

$$\omega_\chi(v) = \sum_{a\in G} \chi(a)\, a \cdot v.$$

Note that $\omega_\chi(v)$ is an eigenvector of the action of $G$ on $V^+$, with eigenvalue $\chi^{-1}$, that is, $a \cdot \omega_\chi(v) = \chi^{-1}(a)\omega_\chi(v)$.

Let us focus in on the special element

$$S_\chi = \omega_\chi(S_1) = \sum_{a\in G} \chi(a) S_a.$$

Since this vector is a linear combination of the $S_a$'s, to show that the $S_a$'s span $V^+$, it suffices to show that the $S_\chi$'s do.

We focus attention on the *even* characters $\chi$, that is, those characters with $\chi(-1) = 1$. (If $\chi(-1) = -1$, $\chi$ is called *odd*.) There are exactly $(p-1)/2$ even characters for $G$, and we will finish the proof if we can show that for the $(p-1)/2$ even characters $\chi$, the vectors $S_\chi$ are non-zero; if so, they will be linearly independent, since they are eigenvectors for distinct characters. Hence they will span $V^+$.

In fact we need only show that the first coordinate of each $S_\chi$ (for $\chi$ even) is non-zero. This is the number

$$s_\chi = (S_\chi)_1 = \sum_{a\in G} \chi(a)P(\langle a/p\rangle).$$

If $\chi$ is the trivial character which is identically 1, then $s_\chi$ is obviously non-zero, so we may assume $\chi$ is one of the $(p-3)/2$ nontrivial even characters of $G$.

Let $\mathbf{B}_k(X)$ denote the $k^{th}$ Bernoulli polynomial. In particular, we have

$$\mathbf{B}_2(X) = X^2 - X + 1/6 = 1/6 - P(X), \text{ or } P(X) = 1/6 - \mathbf{B}_2(X) = \mathbf{B}_2(0) - \mathbf{B}_2(X).$$

Therefore the number $s_\chi$ can be written as

$$s_\chi = \sum_{a\in G} \chi(a)[\mathbf{B}_2(0) - \mathbf{B}_2(\langle a/p\rangle)],$$

and since $\chi$ is nontrivial, $\sum_{a\in G}\chi(a) = 0$, so that

$$s_\chi = -\sum_{a\in G} \chi(a)\mathbf{B}_2(\langle a/p\rangle).$$

Now the definition of the generalized Bernoulli numbers $B_{k,\chi}$ according to Leopoldt (see [**L**, page 37], or [**W**, Proposition 4.1]), gives

$$B_{k,\chi} = p^{k-1}\sum_{a=0}^{p-1} \chi(a)\mathbf{B}_k(\langle a/p\rangle);$$

hence we have

$$s_\chi = \frac{-1}{p}B_{2,\chi}$$

for an even nontrivial character $\chi$ on $G$.

Now the classical theorem that $B_{k,\chi}$ is nonzero when $k$ and $\chi$ have the same parity shows that for every even character $\chi$, $s_\chi \neq 0$, finishing the proof. This theorem is proved by using the associated $L$-series

$$L(s,\chi) = \sum_{n=1}^{\infty} \frac{\chi(n)}{n^s}$$

and noting first that for $n \geq 1$, $B_{n,\chi} = -nL(1-n,\chi)$ (see [**I-R**, Proposition 16.6.2] or [**W**, Theorem 4.2]). Hence it suffices to show that for even $\chi$, $L(1-n,\chi)$

is nonzero. This follows from considering the functional equation for $L$, which for even $\chi$ can be written as (see [**W**, page 29])

$$L(1-s,\chi) = \Gamma(s)\frac{2}{\tau(\chi)}\Big(\frac{p}{2\pi}\Big)^{s} L(s,\overline{\chi})$$

where $\tau(\chi) = \sum_{a=1}^{p} \overline{\chi}(a)e^{2\pi i a/p}$ is a Gauss sum, and is known to be non-zero (it has absolute value $\sqrt{(p)}$, see [**W**, Lemma 4.8]). Therefore to finish we must check that for even $n \geq 2$, $L(n,\overline{\chi}) \neq 0$, and this follows since for any $\chi$, $L(s,\chi)$ is nonzero whenever $\mathbf{Re}(s) > 1$, by considering the Euler product expansion for $L$, which is

$$L(s,\chi) = \prod_{p} \left(1-\chi(p)p^{-s}\right)^{-1}.$$

□

The techniques of the proof given above was inspired by those used in a similar situation in [**R**, appendix to section 5], following [**M-S**]. I am indebted to Miles Reid for leading me in this direction.

The first corollary of the above Proposition is that the unoriented distribution numbers $M_{i,n}(S)$ for a torsion section $S$ of prime order are determined by the linear system of Lemma 5.5. This proves that they are independent of $S$! Moreover, we can easily calculate them from the linear system. We require one lemma concerning the coefficient matrix $\mathbf{P}_p$:

LEMMA 6.2. *For an odd prime $p$, and for any fixed $\alpha = 1, \ldots, (p-1)/2$,*

$$\sum_{i=1}^{(p-1)/2} P(\langle \alpha i/p\rangle) = \frac{p^2-1}{12p}.$$

PROOF. For fixed $\alpha$, the integers $\alpha i$ as $i$ ranges from 1 to $p-1$ themselves range over a complete set of representatives for the nonzero residue classes modulo $p$. Therefore, since $P(\langle \alpha i/p\rangle)$ is an even function of $i$, the set of values of this function for $i = 1, \ldots, (p-1)/2$ are the same as the set of values for the function when $\alpha = 1$. In this range we have $\langle i/p\rangle = i/p$, so that for any $\alpha$, the sum above is simply

$$\sum_{i=1}^{(p-1)/2} (i/p)(1-i/p) = \frac{p^2-1}{12p}$$

as claimed. □

Our main result is the following.

THEOREM 6.3. *Let $S$ be a torsion section of odd prime order $p$. Then*

$$M_{i,p}(S) = 2p/(p^2-1) \quad \text{if } 1 \leq i \leq (p-1)/2,$$

*and*

$$M_{0,p}(S) = 1/(p+1).$$

*If $S$ is a torsion section of order 2, then*

$$M_{1,2}(S) = 1/3 \quad and \quad M_{0,2}(S) = 2/3.$$

PROOF. If $p$ is odd, Lemma 6.2 shows that the matrix $\mathbf{P}_p$ has constant row sums. In other words, $\mathbf{P}_p\mathbf{1} = ((p^2-1)/12p)\mathbf{1}$, where $\mathbf{1}$ is the constant vector with every coordinate 1. Therefore a solution to the linear system $\mathbf{P}_p\mathbf{X} = (1/6)\mathbf{1}$ (of which the vector $\mathbf{M}_p$ of $M_{i,p}(S)$'s is a solution by Lemma 5.5) is given by a constant vector $(2p/(p^2-1))\mathbf{1}$. Since $\mathbf{P}_p$ is invertible, such a solution is unique, showing that for $i = 1, \ldots, (p-1)/2$, we must have $M_{i,p}(S) = 2p/(p^2-1)$. The value of $M_{0,p}$ is then gotten from (5.3).

For $p = 2$, the computations were made at the end of the last section. □

We will use the notation $M_{i,p}$ for $M_{i,p}(S)$ from now on. The computation of $M_{0,p}$ was done previously in [**M-P3**] using a quotient argument.

We view the above theorem as an "equidistribution" result: for a $p$-torsion section $S$, the number of times $S$ meets a component which is "distance $i$" away from the zero component $C_0$, (measured in units of $m_j/p$), counted with multiplicity ($m_j$ for the fiber $F_j$), is independent of $i$, for $i \neq 0$.

As a concrete example, take the elliptic $K3$ surface with 6 singular fibers, having $[m_j] = [1,1,1,7,7,7]$. Such a $K3$ surface exists, (see [**M-P1**]) and has a torsion section $S$ of order 7 (see [**M-P2**]). The component numbers $\{k_j(S)\}$ must be $(0,0,0,a,b,c)$, where we can take the orientations of the $I_7$ fibers so that $0 \le a,b,c \le 3$. The equidistribution property above says that $M_{i,7} = 7/24$ for $i = 1,2,3$; hence we are forced to have $(k_j) = (0,0,0,1,2,3)$ after possibly reordering the fibers. Since $\chi(\mathcal{O}_X) = 2$, the sum of the component numbers $k_j(S)$ must be 6 by Corollary 4.3, so this is consistent. The quadratic relation for $S$ in this case says that $1 \cdot 6/7 + 2 \cdot 5/7 + 3 \cdot 4/7 = 2\chi(\mathcal{O}_X) = 4$, which also checks out. Finally note that any multiple of $S$ (except $S_0$ of course) has component numbers $\{0,0,0,1 \text{ or } 6, 2 \text{ or } 5, 3 \text{ or } 4\}$ (after a possible permutation of the fibers) so that all of the relations above are satisfied for all multiples.

As another example, take the Beauville surface $X_{6321}$, which is a rational elliptic surface ($\chi(\mathcal{O}_X) = 1$) with four singular fibers of types $I_6$, $I_3$, $I_2$, and $I_1$ (see [**Be**]). Such a surface has a finite Mordell-Weil group of order 6 (see [**M-P4**]). Using the relations given here for the component numbers, we can easily deduce them. Let $S_2$ be a torsion section of order 2 and $S_3$ be a torsion section of order 3. The component numbers for $S_2$ must be $(0 \text{ or } 3, 0, 0 \text{ or } 1, 0)$, and since their sum must be 4 by Corollary 4.1, we must have $(k_j(S_2)) = (3,0,1,0)$. Note that these numbers do not change if we change orientations, since $S_2$ has order 2. The component numbers for $S_3$ must be $(0 \text{ or } 2 \text{ or } 4, 0 \text{ or } 1 \text{ or } 2, 0, 0)$, and we may choose orientations so that they are $(0 \text{ or } 2, 0 \text{ or } 1, 0, 0)$. By Corollary 4.3, these numbers must sum to 3, so that we must have $(k_j(S_3)) = (2,1,0,0)$. In particular we have with these orientations that the sum of these sections, which is a section $S_6$ of order 6 generating the Mordell-Weil group of $X$, has component numbers $(5,1,1,0)$. (We may re-orient the fibers to achieve $(k_j(S_6)) = (1,1,1,0)$ if we desire.)

As a final example let us consider the Beauville surface $X = X_{3333}$, which is the Hessian pencil surface obtained by taking a smooth cubic curve $C$ in the plane and considering the pencil formed by $C$ and its Hessian $H$. One obtains after blowing up the 9 base points of this pencil (which are the 9 flexes of $C$) a rational elliptic surface ($\chi = 1$) with four $I_3$ fibers. (One gets the same surface starting from any smooth $C$!). The Mordell-Weil group of $X$ is of order 9, with the zero-section $S_0$ and 8 other sections of order 3. The component numbers for a section are all 0, 1, or 2, and if $S$ is not the zero section, then the component number sum formula Corollary 4.4 says that exactly one $k_j(S)$ is 0, all others are non-zero. We may choose an ordering of the fibers and orientations so that one of the order 3 sections $S_1$ has component numbers $(0,1,1,1)$. Let $S_2$ be another order 3 section, not equal to $2S_1$, so that $S_1$ and $S_2$ generate the Mordell-Weil group. By reordering the fibers we may assume that the component numbers for $S_2$ are $(a,0,b,c)$ where $a,b,c$ are not 0; after reorienting the first fiber we may further assume that $a = 1$. Since the sum $S_3 = S_1 + S_2$ will then have component numbers $(1,1,b+1,c+1)$ and must have three non-zero component numbers, we may (after possibly switching the last two fibers) assume that $b = 2$ and $c = 1$, so that $(k_j(S_2)) = (1,0,2,1)$. This then determines the component numbers for all 8 nonzero sections of $X$.

As an application of these computations of the unoriented distribution numbers $M_{i,p}$ for prime $p$, we obtain a divisibility result. Since each of the numbers $12\chi M_{i,p}(S)$ is an integer, and since $p$ and $p^2-1$ are always relatively prime, we obtain the following.

COROLLARY 6.4. *Let $p$ be an odd prime and suppose that $f$ admits a section of order exactly $p$. Then $(p^2-1)/2$ divides $e = 12\chi$.*

This improves a divisibility result stated in [**M-P3**, page 255] that $p+1$ divides $e = 12\chi$, which was obtained by using $M_{0,p}$ only. Moreover this is sharp; for the elliptic modular surface (associated to the group $\Gamma_1(p)$) which has a section of order $p$, the Euler number is exactly $e = (p^2-1)/2$ (see [**Shm**] or [**C-P**]). In some sense we can view this result as enabling us to "divine" the Euler number of such a universal surface, without constructing it: if it existed, it surely would have minimal $e$.

We also have recovered what Persson and I referred to as the "Fixed Point Rule" in [**M-P1**, Corollary 5.5] (see also [**M-P2**, Lemma 3(c)] and [**M**, Corollaries X.3.3 and X.3.4]). Again these works concentrated on the K3 surface case, which has $\chi(\mathcal{O}_X) = 2$; the statement in this case was that if $S$ is a $p$-torsion section on a K3 elliptic surface, with component numbers $\{k_j\}$, then

$$\sum\{m_j \,:\, k_j \neq 0\} = 24p/(p+1).$$

In general of course we have (using the results of this article) that

$$\begin{aligned}\sum\{m_j\ :\ k_j \neq 0\} &= 12\chi - \sum\{m_j\ :\ k_j = 0\}\\ &= 12\chi - 12\chi\cdot M_{0,p}(S)\\ &= 12\chi(1-1/(p+1))\\ &= 12\chi p/p+1,\end{aligned}$$

which generalizes the Fixed Point Rule as stated in the above articles.

Finally, I would like to mention a "dual" equidistribution property. It concerns not the indices of the components other than the zero component which a $p$-torsion section $S$ hits, but rather the position in $C_0^{(j)}$ which $S$ hits, when it does hit this zero component. Specifically, each $C_0^{(j)}$ can be identified with $\mathbb{C}^*$ by sending the nodes to 0 and $\infty$, and sending the point $S_0 \cap C_0^{(j)}$ to 1. (Note that if one exchanges the nodes, the identification with $\mathbb{C}^*$ changes by the inversion $z \mapsto 1/z$.) Any torsion section $S$ of order $p$ will then hit $C_0^{(j)}$ in a $p^{th}$ root of unity. The equidistribution property in this context is that each $p^{th}$ root of unity occurs equally often, (counted properly, and up to the inversion given by exchanging the nodes).

I do not know how to detect this root of unity using the elementary types of techniques used in this paper. Of course this property, properly formulated, is invariant under base change, and so it would suffice to check it for the modular surfaces. Peter Stiller and I have done this and this work will appear in a forthcoming paper.

## REFERENCES

[Be] A. Beauville: "Les familles stables de courbes elliptiques sur $\mathbb{P}^1$ admettant quatre fibres singulieres". C.R. Acad. Sc. Paris, t. 294, (1982), 657-660.

[C-P] D.A. Cox and W. R. Parry: "Torsion in elliptic curves over $k(t)$". Compositio Math. vol. 41, Fasc. 3, (1980),337-354.

[C-Z] D. Cox and S. Zucker: "Intersection numbers of sections of elliptic surfaces". Inventiones Math. 53, (1979), 1-44.

[G] J. Giraud: "Surfaces d'Hilbert-Blumenthal". In:*Surfaces Algébriques*, Séminaire de Géométrie Algébrique d'Orsay 1976-78, Expose III, Springer Lecture Notes in Mathematics, Vol. 868 (1981), 35-57.

[I-R] K. Ireland and M. Rosen: *A Classical Introduction to Modern Number Theory.* Graduate Texts in Mathematics, Vol. 84, Springer-Verlag (1982).

[K] K. Kodaira: "On compact analytic surfaces II, III". Annals of Math. 77, (1963), 563-626; 78 (1963), 1-40.

[L] S. Lang: *Cyclotomic Fields I and II.* Graduate Texts in Mathematics, Vol. 121, Springer-Verlag (1990).

[M] R. Miranda: *The Basic Theory of Elliptic Surfaces.* Dottorato di Ricerca in Matematica, Dipartimento di Mathematica dell'Universiá di Pisa, EDS Editrice Pisa, (1989).

[M-P1] R. Miranda and U. Persson: "Configurations of singular fibers on elliptic $K3$ surfaces". Math. Zeitschrift, vol. 201 (1989), 339-361.

[M-P2] R. Miranda and U. Persson: "Mordell-Weil Groups of extremal elliptic $K3$ surfaces". In: *Problems in the Theory of Surfaces and their Classification*, F. Catanese, C. Ciliberto, and M. Cornalba, editors. Symposia Mathematica, Vol. XXXII, (1991), 167-192.

[M-P3] R. Miranda and U. Persson: "Torsion groups of elliptic surfaces", Compositio Math., vol. 72 (1989), 249-267.

[M-P4] R. Miranda and U. Persson: "On extremal rational elliptic surfaces". Math. Zeit. 193, (1986), 537-558.

[M-S] D.R. Morrison and G. Stevens: "Terminal Quotient Singularities in Dimensions Three and Four". Proc. of the AMS, Vol. 90, No. 1, (Jan. 1984), 15-20.

[R] M. Reid: "Young Person's Guide to Canonical Singularities". In: *Algebraic Geometry Bowdoin 1985*, Spencer Bloch, Editor. Proceedings of Symposia in Pure Mathematics, Vol. 46, Part 1 (1987), 345-414.

[Sch] C. W. Schwartz: "A Mordell-Weil group of rank 8, and a subgroup of finite index".

[Shm] G. Shimura: *Arithmetic theory of automorphic functions*. Princeton University Press, Princeton (1971).

[Shd1] T. Shioda: "On elliptic modular surfaces". J. Math. Soc. Japan, vol. 24, (1972), 20-59.

[Shd2] T. Shioda: "On the Mordell-Weil Lattices". Commentarii Mathematici, Universitatis Sancti Pauli, vol. 39, no. 2, (1990), 211-240.

[Shd3] T. Shioda: "The Mordell-Weil Lattice of a Rational Elliptic Surface". Commentarii Mathematici, Universitatis Sancti Pauli, vol. 40, no. 1, (1991), 83-99.

[Stl] P.F. Stiller: "The Picard Numbers of Elliptic Surfaces with Many Symmetries". Pacific J. of Math., vol 128, No. 1 (1987), 157-189.

[W] L. C. Washington: *Introduction to Cyclotomic Fields*. Graduate Texts in Mathematics, Vol. 83, Springer-Verlag (1982).

DEPT. OF MATHEMATICS, COLORADO STATE UNIVERSITY, FT. COLLINS, CO 80523
*E-mail address*: miranda@riemann.math.colostate.edu

Contemporary Mathematics
Volume **162**, 1994

# Braid Group Techniques in Complex Geometry III: Projective Degeneration of $V_3$

BORIS MOISHEZON and MINA TEICHER

ABSTRACT. In this paper we present a projective degeneration of the Veronese surface of order 3 into union of planes. Our techniques could be generalized to Veronese surfaces of order n. In our next papers we shall use this degeneration to compute the braid monodromy of $S_3$ (the branch curve of the generic projection $V_3 \to \mathbb{CP}^2$) and the fundamental group $\Pi_1(\mathbb{CP}^2 - S_3, *)$.

## 0. Introduction

This paper is a continuation of our articles, "Braid group techniques in complex geometry I and II," which appeared a few years ago (see [MoTe4], [MoTe5]).

One of the main applications of the braid group technique is the topology of $\mathbb{CP}^2$-complements of cuspidal branch curves corresponding to generic $\mathbb{CP}^2$-projections of algebraic surfaces, and, primarily, to fundamental groups of such complements.

Not much is known about these groups. One can mention the works [Za],[Mo2] and the recent article [Mol]. It is important to note that the earlier works created the wrong impression about the complexity of these fundamental groups, namely that they are "big", in particular, they contain free subgroups with two generators. The braid groups and their close analogs were considered as the typical examples.

In the work [Mo2], the first indications that the complexity of the results in [Za], [Mol] was due to a "non-stable range situation" were obtained. Thus, it is possible that in a "stable range" (for instance, for projective embeddings of algebraic surfaces corresponding to sufficiently large very ample divisors) the groups $\Pi_1(\mathbb{CP}^2 - S)$, $S$ the branch curve of a generic projection, are much simpler than it was generally expected.

To illustrate this, let us consider the following: The classical Veronese surface $V_2$ behaves "non-typically" in different algebraic-geometrical theories. It gives

1991 Mathematics Subject Classification. Primary 20F36, 14J10
Final version.

also an example of a "non-stable range" behavior when one asks about the structure of $\Pi_1(\mathbb{CP}^2 - S_2, *)$, where $S_2$ is the branch curve of a generic projection $V_2 \to \mathbb{CP}^2$. One can prove that $\Pi_1(\mathbb{CP}^2 - S_2, *)$ is "big" in the sense mentioned above (see also [MoTe3]).

Let $V_k$ be the Veronese surface of order $k$, that is the projective embedding of $\mathbb{CP}^2$ corresponding to $k\times$ (projective line). Denote by $S_k$ the branch curve of the generic projection $V_k \to \mathbb{CP}^2$. The "stable range" starts with $V_3$, where the corresponding $\Pi_1(\mathbb{CP}^2 - S_3, *)$ is "small", more precisely, contains a solvable group of finite index.

Thus, the case of $V_3$ is the critical one. In the present work and in its continuation, we consider this important case in detail. First, we present a sequence of projective degenerations of $V_3$ ending in a union of nine planes. This sequence is used in our next paper in this series, "Braid group techniques IV" to get an explicit braid monodromy formula for $S_3$. We start with the braid monodromy related to the degenerate object and regenerate this formula to get the braid mondodromy for $S_3$. Such a formula is needed for the study of $\Pi_1(\mathbb{CP}^2 - S_3, *)$.

Motivated by our results we expect that for many classes of simply-connected algebraic surfaces, the groups $\Pi_1(\mathbb{CP}^2 - S, *)$ ($S$ is the branch curve of a generic $\mathbb{CP}^2$-projection) are extensions of symmetric groups by solvable ones. One should remember, however, that in general the situation could be much more complicated. It is known that many abstract groups (in particular, all finite ones) could be fundamental groups of algebraic surfaces. The corresponding $\Pi_1(\mathbb{CP}^2 - S, *)$'s in such cases will contain subgroups at least as complicated as those fundamental groups. We believe that the right approach is to distinguish different levels of complexity, even in the case of simply-connected algebraic surfaces. The expectation formulated above will be true then for the first level of complexity. It could correspond to algebraic surfaces which are diffeomorphic to "almost completely decomposable 4-manifolds" in the sense of [Mo3], Intoduction. The next level of complexity possibly is represented by simply-connected surfaces with positive index which we constructed in ([MoTe1], [MoTe2]), as well as by simply-connected surfaces discovered by the Chinese school.

## 1. D-Construction and elevated projective degenerations

*Definition: Equivalence of projective embeddings.*

We say that an embedding $k : Z \to \mathbb{CP}^N$ is *equivalent* to an embedding $g : Z \to \mathbb{CP}^M$ if there exists an embedding $\lambda : Z \to \mathbb{CP}^L$ and projections $\mathbb{CP}^L \overset{p_1}{\to} \mathbb{CP}^M$ , $\mathbb{CP}^L \overset{p_2}{\to} \mathbb{CP}^N$ s.t. $p_i$ is regular on $\lambda(Z)$ ($i = 1, 2$), $p_1\lambda = g$, $p_2\lambda = k$. Thus, the following diagram (Figure 1) is commutative.

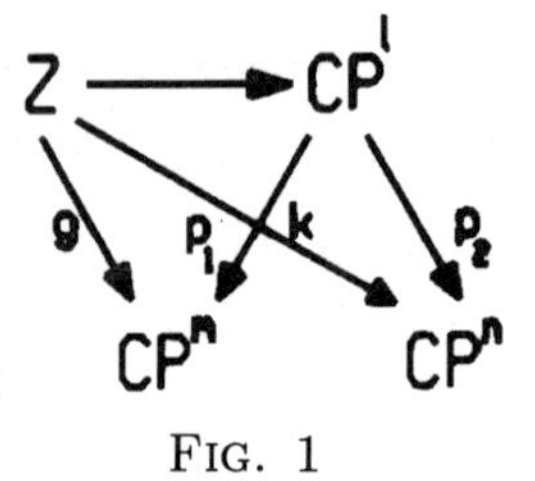


FIG. 1

*Definition: Projective degeneration.*

Let $k : Y \to \mathbb{CP}^n$, $k' : X \to \mathbb{CP}^n$ be projective embeddings. We say that $k'$ is a *projective degeneration* of $k$ if there exists an algebraic variety $V$, with $\pi : V \to \mathbb{C}$ s.t. for every irreducible component $V_i$ of $V$, $\pi(V_i) = \mathbb{C}$ and $\pi^{-1}(0) \simeq X$, $\pi^{-1}(1) \simeq Y$, $\pi^{-1}(1)$ is a generic fiber, and there exists a regular morphism $F : V \to \mathbb{CP}^n \times \mathbb{C}$ s.t. $F(\pi^{-1}(t)) \subset \mathbb{CP}^n \times t$, $F_t = F\big|_{\pi^{-1}(t)} : \pi^{-1}(t) \to \mathbb{CP}^n \times t$ is a projective embedding of $\pi^{-1}(t)$, $F_0 = k'$ and $F_1 = k$ under the identification of $\pi^{-1}(0)$ and $\pi^{-1}(1)$ with $X$ and $Y$, respectively.

*Definition: Elevated projective degeneration.*

Let $g : Y \to \mathbb{CP}^N$ and $g' : X \to \mathbb{CP}^{N'}$ be projective embeddings. We say that $g'$ is an elevated projective degeneration of $g$ if there exist projective embeddings $k, k'$, s.t. $k, k'$ are equivalent to $g, g'$, respectively, and $k'$ is a projective degeneration of $k$.

We want to construct an *elevated projective degeneration* of the Veronese surface into union of planes. Construction-D, which we shall present now, is a technique to produce *elevated projective degeneration* in special cases that include the Veronese case.

*Definition: D-Assumptions.*

Let $Z$ be a projective algebraic variety. Let $g : Z \to \mathbb{CP}^N$ be a projective embedding. Let $G$ be a line bundle on $Z$ defined by a hyperplane section corresponding to $g$. Let $Z = A \cup B$, $A$ and $B$ subvarieties of $Z$ and $C = A \cap B$. We say that $(Z, A, B, g)$ satisfy D-assumptions if:

(i) $B$ is defined by a sheaf of principal ideals.
(ii) The ideal of $C = A \cap B$ on $A$ is equal to the restriction to $A$ of the ideal $B$ on $Z$.
(iii) $C$ does not have a common irreducible component with $A$.
(iv) There exists a base $\alpha_1 \dots \alpha_M$ of $H^0(A, \mathcal{O}_A[G\big|_A - [C]])$ with no common zeroes.

*Remark.* (iii) $\Rightarrow$ Ideal of $A \cap B$ is a principal ideal sheaf which is not a zero divisor, i.e. an invertible ideal sheaf.

*Definition: D-Construction.*

Let $(Z, A, B, g)$ satisfy D-assumptions. Consider $Z \times \mathbb{C}$. Let $t$ be a complex coordinate in $\mathbb{C}$. For $S \subset Z$ let $S_0 = S \times \{0\}$. Let $q : Z \times \mathbb{C} \to Z$, $p : Z \times \mathbb{C} \to \mathbb{C}$ be the natural projections. Let $\phi : W \to Z \times \mathbb{C}$ be the monoidal transformation of $Z \times \mathbb{C}$ with center $B_0$. For $T \subset Z \times \mathbb{C}$, let $T^s = \overline{\phi^{-1}(T - B_0)}$ the strict image of $T$ in $W$. In particular, $(B \times \mathbb{C})^s = \overline{\phi^{-1}(B \times \mathbb{C} - B_0)}$ and $A_0^s = \overline{\phi^{-1}(A_0 - B_0)}$. For $S \subset B$, let $\tilde{S}_0 = (S \times \mathbb{C})^s \cap \phi^{-1}(S \times \{0\})$. In particular, $\tilde{B}_0 = (B \times \mathbb{C})^s \cap \phi^{-1}(B_0)$

**Lemma 1.** (Proof omitted, see Figure 2)

(i) $(B \times C)^s \cap A_0^s = \emptyset$.
(ii) $\phi^{-1}(B_0) = \tilde{B}_0 \cup \phi^{-1}(C_0)$.
(iii) $\phi^{-1}(Z_0) = \tilde{B}_0 \cup \phi^{-1}(C_0) \cup A_0^s$.
(iv) $\tilde{B}_0 \cap A_0^s = \emptyset$.

(v) $\phi|_{(B\times C)^s} : (B\times C)^s \to B\times C$ *is an isomorphism which sends* $\tilde{B}_0$ *to* $B_0$ *and* $\tilde{B}_0 \cap \phi^{-1}(C_0)$ *to* $C_0$.

(vi) $\phi|_{A_0^s} : A_0^s \to A_0$ *is an isomorphism.*

(vii) $\phi|_{\phi^{-1}(C_0}$ $: \phi^{-1}(C_0) \to C_0$ *is a projective line bundle with 2 disjoint cross-sections* $\phi^{-1}(C_0)\cap A_0^s$ *and* $\phi^{-1}(C_0)\cap \tilde{B}_0$

(viii) $A_0^s \cap \phi^{-1}(C_0) \sim C_0$.

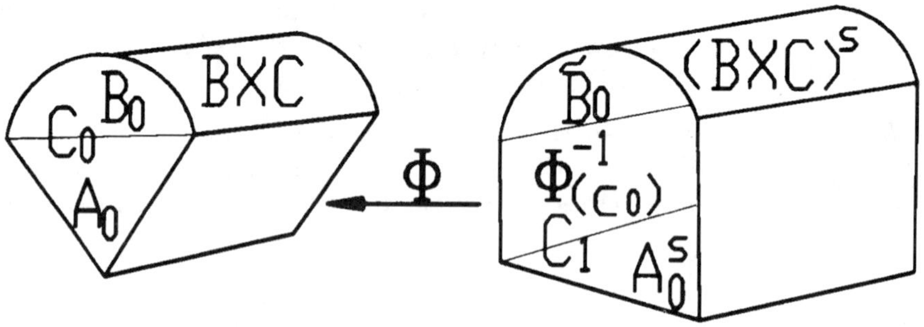


FIG. 2

Now we are going to construct a morphism from $W$ to a "big" projective space.

Let $G' = \phi^* q^*(G) - [\phi^*(B_0)]$ where $[\phi^*(B_0)]$ is the line bundle that corresponds to the pullback of the ideal of $B_0$. For every $\alpha \in H^0(A, \mathcal{O}_A[G|_A - [C]])$ we shall construct here an element $\alpha' \in H^0[W, \mathcal{O}_W[G']]$.

Let $\{U_r\}$ be an affine open cover of $Z$ s.t. the ideal of $B\cap U_r$ in $U_r$ is defined by a single element $\tau_r$ and $G|_{U_r}$ is trivial for each $U_r$. (D-Assumption (i))

Let

$$U_r^1 = \phi^{-1}q^{-1}(U_r) - A_0^s$$
$$U_r^2 = \phi^{-1}q^{-1}(U_r) - (B\times C)^s.$$

Let $u_r = \frac{1}{t}q^*(\tau_r)$ be a meromorphic function on $q^{-1}(U_r)$, which is holomorphic outside of $Z_0$. Evidently, $\frac{1}{u_r}$ is holomorphic outside of $(B\times C)^s$.

Take $\alpha \in H^0(A, \mathcal{O}_A[G|_A - [C]])$, thus $\alpha = \{\alpha_r\}$ where $\alpha_r$ is regular on $U_r \cap A$. W.l.o.g. we can assume that $\alpha_r$ is defined in the whole of $U_r$.

Let

$$\alpha_r^1 = \phi^* q^*(\alpha_r)\phi^*(u_r)$$
$$\alpha_r^2 = \phi^* q^*(\alpha_r).$$

$\alpha_r^1$ is holomorphic in $U_r^1$ and $\alpha_r^2$ is holomorphic in $U_r^2$, where $\{U_r^1, U_r^2\}_r$ is a cover of $W$. It is possible to check that $\{\alpha_r^\varepsilon\}$ is an element of $H^0(W, \mathcal{O}_W[G'])$. Denote this element by $\alpha'$.

For every $\psi \in H^0(Z, \mathcal{O}_Z[G])$ we shall define an element $\psi'$ of $H^0(W, \mathcal{O}_W[G'])$ as follows: $\psi = \{\psi_r\}$, $\psi_r$ regular on $U_r$. Let

$$\psi_r^1 = \phi^* q^*(\psi_r)$$
$$\psi_r^2 = \phi^* q^*(\psi_r)\phi^*\left(\frac{1}{u_r}\right).$$

$\psi_r^1$ is holomorphic on $U_r^1$ and $\psi_r^2$ is holomorphic on $U_r^2$. It is easy to check that $\{\psi_r^\varepsilon\}$ defines an element of $H^0(W, \mathcal{O}_W[G'])$. We call this element $\psi'$.

Let us take a base $\alpha_1, \ldots, \alpha_M$ of $H^0(A, \mathcal{O}_A[G|_A - [C]])$ with no common zeroes on $A$. (It exists under D-assumption (iv).) Let us take a base $\psi_0 \ldots \psi_N$ of $H^0(Z, \mathcal{O}_Z[G])$. Consider the map $f : W \to \mathbb{CP}^{N+M}$ defined by

$$x \xrightarrow{f} (\psi_0'(x); \ldots; \psi_N'(x); \alpha_1'(x); \ldots; \alpha_M'(x)).$$

It is easy to see that the following lemma is true.

**Lemma 2.** (Proof omitted) *$f$ is regular.*

Consider $i : f(\phi^{-1}(Z_0)) \hookrightarrow \mathbb{CP}^{N+M}$. Let $G^{(1)} = i^*\mathcal{O}(1) \left(= \mathcal{O}(1)|_{f(\phi^{-1}(Z_0))}\right)$. The map $i$ is defined by a set of elements of $H^0(f(\phi^{-1}((Z_0), O[G^{(1)}])$. Let $Z^{(1)} = f(\phi^{-1}(Z_0))$, $A^{(1)} = f(A_0^s)$, $B^{(1)} = f(\phi^{-1}(B_0))$, $g^{(1)} =$ inclusion of $Z^{(1)}$ in $\mathbb{CP}^{N^{(1)}}$.

*Definition: D-Construction.*

We say that $(Z^{(1)}, A^{(1)}, B^{(1)}, g^{(1)})$ is obtained from $(Z, A, B, g)$ by the D-construction.

**Proposition 3.** (Proof omitted)

(i) *$g^{(1)}$ is an elevated projective degeneration of $g$. (Compare the image of the special and generic fibers of $W$ under $f$. See also* (iii) *below.)*

(ii) $f(A_0^s \cap \phi^{-1}(B_0)) \cap f(\phi^{-1}(B_0) - A_0^s) = \emptyset$.

(iii) *Let $\pi_1$ be the projection on the first $N+1$ coordinates. Then $\forall t \in \mathbb{C}^1 - \{0\}$, $\phi^{-1}(Z \times t) \simeq Z$ and the map $\pi_1|_{\phi^{-1}(Z\times t)}$ coincide with the embedding $Z \hookrightarrow \mathbb{CP}^N$ that corresponds to $G$.*

(iv) *$\pi_1 f : \tilde{B}_0 \to \mathbb{CP}^N$ coincide with the embedding $\tilde{B}_0 \to \mathbb{CP}^N$ corresponding to $g|_{B_0}$.*

(v) *$f|_{\phi^{-1}(B_0)-A_0^s} : \phi^{-1}(B_0) - A_0^s \to \mathbb{CP}^{N+M}$ is an embedding.*

(vi) *Let $\pi_2$ be the projection on the last $M$ coordinates. Then $\pi_2 f|_{A_0^s} : A_0^s \to \mathbb{CP}^{M-1}$ coincide with the regular map $A \to \mathbb{CP}^{M-1}$ defined by $x \to (\alpha_1(x); \ldots; \alpha_M(x))$. Moreover, $\pi_2|_{f(A)} : f(A) \to \pi_2 f(A)$ is an isomorphism.*

**Lemma 4.** *Let $A$ be a nonsingular projective algebraic surface. Let $C \subset A$ be a nonsingular algebraic curve with $(C)_A^2 = k$. Let $q : A \times \mathbb{C} \to A$ be the natural projection, $A_0 = A \times \{0\}$, $C_0 = C \times \{0\}$. Let $\phi : W \to A \times \mathbb{C}$ be the monoidal transformation with center $C_0$. Let $G$ be a line bundle on $A$ with $\deg G|_C = n$.*

*Let $G'$ be $\phi^*q^*G - [\phi^*C_0]$. Let $A_0^s$ be the strict image of $A_0$; $Z_1 = \phi^{-1}(A_0)$, $C_1 = A_0^s \cap \phi^{-1}(C_0)$. Then*

(i) $Z_1 = A_0^s \cup \phi^{-1}(C_0)$.

(ii) $\phi\big|_{A_0^s} : A_0^s \to A_0^s$ *is an isomorphism,* $\phi\big|_{C_1} : C_1 \xrightarrow{\sim} C_0$ *is an isomorphism.*

(iii) $\phi : \phi^{-1}(C_0) \to C_0$ *is a projective line bundle with cross-section* $C_1$ *and fiber* $\ell$, *and* $(\phi^{-1}(C_0).\ell))_W = -1$.

(iv) $(C_1)^2_{\phi^{-1}(C_0)} = -k$.

(v) $G'\big|_{\phi^{-1}(C_0)} = [C_1 + n\ell]$.

(vi) *If* $G = [nC]$ *then* $G'\big|_{A_0^s} = [(n-1)C_1]$.

*Proof.*

(i)+(ii)+(iii): Since $A$ is nonsingular, $C$ is of codimension 1 in $A$, and $\phi$ is the blowing-up of $A \times \mathbb{C}$ with center $C_0$.

(iv): Consider $C_1$ as a cycle. We can move $C_1$ to be contained only in $\phi^{-1}(C_0)$ or only in $A_0^s$. Thus:

$$(C_1)^2_{\phi^{-1}(C_0)} = A_0^s.C_1$$
$$(C_1^2)_{A_0^s} = (\phi^*(C_0).C_1)_W.$$

By (ii) $C_1 = A_0^s \cap \phi^{-1}(C_0) \overset{\phi}{\rightsquigarrow} C_0 \overset{q}{\rightsquigarrow} C$. Thus:

$$(C_1)^2_{A_0^s} = (C_0)^2_{A_0} = (C)^2_A = k.$$

So $(\phi^*(C_0).C_1)_W = k$. Now $(\phi^*(A_0).C_1)_W = (A_0.\phi_*(C_1))_{A\times\mathbb{C}} = (A_0.C_0)_{A\times\mathbb{C}}$ (by (ii)). $C_0$ can be moved in $A \times \mathbb{C}$ to $C \times \{1\}$, thus $(A_0.C_0)_{A\times\mathbb{C}} = (A_0.C \times \{1\})_{A\times\mathbb{C}}$. But $(A_0.C \times \{1\})_{A\times\mathbb{C}} = 0$ (since they do not intersect), so $(\phi^*A_0.C_1)_W = 0$. Since

$$0 = (\phi^*(A_0).C_1)_W = ((A_0^s + \phi^*C_0).C_1)_W = (A_0^s.C_1)_W + k,$$

we have:

$$(A_0^s.C_1)_W = -k.$$

Thus

$$(C_1)^2_{\phi^{-1}(C_0)} = -k.$$

(v): Since $\phi^{-1}(C_0)$ is a ruled surface (see (iii)), we have $G'\big|_{\phi^{-1}(C_0)} = [xC_1 + y\ell]$ for some $x$ and $y$.

$$\deg G'\big|_{\phi^{-1}(C_0)}\Big|_{\ell} = \left(G'\big|_{\phi^{-1}(C_0)}.\ell\right)_{\phi^{-1}(C_0)} = x(C_1.\ell)_{\phi^{-1}(C_0)} + y(\ell.\ell)_{\phi^{-1}(C_0)} = x$$

(by (iii))

$$\deg G'\big|_{\phi^{-1}(C_0)}\Big|_{C_1} = \left(G'\big|_{\phi^{-1}(C_0)}.C_1\right)_{\phi^{-1}(C_0)} = x(C_1.C_1)_{\phi^{-1}(C_0)} + y(\ell.C_1)_{\phi^{-1}(C_0)}$$
$$= -kx + y$$

(by (iii) and (iv))

$$\begin{aligned} \deg G'\big|_{\phi^{-1}(C_0)}\big|_\ell &= \deg(\phi^* q^*(G) - [q^*(C_0])\big|_\ell \\ &= \deg G\big|_{q\phi(\ell)} - (\phi^{-1}(C_0).\ell)_W \\ &= 0 \qquad\qquad - (-1) = 1 \end{aligned}$$

$$\begin{aligned} \deg G'\big|_{\phi^{-1}(C_0)}\big|_{C_1} = \deg G'\big|_{C_1} &= \deg(\phi^* q^*((G) - [\phi^*(C_0)])\big|_{C_1} \\ &= \deg G\big|_C - (\phi^*(C_0).C_1)_W = n - k \\ \Rightarrow 1 = x,\ y &= (n-k) + kx = n - k + k\cdot 1 = n. \\ \Rightarrow G'\big|_{\phi^{-1}(C_0)} &= [C_1 + n\ell]. \end{aligned}$$

(vi):

$$\begin{aligned} G'\big|_{A_0^s} = \phi^* q^* G - [\phi^*(C_0)]\big|_{A_0^s} &= \left(q\phi\big|_{A_0^s}\right)^* (G\big|_A) - [C_1] \\ &= \left(q\phi\big|_{A_0^s}\right)^* [nC] - [C_1] \\ &= n[C_1] - [C_1] \\ &= [(n-1)C_1] \qquad \square \end{aligned}$$

## 2. Construction of the elevated projective degeneration of the Veronese surface $V_n$ into $n^2$ planes

Consider $\mathbb{CP}^2 \overset{v_n}{\underset{\rightarrow}{\sim}} V_n \subseteq \mathbb{CP}^N$, $N = \binom{n+2}{2} - 1$, defined by: $v_n(x_0; x_1; x_1) = (\ldots; x_0^i x_1^j x_2^k; \ldots)_{i+j+k=n}$. The map $v_n$ is defined by a base of $H^0(\mathbb{CP}^2, nL)$, where $L$ is a line in $\mathbb{CP}^2$. This gives the classical construction of the Veronese surface.

We shall consider a more general definition: Let $\gamma_1 \ldots \gamma_{N'} \in H^0(\mathbb{CP}^2, nL)$ s.t. $\mathbb{CP}^2 \to \mathbb{CP}^{N'-1}$ defined by $x \overset{\sigma}{\to} (\gamma_1(x) \ldots \gamma_{N'}(x))$ is biregular. We still call $\sigma(\mathbb{CP}^2)$ a Veronese surface and denote it by $V_n$. Let $F_n$ be a Hirzebruch surface, namely, a ruled surface with fiber $\ell$ and cross-section $C$ such that $(C)^2_{F_n} = -n$. Let $T_k$, $k \geq 2$ be the embedding of $F_1$ in $\mathbb{CP}^M$ w.r.t. the linear system $|C + k\ell|$. Let $U$ be the embedding of $F_0$ in $\mathbb{CP}^M$ w.r.t. the linear system $|C + \ell|$.

**Lemma 5.** *Let $V_n$ be the Veronese surface.*

$$\sigma : \mathbb{CP}^2 \overset{\sim}{\to} V_n \subseteq \mathbb{CP}^N \qquad N = \binom{n+2}{2} - 1$$

*Let $L$ be a line in $\mathbb{CP}^2$, $C = \sigma(L)$.*
*Let $q : V_n \times \mathbb{C} \to V_n$ be the natural projection.*
*Let $A_0 = V_n \times \{0\}$.*

*Let $\phi : W \to V_n \times \mathbb{C}$ be the monoidal transformation with center $C_0 = C \times \{0\}$.*
*Let $G = [nC]$ (the line bundle corresponding to $nC$).*
*Let $G' = \phi^* q^*(G) - \phi^*(C_0)$.*
*Let $Z_1 = \phi^{-1}(V_n \times \{0\})$, $A_1 = (A_0)^s$, $B_1 = \phi^{-1}(C_0)$, $C_1 = A_1 \cap B_1$.*
*Then:*

(1) $Z_1 = A_1 \cup B_1$.
(2) $\phi\big|_{A_1} : A_1 \overset{\sim}{\to} A_0$ *is an isomorphism,* $q\phi\big|_{A_1} : A_1 \to V_n$ *is an isomorphism and* $C_1 \underset{\to}{\overset{q\phi}{\approx}} C$.
(3) $\phi\big|_{B_1} : B_1 \to C_0$ *is a projective line bundle with cross-section* $C_1$ *and fiber* $\ell$, $(\phi^{-1}(C_0).\ell)_W = -1$.
(4) $(C_1)^2_{B_1} = -1$, $(C_1)^2_{A_1} = 1$
(5) $G'\big|_{C_1} = C_1 + n\ell$.
(6) $G'\big|_{A_1} = (n-1)C_1$.

*Proof.* We apply Lemma 4 with $A = V_n$, $C = \sigma(L)$, $G = [nC]$, $k = (C)^2_{V_n} = (L)^2_{\mathbb{CP}^2} = 1$. The only intersection number that does not appear in Lemma 4 is $(C_1)^2_{A_1}$. By (2), $C_1 \overset{q\phi}{\to} C$ so $(C_1)^2_{A_1} = (C)^2_{V_n} = 1$. □

**Lemma 6.** *In the notation of Lemma 5:*

(i) $(V_n, V_n, C, i)$ *satisfy D-assumptions. Let $f$ be the map introduced in the D-construction. Then*
(ii) $f(A_1)$ *and* $f(B_1)$ *intersect transversally in* $f(C_1)$.
(iii) $f\big|_{A_1 \cup B_1}$ *is an embedding for* $n > 1$.
(iv) $f(A_1) = \begin{cases} V_{n-1} & n > 1 \\ pt & n = 1 \end{cases} \qquad f(B_1) = \begin{cases} T_n & n > 1 \\ V_1 & n = 1. \end{cases}$

*Proof.*
(i) and (ii) are easy to check.
(iii) By Proposition 3(v), $g\big|_{B_1 - C_1}$ is biregular. By Proposition 3(vi), $f\big|_{A_1}$ is induced from the global section of $G'\big|_A$. By Lemma 5(vi) $G'\big|_A = (n-1)C_1$, thus $f\big|_{A_1}$ is actually the Veronese map $v_{n-1}$ and thus it is biregular for $n > 1$. From here it is possible to deduce (iii).
(iv) Since $f\big|_{A_1}$ is $v_{n-1}$, $f(A_1) = V_{n-1}$ for $n > 1$ and point for $n = 1$.
By Lemma 4, $\phi\big|_{B_1} : B_1 \to C_0$ is a projective line bundle with cross-section $C_1$ and fiber $\ell$ s.t. $(C_1)^2_{B_1} = -1$. Thus $B_1$ is a Hirzebruch surface, $F_1$. For $n > 1$ $f\big|_{B_1}$ is biregular and induced from the global sections of $G'\big|_{B_1}$ which is $[C_1 + n\ell]$. Thus $f(B_1)$ is $T_n$ for $n > 1$. For $n = 1$ we get $G'\big|_{B_1} = C_1 + \ell$ with $(C_1 + \ell)^2_{B_1} = 1$ and $(C_1 + \ell).C_1 = 0$. Thus $f(B_1) = V_1$. □

**Lemma 7.** *Let $n > 1$. In the notation of Lemma 6, after applying the D-construction on $(V_n, V_n, C, i)$, we get $(Z^{(1)}, V_{n-1}, T_n, i)$ where $T_n$ intersects $V_{n-1}$ transversally in a curve $C_{n-1}$, $(C_{n-1})^2_{T_n} = -1$ and $(C_{n-1})^2_{V_{n-1}} = 1$.*

*Proof.* From Lemma 5, $A_1 \cap B_1 = C_1$ and $(C_1)^2_{B_1} = -1$, $(C_1)^2_{A_1} = 1$. By Lemma 6, $f\big|_{A_1 \cup B_1}$ is biregular so $(f(C_1)\big|^2_{f(B_1)} = -1$ and $f(C_1)\big|^2_{f(A_1)} = 1$. Denote

$C_{n-1} = f(C_1)$. The result of the D-construction is $(Z^1, f(A_1), f(B_1), i)$. We conclude the proof using Lemma 6. (See Figure 3.)

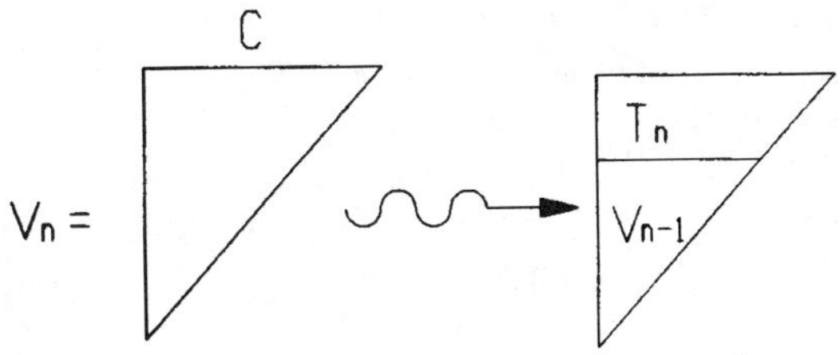


FIG. 3

## 3. DEGENERATION OF $V_3$

We shall construct a sequence of 5 elevated projective degenerations starting with $V_3$ in which the final object is a union of 9 planes. The first 4 degenerations will be obtained by the D-construction.

First degeneration $V_3 \rightsquigarrow$ union of $T_3$ and $V_2$ $(Z^{(0)} \rightsquigarrow Z^{(1)})$

Second degeneration $\rightsquigarrow$ union of $T_3$, $T_2$ and $V_1$ $(Z^{(1)} \rightsquigarrow Z^{(2)})$

Third degeneration $\rightsquigarrow$ union of 2 quadrics, two $V_1$'s and one $T_2$ $(Z^{(2)} \rightsquigarrow Z^{(3)})$

Fourth degeneration $\rightsquigarrow$ union of 3 quadrics and 3 $V_1$'s $(Z^{(3)} \rightsquigarrow Z^{(4)})$

Fifth degeneration $\rightsquigarrow$ each of the quadrics is degenerated to a union of 2 planes $(Z^{(4)} \rightsquigarrow Z^{(5)})$

Sixth degeneration $\rightsquigarrow$ same configuration with real vertices $(Z^{(5)} \to Z^{(6)})$

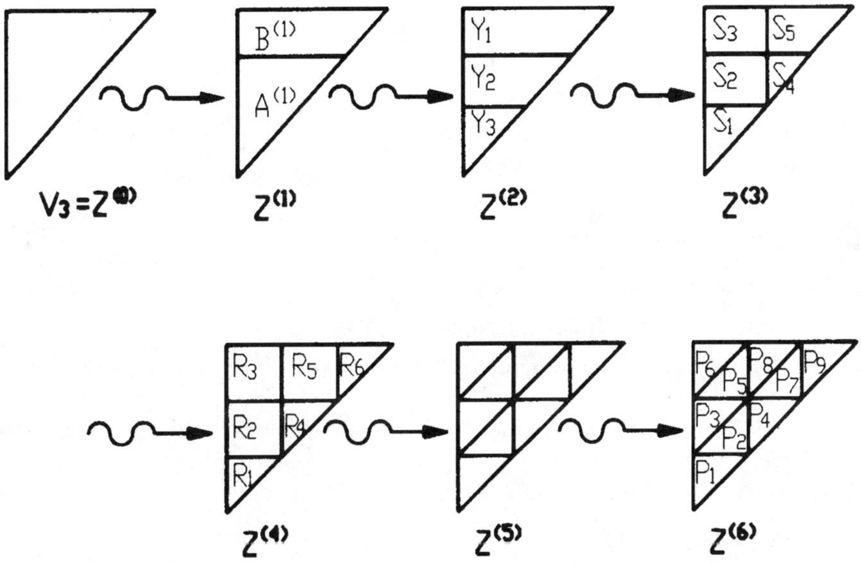


FIG. 4

*Remarks.*

(i) For $i = 1 \dots 4$ we denote by $\phi^{(i)}$ and $f^{(i)}$ the blowing-up and the map $f$ which is used in the corresponding D-construciton.

(ii) $F_0 = \mathbb{P}^1 \times \mathbb{P}^1$, $F_1$ =b.u. of $\mathbb{CP}^2$ in one point.

(iii) Any cross-section $C^+$ of $F_1$ that does not intersect $C$ the minimal cross-section ($C^2 = -1$) has self-intersection 1 and is linear equivalent to $C + \ell$. Let $f$ be the map of $F_1$ to a projective space defined by $|C + \ell|$. Then $f$ contracts $C$ to a point and $f(F_1)$ is $V_1$. Thus, $(f(C^+))^2_{\mathbb{P}^2} = 1$ and $(f(\ell))^2_{\mathbb{P}^2} = 1$ so $f(C^+)$ and $f(\ell)$ are projective lines on $\mathbb{P}^2$. Moreover, $f$ is a blowing-down and on every curve $C'$ s.t. $C.C' = 0$ or 1, $f$ is an isomorphism.

*3.1 First degeneration.*

For unification of future notations we denote $V_3$ by $Z^{(0)}$.

We apply the D-construciton on $(V_3, V_3, C, i)$. By Lemma 7 we get the following projective degeneration of $V_3$, namely $Z^{(1)}$.

$Z^{(1)} = A^{(1)} \cup B^{(1)}$.

$C^{(1)} = A^{(1)} \cap B^{(1)}$.

$A^{(1)} = V_2$, Veronese surface of order 2.

$B^{(1)} = T_3$, projective line bundle with cross-section $C^{(1)}$.

$(C^{(1)})^2_{A^{(1)}} = 1$.

$(C^{(1)})^2_{B^{(1)}} = -1$.

(See abstract configuration of $Z^{(1)}$ in Figure 5.)

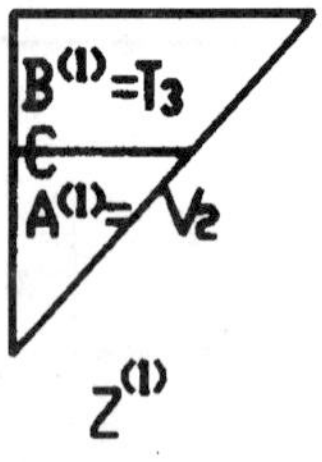


FIG. 5

*3.2 Second degeneration.*

**Lemma 8.**

(i) *$(Z^{(1)}, A^{(1)}, B^{(1)}, i)$ satisfy D-assumptions.*

(ii) *The restriction of the D-construction of the above quadruple restricted to $(A^{(1)}, A^{(1)}, C^{(1)}, i)$ coincides with the D-construction of the last quadruple.*

(iii) *The map $f$ from the D-construction of the above quadruple is an isomorphism on $\widetilde{B_0^{(1)}}$.*

*Proof.* (i) Easy to check.

(ii) Easy to check.

(iii) By Proposition 3(v), $f$ is isomorphic on $\phi^{-1}(B_0^{(1)}) = (A_0^{(1)})^s$ which include $\widetilde{B_0^{(1)}}$.

**Proposition 9.** *After applying the D-construction on $(Z^{(1)}, A^{(1)}, B^{(1)}, i)$ we get the following elevated projective degeneration of $Z^{(1)}$, namely $Z^{(2)}$, with a projective embedding $g^{(2)}$. Then*

$$Z^{(2)} = Y_3 \cup Y_2 \cup Y_1$$

*where*

$$g^{(2)}(Y_1) = V_1, \qquad Y_3 \cap Y_1 = \emptyset.$$

*For $i = 2, 3$:*

*$Y_i$ is a projective line bundle with cross-section $C_{i-1}$ and fiber $\ell_i$, $g^{(2)}(Y_i) = T_i$.*

*$Y_i \cap Y_{i-1} = C_{i-1}$.*

*$(C_{i-1})^2_{Y_{i-1}} = 1$.*

*$(C_i)^2_{Y_i} = -1$. (See Figure 6 for abstract configuration of $Z^{(2)}$.)*

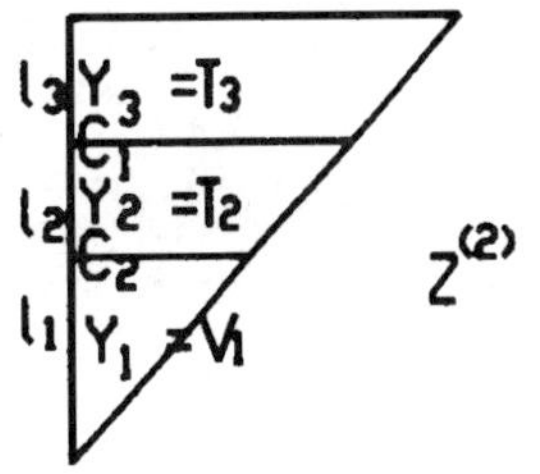


FIG. 6

*Proof.* By Lemma 8(i) we can apply the D-construction on $(Z^{(1)}, A^{(1)}, B^{(1)}, i)$. By Lemma 8(ii) the restriction of such a D-construction to $(A^{(1)}, A^{(1)}, C^{(1)}, i)$ coincides with the D-construction of $(A^{(1)}, A^{(1)}, C^{(1)}, i)$, which is analyzed in Lemma 7. By Lemma 7 from $(A^{(1)}, A^{(1)}, C^{(1)}, i)$ we get $(Y_1 \cup Y_2, Y_1, Y_2, i)$ with $Y_1 = f^{(2)}(A_0^{(1)})^2$, $Y_2 = f^{(2)}\phi^{(2)^{-1}}(C_0^{(1)})$.

$C_1 = Y_1 \cap Y_2$ satisfies

$$(C_1)^2_{Y_1} = 1$$
$$(C_1)^2_{Y_2} = -1$$

and $Y_1 = V_1$ and $Y_2 = T_2$, $Y_2$ is a projective line bundle with cross-section $C_1$ and fiber $\ell_2$.

Let us denote $Y_3 = f^{(2)}(\widetilde{B_0^{(1)}})$. Using Lemma 6 and Proposition 3(v) one can check that $f^{(2)}$ is an embedding. Together with Lemma 1(v) we get $Y_3 \overset{f^{(2)}}{\underset{\to}{\sim}} \widetilde{B_0^{(1)}} \overset{\phi^{(2)}}{\underset{\to}{\sim}} B_0^{(1)} \overset{\sim}{\to} B^{(1)}$. Thus $Y_3 \overset{\sim}{\to} B^{(1)}$ in this isomorphism $C_2 \overset{\sim}{\to} C^{(1)}$, where $C_2 = Y_2 \cap Y_3$, and $Y_3$ is a projective line bundle with cross-section $C_2$ and fiber $\ell_3$. Thus $(C_2)^2_{Y_3} = (C^{(1)})^2_{B^{(1)}} = -1$.

By Lemma 1(vii), $\phi^{(2)^{-1}}(C_0^{(1)} \cap (A_0^{(1)})^2$ and $\phi^{(2)^{-1}}(C_0^{(1)}) \cap \widetilde{B_0^{(1)}}$ are 2 disjoint cross-sections of $\phi^{(2)^{-1}}(C_0^{(1)})$. Applying $f^{(2)}$ on $\phi^{(2)^{-1}}(C_0^{(1)})$ we get that $C_1$ and $C_2$ are 2 disjoint cross-sections of $Y_2$. By Remark (iii), $(C_2)^2_{Y_2} = 1$. □

*3.3 Third degeneration.*

We use here the notations of Proposition 9. Let $A^{(2)} = Z^{(2)}$. Let us choose $\ell_i$ $i = 2, 3$ s.t. $\ell_2.\ell_3 = pt$. Let $\ell_1$ be a line in $Y_1$ s.t. $\ell_1 \ell_2 = pt$. Let $B^{(2)} = \ell_1 \cup \ell_2 \cup \ell_3$ Let $C^{(2)} = B^{(2)}$.

**Lemma 10.**

(i) $(Z^{(2)}, A^{(2)}, B^{(2)}, g^{(2)})$ *satisfy D-assumptions.*

(ii) *After construction of the monidal transformations* $\phi^{(3)}$ *and taking the 0-fiber we get an algebraic variety* $\phi^{(3)^{-1}}(Z_0^{(2)})$ *where:*

$$\phi^{(3)^{-1}}(Z_0^{(2)}) = \bigcup_{i=1}^{3} (Y_i)_0^s \cup \bigcup_{i=1}^{3} \phi^{(3)^{-1}}(\ell_i)_0 = \text{union of 6 surfaces}$$

$$\bigcup_{i=1}^{3} (Y_i)_0^s = (A^{(2)})_0^s \qquad \bigcup_{i=1}^{3} \phi^{(3)^{-1}}(\ell_i)_0 = \phi^{(3)^{-1}}(B_0^{(2)})$$

$$(A^{(2)})_0^s \cap \phi^{(3)^{-1}}(B_0^{(2)}) = \bigcup_{i=1}^{3} ((Y_i)_0^s \cap \phi^{(3)^{-1}}(\ell_i)_0).$$

*Moreover,*

(a) *For every* $i$, $\phi^{(3)^{-1}}(\ell_i)$ *is irreducible and* $\phi^{(3)^{-1}}(\ell_1)_0 \cap \phi^{(3)^{-1}}(\ell_3)_0 = \emptyset$.

(b) *For every* $i$, $(Y_i)_0^s$, *the strict image of* $(Y_i)_0$ *is irreducible.*

(c) *Let* $(\ell_i)_1 = \phi^{(3)^{-1}}(\ell_i)_0 \cap (Y_i)_0^s$. *Then* $\phi^{(3)}|\phi^{(3)^{-1}}((\ell_i)_0) : \phi^{(3)^{-1}}((\ell_i)_0) \to (\ell_i)_0$ *is a projective line bundle with cross-section* $(\ell_i)_1$ *and fiber* $\ell_i^{\text{hor}}$. *We can choose* $\ell_i^{\text{hor}} = \phi^{(3)^{-1}}(\ell_i)_0 \cap \phi^{(3)^{-1}}(\ell_{i+1})_0$ $(i = 1, 2)$.

(d) $((\ell_i)_1)^2_{\phi^{-1}(\ell_i)_0} = \begin{cases} 0 & i = 2, 3 \\ -1 & i = 1. \end{cases}$

(e) $\phi^{(3)^{-1}}(\ell_i)_0 = F_0 \quad i = 2, 3 \qquad \phi^{(3)^{-1}}(\ell_1)_0 = F_1$.

(f) $(Y_i)_0^s \xrightarrow{\sim} Y_i = \begin{cases} F_1 & i = 2, 3 \\ \mathbb{P}^2 & i = 1. \end{cases}$

(g) $(Y_i)_0^s \cap (Y_{i+1})_0^s = (C_i)_0^s \qquad (Y_1)_0^s \cap (Y_3)_0^s = \emptyset$.

(h) $((\ell_i)_1)^2_{(Y_i)_0^s} = \begin{cases} 0 & i = 2, 3 \\ 1 & i = 1. \end{cases}$

(i) $(\ell_i^{\text{hor}})^2_{\phi^{(3)^{-1}}(\ell_i)_0} = 0$.

(j) $(\ell_i^{\text{hor}})_{\phi^{(3)^{-1}}(\ell_{i+1})_0} = 0$.

(k) $((C_i)_0^s)^2_{(Y_i)_0^s} = 1 \quad i = 1, 2$.

(l) $((C_i)_0^s)^2_{(Y_{i+1})_0^s} = -1 \quad i = 1, 2$.

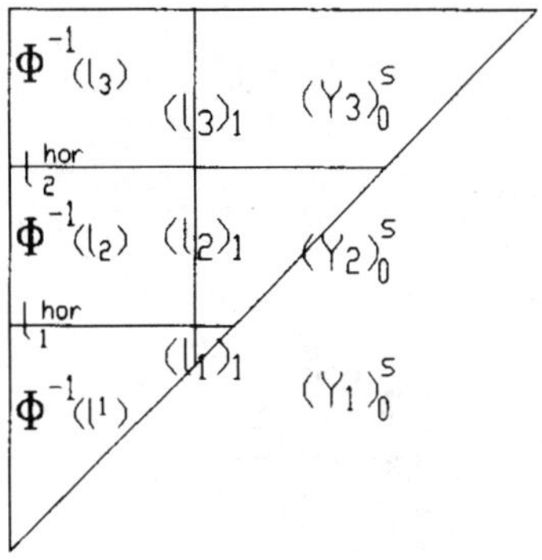

FIG. 7

*(See Figure 7 for abstract configuration of $\phi^{(3)^{-1}}(Z_0^{(2)})$.)*

*Proof.*

(i) We skip the proof.

(ii) $A^{(2)} = \bigcup_{i=1}^3 Y_i \Rightarrow (A_0^{(2)})^s = (\bigcup_{i=1}^3 Y_i)_0^s = \bigcup_{i=1}^3 (Y_i)_0^s$.

$B^{(2)} = \bigcup_{i=1}^3 \ell_i \Rightarrow \phi^{(3)^{-1}}(B_0^{(2)}) = \bigcup_{i=3}^3 \phi^{(3)^{-1}}(\ell_i)_0$.

$\phi^{(3)^{-1}}(Z_0^{(2)}) = (A_0^{(2)})^s \cup \phi^{(3)^{-1}}(B_0^{(2)}) = \bigcup_{i=1}^3 (Y_i)_0^s \cup \bigcup_{i=1}^3 \phi^{(3)^{-1}}(\ell_i)_0$.

(a) Since $\ell_i$ is irreducible and $\ell_1 \cap \ell_3 = \emptyset$.

(b) Since $(Y_i)_0$ is irreducible.

(c) From Lemma 4(iii) applied to $\ell_i \subset Y_i \quad i = 1, 2$.

(d) By Lemma 4(iv)

$$((\ell_i)_1)^2_{\phi^{(3)^{-1}}(\ell_i)_0} = -(\ell_i)^2_{Y_i} = \begin{cases} 0 & i = 2, 3 \\ -1 & i = 1. \end{cases}$$

(e) Since $Y_i$ is a projective line bundle with cross-section $C_i$ with self intersection 0 if $i = 2, 3$ and $-1$ if $i = 1$, we get

$$Y_i \simeq F_0 \quad i = 2, 3$$
$$Y_1 \simeq F_1.$$

(f)(g) Since $\phi^{(3)}$ is an isomorphism on $\bigcup_{i=1}^2 (Y_i)_0^s$ (Lemma 1 (vi)).

(h) Since $\phi^{(3)}$ is an isomorphism on $(Y_i)_0^s$ and the choice of $\ell_i$ on $Y_i \quad i = 2, 3$ (Proposition 9) and the fact that $\ell_1$ is a line on $Y_1 \sim V_1$.

(i) By Lemma 4(iii).

(j) Since $\ell_i^{\text{hor}}$ is also a fiber of $\phi^{(3)^{-1}}(\ell_{i+1})_0$.

(k) (l) By Proposition 9 and $\phi^{(3)}|_{(A^{(2)})_0^s}$ is an isomorphism. □

**Proposition 11.** *After applying the D-construction on $(Z^{(2)}, A^{(2)}, B^{(2)}, g^{(2)})$, we get $Z^{(3)}$, with a projective embedding $g^{(3)}$, where $Z^{(3)} = S_1 \cup S_2 \cup S_3 \cup S_4 \cup S_5$ and $g^{(3)}$-images of $S_1, S_2, S_3, S_4, S_5$ are projective surfaces isomorphic respectively to $V_1, U, U, V_1, T_2$.*

*Denote:*
$m_i = S_i \cap S_{i+2}$ $i = 2, 3$; $m_i^{\text{hor}} = S_i \cap S_{i+1}$ $i = 1, 2, 4$.
*Then*
$(m_i)^2_{S_i} = 0$ $i = 2, 3$, $(m_2)^2_{S_4} = 1$, $(m_3)^2_{S_5} = 0$, $(m_1^{\text{hor}})^2_{S_1} = 1$, $(m_1^{\text{hor}})^2_{S_2} = 0$, $(m_2^{\text{hor}})^2_{S_2} = 0$, $(m_2^{\text{hor}})^2_{S_3} = 0$, $(m_4^{\text{hor}})^2_{S_5} = -1$, $(m_4^{\text{hor}})^2_{S_4} = 1$.
*(See Figure 8 for abstract configuration of $Z^{(3)}$.)*

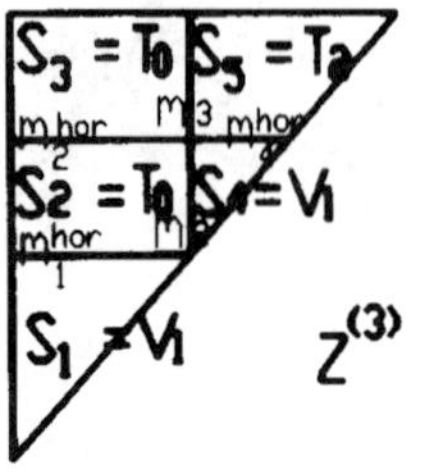


FIG. 8

*Proof.* We apply $f^{(3)}$ that is induced from $g^{(2)}$ on $\phi^{(3)^{-1}}(Z_0^{(2)})$ (See Lemma 10.)

Let $G^{(2)}$ be the line bundle on $Z^{(2)}$ defined by a hyperplane section corresponding to $g^{(2)}$. If follows from Proposition 9 that

$$G^{(2)}\big|_{Y_3} = [C_2 + 3\ell_3]$$
$$G^{(2)}\big|_{Y_2} = [C_1 + 2\ell_2]$$
$$G^{(2)}\big|_{Y_1} = [C_1].$$

Recall: $(G^{(2)})' = \phi^{(3)^*} q^{(3)^*} G^{(2)} - [\phi^{(3)^*} B_0^{(2)}]$.

Thus $(G^{(2)})'\big|_{(Y_3)_0^s} = (C_2)_0^s + 3(\ell_3)_1 - (\ell_2)_1 = (C_2)_0^s + 2(\ell_3)_1$

$(G^{(2)})'\big|_{(Y_2)_0^s} = (C_1)_0^s + 2(\ell_2)_1 - (\ell_2)_1 = (C_1)_0^s + (\ell_2)_1 = (C_2)_0^s$ (Remark iii)

$(G^{(2)})'\big|_{(Y_1)_0^s} = (C_1)_0^s - (\ell_1)_1 = 0$ (on $V_1$ every 2 lines are linear equivalent).

Since $f\big|_{(A^{(2)})_0^s}$ is determined by $(G^{(2)})'\big|_{(A^{(2)})_0^s}$ and $(A^{(2)})_0^s$ is known from Lemma 10, we get that

$f(Y_3)_0^s = T_2$ and $f$ is an isomorphism there
$f(Y_2)_0^s = V_1$ and $f$ is an isomorphism on $(\ell_2)_1$ and $(C_2)_0^s$ (Remark iii)
$f(Y_1)_0^s = \text{point}$.

By Lemma 3(v) $f\big|_{\phi^{(3)^{-1}}(B^{(2)})_0 - (A^{(2)})_0^s}$ is an isomorphism, we get

$$f\big|_{\phi^{(3)^{-1}}(\ell_i)_0 - (Y_i)_0^s} \text{ is an isomorphism.}$$

From above $f\big|_{(Y_3)_0^s}$ is an isomorphism, so

$$f\big|_{\phi^{(3)^{-1}}(\ell_3)_0} \text{ is an isomorphism.}$$

Since $(Y_2)_0^s \cap {\phi^{(3)}}^{-1}(\ell_2) = (\ell_2)_1$, and $f$ is an isomorphism there, we get

$$f\big|_{{\phi^{(3)}}^{-1}(\ell_2)_0} \text{ is an isomorphism.}$$

By the above $f({\phi^{(3)}}^{-1}(\ell_1)_0 \cap (Y_1)_0^s) = f((\ell_1)_1) = pt$, from Lemma 10(e) ${\phi^{(3)}}^{-1}(\ell_1)_0 \simeq F_1$ so $f\big|_{{\phi^{(3)}}^{-1}(\ell_1)_0}$ is a blowing down of $F_1$ to $V_1$ (Remark iii).

We introduce some new notation to the components of $Z^{(3)} = f(Y_0^{(2)})$ and to the intersection lines.

$$\begin{aligned}
S_1 &= f({\phi^{(3)}}^{-1}(\ell_1)_0) & m_2 &= S_2 \cap S_4\\
S_2 &= f({\phi^{(3)}}^{-1}(\ell_2)_0) & m_3 &= S_3 \cap S_5\\
S_3 &= f({\phi^{(3)}}^{-1}(\ell_3)_0) & m_1^{\text{hor}} &= S_1 \cap S_2\\
S_4 &= f((Y_2)_0^s) & m_2^{\text{hor}} &= S_2 \cap S_3\\
S_5 &= f((Y_3)_0^s) & m_4^{\text{hor}} &= S_4 \cap S_5
\end{aligned}$$

From above $f$ is an isomorphism on ${\phi^{(3)}}^{-1}(\ell_2)_0, {\phi^{(3)}}^{-1}(\ell_3)_0, (Y_3)_0^s$. Thus the self-intersection numbers of the intersection curves in these subvarieties are equal to the corresponding intersection numbers before applying $f$. So it remains to compute

$$(m_4^{\text{hor}})^2_{S_4},\ (m_2)^2_{S_4}\ (m_1^{\text{hor}})^2_{S_1}.$$

Since $S_4$ and $S_1$ are $V_1$, the self-intersection of every line on it is 1. Thus the above self-intersections are 1. □

*3.4 Fourth degeneration.*

We use here notations from Proposition 11.

Consider $(Z^{(3)}, A^{(3)}, B^{(3)}, g^{(3)})$ where

$A^{(3)} = S_4 \cup S_5$

$B^{(3)} = S_1 \cup S_2 \cup S_3$

$C^{(3)} = m_3 \cup m_2$

**Lemma 12.**

(i) $(Z^{(3)}, A^{(3)}, B^{(3)}, g^{(3)})$ *satisfy the D-assumptions.*

(ii) *Let $\phi^{(4)}$ be the monoidal transformation that appears in the D-construction of the above quadruple. Then using the notations introduced in section 1 and the presentation of the D-construction, we have:*

$$(\phi^{(4)})^{-1}(Z_0^{(3)}) = \tilde{S}_1 \cup \tilde{S}_2 \cup \tilde{S}_3 \cup {\phi^{(4)}}^{-1}(m_2)_0 \cup {\phi^{(4)}}^{-1}(m_3)_0 \cup (S_4)_0^s \cup (S_5)_0^s$$

*where:*

(a) $(A^{(3)})_0^s = (S_4)_0^s \cup (S_5)_0^s$.

(b) $(B^{(3)})_0 = \tilde{S}_1 \cup \tilde{S}_2 \cup \tilde{S}_3$ *and do not intersect* $(A^{(3)})_0^s$.

(c) $\phi^{(4)^{-1}}(C^{(3)})_0 = \phi^{(4)^{-1}}(m_3)_0 \cup \phi^{(4)^{-1}}(m_2)_0$,
$\phi^{(4)^{-1}}(C^{(3)})_0 \cap \widetilde{(A^{(3)})^s_0} = ((S_4)^s_0 \cap \phi^{(4)^{-1}}(m_2)) \cup ((S_5)^s_0 \cap \phi^{(4)^{-1}}(m_3))$ and
$\phi^{(4)^{-1}}(C^{(3)})_0 \cap \widetilde{(B^{(3)})}_0 = (\widetilde{(S_2)}_0 \cap \phi^{(4)^{-1}}(m_2)_0) \cup ((\widetilde{(S_3)}_0 \cap \phi^{(4)^{-1}}(m_3)_0)$.
(d) $\widetilde{(m_4^{\mathrm{hor}})^s_0} = (S_4)^s_0 \cap (S_5)^s_0$.
(e) $\widetilde{(m_i^{\mathrm{hor}})}_0 = \tilde{S}_i \cap \tilde{S}_{i+1} \quad i = 1, 2$.
*Denote:*
$m_3^{\mathrm{hor}} = \phi^{(4)^{-1}}(m_2)_0 \cap \phi^{(4)^{-1}}(m_3)_0$.
$n_i = \tilde{S}_i \cap \phi^{(4)^{-1}}(m_i)_0 \quad i = 2, 3$.
*Let* $(m_i)_1 = \phi^{(4)^{-1}}(m_i)_0 \cap (S_{i+2})^s_0 \quad i = 2, 3$.
*Then*
(f) $((m_3)_1)^2_{\phi^{(4)^{-1}}(m_3)_0} = 0 \qquad ((m_3)_1)^2_{(S_5)^s_0} = 0$.
(g) $((m_2)_1)^2_{\phi^{(4)^{-1}}(m_2)_0} = -1 \qquad ((m_2)_1)^2_{(S_4)^s_0} = +1$.
(h) $((m_4^{\mathrm{hor}})^s_0)^2_{(S_5)^s_0} = -1$.
(i) $(\widetilde{(m_i^{\mathrm{hor}})}_0)^2_{\tilde{S}_{i+1}} = 0 \quad i = 1, 2 \qquad \widetilde{(m_i^{\mathrm{hor}})}_0)^2_{\tilde{S}_i} = \begin{cases} 0 & i = 2 \\ 1 & i = 1. \end{cases}$
(j) $(m_3^{\mathrm{hor}})^2_{\phi^{(4)^{-1}}(m_3)_0} = 0 \qquad (m_3^{\mathrm{hor}})^2_{\phi^{(4)^{-1}}(m_2)_0} = 0$.
(k) $(n_3)^2_{\phi^{(4)^{-1}}(m_3)_0} = 0 \qquad (n_3)^2_{\widetilde{(S_3)}_0} = 0$.
(l) $(n_2)^2_{\phi^{(4)^{-1}}(m_2)_0} = 1 \qquad (n_2)^2_{\widetilde{(S_2)}_0} = 0$.
<u>and</u>
(m) $\tilde{S}_3 \simeq F_0, \quad \tilde{S}_2 \simeq F_0, \quad \tilde{S}_1 \simeq \mathbb{P}^2$.
(n) $\phi^{(4)^{-1}}(m_3) \simeq F_0$.
(o) $\phi^{(4)^{-1}}(m_2) \simeq F_1$.
(p) $(S_5)^s_0 \simeq F_1, \quad (S_4)^s_0 \simeq \mathbb{P}^2$.
*(See Figure 9 for abstract configuration.)*

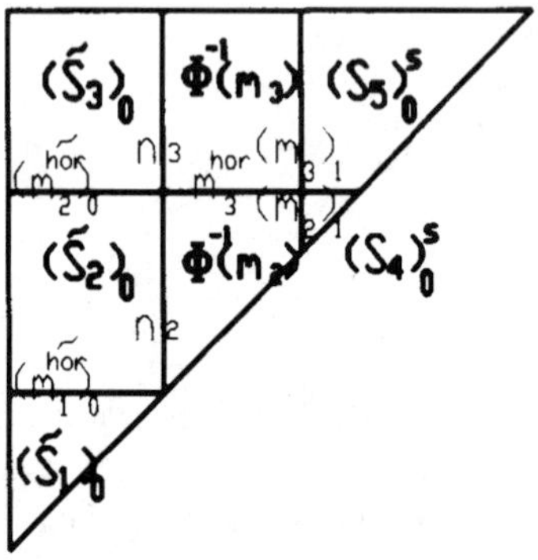

FIG. 9

*Proof.* We skip (i).
$\phi^{(4)^{-1}}(Z_0^{(3)}) = (A^{(3)})^s_0 \cup \phi^{-1}(C^{(3)})_0 \cup \widetilde{(B^{(3)})}_0$ (Lemma 1(iii)).
(a) $A^{(3)} = S_5 \cup S_4$ so $(A^{(3)})^s_0 = (S_5)^s_0 \cup (S_4)^s_0$.
(b) $C^{(3)} = m_3 \cup m_2$ so $\phi^{(4)^{-1}}(C_0^{(3)}) = \phi^{(4)^{-1}}(m_3)_0 \cup \phi^{(4)^{-1}}(m_2)_0$.

(c) $B^{(3)} = S_1 \cup S_2 \cup S_3$ so $\widetilde{(B^{(3)})}_0 = \widetilde{(S_1)}_0 \cup \widetilde{(S_2)}_0 \cup \widetilde{(S_3)}_0$ and do not intersect $(A^{(3)})_0^s$.

(d) The strict image of intersection is the intersection of the strict images.

(e) From the definition of $\tilde{S}$.

(f) By Lemma 4(ii) applied on $m_3 \subset S_5$, we get $((m_3)_1)_{(S_5)_0^s} = (m_3)^2_{S_5} = 0$ and by Lemma 4(iv), we get $((m_3)_1)^2_{\phi^{(4)-1}(m_3)_0} = 0$.

(g) Using the same arguments on $m_2 \subset S_4$, $((m_2)_1)^2_{(S_4)_0^s} = 1$ and $((m_2)_1)^2_{\phi^{(4)-1}(m_2)_0} = -1$.

(p) (h) By Lemma 4(ii) and Lemma 11 ($(m_4^{\text{hor}})^2_{S_5} = -1$), $S_5 \simeq T_2$, $S_1 \simeq V_1$) we get (h) and (p).

(j) (o) (n) By Lemma 4(iii) applied to $m_i \subset S_{i+2}$, ${\phi^{(4)}}^{-1}(m_i)_0$ is a projective line bundle with fiber $m_3^{\text{hor}}$ and cross-section $(m_i)_1$. $(m_i)_1$ has self intersection 0 if $i = 3$ and $-1$ if $i = 2$ (from (f) and (g) above). Thus ${\phi^{(4)}}^{-1}(m_3)_0$ is $F_0$ and ${\phi^{(4)}}^{-1}(m_2)_0$ is $F_1$.

(m) (i) By Lemma 1(v), $\phi^{(4)}$ is an isomorphism on $\tilde{S}_1 \cup \tilde{S}_2 \cup \tilde{S}_3$. Using this and Lemma 11 one gets (i) and (m).

(k) (l) By Remark iii, $(n_2)^2_{\phi^{(4)-1}(m_2)_0}$ is 1. By Lemma 1(v), $\phi^{(4)}$ is an isomorphism on $\widetilde{(S_2)}_0 \cup \widetilde{(S_3)}_0$. Thus the self intersections of $n_2$ and $n_3$ on $\widetilde{(S_2)}_0$ and $\widetilde{(S_3)}_0$ are equal to the self intersection of $m_2$ and $m_3$ on $S_2$ and $S_3$ respectively. □

**Proposition 13.** *Applyng the D-construction on $(Z^{(3)}, A^{(3)}, B^{(3)}, g^{(3)})$, we get an elevated projective degeneration, $Z^{(4)}$, with projective embedding $g^{(4)}$ where $Z^{(4)} = R_1 \cup R_2 \cup R_3 \cup R_4 \cup R_5 \cup R_6$ and the $g^{(4)}$-images of $R_1, R_2, R_3, R_4, R_5, R_6$ are projective surfaces isomorphic respectively to $V_1, U, U, V_1, U, V_1$, and they intersect each other as in Figure 4, where each intersection curve has self intersection 0 on $R_2, R_3, R_5$ and 1 on $R_1, R_4, R_6$.*

*Proof.* Consider ${\phi^{(4)}}^{-1}(Z_0^{(3)})$. In Lemma 12 we presented its factorization to irreducible components.

Let $G^{(3)}$ be the line bundle on $Z^{(3)}$ defined by a hyperplane section corresponding to $g^{(3)}$ (see definition of $g^{(3)}$ in Proposition 11). It follows from Proposition 11 that

$G^{(3)}|_{S_1} = [m_1^{\text{hor}}]$.

$G^{(3)}|_{S_2} = [m_2^{\text{hor}} + m_2]$.

$G^{(3)}|_{S_3} = [m_3^{\text{hor}} + m_3]$.

$G^{(3)}|_{S_4} = [m_4^{\text{hor}}]$.

$G^{(3)}|_{S_5} = [m_4^{\text{hor}} + 2m_3]$.

Let $(G^{(3)})' = \phi^{(4)*} q^{(4)*} G^{(3)} - [\phi^{(4)*}(B_0^{(3)})]$. We apply Lemma 4(v) to $m_i \subset S_{i+2}$ $(i = 2, 3)$ to get:

$G^{(3)'}|_{\phi^{(4)-1}_{(m_i)_0}} = (m_i)_1 + (\deg G^{(3)}|_{m_i}) m_3^{\text{hor}}$.

Now: $\deg G^{(3)}|_{m_i} = G^{(3)}|_{S_{i+2}} m_i = 1$.

Therefore $G^{(3)'}|_{\phi^{(4)-1}(m_i)_0} = (m_i)_1 + m_3^{\text{hor}}$.

Since $f^{(4)}$ is determined by $G^{(3)'}$, we get from Remark (iii) that $f$ is a blowing-down on ${\phi^{(4)}}^{-1}(m_2)_0$ ($\simeq F_1$) and $f^{(4)}({\phi^{(4)}}^{-1}(m_2)_0) = V_1$. Furthermore $f$ is an isomorphism on ${\phi^{(4)}}^{-1}(m_3)_0$ ($\simeq F_0$) and $f^{(4)}({\phi^{(4)}}^{-1}(m_3)_0) = U$.

By Proposition 3(iv) we get that $f^{(4)}$ on $\bigcup_{i=1}^3 \tilde{S}_i$ coincide with $f^{(3)}$on $\bigcup_{i=1}^3 S_i$ ($S_i \sim \tilde{S}_i \quad i = 1,2,3$). Thus the $f^{(4)}$-images of $\tilde{S}_i$ $i = 1,2,3$ are projective surfaces isomorphic respectively to $V_1, U, U$.

By Proposition 3(vi) we get that $f^{(4)}$ on $(S_4)_0^2, (S_5)_0^2$ coincide with the regular map of $S_4, S_5$ induced from $G^{(3)}|_{S_4} - [m_2]$ and $G^{(3)}|_{S_5} - [m_3]$, respectively ($S_i \sim (S_i)_0^2$ $i = 4,5$). Now: $G^{(3)}|_{S_4} - [m_2] = [m_4^{\text{hor}}] - [m_2] = 0$ (since $S_4 \simeq V_1$). Thus $f^{(4)}$ contracts $(S_4)_0^2$ to a point. Now $G^{(3)}|_{S_5} - [m_3] = [m_4^{\text{hor}} + 2m_3] - [m_3] = [m_4^{\text{hor}}] + m_3$. Since $(S_5)_0^s$ is $F_1$ we can use Remark (iii) to conclude that $f$ contracts $(S_5)_0^s$ to $V_1$.

Let us denote:

$$
\begin{aligned}
R_i &= f^{(4)}(\tilde{S}_i) && i = 1,2,3 \\
R_i &= f^{(4)}({\phi^{(4)}}^{-1}(m_{i-2})_0) && i = 4,5 \\
R_6 &= f^{(4)}((S_5)_0^s).
\end{aligned}
$$

Since $f^{(4)}$ is an isomorphism on $\tilde{S}_i$ for $i = 1,2,3$ and on ${\phi^{(4)}}^{-1}(m_3)_0$, the self-intersection numbers of the intersection lines lying on $R_1, R_2, R_3$ and $R_5$ can be deduced from Lemma 12. Since $R_4$ and $R_6$ are obtained from ${\phi^{(4)}}^{-1}(m_2)_0$ and $(S_5)_0^2$ as in Remark (iii), we can use the remark to deduce the other self-intersection numbers. □

*3.5 Fifth degeneration.*

**Proposition 14.** *$Z^{(4)}$ can be regenerated to $Z^{(5)}$ a union of 9 planes as in Figure 4.*

*Proof.* Recall from the previous degeneration, $Z^{(4)} = R_1 \cup \cdots \cup R_6$ with intersection lines and intersection points $a_i \quad i = 3,5,6$ as in Figure 10.1.

We add points $a_i \quad i = 1,2,4,7,8,9,10$ to get $a_1 \ldots a_{10}$ as in Figure 10.2, such that the connecting lines are contained in the corresponding $R_k$, and $\{a_i\}_{i=1}^{10}$ are in the general position. We then get a graph (Figure 10.3) in which squares correspond to projective nonsingular quadrics, triangles to 2-planes and edges on the border of two complexes to an intersection line.

Given such ten points $a_1, \ldots a_{10}$ we want to describe the parameter space of such a configuration with $a_1, \ldots a_{10}$ as vertices. Consider a square of the configuration. Its vertices are $a_{i_1}, a_{i_2}, a_{i_4}$. There exists a unique projective 3-space, $\widetilde{\mathbb{CP}^3}$ that contains $\{a_{i_t}\}_{t=1}^4$. There exist 4 planes in $\mathbb{CP}^3$ $\{H_{i_t}\}_{t=1}^4$ such that $H_{i_t}$ contains $\{a_{i_k}\}_{\substack{k=1\\k\neq t}}^4$ and $H_{i_t}$ does not contain $a_{i_t}$. Let $L_{i_t}$ be the linear form on $\widetilde{\mathbb{CP}^3}$ defining $H_{i_t}$. Consider $Q_{\lambda,\mu} = \lambda L_{i_2} L_{i_3} + \mu L_{i_1} L_{i_4}$. Let $M_{\lambda,\mu}$ be the quadric in $\widetilde{\mathbb{CP}^3}$ defined by $Q_{\lambda,\mu}$. Let $C$ be the union of the 4 lines represented by the edges of the square of Figure 4.

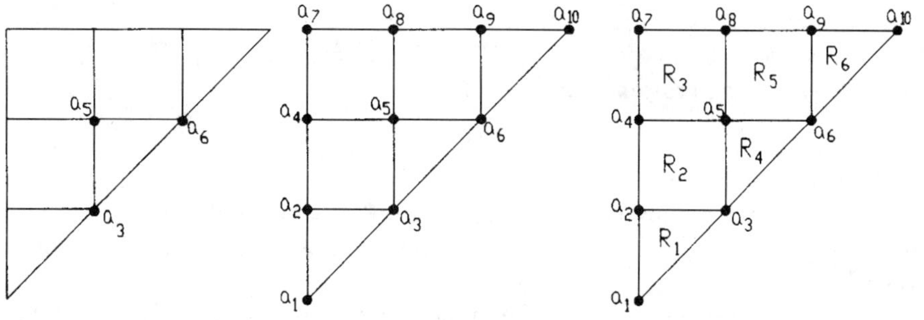


FIG. 10

**Claim.** *Let $M$ be a quadric surface in $\mathbb{CP}^N$. Then $M$ contains $C$ iff $M = M_{\lambda,\mu}$ for some $\lambda, \mu$.*

*Proof.* The "iff" part is trivial. Assume that $M$ contains $C$. There exists a 3-dimensional projective subspace $P$ in $\mathbb{CP}^N$ containing $M$. Then $P$ contains $\{a_{i_t}\}_{t=1}^{4}$. Since $\{a_{i_t}\}$ are in the general position, $P$ must equal $\widetilde{\mathbb{CP}^3}$. Let $Q$ be the quadric equation in $\widetilde{\mathbb{CP}^3}$ defining $M$. Let $M_{\lambda_1,\mu_1}$ be an irreducible member of the family $\{M_{\lambda,\mu}\}$.

Let $H$ denote a hyperplane in $\widetilde{\mathbb{CP}^3}$. Then $Q\big|_{M_{\lambda_1,\mu_1}}$, $Q_{1,0}\big|_{M_{\lambda_1,\mu_1}} \in H^0\left[M_{\lambda_1,\mu_1}, \mathcal{O}_{M_{\lambda_1,\mu_1}}[2H_0]\big|_{M_1}\right]$. The divisor $Q\big|_{M_{\lambda_1,\mu_1}}$ equals $Q.Q_{\lambda_1,\mu_1}$. Now, $Q.Q_{\lambda_1,\mu_1}$ and $C$ are both positive cycles of degree 4 and $Q.Q_{\lambda_1,\mu_1} \supseteq C$. Therefore $Q.Q_{\lambda_1,\mu_1} = C$. Thus $\left(Q\big|_{M_{\lambda_1,\mu_1}}\right) = C$. Similarly $\left(Q_{1,0}\big|_{M_{\lambda_1,\mu_1}}\right) = C$. Thus $\left(Q\big|_{M_{\lambda_1,\mu_1}}\right) = \left(Q_{1,0}\big|_{M_{\lambda_1,\mu_1}}\right)$. Therefore $Q\big|_{M_{\lambda_1,\mu_1}} = \alpha Q_{1,0}\big|_{M_{\lambda_1,\mu_1}}$ for some $\alpha \in \mathbb{C}$. Thus $Q - \alpha Q_{1,0} = \beta Q_{\lambda_1,\mu_1}$. By comparing degrees we get that $\beta \in \mathbb{C}$. Thus $Q$ is a linear combination of $Q_{1,0}$ and $Q_{\lambda_1,\mu_1}$. Thus $Q = Q_{\lambda,\mu}$ for some $\lambda, \mu$ and $M = M_{\lambda,\mu}$ for some $\lambda, \mu$.

We go back to the 10 points $a_1 \ldots a_{10}$. Every three points which are the vertices of a triangle uniquely define the 2-plane. Every 4 points which are the vertex of a sphere uniquely define a $\widetilde{\mathbb{CP}^3}$ which contains them. By the claim a projective quadric that contains the curve represented by the edges is determined by some $(\lambda : \mu)$. If it is irreducible, then $\lambda, \mu \neq 0$. Assume that $R_i \quad i = 2, 3, 5$ are determined by $(\lambda_i, \mu_i)$. Thus any configuration as in Figure 3 is determined by the three pairs $(\lambda_2 : \mu_2)$ $(\lambda_3 : \mu_3)$ $(\lambda_5, \mu_5)$ where $\lambda_i, \mu_i \neq 0$. Thus the parameter space defining all of these configurations can be described by $\mathbb{P}^1 \times \mathbb{P}^1 \times \mathbb{P}^1 - Z$ where $Z$ is a proper subvariety. This space is irreducible and nonsingular. $Z^{(4)}$ corresponds to a point $t_1$ in $\mathbb{P}^1 \times \mathbb{P}^1 \times \mathbb{P}^1 - Z$.

Let $t_0 = ((1 : 0), (1 : 0), (1 : 0) \in \mathbb{P}^1 \times \mathbb{P}^1 \times \mathbb{P}^1$. This point corresponds to $Z^{(5)}$

in a configuration as in Figure 4.

We can take a simple path $\Gamma$ in $\mathbb{P}' \times \mathbb{P}' \times \mathbb{P}'$ from $t_1$ to $t_0$ s.t. $\Gamma \cap Z = t_0$ which gives a degeneration of $Z^{(4)}$ to $Z^{(5)}$. □

*3.6 Sixth degeneration.*

In the last step we are going to degenerate $Z^{(5)}$ to a similar configuration in which $\{a_i\}$ are real and satisfy a growth condition as in Theorem IX.2.1 of our paper [MoTe4].

Every configuration of planes like in figure 5 is determined by the 10 vertices and which are "10 points in the general position in $\mathbb{CP}^N$" and vice versa. The sets of "10 points in the general position in $\mathbb{CP}^n$" form a nonsingular irreducible variety in which two such sets can be connected in a continuous family. Thus $Z^{(5)}$ can be degenerated to $Z^{(6)}$ in which the vertices satisfy the above conditions.

## References


[Mo1] Moishezon, B.G., *Stable branch curves and braid monodromies*, Lecture Notes in Math. **862** (1981), 107-192.

[Mo2] Moishezon, B.G., *On branch curves of cuspidal curves*, J. of Algebraic Geometry **2, no. 2** (1993), 309-384.

[Mo3] Moishezon, B.G., *Complex surfaces and connected sums of complex projective planes*, Lecture Notes in Math. **603** (1977).

[MoTe1] Moishezon, B.G, Teicher, M., *Simply connected algebraic surfaces of positive index*, Inventiones Math. **89** (1987), 601-643.

[MoTe2] Moishezon, B.G, Teicher, M., *Existence of simply-connected algebraic surfaces of general type with positive and zero indices*, Proc. Nat. Ac. Sci., USA **83** (1986), 6665-6666.

[MoTe3] Moishezon, B.G, Teicher, M., *Galois covering in the theory of algebraic groups*, Proceedings in Symposia in Pure Mathematics **46** (1987), 47-65.

[MoTe4] Moishezon, B.G, Teicher, M., *Braid group techniques in complex geometry I: Line arrangements in* $\mathbb{CP}^2$, Contemporary Mathematics **78** (1988), 425-555.

[MoTe5] Moishezon, B.G, Teicher, M., *Braid group techniques in complex geometry I: From arrangements of lines to cuspidal curves*, Lecture Notes in Math. **1479** (1989), 131-179.

[Za] Zariski, O., *On the Poincare group of rational plane curves*, Amer. J. Math. **58** (1936), 607-619.

Contemporary Mathematics
Volume **162**, 1994

# Braid Group Techniques in Complex Geometry IV: Braid Monodromy of the Branch Curve $S_3$ of $V_3 \to \mathbb{CP}^2$ and application to $\Pi_1(\mathbb{CP}^2 - S_3, *)$

BORIS MOISHEZON and MINA TEICHER

ABSTRACT. This work is a continuation of our articles "Braid group techniques in complex geometry I, II, III." We consider here the Veronese surface of order 3, $V_3$ and compute the braid monodromy of $S_3$, the branch curve of $V_3 \to \mathbb{CP}^2$. We accomplish this by using the projective degeneration introduced in "Braid group techniques III" and the algorithms introduced in "Braid group techniques I, II."

## 0. Introduction

This work is the continuation of "Braid group techniques in complex geometry III: Projective degeneration of $V_3$" and of "Braid group techniques in complex geometry I: Line arrangements in $\mathbb{CP}^2$" and "Braid group techniques in complex geometry II: From arrangements of lines to cuspidal curves" that preceded it ([MoTe3], [MoTe1], [MoTe2]). We study here the generic projection $\pi : V_3 \to \mathbb{CP}^2$, and give an explicit description of the braid monodromy of $S_3$, the branch curve of $\pi$.

We use the sequence of projective degenerations
$V_3 = Z^{(0)} \rightsquigarrow Z^{(1)} \dots Z^{(i)} \rightsquigarrow Z^{(i+1)} \rightsquigarrow \dots \rightsquigarrow Z^{(6)}$ of [MoTe3]. Let $S^{(i)}$ be the branch curve of the generic projection $\pi^{(i)} : Z^{(i)} \to \mathbb{CP}^2$, $S^{(i+1)}$ is a degeneration of $S^{(i)}$. The curve $S^{(6)}$ is an arrangement of lines, so we can use Theorem XI.2.1 of [MoTe1] to get an explicit formula of the braid monodromy of $S^{(6)}$. We then use the results of [MoTe2] to get the detailed deduction formulas of the braid monodromy of $S^{(5)}, S^{(4)}, \dots S^{(0)} = S_3$ from the formula of $S^{(6)}$. In our next work, we shall prove the result about the explicit structure of $\Pi_1(\mathbb{CP}^2 - S_3, *)$ using the braid monodromy formula for $S_3$ that we obtained here. In this paper we give the precise formulation of this result.

1991 Mathematics Subject Classification. Primary 20F36, 14J10
Final version.

In fact, what we get here is a braid monodromy factorization of $S_3$. We refer the reader to [MoTe1] for definitions of braid monodromy and braid monodromy factorization. Nevertheless, we recall

*Definition.* Braid monodromy factorization

Let $\varphi : \Pi_1 \to B_m$ be a braid monodromy. Let $\Delta^2$ be the generator of center $(B_m)$. We say that $\Delta^2 = \Pi b_i$ is a braid monodromy factorization if there exists a base $\{\delta_i\}$ of $\Pi_1(\mathbb{C} - M_i u)$ s.t. $b_i = \varphi(\delta_i)$.

## 1. Regeneration of Braid Monodromy

*Definition.* Branch curve of a projection

Let $V$ be an algebraic surface in $\mathbb{CP}^N$ s.t. every irreducible component is of dimension 2. Let $\pi : V \to \mathbb{CP}^2$ be a generic projection. There exists a curve $S \subseteq \mathbb{CP}^2$ and $n \in Z^+$ such that $\forall x \in \mathbb{CP}^2 - S$, $\#\pi^{-1}(x) = n$ and $\forall x \in S$ $\#\pi^{-1}(x) \leqq n$. We call $n = \deg \pi = \deg V$ and $S$ the branch curve of $V$.

*Definition.* Regeneration of a Braid group, of a frame

1) We say that $B_q = B[(D_1, K_1)]$ is a regeneration of $B_p = B[(D_2, K_p)]$ if there exist $A \subset K_p$ $A = \{d_j\}_{j=1}^{q-p}$ and $d_j$ a small disc around each of the points $c_j \in A$ s.t. $d_j \cap d_i = \emptyset$ for $i \neq j$ and $\alpha : D_2 \to D_1$ s.t. $K_q = \alpha(K_p - A) \cup B$ where $B = \cup\{c'_j, c''_j\}$ s.t. $c'_j, c''_j \in \alpha(d_j)$. We call $\alpha$ the diffeomorphism defining the regeneration.

2) We say that a frame $Y_1 \dots Y_{q-1}$ of $B_q$ is a regeneration of a frame $X_1 \dots X_{p-1}$ of $B_p$ if there exists a sequence $y_1 \dots y_{q-1}$ of consecutive paths giving $Y_1 \dots Y_{q-1}$ and a sequence $x_1 \dots x_{p-1}$ giving $X_1 \dots X_{p-1}$ s.t. $\alpha(\cup x_i - \cup d_j) = \cup y_i = \alpha(\cup d_j)$. Let $\rho_j$ be the halftwist in $B_q$ defined the segment $[c'_j, c''_j]$. One can see that $Y_1 \dots Y_{q-1}$ is defined up to conjugation by $\prod\limits_{j=1}^{q-p} \rho_j^{m_j}$ for some $m_j \in \mathbb{Z}$. (See definition of a frame of $B_n$ in [MoTe1].)

3) We say that an embedding $\alpha' : B_\ell \hookrightarrow B_q$ is a regeneration of an embedding $\alpha : B_k \hookrightarrow B_p$ if $\alpha$ is defined by $X_1, \dots, X_{k-1}$ which is part of a frame $X_1, \dots, X_{p-1}$ of $B_p$, if $\alpha'$ is defined by $Y_1 \dots Y_{\ell-1}$ which is part of a frame $Y_1 \dots Y_{q-1}$ and $Y_1 \dots Y_{q-1}$ is a regeneration of $X_1 \dots X_{p-1}$. $\alpha'$ is determined by $\alpha$ up to the conjugation by $\rho$ for some $\rho \in G_1$, when $G_1 = \langle \rho_j \mid \rho_j \in A, \rho_j \in x_t \;\; 1 \leq t \leq k-1 \rangle$.

In [MoTe3] we constructed a series of algebraic varieties $Z^{(i)}$ $i = 0 \dots t$, $Z^{(0)} = V_3$, s.t. $Z^{(i+1)}$ is a degeneration of $Z^{(i)}$ $i = 0 \dots 5$.

For every $Z^{(i)}$ we take a generic projection $\pi^{(i)}$ tp $\mathbb{CP}^2$. We denote by $S^{(i)}$ the corresponding branch curve. One can choose a family of generic projectionss to $\mathbb{CP}^2$ such that $S^{(i+1)}$ is a projective degeneration of $S^{(i)}$.

Let us choose a line in infinity transversal to $S^{(i)}$ $i = 0 \dots 6$ and a system of coordinates $x.y$ in the affine part, $\mathbb{C}^2$, which is in a general position to each $S^{(i)}$ $i = 0 \dots 6$.

Let $m^{(i)} =$ degree of $S^{(i)} = \deg \pi^{(i)}$.

Let $N^{(i)} = \{x \in X-\text{axis} \mid \#\pi^{(i)-1}(x) \leq m^{(i)}\}$.

Let $u$ be a real point in $X$-axis. $N^{(i)} << u\ \forall i$. Let $\mathbb{C}_u = \{(u,y) \mid y \in \mathbb{C}\}$.

Let $K^{(i)} = S^{(i)} \cap \mathbb{C}_u$.

Let $B^{(i)} = B_{m_i}[\mathbb{C}_u, K^{(i)}]$.

Let $\varphi^{(i)} : \prod_1(X\text{-axis - } N^{(i)}, u) \to B_{m_i}[\mathbb{C}_u, K^{(i)}]$ be the braid monodromy with respect to $S^{(i)}, \pi^{(i)}.u$.

We want to compute $\varphi^{(0)}$.

By the construction of $Z^{(6)}$, $K^{(6)}$ is a set of real points. For every point $c$ in $K^{(6)}$ there exists an $i$ s.t. $c$ is replaced in $K^{(j)}$ for $j \leq i$ by 2 points close to each other. In $K^{(0)}$ we end up with twice as many points as in $K^{(6)}$. This implies that $B^{(i)}$ is a natural regeneration of $B^{(i+1)}$. We choose a real frame in $B^{(6)}$ and inductively a frame of $B^{(i)}$ which is a regeneration of the frame of $B^{(i+1)}$. We describe the different $\varphi^{(i)}$ using those frames. We start with $\rho^{(6)}$ and use the forthcoming 3 regeneration rules to obtain a formula for $\varphi^{(0)}$ using the regenerated frame going step by step from $i = 6$ to $i = 0$.

Let $\alpha_i$ be the diffeomorphism defining the regeneration of $B^{(i+1)}$ to $B^{(i)}$. Let $\hat{\alpha}_i = \alpha_6 \circ \cdots \circ \alpha_i$. In what follows, we shall use new geometric realizations for $B_{m_i}$ obtained from $B_{m_i}[\mathbb{C}_u, K^{(i)}]$ using $\hat{\alpha}_i^{-1}$. For instance, when we say that a curve is defining a halftwist in $B^{(i)}$ goes above the real line, we mean that its image under $\hat{\alpha}_i^{-1}$ goes above the real line. We call this geometric realization "a real model."

In each step we compute first the local braid monodromy and L.V.C. for the new singularities, using the regenerated frame. The embedding of the local braid monodromy into the global one is defined up to conjugation that appears in the appropriate regenerated frame. Thus we get global braid monodromy and L.V.C. up to conjugation. (See the definition of L.V.C. in [MoTe1].)

**Regeneration Rules**

In what follows we shall describe what happens to a factor in a factorized expression of a braid monodromy after a regeneration process of the braid group and the frame. We shall consider 3 different cases according to the type of singularity that induced the factor. We shall need the following notations and definitions.

*Definition.* Regenerated halftwist with respect to bushes $J_1, J_2$.

Let $Y_{ij} = H(y_{ij})$ be a halftwist with respect to a path $y_{ij}$ from $q_i$ to $q_j$ in $\mathbb{C}_u \cdot q_i, q_j \in K^{(\ell)}$. Let $d_i, d_j$ be 2 small discs around $q_i, q_j$ respectively, such that $d_i \cap d_j = \emptyset$. Assume that in a regeneration process of $B[\mathbb{C}_u, k^{(\ell)}]$, $q_i$ is replaced by $q_{i_1}, \ldots, q_{i_p}$ in $d_i$ and $q_j$ is replaced by $q_{j_1}, \ldots, q_{j_p}$ in $d_j$. Let $q_i^1 = \partial(d_i) \cap y_{ij}$, $q_j^1 = \partial(d_j) \cap y_{i_j}$, $y'_{ij} = y_{ij} - d_i - d_j$. Let $J_1 = \{\alpha_{i_k}\}_{k=1}^{p}$ be a bush in $d_i$ starting at $q_i^1$ and let $J_2 = \{\beta_{i_\ell}\}_{\ell=1}^{q}$ be a bush in $d_i$ starting at $q_j^1$. Let $y_{i_k.j_\ell} = \alpha_{i_k} y'_{ij} \beta_{j_\ell}^{-1}$. We call $\{Y_{i_k,j_\ell}(J_1, J_2)\}_{k,\ell} = \{H(y_{i_k,j_\ell})\}_{k,\ell}$ the regenerated halftwist with respect to $J_1$ and $J_2$.

*Notation.*

$Z_{a,b} = H(z_{a,b})$ the halftwist with respect to $z_{a,b}$ where $a, b \in K \subseteq \mathbb{C}_u$ and: $z_{a,b}$ = shortest path in $\mathbb{C}_u - K$ connecting $a$ and $b$.

$\rho_j = Z_{jj'}$.

$$Y^{(1)}_{i,j}(J_1, J_2) = Y_{i_1,j_1}(J_1, J_2) \cdot Y_{1_2,j_2}(J_1, J_2).$$
$$Y^{(2)}_{i,j}(J_1, J_2) = \prod_{k=1}^{k} \prod_{\ell=1}^{q} Y^2_{i_k,j_\ell}(J_1, J_2).$$
$$Y^{(3)}_{i,j} = \prod_{m=1}^{-1} (Y_{ij})^3 \rho_j^m.$$
$$(A)_B = A_B = B^{-1}AB.$$

Recall the following definitions that we use in the regeneration rules.

*Definition.* Equality of factorized expressions

Two factorized expressions $g_i \cdot \ldots \cdot g_k$ and $h_i \cdot \ldots \cdot h_k$ of $g$ in a froup $G$ are equal if $g_i = h_i \ \forall 1 \le i \le k$.

*Definition.* Hurwitz move

Let $g_i \cdot \ldots \cdot g_k = h_i \cdot \ldots \cdot h_k$ be two factorized expressions of the same element in a group $G$. We say that $g_i \cdot \ldots \cdot g_k$ is obtained from $h_i \cdot \ldots \cdot h_k$ by a Hurwitz move if $\exists\, 1 \le p \le k-1$ s.t. $g_i = h_i \ i \ne p, p+1$, $g_p = h_p h_{p+1} h_p^{-1}$ and $g_{p+1} = hp$ or $g_p = h_{p+1}$ and $g_{p+1} = h_{p+1}^{-1} h_p h_{p+1}$

*Definition.* Hurwitz equivalence of factorized expressions

Let $g_i \cdot \ldots \cdot g_k = h_i \cdot \ldots \cdot h_k$ be two factorized expressions of the same element in a group $G$. We say that $g_i \cdot \ldots \cdot g_k$ is a Hurwitz equivalent to $h_i \cdot \ldots \cdot h_k$ if $h_i \cdot \ldots \cdot g_k$ is obtained from $h_i \cdot \ldots \cdot h_k$ by a finite number of Hurwitz moves. We denote it by $g_1 \cdot \ldots \cdot g_k \underset{He}{\simeq} h_i \cdot \ldots \cdot h_k$.

*Definition.* A factorized expression invariant under $h$

Let $g = g_i \cdot \ldots \cdot g_k$ be a factorized expression in a group $g$. We say that $g$ is invariant under $h \in G$ if $(g_1)_h \cdot \ldots \cdot (g_k)_h$ is Hurwitz equivalent to $g_i \cdot \ldots \cdot g_k$.

In our process of regeneration, we get doubling in generic points such that a branch point is replaced by 2 branch points, a node is replaced by 2 or 4 nodes and a tangent point is replaced by 3 cusps. The following three rules explain the changes in the braid monodromy factors when these regenerations appear.

*First regeneration rule.* Regeneration of a branch point.

(a) In a regeneration process of a branch point, in the typical fiber, $q_i$ is replaced by $\{q_{i_k}\}_{k=1}^2$ and $q_j$ is replaced by $\{q_{j_\ell}\}_{\ell=1}^2$. Then a factor of the braid monodromy of the form $Y_{ij}$ is replaced in the regeneration process by $Y^{(1)}_{ij}(J_1, J_2)$ for some $J_1, J_2$ where $Y^{(1)}_{i,j}(J_1, J_2) = Y_{i_1,j_1}(J_1, J_2) \cdot Y_{i_2,j_2}(J_1, J_2)$.

(b) Assume $q_{i_k}$ and $q_{i_\ell}$ are real points $k = 1, 2$ $\ell = 1, 2$ and $q_i < q_j$. Let $J_-$ be a bush in $d_i$ below the real line and $J_+$ be a bush in $d_i$ above the real line. Then for every bush $J_1$ in $d_i$ and every bush $J_2$ in $d_j$ there exists $m_1, m_2 \in \mathbb{Z}$ such that $Z_1 = (Z_-)Z^{m_1}_{i_1,i_2}$ and $Z_2 = (Z_+)Z^{m_2}_{j_1,j_2}$. Denote $Y^{(1)}_{i,j}(m_1, m_2) = Y^{(1)}_{i,j}(Z_1, Z_2)$, and $Y^{(1)}_{i,j} = Y^{(1)}_{i,j}(0,0) = Y^{(1)}_{i,j}(Z_-, Z_+)$.

(c) $Y^{(1)}_{i,j}(m_1, m_2)$ is Hurwitz equivalent as a factorized expression to $Y^{(1)}_{i,j}(0, m_2 - m_1)$.

(d) Let us denote $Y^{(1)}_{i,j}(m) = Y^{(1)}_{i,j}(0, m)$. Then $Y_{i,j}$ is replaced in the regeneration process by $Y^{(1)}_{i,j}(m)$ for some $m$.

(e) $Y^{(1)}_{i,j}(m)$ is invariant under $(Z_{i_1i_2}Z_{j_1j_2})^r$ $\forall r \in \mathbb{Z}$.

*Proof.*
(a) Lemma 3.1 of [MoTe2].
(b) Simple analysis of 2 points punctured disc.
(c) Lemma 3.1 as in (a), second part.
(d) From (a) and (c).
(e) By (c).

*Second regeneration rule.* Regeneration of a node.
(a) In our regeneration process of a node, the element $q_i$ of the typical fiber is replaced by $\{q_{i_k}\}_{k=1}^p$ and $q_j$ is replaced by $\{q_{j_\ell}\}_{\ell=1}^q$. Then a factor of the form $Y^{(2)}_{i,j}$ will be replaced by a factorized expression of the form $Y^{(2)}_{i,j}(J_1J_2)$ where $Y^{(2)}_{i,j}(J_1J_2) = \prod\limits_{k=1}^{p}\prod\limits_{\ell=1}^{q} Y^{(2)}_{i_k,j_\ell}(J_1J_2)$.
(b) Up to a Hurwitz equivalence of factorized expression $Y^{(2)}_{i_j,j}(J_1J_2)$ is independent of the choice of $J_1$ and $J_2$ and by abuse of the notation we denote it by $Y^{(2)}_{i,j}$.
(c) If in a regeneration process $q_i$ is replaced by $q_i$ and $q_{i'}$ then $Y^{(2)}_{i,j}$ is replaced by $Y^{(2)}_{ii',j}$, and if also $q_j$ is replaced by $q_j$ and $q_{j'}$, $Y^{(2)}_{i,j}$ is replaced by $Y^{(2)}_{ii',jj'}$. Moreover, $Y^{(2)}_{ii',j}$ is invariant under $Z^p_{ii'}$, and $Y^{(2)}_{ii',jj'}$ is invariant under $Z^p_{ii'}Z^q_{jj'}$.

*Proof.*
(a) (b): Proposition 3.2 of "Braid group techniques II," [MoTe2].
(c): By (a), (b).

*Third regeneration rule.*
(a) In our regeneration process of a tangent point, in the typical fiber, $q_j$ is replaced by $q_j$ and $q_{j'}$ in $d_j(\subseteq \mathbb{C}_u)$ and a factor of the form $Y^4_{i,j}$ in the braid monodromy factorized expression is replaced by $Y^{(3)}_{i,jj'}(m)$ for some bush $J_2$ in $d_j$, where
$Y^{(3)}_{i,jj'}(m) = (Y^3_{ij})_{\rho_j^{m+1}} \cdot (Y^3_{ij})_{\rho_j^m} \cdot (Y^3_{ij})_{\rho_j^{m-1}}$.
(b) $Y^{(3)}_{i,jj'}(m)$ is independent of the choice of $m$.
(c) $Y^4_{i,j}$ is replaced by $Y^{(3)}_{i,jj'} = Y^{(3)}_{i,jj'}(0)$.
(d) $Y^{(3)}_{i,jj'}$ is invariant under $Z^r_{jj'}$.

*Proof.*
(a) A degree 4-element comes from a tangency point in the branch curve.

By Lemma 3.3 of ([MeTe2]) in the regeneration process, the point $q_j$ is replaced by $q_j, q_j'$, the tangency point is degenerated to 3 cusps, and in the braid monodromy formula, the degree 4 element is replaced by 3 cubes of halftwists as follows: $(Y_{ij})_{\rho_j^{m+1}} \cdot (Y_{ij})_{\rho_j^m} \cdot (Y_{ij})_{\rho_j^{m-1}}$.
(b) Again by Lemma 3.3, as above.
(c) By (a) and (b).
(d) By (b).

## 2. Local braid monodromies of the branch curve of generic projection to $\mathbb{CP}^2$ of a Veronese surface of order 3

The configuration of Fig. 1 describes $Z^{(0)}$ to $Z^{(6)}$ (see [MoTe3]). We denote there the numeration of intersection lines on $Z^{(6)}$, $\{\hat{L}_i\}_{i=1}^9$ and their intersection points $\{a_j\}$. We also denote the irreducible components of $Z^{(5)}$ by $\{P_i\}_{i=1}^9$ and of $Z^{(4)}$ by $\{R_i\}_{i=1}^6$. We have generic projections $\pi^{(i)} : Z^{(i)} \to \mathbb{CP}^2$, $S^{(i)}$ the branch curves in $\mathbb{CP}^2$ and $\varphi^{(i)}$ their braid monodromy. We start with $\varphi^{(6)}$.

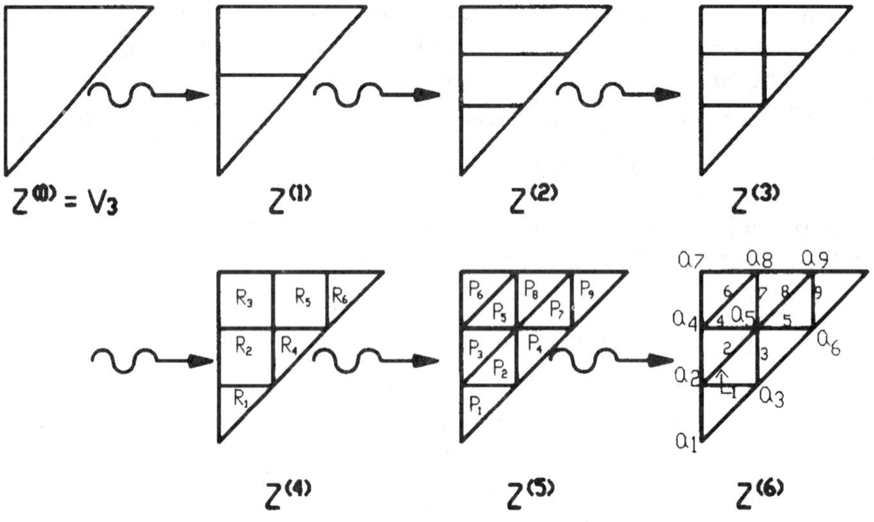


FIG. 1

Let $L_j = \pi^{(6)}(\hat{L}_j)$ $j = 1 \dots 9$. So, $S^{(6)} = \bigcup\limits_{j=1}^{9} L_j$ = union of 9 lines. Let

$$v_j = \begin{cases} \pi^{(6)}(a_{j+1}) & j \leq 5 \\ \pi^{(6)}(a_{j+2}) & j \leq 6.7, \end{cases}$$

$v_1 \dots v_7$ are the singular points of $S^{(6)}$.

Let $C$ be the union of all lines connecting pairs of the $v_j$'s. $S^{(6)}$ is a subcurve of $C$.

Theorem IX. 2.1 of [MoTe1] gives us a description of braid monodromy for $C : \Delta_C^2 = \prod\limits_{\gamma=7}^{1} C_v \Delta_v^2$ with an appropriate description of the L.V.C. We use this formula to obtain a description of $\varphi^{(6)}$ by deleting all factors that involve lines that do not appear in $S^{(6)}$. Thus we get

$$\Delta_{S^{(6)}}^2 = \prod_{j=7}^{1} \tilde{C}_j \tilde{\Delta}_j^2$$

We want to describe each factor separately.

Let $q_i = L_i \bigcap \mathbb{C}_u$. $q_i$ be real.

We shall use the following notations for paths and braids in $B[\mathbb{C}_u, K]$ :

$z_{ij}$ = path connecting $q_i$ and $q_j$ below a real line.
$\overline{z}_{ij}$ = path connecting $q_i$ and $q_j$ above a real line.
$\overset{(a)(b)}{z_{ij}}$ = path connecting $q_i$ and $q_j$ almost always below a real line, but going below $a$ and $b$.
$\underset{(a)(b)}{\overline{z}_{ij}}$ = path connecting $q_i$ and $q_j$ almost always above a real line, but going below $a$ and $b$.
$Z_{ij} = H(z_{ij})$.
$\overline{Z}_{ij} = H(\overline{z}_{ij})$.
$\overset{(a)(b)}{Z_{ij}} = H(\overset{(a)(b)}{z_{ij}})$.
$\underset{(a)(b)}{\overline{Z}_{ij}} = H(\underset{(a)(b)}{\overline{z}_{ij}})$.

By Theorem IX.2.1 of [MoTe1] we get:

$$\tilde{C}_1 = D_1 \cdot D_2 = \mathrm{Id}$$
$$\tilde{C}_2 = D_3 \cdot = \mathrm{Id}$$
$$\tilde{C}_3 = D_4 \cdot D_6 = Z^2_{14} \prod_{i=1,2,3,5} \overline{Z}^2_{i6}$$
$$\tilde{C}_4 = D_5 \cdot D_7 \cdot D_8 = \overline{Z}^2_{15} \cdot \underset{(6)}{\overline{Z}^2_{17}} \cdot \overline{Z}^2_{18} \overline{Z}^2_{68}$$
$$\tilde{C}_5 = D_9 = \prod_{\substack{i=1 \\ i \neq 5}}^{7} \overline{Z}^2_{i9}$$

For $j \neq 4$ : $\tilde{\Delta}^2_j = \underline{Z}^2_{j_1,j_2}$ for $L_{j_1}$ and $L_{j_2}$ s.t. $v_j = L_{j_1} \cap L_{j_2}$.

For $j = 4$ the L.V.C. corresponding to the embedding $B_6 \hookrightarrow B_9$ is defined as follows: We take a local real frame of $B_6$ and embed it in $B_9$ by a sequence of consecutive simple paths connecting the points 2, 3, 4, 5, 7, 9 below the real line.

For every $k = 1 \dots 7$ we shall take a small neighborhood of $v_k$ and analyse the local braid monodromy of $S^{(j)}$ which come from singularities of this neighborhood, from $j = 6$ to $j = 0$.

**Lemma 1.** *Let $u_i = L'_\alpha \cap L_\beta$ $\alpha < \beta$, $i \neq 4$, then the local braid monodromy factorization of $S^{(0)}$ in a neighborhood of $v_i$ is*

(a) $\left(Z^3_{\alpha'\beta}\right)_{\rho_\alpha} Z^3_{\alpha\beta} \left(Z^3_{\alpha'\beta}\right)_{\rho_1^{-1}} \tilde{\underline{Z}}_{\beta\beta'(\alpha)}$ $\quad i = 1, 2, 3, 5.$

(b) $\left(Z^3_{\alpha'\beta}\right)_{\rho_\beta} Z^3_{\alpha'\beta} \left(Z^3_{\alpha'\beta}\right)_{\rho_\beta^{-1}} \tilde{\underline{Z}}_{\alpha\alpha'(\beta)}$ $\quad i = 6, 7$

*where*

$\tilde{\underline{z}}_{\beta\beta'(\alpha)} =$ 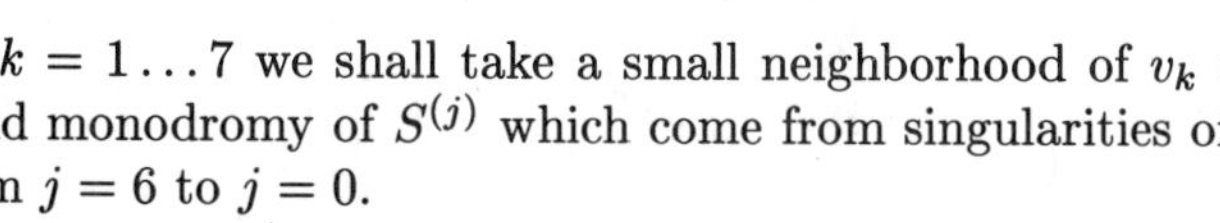


$\tilde{\underline{z}}_{\alpha\alpha'(\beta)} =$ 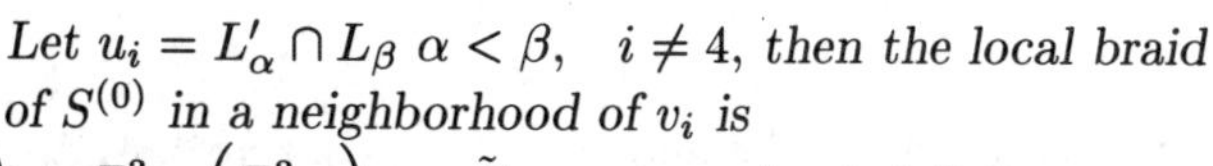

*Proof.* We start with the following

**Claim.** *Let $V$ be a projective algebraic surface. Let $D'$ be a curve in $V$. Let $f : V \to \mathbb{CP}^2$ be a generic projection. Let $S \subseteq \mathbb{CP}^2$, $S' \subset V$ be the branch curve of $f$ and the corresponding ramification curve. Assume $S'$ interset $D'$ in $a'$. Let $D = f(D')$, $a = f(a')$. Assume that there exist neighborhoods of $a$ and $a'$ such that their $f|_{S'}$ and $f|_{D'}$ are isomorphisms. Then $D$ is tangent to $S$ at $a$.*

*Proof of the claim.* Let us choose tangent vectors $e_1$ and $e_2$ in $V$ at $a'$ such that one is tangent to $S'$ and one is tangent to $D'$. Since $a' \in S'$ $df_{a_1}$ is not an isomorphism at $a'$ and thus $df_{a'}(e_1)$ and $df_{a'}(e_2)$ are colinear. Thus, $S$ and $D$ are tangent at $a$. □

We shall prove the lemma for $v_1 = L_1 \cap L_2$. After the first regeneration we get $\pi^{(4)} : Z^{(4)} \to \mathbb{CP}^2$ where above a neighborhood of $v_1$ we have a quadric $R_2$ and a plane $R_1$ intersecting in a line $\hat{L}_1$, $(\hat{L}_1 \cdot \hat{L}_1)_{R_2} = 1$ and $S'$ the ramification curve of $\pi^{(4)} : R_2 \to \mathbb{CP}^2$ $S'$ is linear equivalent to $\hat{L}_1 + \hat{L}_3$ ($\hat{L}_3 = R_2 \cap R_4, \hat{L}_1 \cdot \hat{L}_3 = 1$). Thus $S' \cdot \hat{L}_1 = 1$. By the claim the branch curve in $\mathbb{CP}^2$ of $\pi^{(4)}|_{R_2}$ is a conic tangent to $L_1 = \pi^{(6)}(\hat{L}_1) = \pi^{(4)}(\hat{L}_1)$. Thus, in a neghborhood of $v_1$, $S^{(4)}$ looks like in Figure 2. We use the algorithm introduced in Proposition 1.5 of ([MoTe2]) to compute the braid monodromy of such configuration. We use the following table (table 1):

| | $\varepsilon$ | $\delta$ | |
|---|---|---|---|
| tangent point | 4 | $\Delta^2\langle 1,2\rangle$ | $Z^4\langle 1,2\rangle = Z^4_{12}$ |
| branch point | 1 | | $\tilde{Z}_{22'} = H(\tilde{z}_{22'})$ |

table 1

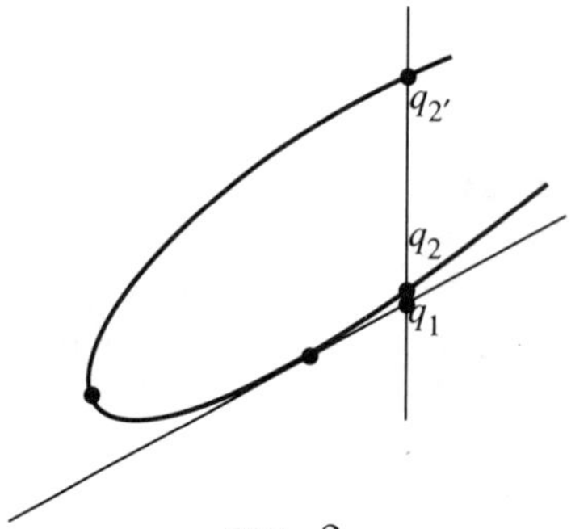


FIG. 2

to get the braid monodromy of the above configuration to be $Z^4_{12}\ \tilde{Z}_{22'}$.

When we regenerate from $S^{(4)}$, to $S^{(2)}$ the part of $S^{(i)}$ $i = 4,3,2$ around $v_1$ does not change from the point of view of braid monodromy. When regenerating $S^{(2)}$ to $S^{(1)}$ in the neighborhood of the tangent point above, we are locally in the situation described by the third regeneration rule (3 cubes Lemma) where $q_1$ (from fig. 2) is replaced by $q_1$ and $q_{1'}$ which are close to each other, and the factor $Z^{(4)}_{12}$ corresponding to this tangent point is replaced by the following expression: $Z^{(3)}_{11',2} = \left(Z^{(3)}_{1'2}\right)_{\rho_1} Z^{(3)}_{1'2} \left(Z^{(3)}_{1'2}\right)_{\rho_1^{-1}}$ where $\rho_1 = Z_{11'}$ and $\tilde{Z}_{22'}$ is replaced by a $\tilde{\underline{Z}}_{22'(1)}$ braid similar to it but going around $q_1$ and $q_{1'}$. When regenerating from

$S^{(1)}$ to $S^{(0)}$ the braid monodromy does not change and so we get the Lemma for $v_1$. The proofs for the other $v_i$'s $i \neq 4$ are similar. $\square$

**Lemma 2.**
(i) $Z^{(3)}_{11',2} = Z^2_{1'2} Z^2_{12} Z^2_{1'2} Z^2_{12} Z_{11'}$
(ii) $\Delta^{(2)}_4 = \overline{Z}^{(2)}_{1'2'} \underline{Z}^2_{12} \overline{Z}^2_{1'2'} Z^2_{12} Z^4_{11'} Z^4_{22'}$
(iii) $\left(\overline{Z}_{1'2'} Z_{12}\right)_{Z^t_{22'}} = \overline{Z}_{1'2'} Z_{11'} \Rightarrow t = 0$.

*Proof.* In order to simplify the proof, we denote $X_1 = Z_{1'2}$, $X_2 = Z_{11'}$.

Thus:

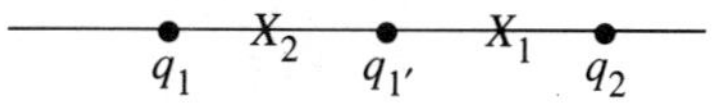


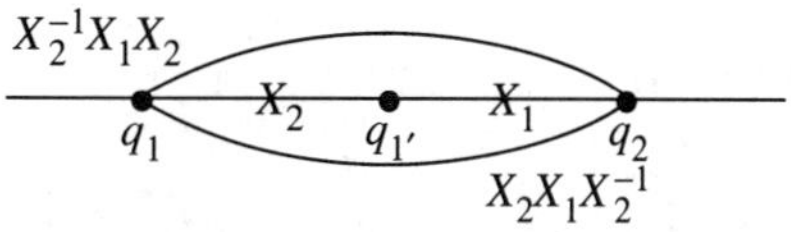


Then $(Z_{1'2})_{\rho_1} = X_2^{-1}X_1X_2$ a $(Z_{1'2})_{\rho_1^{-1}} = X_2X_1X_2^{-1}$. In what follows, we use $\langle A, B\rangle = 1 \Rightarrow A^{\pm 1}BA = BAB^{\pm 1}$. Thus $X_2^{-1}X_1X_2 = X_1X_2X_2^{-1}$, $X_1X_2X_1 = X_2X_1X_2$ and $X_2X_1X_2^{-1} = X_1^{-1}X_2X_1$.

(i)
$$\begin{aligned} Z^{(3)}_{11',2} &= \left(X_2^{-1}X_1X_2\right)^3 \; X_1^3 \; \left(X_2X_1X_2^{-1}\right)^3 \\ &= \left(X_1X_2X_1^{-1}\right)^3 \; X_1^3 \; \left(X_1^{-1}X_2X_1\right)^3 \\ &= X_1X_2^3X_1X_2^3X_1 \end{aligned}$$

Now:

$$\begin{aligned} Z^2_{12}Z^2_{1'2}Z^2_{12}Z^2_{1'2}Z^2_{11'} &= X_2^{-1}X_1^2X_2X_1^2X_2^{-1}X_1^2X_2X_1^2X_2 \\ &= X_1X_2^2X_1^{-1}X_1^2X_1X_2^2X_1^{-1}X_1^2X_2 \\ &= X_1X_2^2X_1^2X_2^2X_1X_2. \end{aligned}$$

Now:

$$\begin{aligned} X_1X_2^3X_1X_2^3X_1 &= X_1X_2^2X_1^2X_2^2X_1X_2 \Leftrightarrow \\ X_2X_1X_2^3X_1 &= X_1^2X_2^2X_1X_2 \qquad \Leftrightarrow \\ X_2X_1X_2^3X_1 &= X_1^2X_2X_2X_1X_2 \qquad \Leftrightarrow \\ X_2X_1X_2^3X_1 &= X_1^2X_2X_1X_2X_1 \qquad \Leftrightarrow \\ X_2X_1X_2^2 &= X_1^2X_2X_1 \qquad \Leftrightarrow \\ X_2X_1X_2 \cdot X_2 &= X_1 \cdot X_1X_2X_1 \qquad \Leftrightarrow \\ X_1X_2X_1 \cdot X_2 &= X_1 \cdot X_2X_1X_2 \qquad \square \end{aligned}$$

(ii) This can be verified by direct computation in an abstract $B_4$. We prove it using our algorithm from [MoTe1] for computing braid monodromy in the special case of 2 conics in general position defined over the reals.

(iii) Assume to the contrary that $t \neq o$. By conjugation with $Z_{22'}^{-t}$ we can assume $t > 0$. Let $\{\gamma_1, \gamma_{1'}, \gamma_2, \gamma_{2'}\}$ be a good geometric base of $\Pi_1\left(D - \{q_1, q_{1'}, q_2, q_{2'}\}\right)$. We show that $\overline{Z}_{1'2'}Z_{12}$ acts on $\gamma_1$ differently than $\left(\overline{Z}_{1'2'}Z_{12}\right)_{Z_{22'}^t} = Z_{22'}^{-t}\overline{Z}_{1'2'}Z_{12}Z_{22'}^t$.

$(\gamma_1)\overline{Z}_{1'2'} = \gamma_1$

$(\gamma_1)Z_{12} = \gamma_2$

$(\gamma_1)Z_{22'} = \gamma_1$

$(\gamma_2)Z_{22'}^t = \begin{cases} (\gamma_{2'}\gamma_2)^{\tau}\gamma_2(\gamma_{2'}\gamma_2)^{-\tau} & t = 2\tau \\ (\gamma_{2'}\gamma_2)^{\tau}\gamma_{2'}(\gamma_{2'}\gamma_2)^{-\tau} & t = 2\tau + 1 \end{cases}$

Thus, $(\gamma_1)\overline{Z}_{1'2'}Z_{12} = \gamma_2$. On the other hand,

$$(\gamma_1)\left(\overline{Z}_{1'2'}Z_{12}\right)_{Z_{22'}^t} = \begin{cases} (\gamma_{2'}\gamma_2)^{\tau}\gamma_2(\gamma_{2'}\gamma_2)^{-\tau} & t = 2\tau \\ (\gamma_{2'}\gamma_2)^{\tau}\gamma_{2'}(\gamma_{2'}\gamma_2)^{-\tau} & t = 2\tau + 1 \end{cases}$$

which does not equal $\gamma_2$ for $t \neq 0$. □

**Lemma 3.** *In the notations of Lemma 1,* $A_i = \Delta_4^2\rho_\alpha^{-1}\rho_\beta^{-1}$ $i \neq 4$.

*Proof.* We shall prove it for $i = 1$.

**Claim.** $\overline{Z}_{2',11'}^{(2)} \cdot Z_{22'} = Z_{2',11'}^{(2)} \cdot \tilde{Z}_{22'}$.

*Proof of the claim.*

$$\begin{aligned} \overline{Z}_{2',11'}^{(2)} \cdot Z_{22'} &= Z_{22'}\left(\overline{Z}_{2',11'}^{(2)}\right)_{Z_{22'}} \\ &= Z_{22'}Z_{2,11'}^{(2)} \\ &= Z_{2,11'}^{(2)}(Z_{22'})_{Z_{2,11'}^{(2)}} \\ &= Z_{2,11'}^{(2)} \cdot \tilde{Z}_{22'}. \end{aligned}$$

The last equality is induced from the fact that $Z_{2,11'}^{(2)}$ is actually the braid obtained from the motion of $q_2$ around $q_1$ and $q_{1'}$ counterclockwise. □

We go now to Lemma 3. By induction on Corollary V.2.5 of [MoTe1] we have

$$\Delta_4^2 = Z_{2,11'}^{(2)}\overline{Z}_{2',11'}^{(2)}Z_{22'}^2Z_{11'}^2.$$

By the claim:

$$\Delta_4^2 = Z_{2,11'}^{(2)}Z_{2,11'}^{(2)}\tilde{Z}_{22'}Z_{22'}Z_{11'}^2.$$

Since $Z_{11'}^2$ commutes with $Z_{22'}$ and $\tilde{Z}_{22'}$ we get:

$$\Delta_4^2 = Z_{2,11'}^{(2)}Z_{2,11'}^{(2)}Z_{11'}\tilde{Z}_{22'}Z_{22'}Z_{11'}.$$

By Lemma 2:

$$\Delta_4^2 = Z_{2,11'}^{(3)}\tilde{Z}_{22}Z_{22'}Z_{11'}.$$

By Lemma 1:

$$\Delta_4^2 = A_1Z_{22'}Z_{11'}. \quad \square$$

**Corollary 4.** *Let $H_i$ be the local braid monodromy of $S^{(0)}$ around $v_i$ $i \neq 4$, then*

$H_1 = Z^{(3)}_{11',2} \cdot \tilde{Z}_{22'(1)}$
$H_2 = Z^{(3)}_{11',3} \cdot \tilde{Z}_{33'(1)}$
$H_3 = Z^{(3)}_{44',6} \cdot \tilde{Z}_{66'(4)}$
$H_5 = Z^{(3)}_{55',9} \cdot \tilde{Z}_{99'(5)}$
$H_6 = Z^{(3)}_{6',77'} \cdot \tilde{Z}_{66'(7)}$
$H_7 = Z^{(3)}_{8',99'} \cdot \tilde{Z}_{88'(9)}$

*Proof.* By Lemma 1.

We want to compute the braid monodromy factors of $\varphi^{(0)}$ that come from singularities in a neighborhood of $v_4$. We start with $\rho^{(4)}$, using Proposition 2.4 of [MoTe2]. We notice that $\rho^{(3)} = \rho^{(4)}$ there because the third regeneration does not change a neighborhood of $v_4$. Then we compute $S^{(2)}$ and $\varphi^{(2)}$ in a neighborhood of $v_4$ and using proposition 3.2 of [MoTe2]. We notice that $\rho^{(1)} = \rho^{(2)}$ there and finally we get a factorized expression for $\rho^{(0)}$.

**Lemma 5.** *In a neighborhood of $v_4$ the local braid monodromy of $S^{(4)}$ around $v_4$ is given by*

$$(4) \quad Z^4_{2'3} \cdot \tilde{Z}_{88'} \cdot Z^2_{48'} \cdot (Z^2_{38'})^\bullet \cdot \overline{Z}^4_{58} \cdot (Z^2_{48}) \cdot (Z^2_{38})^\bullet \cdot (\Delta^2\langle 3,4,5,7\rangle)^\bullet \cdot Z^4_{78}$$
$$(Z^2_{2'i}\ i = 8'875)^\bullet \overline{Z}^4_{2'4} \cdot Z^2_{2i}\ \ i = 8'875 \cdot \tilde{Z}_{22'}\cdot$$

*where $(\ \ )^\bullet$ = conjugation by $Z^2_{2'3}Z^{-2}_{78}$, where the braids $\tilde{Z}_{22'}$ and $\tilde{Z}_{88'}$ are determined by*

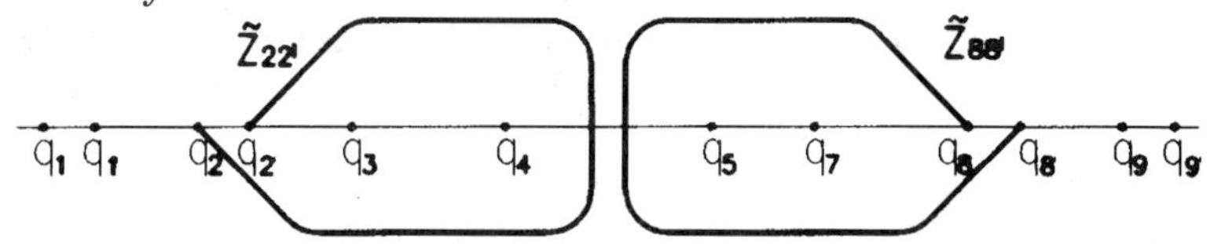

*and the L.V.C. corresponding to $\Delta^2\langle 3,4,5,7\rangle$ is given by:* 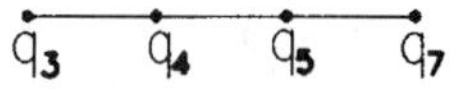

*Proof.* Let $\hat{L}_i$ $i = 2,3,4,5,7,8$ be the lines in $Z^{(6)}$ and in $Z^{(5)}$ (abuse of notation) intersecting in a point $a_5$. Let $L_i = p^{(6)}(\hat{L}_i)$. Then $L_i$ $i = 2,3,4,5,7,8$ intersect in $v_4$. $S^{(6)}$ (and $S^{(5)}$) around $v_4$ is $\bigcup L_i$ $i = 2,\dots,8$ $i \neq 4$. After the first regeneration, $S^{(4)}$ in a neighborhood of $v_4$ looks as follows: The lines $L_2$ and $L_8$ are replaced by 2 conics $Q_2$ and $Q_8$ where $Q_2$ is tangent to $L_3$ and $L_4$, and $Q_8$ is tangent to $L_5$ and $L_7$. This follows from the fact that $P_2 \bigcup P_3$ is regenerated to $R_2$ and the tangency follows from the claim in Lemma 1. The same argument applies to $Q_8$ replacing $L_8$. Thus in a neighborhood of $v_4$, $S^{(4)}$ looks like the configuration from Proposition 2.4 of [MoTe2] with a change of indices and therefore we obtain our lemma. □

**Lemma 6.** *Let $w = L_3 \cap L_4 \cap L_5 \cap L_7$ in $S^{(4)}$.*

(a) *In a small enough neigborhood $U_2$ of $w$ $T^{(2)} = S^{(2)} \cap U_2$ looks like Figure 3 (i.e., the singularities of $T^{(2)}$ are 4 nodes 4 tangency points and 2 branch points).*

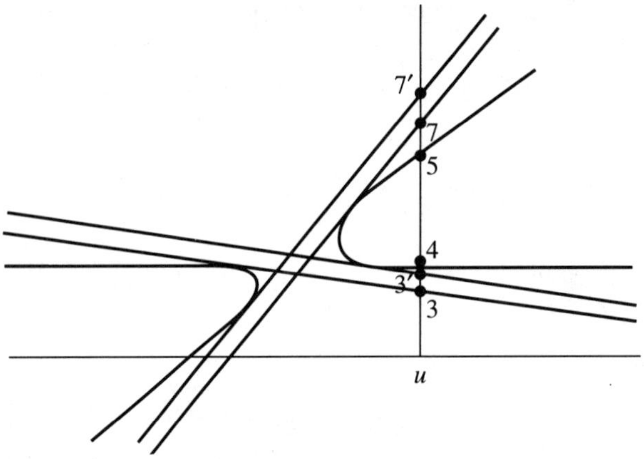


FIG. 3

(b) *The local braid monodromy of* $S^{(2)}$ *in that neighborhood is presented by* $F_1(F_1)_{\rho^{-1}}$ *where* $F_1 = (Z_{3'4})^4(Z_{57})^4\tilde{Z}^4_{45}(\overset{(4)}{Z}_{3'7})^2(\overline{Z}_{3'7})^2$ *and* $\rho = A_{33'}Z_{77'} = \rho_3\rho_7$ *and*

$\tilde{z}_{45} =$

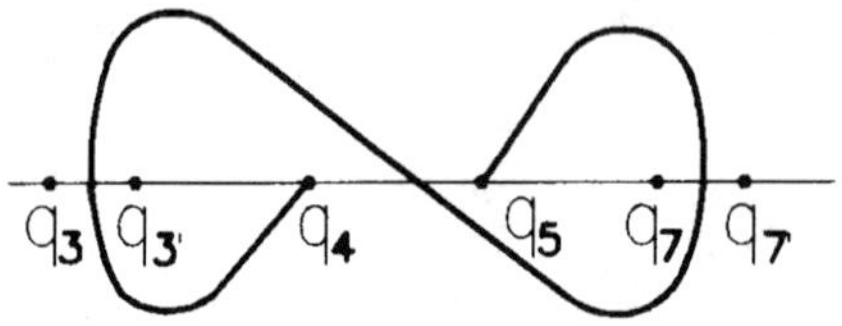


*Proof.*

(a) Let $\hat{w} = R_2 \cap R_3 \cap R_4 \cap R_5$ in $Z^{(4)}$.

For local analysis we can use holomorphic coordinates in a neighborhood $\hat{U}$ of $\hat{w}$ in $\mathbb{CP}^n$. This allows us to consider $\hat{U}$ as a subvariety of a neighborhood of the origin in $\mathbb{C}^4$ with coordinates $X, Y, Z, T$ defined by the following system of equations.

$$\hat{U} : \begin{cases} XT = 0 \\ YZ = 0 \end{cases}$$

By abuse of notation, $\hat{L}_3$ is called the $X$-axis (see Figure 4), and then

$R_2 = \{Z = 0,\ T = 0\}$
$R_3 = \{Z = 0,\ X = 0\}$
$R_4 = \{T = 0,\ Y = 0\}$
$R_5 = \{X = 0,\ Y = 0\}$

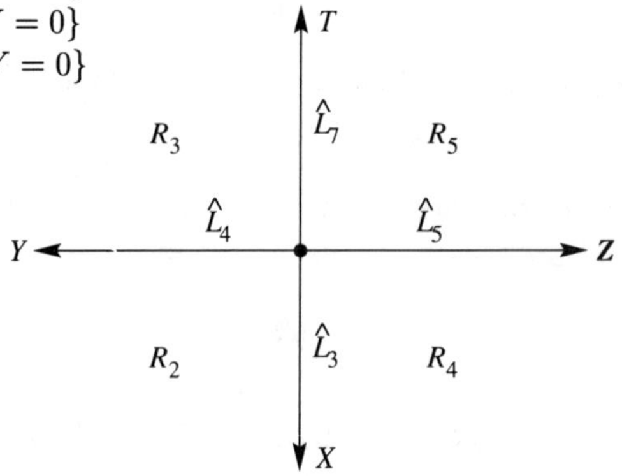


FIG. 4

Let us take the following new coordinates:
$\tau = Z$
$z = Y + Z$
$y = T - Y - 2Z$
$x = X + T - 2Y - 3Z$
i.e., $Z = \tau$
$Y = z - \tau$
$T = y + z + \tau$
$X = x - y + z$

To study generic projections of $\hat{U}$ and its regeneration to $\mathbb{CP}^2$ and $\mathbb{CP}^1$ we can use our projection of $\mathbb{C}^4$ to the $x$-$y$ plane and to the $x$-axis. Also, here we denote the $x$-$y$ projection of $\hat{L}_i$ by $L_i$.

Let $\hat{U}^{(2)}$ be the regeneration of $\hat{U}$ (as a part of $Z^{(2)}$). Then $\hat{U}^{(2)}$ can be described as follows:

$$\tau(z - \tau) = \varepsilon$$

$$(y + z + \tau)(x - y + z) = 0$$

and thus, $\hat{U}^{(2)}$ is the union of two irreducible surfaces:

$$W_1 : \tau(z - \tau) = \varepsilon \qquad y + z + \tau = 0$$
$$W_2 : \tau(z - \tau) = \varepsilon \qquad x - y + z = 0.$$

The $x$-$y$ projection of $W_1 \cap W_2$ is given by a conic: $(-2y + x)(3y - 2x) = u$ (This corresponds to $L_4 \cup L_5$). The branch curve of the $W_1 \to x$-$y$ plane is given by 2 parallels: $y = \pm\sqrt{8\epsilon}$. (This corresponds to the doubling of $L_3$.) The branch curve of the $W_2 \to x$-$y$ plane is given by 2 parallel lines $y - x = \pm\sqrt{2\varepsilon}$. (This corresponds to the doubling of $L_7$.) Thus $S^{(2)}$ in a neighborhood $U^{(2)}$ around $w$ is:

$$\{(2y + x)(3y - 2x) = u\} \bigcup \{y = \pm\sqrt{8\varepsilon}\} \bigcup \{y - x = \pm\sqrt{2\varepsilon}\}$$

where each of the lines in $S^{(2)}$ around $w$ is tangent to the conic. Thus it appears as in the formulation of the lemma.

(b) By proposition 2.5 of [MoTe2]. □

**Lemma 7.** *The local braid monodromy for $S^{(2)}$ around $v_4$ is obtained from formula (4) by the following replacements:*

(i) *Consider each $(\ \ )^\bullet$ as a conjugation by the braid induced from the following motion:*

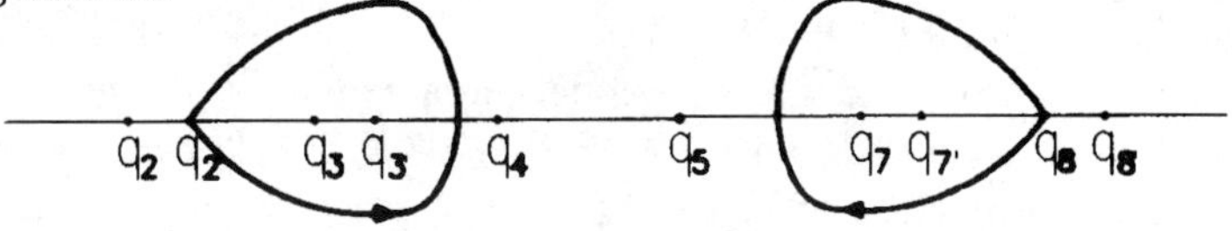


(ii) $\Delta^2\langle 3,4,5,7\rangle$ *is replaced by* $(F_1(F_1)_{\rho^{-1}})^\bullet$.

(iii) *Each of the degree 4 factors in (4) that involves index 3 or 7 is replaced by 3 cubes as in the regeneration rule III.*

(iv) *Each of the degree 2 factors in (4) that involves index 3 or 7 is replaced by 2 degree 2 factors where 3 and 3' are replacing 3 and 7, 7' replacing 7.*

*We call the formula that was obtained formula* (2).

*Proof.* In the regeneration $S^{(4)} \rightsquigarrow S^{(2)}$, $\hat{L}_3$ and $\hat{L}_7$ are replaced in 2 conics, each of which intersect the typical fiber twice in 2 points that are close to $\hat{L}_3 \cap C_{v_4}$ and $\hat{L}_7 \cap C_{v_4}$ respectively, namely, $q_3, q_{3'}, q_7, q_{7'}$.

(i) By the second regeneration rule, $Z^2_{2'3} Z^2_{78}$ should be replaced by $Z^2_{2'3} Z^2_{2'3'} Z^2_{78} Z^2_{7'8}$ which equals the braid induced from that motion.

(ii) By Lemma 6.

(i) By the third regeneration rule.

(iv) By the second regeneration rule. □

**Lemma 8.** *In the notations of Lemma* 6*: Let $T^{(0)}$ be the curve obtained from $T^{(2)}$ in the regeneration $Z^{(2)} \rightsquigarrow Z^{(0)}$.*

(i) *Then the local braid monodromy of $T^{(0)}$ is $\hat{F}_1 \cdot \hat{F}_2$ where*

$$\hat{F}_1 = Z^{(3)}_{3,44'} Z^{(3)}_{55',7} \alpha^{(1)}(m_1) \overset{(4)}{Z}{}^2_{3'7} \overline{Z}^2_{3'7'}$$

$$\hat{F}_2 = \left(Z^{(3)}_{3,44'}\right) \rho^{-1} \left(Z^{(3)}_{55',7}\right)_{\rho^{-1}} \left(\alpha^{(1)}(m_2)\right)_{\rho^{-1}} \left(\underset{\sim}{\overset{(4)}{Z}}{}^2_{3'7}\right)_{\rho^{-1}} \left(\overline{Z}_{3'7'}\right)^2 \rho^{-1}.$$

*$m_1, m_2$ arise from the first regeneration rule applied on $\alpha^{(1)}$ and $(\alpha^{(1)})_{\rho^{-1}}$ respectively, $\alpha = \tilde{Z}_{45}$, $\rho = Z_{33'} Z_{77'}$.*

(ii) *The singularities of the $x$-projections of $T^{(0)}$ are those arising from $T^{(2)}$ by the regeneration rules, namely,*

*4 nodes that exist in $T^2$,*

*$3 \times 4$ cusps arising from 4 tangency points in $T^2$,*

*$2 \times 2$ branch points for 2 branch points of $T^2$.*

(iii) *The braid monodromy of $T^{(0)}$ is $\hat{F}_1 \hat{F}_2$.*

(iv) $\hat{F}_1 \hat{F}_2 = \Delta_8^{-2} Z^{-2}_{33'} Z^{-2}_{77'} Z^{-2}_{44'} Z^{-2}_{55'}$

*Proof.* We use the notations of Lemma 6 (formulation and proof).

(i) In the regeneration process, $Z^{(2)} \rightsquigarrow Z^{(0)}$, $W_1 \cup W_2$ becomes irreducible with equations $ZY = \varepsilon$, $XT = \delta$ where the branch curve in $\mathbb{CP}^2$ of its projection from $U_2$ to $\mathbb{CP}^2$ is a regeneration of $T^{(2)}$ such that the conic is doubled in the generic point. We apply the regeneration rules (I, III) on the local braid monodromy of $T^{(2)}$ on each factor that involves indices 4 or 5, to get $\hat{F}_1 \cdot \hat{F}_2$. (Note that $(\alpha^{(1)}(m_2))_\rho = (\alpha)^{(1)}_\rho(m_2))$.

(ii) Let $V$ be an affine surface in $\mathbb{C}^4$ defined by $YZ = \varepsilon$ $XT = \delta$. Let $\overline{V}$ be its projection closure in $\mathbb{CP}^4$ defined by $YZ = \varepsilon W^2$, $XT = \delta W^2$. It is easy to check that the only singularities of $\overline{V}$ are 4 non-degenerate quadratic singularities at $\infty$, namely the intersection of $\hat{L}_i$ with $\{W = 0\}$ $i = 3, 4, 5, 7$. Let $S_V$ be the branch curve of the $x$-$y$ projection of $V$. Let $\overline{S}_V$ be its projective closure. Each of the singularities of $\overline{V}$ corresponds to a singularity of $\overline{S}_V$, namely, the intersection of $L_i$ with the line in infinity $i = 3, 4, 5, 7$. Thus, $\overline{S}_V$ has 4 nodes at $\infty$. $T^{(0)}$ can be identified with a part of $S_V$. In particular, each singularity of the $x$-projection of $T^{(0)}$ corresponds to a singularity of the $x$-projection of $S_V$. We claim that there are no more singularities of $\overline{S}_V$ besides those mentioned above: the 4 nodes at

$\infty$ and those that come from $T^{(0)}$. We prove it by comparing degrees of the factorized expressions given by the braid monodromy of $S_V$.

The degree of $S_V$ is 8, degree $\Delta_8^2 = 8 \cdot 7 = 56$. The degree of the product in $\Delta_8^2$ of the factors that come from the nodes at $\infty$ is $4 \cdot 2 = 8$. The singularities of $T^{(0)}$ that come from the regenerating singularity of $T^{(2)}$ contributes to the factorization of $\Delta_8^2$ factors whose added degrees are $12 \cdot 3 + 4 \cdot 2 + 4 \cdot 1 = 48$. Thus, we have factors together whose added degrees are $48 + 8 = 56$ and there cannot be any additional singularities of the $x$-projections of $T^{(0)}$ which prove part (ii). Moreover, $S_V$ has no additional singularities, and $\overline{S}_V$ has only the 4 nodes at $\infty$.

(iii) Since $T^{(0)}$ has no additional singularities than those coming from regenerating $T^{(0)}$, these contribute all the factors of $\hat{F}_1 \cdot \hat{F}_2$.

(iv) To prove this point, we have to analyse the contribution of the nodes at $\infty$ to the braid monodromy of $\overline{S}_V$.

We move the line in infinity such that $\overline{S}_V$ is transversal to it. Denote the new affine part of $\overline{S}_V$ by $S'_V$. Then the nodes of $\overline{S}_V$ are (transversal) intersection points of $L_i$ ($i = 3, 4, 5, 7$) with a line at $\infty$. We want to compute the braid monodromy of $S'_V$. In the regeneration process that created $S'_V$ these lines were doubled so that the L.V.C. corresponding to the nodes of $S'_V$ close to $\infty$ are $z_{33'}, z_{44'}, z_{55'}, z_{77'}$ and the contribution of those nodes to the braid monodromy of $S_{V'}$ is $Z^2_{33'}, Z^2_{44'}, z^2_{55'}, z^2_{77'}$. We get the following formula for the braid monodromy of $S_{V'}$ :

$$\Delta_8^2 = \hat{F}_1 \hat{F}_2 Z^2_{33'}, Z^2_{44'}, z^2_{55'}, z^2_{77'} \quad \square$$

**Lemma 9.** *Let $m_1, m_2$ be the integers that appear in $\hat{F}_1, \hat{F}_2$, respectively. Then $m_1 = m_2$.*

*Proof.* Let $G = \{b \in B_8[D, \{3, 3', 4, 4', 5, 5', 7, 7'\}] \mid b\{3, 3'7, 7'\} = \{3, 3', 7, 7'\}\}$. Each $b \in G$ stabilizes $\{4, 4', 5, 5'\}$ and thus can be considered as a braid in $B_4[D, \{4, 4', 5, 5'\}]$. Therefore there exists a natural homomorphism $\beta : G \to B_4[D, \{4, 4', 5, 5'\}]$. We shall refer to $\beta$ as a forgetting homomorphism.

It is obvious that:

(i) $\Delta_8^2 \in G$, $\beta(\Delta_8^2) = \Delta_4^2$.

(ii) $H(\sigma)^2 \in G\ \forall \sigma$.

(iii) $H(\sigma) \in G$ for $\sigma$ s.t. $\sigma \cap \{3, 3', 7, 7'\} = \emptyset$; moreover, $\beta(H(\sigma)) = H(\sigma)$.

(iv) $H(\sigma) \in G$ for $\sigma$ s.t. $\sigma \cap \{4, 4', 5, 5'\} = \emptyset$; moreover, $\beta(H(\sigma)) = I_d$.

By Lemma 8(iv), $\Delta_8^2 = \hat{F}_1 \hat{F}_2 Z^2_{44'} Z^2_{55'} Z^2_{33'} Z^2_{77'}$. We know that $\Delta_8^2, Z^2_{ii'} \in G \Rightarrow \hat{F}_1 \cdot \hat{F}_2 \in G$. So we can apply $\beta$ on each of the factors in the equation: $\beta(\Delta_8^2) = \Delta_4^2$, $\beta(Z^2_{44'}) = Z^2_{44'}$, $\beta(Z^2_{55'}) = Z^2_{55'}$ and $\beta(Z^2_{33'}) = \beta(Z^2_{77'}) = 1$. We still have to compute $\beta(\hat{F}_1 \cdot \hat{F}_2)$. $\hat{F}_1 = Z^{(3)}_{3',44'}\, Z^{(3)}_{55',7}\, \alpha^{(1)}(m_1) \overset{(4,4')}{\underline{Z}}{}^2_{3'7} \overline{Z}^2_{3'7'}$.

By Lemma 2, $Z^{(3)}_{3',44'} \in G$ and $\beta(Z^{(3)}_{3',44'}) = Z_{44'}$. In the same way $\beta(Z^{(3)}_{55',7}) = Z_{55'}$. Clearly $\beta\left(\overset{(4,4')}{\underline{Z}}{}^2_{3'7} \overline{Z}^2_{3'7'}\right) = 1$ (by (iv)).

We shall now compute $\beta(\alpha^{(1)}(m_1))$. $\alpha^{(1)}(m_1) = (\alpha_1 \alpha_2)(m_1)$ where $\alpha_1, \alpha_2$ are the halftwists corresponding to the curves from Figure 5 and Figure 6, respectively.

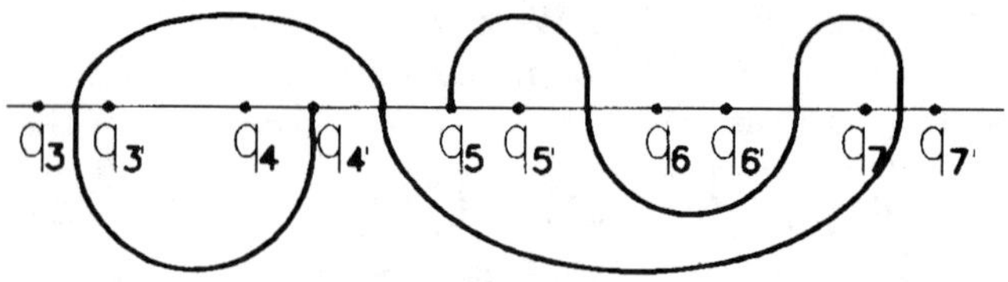


FIG. 5

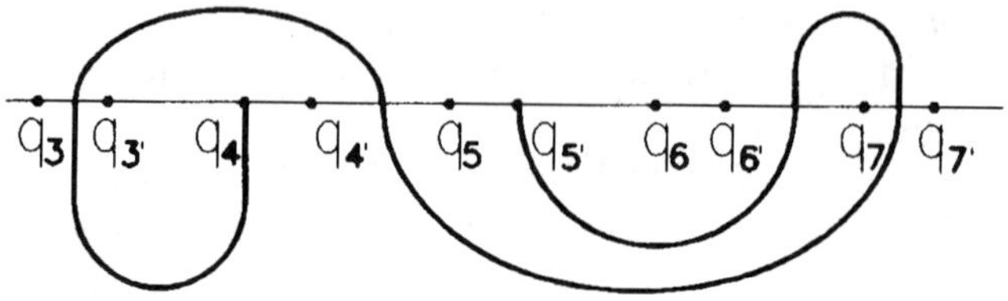


FIG. 6

Thus $\beta(\alpha_1)$ and $\beta(\alpha_2)$ are the halftwists corresponding to the curves from fig. 7 and fig. 8, respectively.

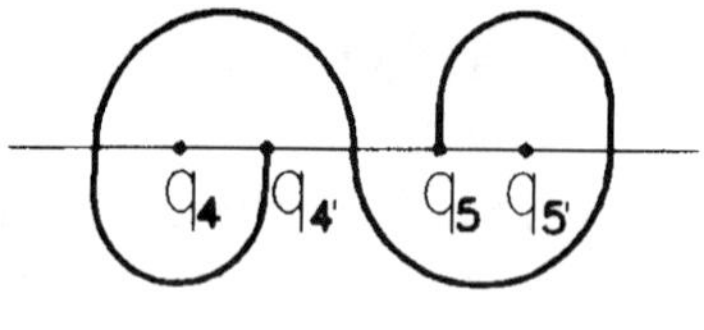


FIG. 7

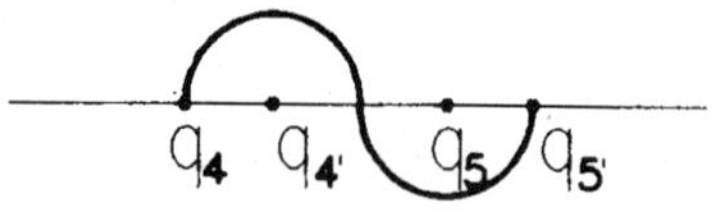


FIG. 8

Let $d_1 = \underline{Z}_{4'5'}$, $d_2 = \underline{Z}_{45}$. It is easy to see that $\beta(\alpha_1) = (d_1)_{Z_{55'}Z^2_{44'}}$, $\beta(\alpha_2) = (d_2)_{Z_{55'}Z^2_{44'}}$. Thus, $\beta(\alpha_1\alpha_2) = (d_1d_2)_{Z_{55'}Z^2_{44'}}$. Now $\alpha^{(1)}(m_1) = (\alpha_1\alpha_2)(m_1) = (\alpha_1\alpha_2)_{Z^{m_1}_{55'}}$ Thus, $\beta(\alpha^{(1)}(m_1)) = \beta((\alpha_1\alpha_2)_{Z^{m_1}_{55'}}) = \beta(\alpha_1\alpha_2)_{Z^{m_1}_{55'}} = (d_1d_2)_{Z^{m_1+1}_{55'}Z^2_{44'}}$.

By Lemma 3.1 of [MoTe2] $d_1d_2$ is invariant under conjugation by $Z_{55'}Z_{44'}$, i.e.,$d_1d_2 \underset{\text{He}}{\simeq} (d_1)_{Z_{44'}Z_{55'}}(d_2)_{Z_{44'}Z_{55'}}$. Therefore, $d_1d_2 = (d_1)_{Z_{44'}Z_{55'}}(d_2)_{Z_{44'}Z_{55'}} = (d_1d_2)_{Z_{44'}Z_{55'}}$. Thus $d_1d_2$ commutes with $Z_{44'}\cdot Z_{55'}$. Therefore, $\beta(\alpha^{(1)}(m_1)) = (d_1d_2)_{Z^{m_1-1}_{55'}}$ and $\beta(\hat{F}_1) = Z_{44'}Z_{55'}\ \ (d_1d_2)_{Z^{m_1-1}_{55'}}$. Now,

$$\hat{F}_2 = (Z^{(3)}_{3',44'})_{\rho^{-1}}(Z^{(3)}_{55',7})_{\rho^{-1}}(\alpha^{(1)}(m_2))_{\rho^{-1}}(\overset{(4,4')}{\underline{Z}}{}^2_{3'7}\overline{Z}^2_{3'7'})_{\rho^{-1}} \quad \rho = Z_{33'}Z_{77'}.$$

All factors except $(\alpha^{(1)}(m_2))_{\rho^{-1}}$ that appear in $\hat{F}_2$ are conjugation by $\rho$ of factors that appear in $\hat{F}_1$. Since $\beta(\rho)^{-1} = \beta(Z^{-1}_{33'}Z^{-1}_{77'}) = \text{Id}$, $\beta$ gives the same value to a factor conjugated by $\rho$ as the factor itself. So the only difference between $\beta(\hat{F}_1)$ and $\beta(\hat{F}_2)$ comes from $\alpha^{(1)}(m_2)_{\rho^{-1}}$. But $\beta(\alpha^{(1)}(m_2)_{\rho^{-1}}) = \beta(\alpha^{(1)}(m_2))$ which equals $(d_1d_2)_{Z^{m_2-1}_{55'}}$ for the same arguments as before. Thus $\beta(\hat{F}_2) =$

$Z_{44'}Z_{55'}(d_1d_2)_{Z_{55'}^{m_2-1}}$.

We apply $\beta$ on both sides of the formula from Lemma 8(iv) to get $\Delta_4^2 = Z_{44'}Z_{55'}\ (d_1d_2)_{Z_{55'}^{m_1-1}}Z_{44'}Z_{55'}(d_1d_2)_{Z_{55'}^{m_2-1}}Z_{44'}^2Z_{55'}^2$.

Since $[d_1d_2, Z_{44'}Z_{55'}]$, we get $\Delta_4^2 = (d_1d_2)_{Z_{55'}^{m_1-1}}(d_1d_2)_{Z_{55'}^{m_2-1}}Z_{44'}^4Z_{55'}^4$. We conjugate both sides by $Z_{55'}^{1-m_2}$ to get $\Delta_4^2 = (d_1d_2)_{Z_{55'}^{m_1-m_2}}d_1d_2Z_{44'}^4Z_{55'}^4$. By Lemma 2(ii) $\Delta_4^2 = d_1d_2d_1d_2Z_{44'}^4Z_{55'}^4$. Thus $(d_1d_2)_{Z_{55'}^{m_1-m_2}} = d_1d_2$. By Lemma 2(iii) it follows that $m_1 - m_2 = 0$. □

**Corollary 10.** *The local braid monodromy of $S^{(0)}$ in $U_2$ is $\hat{F}_1(\hat{F}_1)_{\rho^{-1}}$. Applying regeneration rules on $F_1(F_1)_{\rho_1}$, we get $\hat{F}_1(\hat{F}_1)_{\rho_1}$.*

*Proof.* Since $m_2 = m_1$, $\hat{F}_2 = (\hat{F}_1)_{\rho^{-1}}$. Since $S^{(0)} = T^{(0)} \cap U_2$, we get the corollary by Lemma 8(iii). □

**Lemma 11.** *The local braid monodromy of $S^{(0)}$ around $v_4$ denoted $H_4$ equals:*

$$H_4 = Z_{2',33'}^{(3)}\tilde{Z}_{88'}Z_{44',8'}^{(2)}\left(Z_{33',8}^{(2)}\right)^{\bullet}\overline{Z}_{55',8}^{(3)}\left(Z_{44',8}^{(2)}\right)^{\bullet}\left(Z_{33',8}^{(2)}\right)^{\bullet}\hat{F}_1(\hat{F}_1)_{\rho^{-1}}$$

$$\cdot Z_{77',8}^{(3)}\ Z_{2',i}^2, i = 8, 8, 7', 7, 5', 5\quad \overline{Z}_{2',44'}^{(3)}\ Z_{2i}^2\ \ i = 8', 8, 7', 7, 5', 5\quad \tilde{Z}_{22'}$$

*where ( )$^{\bullet}$ = conjugate by the braid induced from the motion:*

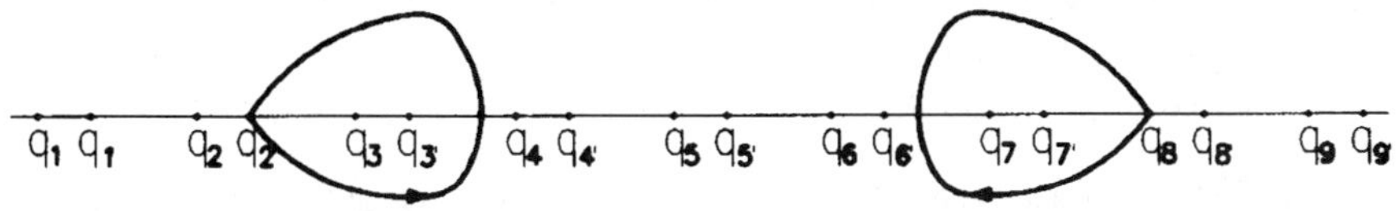

*and $Z_{22'}$, $Z_{88'}$ correspond to:*

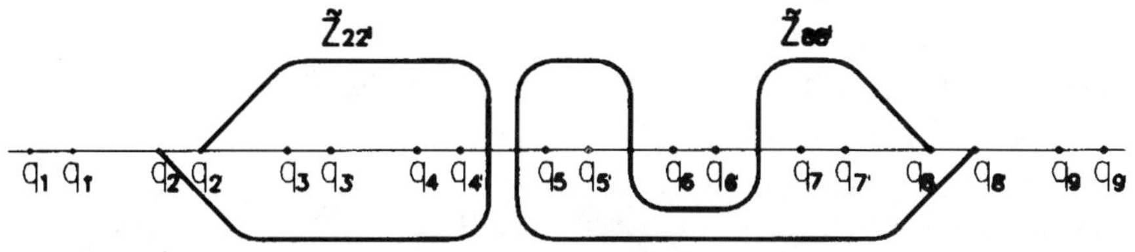

*Proof.* In the last step of regeneration, $L_4$ and $L_5$ are replaced by conics and in the typical fiber $q_4$ is replaced by $q_4$ and $q_{4'}$, and $q_5$ is replaced by $q_5$ and $_{5'}$. To obtain a formula for a local braid monodromy of $S^{(0)}$ around $v_4$ we have to take formula (2) and make the following changes:

(i) ( )$^{\bullet}$ is conjugation by the braid induced from the motion as in the formulation of the Lemma.

(ii) Replace $(F_1)(F_1)_{\rho^{-1}}$ by $\hat{F}_1(\hat{F}_1)_{\rho^{-1}}$.

(iii) Each of the degree 4 factors in (2) that involves index 4 or 5 is replaced by 3 cubes according to the third regeneration rule.

(iv) Each of the degree 2 factors outside of $F_1(F_1)_{\rho^{-1}}$ that involve indexes 4 or 5 is replaced by 2 degree 2 factors as in the third regeneration rule. □

**Lemma 12.**

(i) $\Delta^2_{18} = \prod\limits_{i=7}^{1} C'_i\tilde{H}_i$ *is a braid monodromy factorization for* $S^{(0)}$ *where* $C'_i$ *is* $\tilde{C}_i$ *after applying on it the second regeneration rule and* $\tilde{H}_i = (H_i)_{h_i}$ *for* $h_i \in \langle Z_{jj'} \mid v_i \in L_j\rangle$

(ii) $H_4 = \prod\limits_{\substack{i=2\\ i\neq 6}}^{8} Z^{-1}_{ii'} \cdot \Delta^2_{12}$

*Proof.*

(i) Using Proposition VI.2.1 of [MoTe1] on $S^{(0)}$ we get that

$$\Delta^2_{18} = \prod_{i=1}^{7} C'_i\tilde{H}_i \prod b_i$$

where $C'_i$ is $\tilde{C}_i$ after applying the second regeneration rule, $\tilde{H}_i = (H_i)_{h_i}$ for some $h_i \in \langle Z_{jj'} \mid v_i \in L_j\rangle$ determined by the regeneration of the embedding $B_k \hookrightarrow B_9$ to $B_{2k} \hookrightarrow B_{18}$ where $k = 2$ if $i \neq 4$ and $k = 6$ if $i = 4$, $b_i$ are factors corresponding to singularities that are not covered by $\prod\limits_{i=1}^{7} C_iH'_i$, and each $b_i$ is of the form $Y_i^{t_i}$, $Y_i$ is a positive halftwist and $0 \leq t_i \leq 3$. We compute degrees of all factors involved.

As always, $\deg \Delta^2_{18} = 18 \cdot 17 = 306$.

On the other hand, $\deg \tilde{H}_i = 10 \quad i \neq 4$, and thus $\deg \prod\limits_{i\neq 4} \tilde{H}_i = 60$.

Now, $\prod C'_i$ has 15 factors each of degree 2. After regeneration each of the degree 2 factors give rise to 4 degree 2 factors: $\tilde{Z}^2_{\alpha\beta} \rightsquigarrow Z^2_{\alpha\beta}Z^2_{\alpha\beta'}Z^2_{\alpha'\beta}Z^2_{\alpha'\beta'}$. Thus, $\deg \prod_i C'_i = 15.8 = 120$.

Now deg(degree 1 factors in $H_4$ outside of $\hat{F}_1(\hat{F}_1)_{\rho^{-1}}) = 2$.

deg(squares in $H_4$ outside of $\hat{F}_1(\hat{F}_1)_{\rho^{-1}}) = 20 \cdot 2 = 40$.

deg(cubes in $H_4$ outside of $\hat{F}_1(\hat{F}_1)_{\rho^{-1}}) = 4 \cdot 9 = 36$.

$\deg \hat{F}_1 = \deg(\hat{F}_1)_{\rho^{-1}} = 24 \Rightarrow \deg \hat{F}_1(\hat{F}_1)_{\rho^{-1}} = 48$.

Thus, $\deg \tilde{H}_4 = \deg H_4 = 48 + 36 + 40 + 2 = 126$.

Therefore, $\deg \prod b_i = 306 \cdot (60 + 120 + 126)^{-1} = 1$. Since $\forall i, b_i$ is a positive power of a positive halftwist, we get $b_i = 1\ \forall i$.

(ii) Let

$$G = \left\{b \in B_{18}[D, ii' \ i = 1 \dots 9] \mid b\{1,1',6,6',9,9'\} = \{1,1',6,6',9,9'\}\right\}.$$

Let $\nu : G \to B_{12}[D, \{ii' \ i = 2 \dots 8 \ i \neq 6\}]$ be a forgetting homomorphism as in Lemma 8. Clearly $\nu(\Delta^2_{18}) = \Delta^2_{12}$. Each of the factors in $C'_i$ contains one of the indexes 1, 6 or 9. Therefore, $\nu(C'_i) = 1\ \forall i$. Now $\nu(Z^{-1}_{11'}) = 1$, $\nu(Z^{-1}_{22'}) = Z^{-1}_{22'}$, $\nu(\Delta^2_{11'22'}) = \Delta^2_{22'} = Z^2_{22'}$. Thus, by Lemma 3, $\nu(H_1) = Z_{22'}$. Since $h_1 \in \langle Z_{11'j}, Z_{22'}\rangle$, $h_1$ commutes with $Z_{22'}$ and thus $\nu(\tilde{H}_1) = Z_{22'}$. Similarly, $\nu(\tilde{H}_2) = Z_{33'}$, $\nu(\tilde{H}_3) = Z_{44'}$, $\nu(\tilde{H}_5) = Z_{55'}$, $\nu(\tilde{H}_6) = Z_{77'}$, $\nu(\tilde{H}_7) = Z_{88'}$. Thus,

$\Delta_{12}^2 = \prod_{i=8,7,5} Z_{ii'} \cdot \nu(\tilde{H}_4) \cdot \prod_{i=4}^{2} Z_{ii'}$. Therefore $\nu(\tilde{H}_4) = \Delta_{12}^2 \cdot \prod_{\substack{i=2\\ i\neq 6}}^{8} Z_{ii'}^{-1}$. Since $h_4$ commutes with $\Delta_{12}^2 \prod_{\substack{i=2\\ i\neq 6}}^{8} Z_{ii'}^{-1}$, $\nu(H_4) = \nu(\tilde{H}_4)$ and since 1,6,9 do not appear in $H_4$, $\nu(H_4) = H_4$. □

## 3. Invariance Properties

Invariance properties are results in which we prove that the braid monodromy factorization of $\Delta_{18}^2$ obtained in Lemma 12 is invariant under certain elements of $B_{18}$. Recall that $g = g_1 \dots g_k$ is invariant under $h$ if $g_1 \dots g_k$ is Hurwitz equivalent to $(g_1)_h \dots (g_k)_h$. Invariance properties are important in view of the following Lemma:

**Lemma VI.4.2** [MoTe1]. *If a braid monodromy factorization of $\Delta_{18}^2$ is invariant under $h$ then the equivalent factorization is also a braid monodromy factorization.*

Since every factor of a braid monodromy factorization induces a relation on $\Pi_1(\mathbb{C}^2 - S^{(0)})$ by proving invariant properties, we get more information on $\Pi_1(\mathbb{C}^2 - S^{(0)})$. The braid monodromy factorization obtained above and the invariance properties that we shall prove now were the starting point of our computation of $\Pi_1(\mathbb{C}^2 - S^{(0)})$ which is formulated at the end of the paper.

We shall start by quoting the invariance properties stated in the regeneration rules, we shall prove Chakiri's Lemma and then we shall prove the specific invariance properties for $\Delta_{18}^2$.

From the Regeneration rules we obtain the following

**Invariance rules**

| | |
|---|---|
| Invariance rule I: | $Y_{ij}^{(1)}(m)$ is invariant under $(Z_{ii} Z_{jj'})^q$. |
| Invariance rule II: | $Y_{i,jj'}^{(2)}$ is invariant under $Z_{jj'}^q$ |
| | $Y_{ii',jj'}^{(2)}$ is invariant under $Z_{ii'}^p Z_{jj'}^q$. |
| Invariance rule III: | $Y_{i,jj'}^{(3)}(m)$ is invariant under $Z_{jj'}^q$. |

**Chakiri's Lemma.** *Let $g = g_1 \dots g_k$ be a factorized expression in a group $H$. Then $g_1 \dots g_k$ is invariant under $g^m$ for any $m \in \mathbb{Z}$.*

*Proof.* It is enough to prove it for a free group $H$ and free generators $g_1 \dots g_k$ and for $m = 1$. Let $h$ be the inner automorphism in $H$ defined by conjugation by $g$. Then $(g_1)_g \dots (g_k)_g = (g_1 \dots g_k)_g = (g)_g = g = g_1 \dots g_k$. Since $h(g_i) = (g_i)_g$, $\{(g_i)_g\}$ are also free generators, we can use the final Artin theorem I.4.1 of [MoTe1] to conclude that $(g_1)_g \dots (g_k)_g$ is equivalent to $g_1 \dots g_k$. □

**Invariance Remarks**

(i) To prove invariance of $g_1 \dots g_k$ under $h$ it is enough to prove that $g_1 \dots g_t$ is invariant under $h$ and $g_{t+1} \dots g_k$ is invariant under $h$. Thus we can divide a factorization into subfactorizations and prove invariance on each part separately.

(ii) A factorized expression of one element $g_1$ is invariant under $h$ iff $g_1$ commutes with $h$.

(iii) If the product of elements that commutes with $h$ is invariant under $h$, the relevant factorizations are actually equal.

(iv) If $\sigma_1 \cap \sigma_2 = \emptyset$ then $H(\sigma_1)$ commutes wtih $H(\sigma_2)$.

(v) If $g$ is invariant under $h_1$ and $h_2$ then $g$ is invariant under $h_1 h_2$.

**Lemma 13.** *For $i \neq 4$ and $v_i = L_\alpha \cap L_\beta$, $\alpha < \beta$. $H_i$ is invariant under $Z^p_{\alpha\alpha'} Z^q_{\beta\beta'}$ $\forall p, q \in \mathbb{Z}$.*

*Proof.* We shall only prove it for $i = 1$. Then $\alpha = 1$, $\beta = 2$. Denote $p_i = Z_{11'}$ $p_2 = Z_{22'}$.

Case 1: $q = 0$.

$H_1 = Z^{(3)}_{1'2} \cdot \tilde{Z}_{22'}$

$\tilde{Z}_{22}$ commutes with $Z_{11'}$ because they are halftwists on non-intersecting paths (remark (iv)). By remark (ii) and (v), $\tilde{Z}_{22'}$ is invariant under $Z^p_{11'}$.

$Z^{(3)}_{1'2}$ is invariant under $Z_{11'}$ by Invariance rule III. Thus by Remark (i) $H_1$ is invariant under $Z^p_{11'}$.

Case 2: $p = q$.

By Lemma 3 $H_1 = \Delta^2_4 Z^{-1}_{22'} Z^{-1}_{11'}$.

By Chakiri's Lemma, $H_1$ as a factorized expression is invariant under $(\Delta^2_4 Z^{-1}_{22'} Z^{-1}_{11'})^r$ for every $r \in Z$. Since $\Delta^2_4$ commutes with every element, we get that $H_1$ is invariant under $(Z^{-1}_{22'} Z^{-1}_{11'})^r$. Then $H_1$ is invariant under $(Z^{-1}_{22'} Z^{-1}_{11'})^{-p} = Z^p_{22'} Z^p_{11'}$.

Case 3: $p, q$ arbitrary.

$(H_1)_{Z^p_{11'} Z^q_{22'}}$ is equivalent to $(H_1)_{Z^{p-q}_{11'}}$ by case 2. But $(H_1)_{Z^{p-q}_{11'}}$ is equivalent to $H_1$ by case 1. $\square$

**Corollary 14.** *Complete invariance for $H_i$, $i \neq 4$. $H_i$ is invariant under $\prod_{j=1}^{9} Z^{m_j}_{jj'}$ $\forall m_j \in \mathbb{Z}$.*

*Proof.* We shall prove it for $i = 1$. By Lemma 13, $H_1$ is invariant under $Z^{m_1}_{11'} Z^{m_2}_{22'}$. For $j \neq 1, 2$, $Z_{jj'}$ commutes with any of the factors of $H_1$. Thus $H_1$ is invariant under $\prod_{j=1}^{9} Z^{m_j}_{jj'}$. $\square$

**Lemma 15.** *$H_4$ is invariant under $(Z_{22'} Z_{88'})^p (Z_{33'} Z_{77'})^q (Z_{44'} Z_{55'})^r$ $\forall p, q, r \in \mathbb{Z}$.*

*Proof.*

Case 1: $q = r = p$.

By Lemma 12, $H_4 = \prod_{\substack{i=2 \\ i \neq 6}}^{8} Z^{-1}_{ii'} \cdot \Delta^2_{12}$. Thus by Chakiri's Lemma, $H_4$ is invariant under $\left(\prod\limits_{\substack{i=2\\i\neq 6}}^{8} Z_{ii'}^{-1}\Delta_{12}^2\right)^{-p}$. Since $\Delta_{12}^2$ is a central element, $H_4$ is invariant under $(Z_{22'}Z_{88'})^p(Z_{33'}Z_{77'})^p(Z_{44'}Z_{55'})^p$.

Case 2: $p=0$.

Denote $\varepsilon = (Z_{33'}Z_{77'})^q(Z_{44'}Z_{55'})^r$. We want to prove that $H_4$ is invariant under $\varepsilon$.

Step 1: Product of the factors outside of $\hat{F}_1(\hat{F}_1)_{\rho^{-1}}$

$\tilde{Z}_{22}$ and $\tilde{Z}_{88'}$ commutes with $\varepsilon$.

$Z^{(3)}_{2',33'}$ $Z^{(3)}_{55',8}$ $Z^{(3)}_{77',8}$ $Z^{(3)}_{2',44'}$ are invariant under $\varepsilon$ by Invariance rule III.

$(\ )^\bullet$ is conjugation by a braid which is invariant under $\varepsilon$.

The degree 2 factors outside of $\hat{F}_1(\hat{F}_1)_{\rho^{-1}}$ are of the form $Z^{(2)}_{\alpha'\alpha,\beta}$ where $\beta=2$ or 8. By Invariance rule II, $Z^{(3)}_{\alpha,\alpha',\beta}$ is then invariant under $Z_{\alpha\alpha'}$ and since the other halftwists in $\varepsilon$ commutes with $Z_{\alpha,\beta}$ and $Z_{\alpha',\beta}$, we get that $Z^{(2)}_{\alpha\alpha',\beta}$ is invariant under $\varepsilon$.

Step 2: $\hat{F}_1(\hat{F}_1)_{\rho^{-1}}$

In order to prove that $\hat{F}_1(\hat{F}_1)_{\rho^{-1}}$ is invariant under $\varepsilon$ we consider the following subcases:

| | | |
|---|---|---|
| Case 2.1 | $q=0$ | $\varepsilon=(Z_{44'}Z_{55'})^r$ |
| Case 2.2 | $r=0\ q=1$ | $\varepsilon=Z_{33'}Z_{77'}$ |
| Case 2.3 | $q=2q'$ | $\varepsilon=(Z_{33'}Z_{77'})^{2q'}\cdot(Z_{44'}Z_{55'})^r$ |
| Case 2.4 | $q=2q'+1$ | $\varepsilon=(Z_{33'}Z_{77'})^{2q'+1}\cdot(Z_{44'}Z_{55'})^r$ |

Case 2.1 $\varepsilon = (Z_{44'}Z_{55'})^r$.

$Z^{(3)}_{3',44'}$ is invariant under $Z_{44'}$ (Invariance rule III) and commutes with $Z_{55'}$ (Invariance remark iv). For the same reasons $Z^{(3)}_{55',7}$ is invariant under $Z_{55'}$ and commutes with $Z_{44'}$.

Thus $Z^{(3)}_{3,44'}Z^{(3)}_{55',7}$ is invariant under $\varepsilon$ (Invariance remark (i) and (v)). $\overset{(4)}{\underline{Z}}{}^2_{3'7}$ and $\overline{Z}^2_{3'7'}$ commutes with $Z_{44'}$ and $Z_{55'}$ and thus with $\varepsilon$. $\alpha^{(1)}(m_1) = (\alpha_1\alpha_2)_{Z^{m_1}_{55'}}$ is invariant under $(Z_{44'}Z_{55'})^r$ by Invariance rule I. Thus $\hat{F}_1$ is invariant under $\varepsilon = (Z_{44'}Z_{55'})^r$. Since $\rho^{-1} = Z^{-1}_{33'}Z^{-1}_{77'}$, $\rho^{-1}$ commutes with $\varepsilon$ and thus $(F_1)_{\rho^{-1}}$ is also invariant under $\varepsilon$. Thus $(\hat{F}_1)(F_1)_{\rho^{-1}}$ is invariant under $\varepsilon$.

Case 2.2 $\varepsilon = Z_{33'}Z_{77'}$, $\varepsilon=\rho$.

To prove that $\hat{F}_1(\hat{F}_1)_{\rho^{-1}}$ is invariant under $\rho$, it is necessary to prove that $\hat{F}_1(\hat{F}_1)_{\rho^{-1}}$ is Hurwitz equivalent to $(\hat{F}_1)_\rho\hat{F}_1$. Since $AB$ is Hurwitz equivalent to $BA_B$, it is enough to prove that $(\hat{F}_1)(\hat{F}_1)_{\rho^{-1}}$ is Hurwitz equivalent to $\hat{F}_1((\hat{F}_1)_\rho)_{\hat{F}_1}$. Thus it is enough to prove that $(\hat{F}_1)_{\rho^{-1}}$ is Hurwitz equivalent to $((\hat{F}_1)_\rho)_{\hat{F}_1}$ or that $((\hat{F}_1)_{\rho^{-1}})_{\hat{F}_1^{-1}}$ is Hurwitz equivalent to $(\hat{F}_1)_\rho$.

By Lemma 8(iv), $\hat{F}_1 = \Delta_8^2\rho^{-2}\ Z^{-2}_{44'}Z^{-2}_{55'}(\hat{F}_1^{-1})_{\rho^{-1}}$. Thus $(\hat{F}_1)^{-1} = (\hat{F}_1)_{\rho^{-1}}(Z_{44'}Z_{55'})^2\rho^2\Delta_8^2$. Thus,

$((F_1)_{\rho^{-1}})_{\hat{F}_1^{-1}} = ((\hat{F}_1)_{\rho^{-1}})_{(\hat{F}_1)_{\rho^{-1}}(Z_{44'}Z_{55'})^2\rho^2\Delta_8^{-2}} =$ (as factorized expressions) $= ((\hat{F}_1)_{\rho^{-1}})_{(\hat{F}_1)_{\rho^{-1}}(Z_{44'}Z_{55'})^2\rho^2} \underset{He}{\simeq}$ (by Chakiri's Lemma) $\underset{He}{\simeq} ((\hat{F}_1)_{\rho^{-1}})_{(Z_{44'}Z_{55'})^2\rho^2}$ $= ((\hat{F}_1)_{(Z_{44'}Z_{55'})^2})_\rho \underset{He}{\simeq}$ (by Case 2.1) $\underset{He}{\simeq} ((\hat{F}_1)_\rho$.

Case 2.3 $\varepsilon = (Z_{33'}Z_{77'})^{2q'}(Z_{44'}Z_{55'})^r$.

By Lemma 8(iv), $(\hat{F}_1)(\hat{F}_1)_{\rho^{-1}} = \Delta_8^2(Z_{33'}Z_{77'})^{-2}(Z_{44'}Z_{55'})^{-2}$. By Chakiri's Lemma $\hat{F}_1(\hat{F}_1)_{\rho^{-1}}$ is invariant under $(\Delta_8^2(Z_{33'}Z_{77'})^{-2}(Z_{44'}Z_{55'})^{-2})^{-q'}$ and thus under $(Z_{33'}Z_{77'})^{2q'}(Z_{44'}Z_{55'})^{2q'}$. By Case 2.1, $\hat{F}_1(\hat{F}_1)_{\rho^{-1}}$ is invariant under $(Z_{44'}Z_{55'})^{r-2q'}$. By Invariance remark(v), $(\hat{F}_1)(F_1)_{\rho^{-1}}$ is invariant under $(Z_{33'}Z_{77'})^{2q'}(Z_{44'}Z_{55'})^{2q'}(Z_{44'}Z_{55'})^{r-2q'} = \varepsilon$.

Case 2.4 $\varepsilon = (Z_{33'}Z_{77'})^{2q'+1}(Z_{44'}Z_{55'})^r$.

By Cases 2.2, 2.3 and Invariance remark(v).

Thus $\hat{F}_1(\hat{F}_1)_{\rho^{-1}}$ is invariant under $(Z_{33'}Z_{77'})^q(Z_{44'}Z_{55'})^r$ $\forall q, r \in Z$. This finishes Step 2 of Case 2 and thus Case 2 is complete.

Case 3 $p, q, r$ arbitrary $\varepsilon = (Z_{22'}Z_{88'})^p(Z_{33'}Z_{77'})^q(Z_{44'}Z_{55'})^r$.

By Case 1, $H_4$ is invariant under $(Z_{22'}Z_{88'})^p(Z_{33'}Z_{77'})^p(Z_{44'}Z_{55'})^p$.

By Case 2, $H_4$ is invariant under $(Z_{33'}Z_{77'})^{q-p}(Z_{44'}Z_{55'})^{r-p}$.

By Invariance remark(v), $H_4$ is invariant under $\varepsilon$. □

**Lemma 16.** *Complete invariance of $C_i'$*

*For every $i$ and $\forall m_j \in \mathbb{Z}$ $j = 1 \dots 9$, $C_i'$ is invariant under* $\varepsilon = \prod_{j=1}^{9} Z_{jj'}^{m_j}$.

*Proof.* We apply Invariance rule II and Invariance remark(iv) on each of the factors of $C_i'$ of the form $Y_{ii',jj'}^{(2)}$. □

## 4. Global braid monodromy

We start with the following remark.

*Remark.* If $\Delta_{18}^2 = \prod b_i$ is a braid monodromy factorization, then $\Delta_{18}^2 = \prod (b_i)_h$ is also a braid monodromy factorization for any $h \in B_{18}$, using a different realization to the real model.

We recall that $H_4$ is a product of factors including $\alpha^{(1)}(m)$ where $m$ is an integer that is not determined.

*Definition.* $\underline{H_4(t) \quad t \in \mathbb{Z}}$.

$H_4(t)$ = a factorized expression where all factors but $\alpha^{(1)}(m)$ equal to the factors of $H_4$ and $\alpha^{(1)}(m)$ is replaced by $\alpha^{(1)}(t)$.

According to this definition $H_4 = H_4(m)$.

**Proposition 17.** *Global braid monodromy factorization for $S^{(0)}$ :*

$\Delta_{18}^2 = \prod_{j=2}^{1} C_j'H_j'$ *is a braid monodromy factorization where*

$$H_j' = \begin{cases} H_j & j \neq 4 \\ H_4(0) & j = 4 \end{cases}$$

*Proof.* By Lemma 12, $\Delta^2_{18} = \prod\limits_{j=7}^{1} C'_j(H_j)_{h_j}$ is a braid monodromy factorization for some $h_j \in \langle Z_{jj'} \mid j = 1\dots 9\rangle$. We conjugate the above factorized expression by $h_4^{-1}$ to get $\Delta^2_{18} = \prod (C'_j)_{h_4^{-1}}(H_j)_{h_j h_4^{-1}}$. By the remark opening this section, this factorization is also a braid monodromy factorization. Since we have complete invariance for $C'_j$ and for $H_j$ $j \neq 4$, we get that $\prod (C_j)_{h_4^{-1}}(H_j)_{h_j h_4^{-1}}$ is Hurwitz equivalent to $\prod C'_j H_j$. By Lemma VI.4.2. of [MoTe1] (quoted earlier) $\prod\limits_{j=7}^{1} C'_j H_j$ is also a braid monodromy factorization. We conjugate this expression by $Z_{55'}^{-m}$ to get the following braid monodromy factorization: $\Delta^2_{18} = \prod (C'_j)_{Z_{55'}^{-m}}(H_j)_{Z_{55'}^{-m}}$. Since we have complete invariance for $C'_j$ and for $H_j$ $j \neq 4$, we get the following braid monodromy factorization:

$$\Delta^2_{18} = \prod C'_j H_j{}''$$

where

$$H''_j = \begin{cases} H_j & j \neq 4 \\ (H_4)_{Z_{55'}^{-m}} & j = 4 \end{cases}. \quad \square$$

**Claim.** $H''_4 \underset{He}{\simeq} H'_4$.

Every factor $P$ in $H_4(m)$ that does not involve the indices 5 or 5' satisfies $P_{Z_{55'}^{-m}} = P$. By definition $(\alpha^{(1)}(m))_{Z_{55'}^{-m}} = \alpha^{(1)}(0)$. Every factor in $H_4(m)$ outside of $\alpha^{(1)}(m)$ that involves the indices 5 or 5' is of the form $Z^{(3)}_{55'jj}$ or $Z^{(2)}_{55'jj}$. By Invariance rules II and III, $(Z^{(i)}_{55'jj})_{Z_{55'}^{-m}} \underset{He}{\simeq} Z^{(i)}_{55'jj}$ for $i = 2, 3$. Thus $(H_4(m))_{Z_{55'}^{-m}} \underset{He}{\simeq} H_4(0)$. Thus $H''_4 \underset{He}{\simeq} H'_4$.

Since $H''_j = H'_j$ for $j \neq 4$, we have $\prod\limits_{j=7}^{1} C'_j H''_j \underset{He}{\simeq} \prod\limits_{j=7}^{1} C'_j H'_j$. Thus $\prod\limits_{j=7}^{1} C'_j H'_j$ is a braid monodromy factorization.

*Definition.* $\underline{\varepsilon_{18}}$

$\varepsilon_{18} = \prod\limits_{j=7}^{1} C'_j H'_j$ as a factorized expression.

**Proposition 18.** *$\varepsilon_{18}$ is invariant under*

$$Z_{11'}^{m_1}(Z_{22'}Z_{88'})^{m_2}(Z_{33'}Z_{77'})^{m_3}(Z_{44'}Z_{55'})^{m_4}Z_{66'}^{m_6}Z_{99'}^{m_9}\ \forall m_i \in \mathbb{Z}.$$

*Proof.* Lemmas 14, 15, 16. $\square$

*Definition.* <u>Complex conjugate of a halftwist, L.V.C.</u>

Let $Y = H(\sigma)$. We say that $\overline{Y}$ is a complex conjugate of $Y$ if $\overline{Y} = H(\overline{\sigma})$ where $\overline{\sigma}$ = simple path obtained by complex conjugation of $\sigma$. Let $L = \{\ell_i\}$ be a L.V.C. We say that $\overline{L}$ is a complex conjugate of $L$ if $\overline{L} = \{\overline{\ell}_i\}$ where $\overline{\ell}_i$ are complex conjugates of $\ell_i$.

**Lemma 19.** $\Delta_{18}^2 = \prod_{j=7}^{1} \overline{C}_j' \overline{H}_j'$ *where* $\overline{C}_j'$ *and* $\overline{H}_j'$ *are factorized expressions obtained from* $C_j'$ *and* $H_j'$ *by complex conjugation of the corresponding halftwists.*

*Proof.* All *explicit* curves which we used to analyze regeneration of braid monodromy are defined over the reals. We can change in all our argument and choices the role of the upper and lower halfplanes in the $x$-axis and in the typical fiber $S_n$ by simultaneous complex conjugation of the $x$-axis and the $y$-axis. Then in every intermediate step we would get explicit formulas for braid monodrmy which we obtained from the original formulas by reversing the order of the factors and by replacing each L.V.C. by complex conjugation. In the final step we get another braid monodromy factorization $\Delta_{18}^2 = \prod_{j=1}^{7} \overline{C}_j' \overline{H}_j'$. □

*Remark.* In what follows we use the braid monodromy formula to obtain relations on $\Pi_1(\mathbb{C}^2 - S^{(0)})$. We will use $C_j', H_j'$ or $\overline{C}_j', \overline{H}_j'$ as necessary.

## 5. Applications: Results about $\Pi_1(\mathbb{C}^2 - S^{(0)})$ and $\Pi_1(\mathbb{CP}^2 - S^{(0)})$

In this section we shall state two applications of the results in sections 3 and 4. The proof of these theorems will appear in the fifth paper of this series: "Braid group techniques in complex geometry, V: Fundamental group of the complement of the branch curve of $V_3$."

In order to phrase those results we need a few definitions.

*Definition.* Transversal halftwists

The halftwists $H(\sigma_1)$ and $H(\sigma_2)$ will be called *transversal* if $C_1$ and $C_2$ intersect transversally in one point which is not an end point of either of the $\sigma_i$'s.

*Definition.* $\tilde{B}_n, \tilde{X}_i$

Let $T_n$ be the subgroup of $B_n$ normally generated by $[X, Y]$ for $X, Y$ transversal halftwists. $\tilde{B}_n$ is the quotient of $B_n$ modulo $T_n$. We choose frame $X_i$ of $B_n$. We denote their images in $\tilde{B}_n$ by $\tilde{X}_i$.

**Definition-Proposition.** $G_0(n), \tau$

*Let* $A_{n-1}$ *be the free abelian group on* $w_1, \ldots, w_{n-1}d$. *Let us define a* $\mathbb{Z}/2$ *skew-symmetric form on* $A_{n-1}$ *as follows:*

$$w_i \cdot w_j = \begin{cases} 1 & (i-j) = 1 \\ 0 & \text{otherwise.} \end{cases}$$

*There exists a unique central extension* $G_0(n)$, *of* $\mathbb{Z}/2$ *by* $A_{n-1}$, *with generators* $u_1 \ldots u_{n-1}$ *that satisfies*

$$1 \to \mathbb{Z}/2 \xrightarrow{b} G_0(n) \xrightarrow{a} A_{n-1} \to 1$$

$$a(u_i) = w_i$$

$$[u_i, u_j] = b(w_i \cdot w_j) = \begin{cases} \tau & |i-j| = 1 \\ 0 & \textit{otherwise.} \end{cases}$$

*We always consider $G_0(n)$ with the standard $\tilde{B}_n$ action as follows:*

$$(u_i)\tilde{X}_k = \begin{cases} u_i^{-1}\tau & k = i \\ u_k u_i & |i-k| = 1 \\ u_i & |i-k| \geq 2 \end{cases} .$$

*Remark.* $A_b(G_0(n)) = A_{n-1}, \quad G_0(n)' = \{\tau, 1\}(\simeq \mathbb{Z}/2)$

*Definition.* $\underline{\tilde{\psi}_n, \tilde{P}_n, \tilde{P}_{n,0}, c}$

Let $\psi_n$ be the standard homomorphisms $B_n \overset{\psi_n}{\to} S_n(=$ symmetric group) and $B_n \overset{A_b}{\to} \mathbb{Z}$. Since $\psi_n([X,Y]) = 1$, $\psi_n$ induces a homomorphism $\tilde{\psi}_n : \tilde{B}_n \to S_n$, $A_b$ induces a homomorphism $\tilde{B}_n \overset{A_b}{\to} \mathbb{Z}$. $\tilde{P}_n = \ker \tilde{\psi}_n$, $\tilde{P}_{n,0} = \ker \tilde{\psi}_n \cap \ker \tilde{A}_b$, $c = [\tilde{X}_1, \tilde{X}_2]$.

Recall from [Mo1] that $c = [\tilde{X}_1, \tilde{X}_2] = [\tilde{X}_i, \tilde{X}_{i+1}] = \cdots = [\tilde{X}_{n-2}, \tilde{X}_{n-1}]$. Moreover, $(\tilde{P}_n') = (\tilde{P}_{n,0})' = \{1, c\} \simeq \mathbb{Z}/2$. $Ab(\tilde{P}_n) =$ free abelian group on $n$ generators. $\tilde{B}_n$ acts on $\tilde{P}_{n,0}$ by conjugation. $\tilde{P}_{n,0}$ with this action is isomorphic to $G_0(n)$ with the standard $\tilde{B}_n$ action defined before. There exists a series: $1 \subseteq (\tilde{P}_{n,0})' \subseteq \tilde{P}_{n,0} \subseteq \tilde{P}_n \subset \tilde{B}_n$ s.t. $B_n/\tilde{P}_n = S_n$, $P_n/\tilde{P}_n \simeq \mathbb{Z}$, $\tilde{P}_{n,0}/(\tilde{P}_{n,0})' \simeq A_{n-1} = A_b G_0(n)$, $(\tilde{P}_{n,0})' \simeq \mathbb{Z}_2$.

Consider $\tilde{B}_9 \rtimes G_0(9)$ with respect to the standard $\tilde{B}_9$ action on $G_0(9)$.

*Definition.* $\underline{v_1, N_9, G_9, \tilde{\psi}_9 : G_9 \to S_9, Ab_9, H_9, H_{9,0}}$

$v_1 = (\tilde{X}_2\tilde{X}_1\tilde{X}_2^{-1})^2\tilde{X}_2^{-2}$

$N_9 =$ normal subgroup generated by $c\tau^{-1}$, $(u_1 v_1^{-1})^3$

$G_9 = \tilde{B}_9 \rtimes G_0(9)/N_9$

$\tilde{\psi}_9 : G_9 \to S_9 \quad \tilde{\psi}_q(\alpha, \beta) = \tilde{\psi}_9(\alpha)$

$Ab_9 : G_q \to \mathbb{Z} \quad AB_9(\alpha, \beta) = \tilde{A}b(\alpha)$

$H_9 = \ker \tilde{\psi}_9$

$H_{9,0} = \ker \tilde{\psi}_9 \cap \ker Ab_9$

**Proposition 20.** *There exists a series* $1 \subseteq H_{9,0}' \subset H_{9,0} \subset H_9 \subset G_9$, *where* $H_{9,0}' = H_9' = \{1, c\} \cong \mathbb{Z}/2$, $H_{9,0}/H_{9,0}' \simeq (\mathbb{Z} + \mathbb{Z}/3)^8$, $H_9/H_{9,0} \simeq \mathbb{Z}$, $G_9/H_9 \simeq S_9$.

**Proposition 21.** *Let $\mathbb{C}^2$ be a big "affine piece" of $\mathbb{CP}^2$. Let $S = S^{(0)} \cap \mathbb{CP}^2$. Let $G = \Pi_1(\mathbb{C}^2 - S)$. Then $G \cong G_9$ s.t. $\psi : G \to S_9$ is compatible with $\tilde{\psi}_9 : G_q \to S_9$.*

**Proposition 22.** *Let $\bar{G} = \Pi_1(\mathbb{CP}^2 - S^{(0)})$. Consider $v_1$ as an element of $G_9$. Then there exist $w_0 \in H_{9,0}$ s.t. $\overline{G} \simeq \overline{G}_9$ where $\overline{G}_9 = G_9/\langle v_1^{18} w_0 \rangle$. Let $\overline{H}_9$ and $\overline{H}_{9,0}d$ be the images of $H_{9,0}$ and $H_q$ in $\overline{G}_9$. Then $\overline{H}_{9,0}' = \overline{H}_q'$ and*

$$1 \subseteq \overline{H}_{9,0}' \subseteq \overline{H}_{9,0} \subseteq \overline{H}_q \subset \overline{G}_q$$

*where* $\overline{G}_9/\overline{H}_9 \simeq S_q$, $\overline{H}_9/\overline{H}_{9,0} \simeq \mathbb{Z}_q$, $\overline{H}_{9,0}/\overline{H}_{9,0}' \simeq (\mathbb{Z} + \mathbb{Z}/3)^8$, $\overline{H}_{9,0}' \cong \mathbb{Z}/2$.

## References

[MoTe1] Moishezon, B.G, Teicher, M., *Braid group techniques in complex geometry I: Line arrangements in* $\mathbb{CP}^2$, Contemporary Mathematics **78** (1988), 425-555.

[MoTe2] Moishezon, B.G, Teicher, M., *Braid group techniques in complex geometry II: From arrangements of lines to cuspidal curves,*, LNM **1479** (1989), 131-179.

[MoTe3] Moishezon, B.G, Teicher, M., *Braid group techniques in complex geometry III: Projection degeneraton of* $V_3$, (in this volume).

[Mo1] Moishezon, B.G., *On branch curves of cuspidal curves*, to appear.

Department of Mathematics, Columbia University, New York, N.Y. 10022 U.S.A.
Department of Mathematics and Computer Science, Bar-Ilan University, Ramat-Gan 52900 Israel

Contemporary Mathematics
Volume **162**, 1994

# On the Irregularity of Cyclic Coverings of the Projective Plane

FUMIO SAKAI

*Dedicated to Professor S.Kobayashi on his 60th birthday*

ABSTRACT. We discuss the irregularity of cyclic coverings of the projective plane. We prove that if the degree $n$ of the covering is a power of a prime number, then the irregularity is less than or equal to $(n-1)(r-1)/2$, where $r$ is the number of the irreducible components of the branch curve. This is a generalization of an old result of Zariski to the case in which the branch curve is reducible. In the proof, the Alexander polynomial of the branch curve plays an important role.

## 1. Introduction

The aim of this note is to give a survey on the irregularity of cyclic coverings of the projective plane $\mathbb{P}^2$. Let $f(x,y)$ be a polynomial of degree $d$ over $\mathbb{C}$. Let us consider the cyclic multiple plane:

$$z^n = f(x,y).$$

Factor $f$ into a product of irreducible polynomials: $f = f_1^{m_1} \cdots f_r^{m_r}$. We assume that the condition: $\mathrm{GCD}(n, m_1, ..., m_r) = 1$ is satisfied. This is nothing but the condition that the above surface is irreducible. We pass to the projective model. Let $\tilde{f}(x_0, x_1, x_2)$ be the homogeneous polynomial associated to $f$ so that $\tilde{f}(1,x,y) = f(x,y)$. Let $C$ be the plane curve defined by the equation: $\tilde{f} = 0$. Let $C_i$ be the irreducible component $\tilde{f}_i = 0$. Let $L$ denote the infinite line: $x_0 = 0$. Define $e$ to be the smallest integer with the condition: $e \geq d/n$. Set $m_0 = ne - d$. Note that $m_0 = 0$ if and only if $n$ divides $d$. Let $W_n$ be the normalization of the weighted hypersurface: $x_3^n = x_0^{m_0}\tilde{f}(x_0, x_1, x_2)$ in $\mathbb{P}(1,1,1,e)$. The covering map $W_n \to \mathbb{P}^2$ ramifies over $C$ in case $m_0 = 0$ or over $C \cup L$ in case $m_0 \neq 0$. Let

1991 *Mathematics Subject Classification.* Primary 14E20, 32S55; Secondary 14J17, 14J70.
This paper is in final form and no version of it will be submitted for publication elsewhere.

$\pi : X_n \to W_n$ be a desingularization. Let $\varphi : X_n \to \mathbb{P}^2$ be the composed map. The irregularity $q(X_n)$ of $X_n$ has three equivalent expressions:

$$q(X_n) = \dim H^1(X_n, \mathcal{O}) = \dim H^0(X_n, \Omega^1) = \frac{1}{2} \dim H^1(X_n, \mathbb{R}).$$

In general, the inequality:

$$2q(X_n) \leq \sum_{i=0}^{r} d_i(n - n_i) - 2(n-1)$$

holds, where $n_i = \mathrm{GCD}(n, m_i)$, $d_i = \deg f_i$ and $d_0 = 1$. To see this, let $\Gamma \in X_n$ be the inverse image of a general line on $\mathbb{P}^2$. The assertion follows from the fact that $H^1(X_n, \mathcal{O})$ injects into $H^1(\Gamma, \mathcal{O})$.

In [**Z1**], Zariski showed that if $n$ is a power of a prime number, then $q(X_n) = 0$ provided that $C$ is irreducible. We shall generalize this result to the case in which the curve $C$ is reducible. Namely, we prove the inequality:

$$2q(X_n) \leq (n-1)(r-1),$$

under the assumption that $n$ divides $d$ or that $f$ is a power of a primitive polynomial. See Theorem 7 in §4. My personal motivation to this result is its application to the analysis of singular plane curves. Cf. [**S**].

In order to study the irregularity of cyclic coverings of the projective plane $\mathbb{P}^2$, three different approaches have been developed. The first is the interpretation of the irregularity as the dimension of the vector space of certain rational differential forms, cf. Zariski [**Z2**], Esnault [**E**], Zuo [**Zu**], Artal-Bartolo [**A**], Vaquié [**V**]. The second is the observation of the action of the cyclic group $\mathbb{Z}_n$ on the Albanese variety, cf. de Franchis [**dF**], Comessatti [**C**], Khashin [**K**], Catanese-Ciliberto [**CC**]. The third is a topological one, which studies the infinite cyclic covering of the complement of the branch curve in $\mathbb{C}^2$, cf. Zariski [**Z1**], Libgober [**L1**], Randell [**R**], Kohno [**Ko**], Loeser-Vaquié [**LV**], Dimca [**D**]. In §2, §3, we quickly review the first and second approaches. In §4, we discuss the third approach and prove the above inequality.

*Acknowledgment.* During the preparation of this article, I visited the University of Notre Dame and Illinois University at Chicago. I would like to thank A.Sommese, L.Ein and A.Libgober for their hospitality and for helpful conversations. I would also like to thank the organizers of the Conference in L'Aquila for their kind invitation and for providing a very pleasant atmosphere.

## 2. Differential forms

Let $\psi : S \to \mathbb{P}^2$ be a composition of blow-ups so that the inverse image of $C \cup L$ has normal crossings. Write

$$\psi^*(\sum_{i=0}^{r} m_i C_i) = \sum \nu_j D_j.$$

Here we set $C_0 = L$. We understand that if $j \leq r$, $D_j$ is the strict transform of $C_j$ and $\nu_j = m_j$ and that for $j > r$, $D_j$ is exceptional for $\psi$. Since $\psi^*(\sum_{i=0}^{r} m_i C_i) \in |n\psi^*\mathcal{O}(e)|$, one can construct an $n$-fold covering of $S$, which ramifies over $\cup D_j$. Let $W_n'$ denote its normalization. Up to birational equivalence, we have the commutative diagram:

$$\begin{array}{ccccc} W_n & \longleftarrow & W_n' & \longleftarrow & X_n \\ \downarrow & & \downarrow & & \\ \mathbb{P}^2 & \longleftarrow & S & & \end{array}$$

Let $\phi : X_n \to S$ be the composed map. Set $\zeta = e^{2\pi i/n}$. The eigenspace decomposition of the structure sheaf $\mathcal{O}_{X_n}$ has the following consequence:

PROPOSITION 1 (ESNAULT [**E**]). *In this situation, we have*

$$H^0(X_n, \mathcal{O}_{X_n})^{\zeta^i} \cong H^0(S, \mathcal{L}^{(i)^{-1}}),$$

*where* $\mathcal{L}^{(i)} = \psi^*\mathcal{O}(ie) \otimes \mathcal{O}(-\sum [i\nu_j/n] D_j)$.

As for the eigenspace decomposition of the sheaf $\Omega^1$, we have

PROPOSITION 2 ([**E, Zu**]). *The isomorphism:*

$$H^0(X_n, \Omega^1)^{\zeta^i} \cong H^0(S, \Omega^1(\log D(i)) \otimes \mathcal{L}^{(i)^{-1}})$$

*holds, where* $D(i) = \sum (i\nu_j - n[i\nu_j/n]) D_j$.

*Remark* 1. Note that $D_j \not\subset D(i)$ if and only if $n | i\nu_j$.

The Bogomolov type vanishing theorem gives the following criterion for the vanishing of the irregularity.

THEOREM 1 ([**E, Zu**]). *If $D(i)$ is big for all $i$, then $q(X_n) = 0$.*

PROOF. If $H^0(X_n, \Omega^1)^{\zeta^i} \neq 0$, then one finds that $\mathcal{L}^{(i)} \hookrightarrow \Omega^1(\log D(i))$, which is impossible if $D(i)$ is big, since $D(i) \in |(\mathcal{L}^{(i)})^{\otimes n}|$. □

Since $q(X_n) = p_g(X_n) + 1 - \chi(\mathcal{O}_{X_n})$, one gets the irregularity $q(X_n)$ if one knows both $p_g(X_n)$ and $\chi(\mathcal{O}_{X_n})$.

PROPOSITION 3. $H^0(X_n, \Omega^2)^{\zeta^i} \cong H^0(S, \Omega^2(\log D(i)) \otimes \mathcal{L}^{(i)^{-1}})$.

On the other hand, one has the following formula for the term $\chi(\mathcal{O}_{X_n})$.

PROPOSITION 4.

$$\chi(\mathcal{O}_{X_n}) = \sum_{i=0}^{n-1} \chi(\mathcal{O}_{\mathbb{P}^2}(-(ie - \sum [i\nu_j/n] d_j))) - \dim R^1\pi_*\mathcal{O}_{X_n}.$$

PROOF. Taking the direct image sheaf, we see that

$$\chi(\mathcal{O}_{X_n}) = \chi((\psi \circ \phi)_* \mathcal{O}_{X_n}) - \dim R^1(\psi \circ \phi)_* \mathcal{O}_{X_n}.$$

We have

$$(\psi \circ \phi)_* \mathcal{O}_{X_n} \cong \psi_*(\mathcal{L}^{(i)^{-1}}) \cong \mathcal{O}(-(ie - \sum [i\nu_j/n] d_j)),$$

and

$$\dim R^1(\psi \circ \phi)_* \mathcal{O}_{X_n} = \dim R^1 \pi_* \mathcal{O}_{X_n}.$$

□

PROBLEM. Discuss those line arrangements $C$ such that $X_n$ has positive irregularity for some $n$. Cf. [**A**].

*Remark* 2. See also [**Z2, L3, V**] for somewhat different approaches.

## 3. Albanese map

Let $X_n$ be a non-singular model of a cyclic multiple plane as defined in Introduction. We denote by $G$ the cyclic group $\mathbb{Z}_n$ and let $\sigma$ be its generator. Suppose $q(X_n) > 0$. Let $\alpha : X_n \to \mathrm{Alb}(X_n)$ be the Albanese map. The group $G$ acts on $X_n$ and naturally on $\mathrm{Alb}(X_n)$. We say that $X_n$ *factors through a pencil* if there is a rational map from $X_n$ onto a $k$-fold cyclic covering $\Gamma_k$ of $\mathbb{P}^1$ for some $k$ ($\geq 2$) dividing $n$ such that the group actions are compatible. In [**CC**], this type of situation is called a partial pull-back from a curve. See also Khashin [**K**].

EXAMPLE 1. Let $\Gamma_k$ be a k-fold cyclic covering of $\mathbb{P}^1$ defined by the equation: $y_2^k = \prod(b_i y_0 - a_i y_1)^{\ell_i}$. Assume $k | \sum \ell_i$. Let $\phi : \mathbb{P}^2 \to \mathbb{P}^1$ be a rational map given by a pair $(G, H)$ where both $G$ and $H$ are homogeneous polynomials of degree $\ell$. If $n | \ell \cdot \sum \ell_i$ and if $k | n$, then the surface $X_n$ defined by the equation: $x_3^n = \prod(b_i G(x_0, x_1, x_2) - a_i H(x_0, x_1, x_2))^{\ell_i}$ factors through $\Gamma_k$.

PROPOSITION 5. *If the Albanese image $\alpha(X_n)$ is a curve, then $X_n$ factors through a pencil.*

PROOF. Set $\Gamma = \alpha(X_n)$. The group $G$ also acts on $\Gamma$. We infer that $\Gamma/G$ is isomorphic to $\mathbb{P}^1$, because there exists a rational map from $\mathbb{P}^2$ onto it. □

PROPOSITION 6. *Suppose that there exist two linearly independent holomorphic one forms $\omega$, $\omega'$ such that $\sigma^*\omega = \lambda\omega$, $\sigma^*\omega' = \lambda^{-1}\omega'$ for some $\lambda$. Then the Albanese image $\alpha(X_n)$ is a curve.*

PROOF. By hypothesis, we find that $\sigma^*(\omega \wedge \omega') = \omega \wedge \omega'$. So $\omega \wedge \omega'$ must be a pull-back of a holomorphic 2-form on $\mathbb{P}^2$, hence $\omega \wedge \omega' = 0$. The assertion follows from the Castelnuovo-de Franchis theorem. □

PROPOSITION 7. *Suppose that there exists an n-th root of unity* $\lambda$ ($\lambda \neq \pm 1$) *such that* $\sigma^*\omega = \lambda\omega$ *for all* $\omega \in H^0(X_n, \Omega^1)$. *Then* $\lambda$ *can take one of the values* $\pm i$, $\pm\rho$, $\pm\rho^2$ *where* $\rho = e^{2\pi i/3}$. *Furthermore,*

$$\mathrm{Alb}(X_n) \cong E_\lambda^q,$$

*where* $E_\lambda$ *is the elliptic curve* $\mathbb{C}/\mathbb{Z} \oplus \mathbb{Z}\lambda$.

PROOF. Cf. Comessatti [C]. □

THEOREM 2 (DE FRANCHIS [**dF**]). *If* $q(X_2) > 0$, *then* $X_2$ *factors through a pencil.*

PROOF. . In case $n = 2$, one must have $\sigma^*\omega = -\omega$ for all $\omega \in H^0(X_n, \Omega^1)$. So the assertion follows from Propositions 5 and 6. □

THEOREM 3 (COMESSATTI [**C**]). *If* $q(X_3) > 0$ *and if the Albanese image of* $X_3$ *is a surface, then* $\mathrm{Alb}(X_3) \cong E_\rho^q$.

PROOF. This follows from Propositions 6 and 7. □

We can prove this type of results for the cases $n = 4, 6$, which were also proved by Catanese and Ciliberto [**CC**].

THEOREM 4. *If* $q(X_4) > 0$, *then either* Alb $(X_4) \cong E_i^q$ , *or* $X_4$ *factors through a pencil.*

PROOF. If $H^0(X_4, \Omega^1)^{(-1)} \neq 0$, then the surface: $x_3^2 = x_0^{m_0}\tilde{f}$ factors through a pencil, so does $X_4$. In case $H^0(X_4, \Omega^1)^{(-1)} = 0$, by Propositions 6 and 7, we see that either $X_4$ factors through a pencil or $\mathrm{Alb}(X_4) \cong E_i^q$. □

EXAMPLE 2. $z^4 = (y^2 - 2x^3)x^2(x^2+1)^2(y+2x)$. In this case, $X_4 \cong E_i^2$.

## 4. Alexander polynomials

Set $U = \mathbb{C}^2 \backslash \{f = 0\} = \mathbb{P}^2 \backslash C \cup L$. Write $U_n = \varphi^{-1}(U) \subset X_n$. We see that $\varphi : U_n \to U$ is an unramified covering of degree $n$. We have a commutative diagram:

$$\begin{array}{ccc} U_n & \xrightarrow{\varphi} & U \\ \downarrow & & \downarrow f \\ \mathbb{C}^* & \xrightarrow[z \to z^n]{} & \mathbb{C}^* \end{array}$$

The idea of the topological approach is to calculate the first Betti number of $X_n$ through that of $U_n$. Define the boundary contribution term:

$$B.C. = b_1(U_n) - b_1(X_n).$$

PROPOSITION 8. $B.C. \geq r$.

This follows from the following:

PROPOSITION 9 (ESNAULT [**E**], HIRONAKA [**Hi**]). *Let $S$ be a smooth projective surface and let $D = D_1 \cup \cdots \cup D_n$ be a divisor having simple normal crossings on $S$. Then*

$$b_1(S \backslash D) - b_1(S) = n - \rho(D),$$

*where* $\rho(D) = \dim\{\sum \mathbb{R}[D_i]\}$ *in* $NS(S) \otimes \mathbb{R}$.

PROOF. One can deduce this from the Residue sequence:

$$0 \to H^0(S, \Omega^1) \to H^0(S, \Omega^1(\log D)) \to H^0(\hat{D}, \mathcal{O}) \to H^1(S, \Omega^1)$$

□

EXAMPLE 3. If $f$ is reduced and if $L$ meets $C$ transversely, then $B.C. = r$. Cf. [**L1**].

*Remark* 3. In the first version of this note, I claimed the formula:

$$B.C. = \#\{\text{the irreducible components of } \varphi^{-1}(C \cup L)\} - 1.$$

But this is not true in general. I would like to thank Prof.V.Kulikov for pointing out a counter example.

We now construct an infinite cyclic covering $\tilde{U}$ of $U$. For this, define $\tilde{U}$ to be a connected component of the pull-back of $f : U \to \mathbb{C}^*$ via $\exp : \mathbb{C} \to \mathbb{C}^*$.

$$\begin{array}{ccc} \tilde{U} & \xrightarrow{\Phi} & U \\ {\scriptstyle f_\infty}\downarrow & & \downarrow{\scriptstyle f} \\ \mathbb{C} & \xrightarrow[\exp]{} & \mathbb{C}^* \end{array}$$

It is well known that $H_1(U, \mathbb{Z}) = \mathbb{Z}^r$, which is generated by the meridian loops $\gamma_i$ around $C_i$. The map $f_* : \pi_1(U) \to \pi_1(\mathbb{C}^*) = \mathbb{Z}$ factors through $H_1(U, \mathbb{Z})$ and it sends $[\gamma_1]^{s_1} \cdots [\gamma_r]^{s_r}$ to $\sum m_i s_i$. It turns out that $\tilde{U}$ is nothing but the quotient of the universal covering of $U$ by the kernel of the above homomorphism. Cf. [**D**, p. 107].

Let $T$ be the deck transformation on $\tilde{U}$ corresponding to the above infinite cyclic covering. The transformation $T$ induces a linear transformation $T_* : H_1(\tilde{U}) \to H_1(\tilde{U})$. We have the exact sequence ([**M2**]):

$$\cdots \to H_1(\tilde{U}) \xrightarrow{T_* - I} H_1(\tilde{U}) \to H_1(U) \to \cdots$$

Since $H_1(U, \mathbb{Z}) = \mathbb{Z}^r$, we infer that the sequence:

(1) $$H_1(\tilde{U}, \mathbb{Z})_0 \xrightarrow{T_* - I} H_1(\tilde{U}, \mathbb{Z})_0 \to \mathbb{Z}^{r-1} \to 0$$

is exact, where $H_1(\tilde{U}, \mathbb{Z})_0 = H_1(\tilde{U}, \mathbb{Z})/\mathrm{Tor}$.

DEFINITION. Under the assumption that $H_1(\tilde{U}, \mathbb{C})$ is finite dimensional, the *Alexander polynomial* of $f$ is defined as follows (cf. [**L1**]):

$$\Delta_f(t) = \det(tI - T_*).$$

Since $T_*$ is defined on $H_1(\tilde{U}, \mathbb{Z})_0$, we infer that $\Delta_f(t) \in \mathbb{Z}[t]$. It follows from (1) that $\Delta_f(t) = (t-1)^{r-1} \cdot G(t)$ but $G(1) \neq 0$.

EXAMPLE 4. Suppose that $f(x, y)$ is weighted homogeneous. Let $(a, b)$ be the weights of $(x, y)$ and let $N$ be the degree of $f$ as a weighted homogeneous polynomial. Then $U \to \mathbb{C}^*$ is a fibre bundle, of which fibre is $F = \{(x, y) | f(x, y) = 1\}$. Set $\xi = e^{2\pi i/N}$. Let $h : F \ni (x, y) \to (\xi^a x, \xi^b y) \in F$ be the monodromy map and we denote by $h_*$ the induced linear map on $H_1(F, \mathbb{C})$. In this case, $H_1(\tilde{U}) \cong H_1(F)$ and $\Delta_f(t) = \det(tI - h_*)$. Clearly, the origin $p$ is the only singularity of the affine curve $f = 0$ and $\det(tI - h_*)$ is known to be the local Alexander polynomial $\Delta_p(t)$ of $p$ (cf. [**M1**]).

DEFINITION. In case $N = \dim H_1(\tilde{U}, \mathbb{C}) < \infty$, let $e_j(t)$, $j = 1, ..., N$, be the elementary divisors of $tI - T_*$. Set

$$N(n, T_*) = \sum \#\{\text{distinct } n\text{-th roots of unity which are roots of } e_j(t)\}.$$

THEOREM 5. *If* $\dim H_1(\tilde{U}, \mathbb{C}) < \infty$, *then*

$$2q(X_n) = 1 + N(n, T_*) - B.C.$$

PROOF. . We have the following exact sequence (cf. [**SS**]):

$$\cdots \to H_1(\tilde{U}) \xrightarrow{T_*^n - I} H_1(\tilde{U}) \to H_1(U_n) \to \cdots$$

We infer from this that $b_1(U_n) = 1 + \dim \operatorname{Ker}(T_*^n - I)$. We see easily that $N(n, T_*) = \dim \operatorname{Ker}(T_*^n - I)$. □

COROLLARY. *If* $T_*$ *is of finite order, then*

$$2q(X_n) = 1 + \#\{n\text{-th roots of unity which are roots of } \Delta_f(t)\} - B.C.$$

DEFINITION. We say that $f$ is *primitive* if the general fibre $f^{-1}(a)$ is irreducible. Cf. Suzuki [**Su**]. It is well known that if $f$ is not primitive, then one can find a primitive polynomial $g(x, y)$ and a polynomial $\theta(u)$ such that $f(x, y) = \theta(g(x, y))$. Write $s = \#\{\theta^{-1}(0)\}$. Clearly, $r \geq s$.

*Remark* 4. (i) Note that $s = 1$ if and only if $f$ is a power of a primitive polynomial. (ii) If $s \geq 2$, then the infinite line $L$ cannot meet $C$ transversely. (iii) If $f$ is not primitive, then $X_n$ factors through a pencil.

PROPOSITION 10. *The vector space* $H_1(\tilde{U}, \mathbb{C})$ *is finite dimensional if and only if* $f$ *is a power of a primitive polynomial.*

PROOF. Suppose first that $f$ is primitive. The general fibre of the fibration $f_\infty : \tilde{U} \to \mathbb{C}$ is irreducible. By Lemma 7 in [**Su**], we see that $\dim H_1(\tilde{U}, \mathbb{C}) \leq \dim H_1(\text{a general fibre}, \mathbb{C}) < \infty$. Assume now that $f$ is not primitive. Writing $f = \theta \circ g$ as above, we set $\theta^{-1}(0) = \{a_1, ..., a_s\}$. Define $V = \mathbb{C} \backslash \{a_1, ..., a_s\}$. We have the diagram:

$$\begin{array}{ccc} \tilde{U} & \longrightarrow & U \\ \downarrow & & \downarrow g \\ \tilde{V} & \longrightarrow & V \\ \downarrow & & \downarrow \theta \\ \mathbb{C} & \longrightarrow & \mathbb{C}^* \end{array}$$

If $s \geq 2$, it is easy to see that $\dim H_1(\tilde{V}, \mathbb{C}) = \infty$, hence $\dim H_1(\tilde{U}, \mathbb{C}) = \infty$. If $s = 1$, then $\tilde{V} = \mathbb{C}$ and so $\dim H_1(\tilde{U}, \mathbb{C}) < \infty$. □

*Remark* 5. In case $r = 1$, this fact was pointed out in [**L1**].

Now we come to Zariski's result.

THEOREM 6 (ZARISKI [**Z1**]). *Suppose $r = 1$. If $n = p^a$ ($p$ is a prime number), then $q(X_n) = 0$.*

PROOF. Since $r = 1$, we infer from (1) that $\Delta_f(1) = \det(I - T_*) = \pm 1$. If a primitive $p^i$-th root of unity ($1 \leq i \leq a$) is a root of the integral polynomial $\Delta_f(t)$, then $\Delta_f(t)$ must be divided by the cyclotomic polynomial $\Phi_{p^i}(t)$. Since $\Phi_{p^i}(1) = p$, this is impossible. □

We can generalize this result to the case in which $C$ is reducible.

THEOREM 7. *Assume that $f$ is a power of a primitive polynomial, or that $n$ divides $d$. If $n = p^a$ ($p$ is a prime number), then*

$$2q(X_n) \leq (n-1)(r-1).$$

PROOF. Assume $f = g^\ell$, where $g$ is primitive. By Proposition 10, $N = \dim\, H_1(\tilde{U}, \mathbb{C}) < \infty$. Let $d_j(t)$ (resp. $d_j$) be the GCD of all $j$-minors of the matrix $tI - T_*$ (resp. $I - T_*$). By the exact sequence (1), we see that the elementary divisors of $I - T_*$ are $1, ..., 1, \overbrace{0, ...0}^{r-1}$. We infer that $d_j = 1$ for $j \leq N - (r-1)$ and $d_j = 0$ for $j > N - (r-1)$. Since $d_j(1) | d_j$, we find that $d_j(1) = \pm 1$ for $j \leq N - (r-1)$ and $d_j(1) = 0$ for $j > N - (r-1)$. As in the proof of Theorem 6, any primitive $p^i$-th root of unity other than 1 cannot be a root of $d_j(t)$ for $j \leq N - (r-1)$ . Let $e_1(t), ..., e_N(t)$ be the elementary divisors of $tI - T_*$. We know that $d_j(t) = b_j e_1(t) \cdots e_j(t)$, $b_j \in \mathbb{Q}$. Thus any primitive $p^i$-th root of unity other than 1 cannot be a root of $e_j(t)$ for $j \leq N - (r-1)$. It follows that $N(n, T_*) \leq n(r-1)$. Since $B.C. \geq r$, we conclude that $b_1(X_n) \leq (n-1)(r-1)$.

Assume $n|d$. Since the infinite line $L$ does not appear in the branch locus of $X_n \to \mathbb{P}^2$, by taking a suitable line as the infinite line, we may assume that $s = 1$. □

COROLLARY. *If $n = 2$, $r = 2$ and $d$ is even, then $q(X_2) = 0$.*

DEFINITION. Set $\tilde{F} = \{(x_0, x_1, x_2) \in \mathbb{C}^3 | \tilde{f}(x_0, x_1, x_2) = 1\}$. Since $\tilde{f}$ is homogeneous, $\tilde{f} : \mathbb{C}^3 \backslash \{\tilde{f} = 0\} \to \mathbb{C}^*$ is a fibre bundle. The typical fibre is $\tilde{F}$. Letting $\eta = e^{2\pi i/d}$, we have the *monodromy transformation*:

$$\tilde{h} : \tilde{F} \ni (x_0, x_1, x_2) \longrightarrow (\eta x_0, \eta x_1, \eta x_2) \in \tilde{F}.$$

It induces a linear transformation $\tilde{h}_* : H_1(\tilde{F}, \mathbb{Z}) \to H_1(\tilde{F}, \mathbb{Z})$. Define

$$\Delta_C(t) = \det(tI - \tilde{h}_*) \in \mathbb{Z}[t],$$

which is called the *Alexander polynomial* of the plane curve $C$. Cf. [**R, D**]. We remark that deg $\Delta_C(t) = \dim H_1(\tilde{F}, \mathbb{C})$.

THEOREM 8. *Assume that $f$ is reduced and that the infinite line $L$ is in a general position. Then the equality:*

$$\Delta_f(t) = \Delta_C(t)$$

*holds. Furthermore, we have*

$$2q(X_n) = 1 + \#\{n\text{-th roots of unity which are roots of } \Delta_C(t)\} - B.C.$$

PROOF. Cf. [**R, D**]. We see that $U \cong (\mathbb{C}^3 \backslash \{\tilde{f} = 0\}) \cap \{x_0 = 1\}$. The affine version of the Lefschetz theorem ([**H**]) asserts that $\pi_1(\mathbb{C}^3 \backslash \{\tilde{f} = 0\}) \to \pi_1(U)$ is an isomorphism. It follows that $H_1(\tilde{U}, \mathbb{Z}) \cong H_1(\tilde{F}, \mathbb{Z})$. The transformation $T_*$ corresponds to $\tilde{h}_*$. □

COROLLARY. *Under the same hypothesis, if* $\mathrm{GCD}(n, d) = 1$, *then* $q(X_n) = 0$.

PROOF. By hypothesis, we find that $b_1(U_n) = r - 1$ and $B.C. = r$. □

COROLLARY (VAQUIÉ [**V**]). *Under the same hypothesis, if $\pi_1(\mathbb{P}^2 \backslash C)$ is an abelian group, then $q(X_n) = 0$ for any $n$.*

PROOF. Since $\tilde{F} \to \mathbb{P}^2 \backslash C$ is a $d$-fold unramified covering, one has

$$1 \to \pi_1(\tilde{F}) \to \pi_1(\mathbb{P}^2 \backslash C) \to \mathbb{Z}/d\mathbb{Z} \to 1 \quad \text{(exact)}$$

If $\pi_1(\mathbb{P}^2 \backslash C)$ is abelian, then we infer from this that $H_1(\tilde{F}, \mathbb{C}) \cong H_1(\mathbb{P}^2 \backslash C, \mathbb{C})$. As a consequence, we have $\Delta_C(t) = (t-1)^{r-1}$. The assertion then follows from Theorem 8. □

We quote two divisibility theorems of the Alexander polynomials. See also [**Ko**].

THEOREM 9 (LIBGOBER [**L1**]). *Suppose $f$ is irreducible. Then*

$$\Delta_f(t) \mid \prod_{\tilde{p}} \Delta_{\tilde{p}}(t),$$

*where $\tilde{p}$ moves all local branches of* $\mathrm{Sing}(C \cup L)$.

THEOREM 10 (DIMCA [**D**]). *Suppose $f$ is reduced. Then*

$$\Delta_C(t) \mid \prod_{p \in \mathrm{Sing}(C)} \tilde{\Delta}_p(t),$$

*Remark* 6. Loeser-Vaquié [**LV**] proved a formula of $\Delta_C(t)$ in terms of $H^1$ of some sheaf on $\mathbb{P}^2$.

COROLLARY (ZARISKI [**Z2**]). *Suppose $L$ is in a general position. If $C$ has only nodes and ordinary cusps as its singularities, then $q(X_n) = 0$ unless $6|n$ and $6|d$.*

PROOF. We know that $\Delta_p(t) = t - 1$ if $p$ is a node, $= t^2 - t + 1$ if $p$ is an ordinary cusp. Thus $\Delta_C(t) = (t-1)^{r-1}(t^2 - t + 1)^{\ell}$ for some $\ell$. In view of Theorem 8, the assertion follows from this. □

*Remark* 7. The assumption that $L$ is in a general position is necessary in the above result. Let us consider the polynomial $f = (x+y)(x+y+1)$. In this case, we find that $q(X_3) = 1$.

*Remark* 8. We refer to [**D, L2, L3, V**] for the higher dimensional generalization.

*Remark* 9. We remark that Kulikov [**Ku1, Ku2**] has recently obtained related results on Alexander polynomials of plane curves.

## REFERENCES


[A] E. Artal-Bartolo, *First Betti number of the Milnor fiber of the cone over a projective plane curve and position of its singular points*, preprint.

[CC] F. Catanese and C. Ciliberto, *On the irregularity of cyclic coverings of algebraic surfaces*, Geometry of Complex Projective Varieties, Seminars and Conferences **9**, Mediterranean Press, Rende, 1993, pp. 89–115.

[C] A. Comessatti, *Sui piani tripli ciclici irregolari*, Rend. Circ. Mat. Palermo **31** (1911), 293–310.

[dF] M. de Franchis, *I piani doppi dotati di due o più differenziali totali di prima specie*, Rend. Accad. Lincei **13** (1904), 688–695.

[D] A. Dimca, *Singularities and topology of hypersurfaces*, Universitext, Springer-Verlag, Berlin and New York, 1992.

[E] H. Esnault, *Fibre de Milnor d'un cône sur une courbe plane singuliere*, Invent. Math. **68** (1982), 477–496.

[H] H. Hamm, *Lefschetz theorems for singular varieties*, Proc. Symp.Pure Math. **40** (1983), 547–557.

[Hi] E. Hironaka, *Polynomial periodicity for Betti numbers of covering surfaces*, Invent. Math. **108** (1992), 289–321.

[K] S. I. Khashin, *The irregularity of double surfaces*, Math. Notes **33** (1983), 233–235.

[Ko] T. Kohno, *An algebraic computation of the Alexander polynomial of a plane algebraic curve*, Proc. Japan Acad. Ser. A **59** (1983), 94–97.

[Ku1] V.Kulikov, *The Alexander polynomials of plane algebraic curves*, Izv. RAN, Ser. Math. (in Russian) **57** (1993), 122-130.

[Ku2] ———, *The Alexander polynomials of plane algebraic curves. II*, preprint.

[L1] A. Libgober, *Alexander polynomial of plane algebraic curves and cyclic multiple planes*, Duke Math. J. **49** (1982), 833–851.

[L2] ———, *Homotopy groups of the complements to singular hypersurfaces. II*, preprint.

[L3] ———, *Position of singularities of hypersurfaces and the topology of their complements*, preprint.

[LV] F. Loeser and M. Vaquié, *Le polynôme d'Alexander d'une courbe plane projective*, Topology **29** (1989), 163–173.

[M1] J. Milnor *Singular points of complex hypersurfaces*, Ann. Math. Studies **61**, Princeton Univ. Press 1968.

[M2] ———, *Infinite cyclic coverings*, Topology of manifolds, Prindle, Weber and Schmidt, Boston, 1968, pp. 115–133.

[R] R. Randell, *Milnor fibers and Alexander polynomials of plane curves*, Proc. Symp. Pure Math. **40** (1983), 415–419.

[S] F. Sakai, *Singularities of plane curves*, Geometry of Complex Projective Varieties, Seminars and Conferences **9**, Mediterranean Press, Rende, 1993, pp. 257–273.

[SS] Y. Shinohara and D. W. Sumners, *Homology invariants of cyclic coverings with applications to links*, Trans. A. M. S. **163** (1972), 101–121.

[Su] M. Suzuki, *Sur les opérations holomorphes du groupe additif complexe sur l'espace de deux variables complexes*, Ann. Sci. Éc. Norm. Sup. **10** (1977), 517-546.

[V] M. Vaquié, *Irrégularité des revêtements cycliques*, preprint.

[Z1] O. Zariski, *On the linear connection index of the algebraic surfaces* $z^n = f(x, y)$, Proc. Nat. Acad. Sci. **15** (1929), 494–501.

[Z2] ———, *On the irregularity of cyclic multiple planes*, Ann. of Math. **32** (1931), 485–511.

[Zu] K. Zuo, *Kummer-Überlagerungen algebraischer Flächen*, Bonner Math. Schrift. **193** (1989).

DEPARTMENT OF MATHEMATICS, FACULTY OF SCIENCE, SAITAMA UNIVERSITY, SHIMO-OKUBO 255, URAWA 338, JAPAN

*E-mail address*: fsakai@rimath.saitama-u.ac.jp

Contemporary Mathematics
Volume **162**, 1994

# Counting rational Points on Cubic Surfaces

Sir Peter Swinnerton-Dyer

1. Let $V$ be a Fano variety defined over an algebraic number field $K$, and let $h = h(P)$ be a height function on $V(K)$ associated with the class of anti-canonical divisors. Let

$$N(X) = N(X, V, h)$$

be the number of points of $V(K)$ of height less than $X$. In a series of papers, of which [1] and [3] are already in print, Manin has considered the behaviour of $N(X)$ as $X \to \infty$. It can happen that most of the points of $V(K)$ are concentrated on finitely many subvarieties; thus it is usually better to consider, instead of $V$, the open subset $W$ obtained by deleting finitely many exceptional subvarieties. Manin has conjectured that for a suitable definition of these exceptional subvarieties

$$N(X, W) \sim CX(\log X)^{r-1} \tag{1}$$

for some constant $C$, where $r$ is the rank of $\operatorname{Pic} V$. He has also introduced the alternative counting function

$$H(s) = H(s, W) = \sum (h(P))^{-s},$$

the sum being taken over all points $P$ of $W(K)$. Under modest assumptions on the well-behavedness of $N(X)$ or $H(s)$, the conjecture (1) is equivalent to

$$\lim (s-1)^r H(s) = (r-1)! C \tag{2}$$

with the same value of $C$ as in (1), where the limit is taken as $s \to 1$ from above. This formulation raises questions about the analytic continuability of $H(s)$, the answers to which may well depend on the choice of the height function $h(P)$; but in contrast with the case of abelian varieties, there is at present no reason to think that there exists a canonical height on Fano varieties.

Manin has shown no interest in the value of $C$, but Heath-Brown [4] has given a heuristic formula for $C$, with significant supporting evidence, in the particular case when $V$ is a cubic surface defined over $\mathbf{Q}$ and given by an equation of the form

$$x^3 + y^3 + z^3 = kt^3 \tag{3}$$

Mathematics Subject Classification. Primary 11D25.
This paper is in final form and no version of it will be submitted for publication elsewhere.

where $k$ is integral and not a cube. His height function is

$$h(P) = \max(|x|, |y|, |z|) \tag{4}$$

where $P = (x, y, z, t)$ and $x, y, z, t$ are integers with no common factor. The first object of this paper is to recast Heath-Brown's formula into a shape in which it can plausibly be generalized.

Heath-Brown argues as follows. If we formally apply the Hardy-Littlewood method to estimating $N(X)$ for (3), the leading term will have the form $\sigma X$ where

$$\sigma = \sigma_\infty \prod \sigma_p. \tag{5}$$

Here $\sigma_\infty$ is the real density of solutions, given by

$$\sigma_\infty = \frac{1}{3}k^{-\frac{1}{3}} \int_{-1}^{1}\int_{-1}^{1}\int_{-1}^{1} \frac{dx\,dy\,dz}{(x^3+y^3+z^3)^{2/3}} \tag{6}$$

and $\sigma_p$ is the density of $p$-adic solutions, given by

$$\sigma_p = \lim p^{-3n} N^*(p^n) \tag{7}$$

where $N^*(p^n)$ is the number of primitive solutions mod $p^n$ of

$$x^3 + y^3 + z^3 \equiv kt^3 \qquad \bmod p^n.$$

If $p \nmid 3k$ then

$$\sigma_p = p^{-3}(p-1)N_p \tag{8}$$

where $N_p$ is the number of solutions of (3) in $\mathbf{F}_p$, the finite field of $p$ elements.

There are two complications to this approach. The first is that there seems little prospect of proving that the error term in the Hardy-Littlewood method is in fact $o(X)$, even if this is actually true; hence the best we can hope for from this line of argument is a heuristic formula for $N(X)$. The second is that, as the explicit formulae for $\sigma_p$ show, the product $\prod \sigma_p$ is not absolutely convergent. To cope with this, Heath-Brown proceeds as follows.

Write $K = \mathbf{Q}(k^{1/3})$, and for convenience define a prime $p \nmid 3k$ to be *split* if $p \equiv 1 \bmod 3$ and $p$ splits in $K/\mathbf{Q}$, and *non-split* if $p \equiv 1 \bmod 3$ and $p$ does not split in $K/\mathbf{Q}$; these correspond to $k$ being a cube or a non-cube mod$p$. (If $p \equiv 2 \bmod 3$ and $p \nmid 3k$, then $p$ is a product of two primes in $K$.) The formula for $\sigma_p$ can be derived from that for $N_p$ provided $p \nmid 3k$; and the latter is

$$N_p = \begin{cases} p^2 + 7p + 1 & \text{if } p \text{ is split,} \\ p^2 - 2p + 1 & \text{if } p \text{ is non-split,} \\ p^2 + p + 1 & \text{if } p \equiv 2 \bmod 3. \end{cases}$$

On the other hand, the local factor of the zeta-function $\zeta_K(s)$ for $p \nmid 3k$ has the form

$$\begin{array}{ll} (1-p^{-s})^{-3} & \text{if } p \text{ is split,} \\ (1-p^{-3s})^{-1} & \text{if } p \text{ is non-split,} \\ (1-p^{-s})^{-1}(1-p^{-2s})^{-1} & \text{if } p \equiv 2 \bmod 3. \end{array}$$

At $s = 1$, these factors when cubed and multiplied by $(1 - p^{-s})^2$ are just what is needed to render $\prod \sigma_p$ absolutely convergent. Hence we can formally replace (5) by

$$\sigma = \sigma_\infty a b^3 \prod_1 \prod_2 \prod_3 \tag{9}$$

where $a$ is a factor coming from the primes $p$ with $p \mid 3k$,

$$b = \lim(s-1)\zeta_K(s) = \lim \zeta_K(s)/\zeta(s) \quad \text{as} \quad s \to 1$$

can be evaluated from the class number formula,

$$\begin{aligned} \textstyle\prod_1 &= \textstyle\prod(1-p^{-1})^7(1+7p^{-1}+p^{-2}) && \text{over } p \text{ which are split,} \\ \textstyle\prod_2 &= \textstyle\prod(1-p^{-3})^3 && \text{over } p \text{ which are non-split,} \\ \textstyle\prod_3 &= \textstyle\prod(1-p^{-2})^3(1-p^{-3}) && \text{over } p \equiv 2 \bmod 3. \end{aligned}$$

Of course, each of these last three products is restricted to $p \nmid 3k$. Each of these three products is absolutely convergent, so if we replace (5) by (9) we have a well-defined formula amenable to computing.

If we take the value of $C$ in (1) to be $\sigma$ defined in this way, then the numerical evidence for $k = 2, 3$ and $N \leq 1000$ fits well; the error seems to be oscillatory and of magnitude roughly $X^{1/2}$, which is the best one could realistically hope for. Moreover Heath-Brown has proved a formula which could be described as the average of the estimate over a range of values of $k$.

2. We now turn to the linked questions of putting Heath-Brown's formula into a more canonical form and generalizing it. The picture which the reader should have in mind at this stage is that of a nonsingular cubic surface $V$ defined over $\mathbf{Q}$ and given by an equation

$$f(x, y, z, t) = 0.$$

To define a height function $h(P)$ on $V(\mathbf{Q}$, we start from a finite set of linear forms $L_i$ which do not all vanish at any point of $\overline{V}$. For each prime $v$, finite or infinite, and each point $(x, y, z, t)$ of $\mathbf{A}^4(\mathbf{Q})$ write

$$h_v(x, y, z, t) = \max_i(|L_i(x, y, z, t)|_v). \tag{10}$$

If $P$ is a point of $\mathbf{P}^3(\mathbf{Q})$, choose any representative $(x, y, z, t)$ of it and define its height to be

$$h(P) = \prod_v h_v(x, y, z, t); \tag{11}$$

this product is well-defined because almost all factors are 1, and $h(P)$ does not depend on the choice of representative of $P$, by the product formula for valuations. There are more general notions of height, but there seems no need to introduce them here. Note also that Heath-Brown counts primitive solutions of (3) whereas we wish to count points on the associated projective variety; hence we must insert a factor $\frac{1}{2}$ into his formula.

The only possible generalization of (6) is

$$\sigma_\infty = \int\int\int \frac{dx\,dy\,dz}{|\partial f/\partial t|}, \tag{12}$$

where the integral is to be taken over the subset of $\mathbf{A}^4(\mathbf{R})$ defined by

$$f(x,y,z,t)=0, \qquad h_\infty(x,y,z,t)\leq 1.$$

The integrand here is the Poincaré residue, and has much better invariance properties than appear from the expression given for it here. But our eventual conjecture has to be invariant under linear transformation of variables and under multiplication of $f$ by a non-zero constant $c$; and these actions respectively multiply (12) by $|J|$ and $|c|^{-1}$, where $J$ is the Jacobian of the transformation. In consequence, we are virtually forced to replace the definition of $\sigma_p$ in (7) by a formula which has as one factor

$$\iiint \frac{dx dy dz}{|\partial f/\partial t|_p} \tag{13}$$

where the integral is to be taken over the subset of $\mathbf{A}^4(\mathbf{Q}_p)$ defined by

$$f(x,y,z,t)=0, \qquad h_p(x,y,z,t)\leq 1. \tag{14}$$

For almost all $p$, the second condition (14) is equivalent to $x,y,z,t$ in $\mathbf{Z}_p$ and hence for almost all $p$ the value of the integral (13) is $p^{-2}N_p$. In view of (8), this leads us to replace the definition of $\sigma_p$ by

$$\sigma_p=(1-p^{-1})\iiint \frac{dx dy dz}{|\partial f/\partial t|_p} \tag{15}$$

taken over the set defined by (14). For the special surface (3) with $k$ cube-free, the value of $\sigma_p$ given by (15) is the same as that given by (7) except when $p=3$; in this case we have picked up a factor 3, but we shall find below that we have a natural need for such a factor.

We have now to produce convergence factors for the product $\prod \sigma_p$, to replace the rather ad hoc introduction of $\zeta_K(s)$. We have

$$N_p=p^2+\sum \lambda_{ip}+1$$

whenever $V$ has good reduction at $p$, where the $\lambda_{ip}$ are the seven eigenvalues of Frobenius acting on $\operatorname{Pic}\overline{V}$. If

$$L(V,s)=\prod_p\prod_i(1-\lambda_{ip}p^{-s})^{-1}$$

is the $L$-series derived from the 2-cohomology of $V$, with the correct factors at the primes of bad reduction, we are thus led to replace (9) by

$$\sigma=\frac{1}{6}\sigma_\infty\left(\prod \sigma_p^*\right)\lim(L(V,s)/(s-2)) \tag{16}$$

where the limit is taken as $s\to 2$ and

$$\sigma_p^*=\prod_i(1-\lambda_{ip}p^{-2})\iiint \frac{dx dy dz}{|\partial f/\partial t|_p}.$$

For the primes $p$ at which $V$ has bad reduction, there will in general be less than seven of the $\lambda_{ip}$; the correct recipe in this case was given by Serre and Deligne ([6], see for example [5, pp.57-59]), and the right point of view is that these are the factors which give $L(V,s)$ a functional equation of the correct type.

For the special surface (3), the latest formula for $\sigma$ is still the same as Heath-Brown's formula (9) except for the factor $\frac{1}{2}$ whose introduction was explained at the beginning of this section. To show this, it is enough to check that

$$L(V,s) = \zeta_K^3(s-1)/\zeta^2(s-1).$$

It is easy to verify this for the factors associated with primes $p$ which do not divides $3k$; the remaining factors then have to fit correctly, because of the functional equations satisfied by the three functions involved.

3. The product $\sigma_\infty \prod \sigma_p^*$ can be written as a single integral over the subset of adelic affine 4-space defined by

$$f(x,y,z,t) = 0, \qquad h_v(x,y,z,t) \leq 1 \quad \text{(all } v\text{)} \tag{17}$$

with respect to a suitably weighted measure. As Heath-Brown has remarked, in this form it appears to take no account of the failure of weak approximation — that is, of the fact that in general $V(\mathbf{Q})$ will not be dense in $V(\mathbf{A})$ in the adelic topology. But there are at least two indications that this cannot be the right way to proceed, and that it is necessary to take the adelic integral not over the whole set (17) but over the intersection of that set with the closure of the affine image of $V(\mathbf{Q})$. We note the incidental advantage that this change gets rid of the mysterious factor $\frac{1}{3}$ in (16); for if $V$ is given by (3) with $k$ not a cube, weak approximation always fails and the closure of $V(\mathbf{Q})$ in (17) is always one-third of (17) in a sense that can easily be made precise.

Consider first the case when the Hasse principle fails, so that $V(\mathbf{Q})$ is empty though all the $V(\mathbf{Q}_p)$ are non-empty. Many such examples are known; see for example [2] and the references given there. In such a case we have $C = 0$ in (1) and (2). But the only way of obtaining a zero value from the kind of formula which we have been developing is by taking the adelic integral over the null set; for we know too much about the behaviour of $L(V,s)$ to hope that the necessary zero factor can be derived from it.

Moreover, even if one were seeking a formula for $C$ valid only when $V(\mathbf{Q})$ is not empty, taking the integral over the whole of (17) would lead to inconsistencies as $h$ varies. We can illustrate this by considering

$$x^3 + y^3 + z^3 = 3t^3. \tag{18}$$

It is known ([4, Theorem 1]) that for every primitive solution of (18) in $\mathbf{Z}$ either $x \equiv y \equiv z \bmod 9$ or one of $x, y, z$ is divisible by 9; this is clearly not true of all primitive solutions in $\mathbf{Z}_3$. It follows that every primitive solution in $\mathbf{Z}$ satisfies

$$x + y + z - 3t \equiv 0 \mod 9$$

but this is not true of all primitive solutions in $\mathbf{Z}_3$. Now let $h'$ be the height on $V(\mathbf{Q})$ obtained from the linear forms $x, y, z$, and $h''$ the height obtained from the linear forms $x, y, z, (x+y+z-3t)/9$. We have $h' = h''$ on $V(\mathbf{Q})$, and $h'_v \le h''_v$ on $V(\mathbf{A}$; but for $v = 3$ we do not always have equality here. Hence in an obvious notation, we must have $C' = C''$ in (1) and (2); but any formula for $C$ which involved integration over the whole set (17) would imply $C' > C''$.

Even if we go back to a naive height such as (4), the same problems remain. For there are cubic surfaces $V$ with the following property: $V(\mathbf{R})$ has two connected components $W_1$ and $W_2$; weak approximation holds on $W_1$, but there are no rational points on $W_2$. (The only example I know contains a rational line and therefore falls outside the hypothesis $r = 1$; but I am sure that examples with $r = 1$ do exist.) We can then choose naive heights $h'$ and $h''$ with $h' \le h''$, with equality everywhere on $W_1$ but not everywhere on $W_2$. As in the previous paragraph, we have $C' = C''$, but any formula for $\sigma_\infty$ which involves integration over part of $W_2$ as well as part of $W_1$ would imply $C' > C''$.

In the light of this, it is natural to ask what is known or conjectured about the closure of $V(\mathbf{Q})$ in $V(\mathbf{A})$. The main conjecture is that the only obstruction to weak approximation is the Brauer-Manin obstruction, which is certainly computable for any particular $V$. Regardless of this, it is known that each $V(\mathbf{Q}_v)$ can be partitioned into a finite union of sets $W_{iv}$ with the following properties:

(i) If $V$ has good reduction at $v$ then there is only one $W_{iv}$.

(ii) If $v = \infty$ then the $W_{i\infty}$ are the connected components of $V(\mathbf{R})$.

(iii) $V(\mathbf{Q})$ is either dense in or disjoint from each $\prod W_{iv}$.

The rule for constructing the $W_{ip}$, for a prime $p$ at which $V$ has bad reduction, is as follows. Two points of $V(\mathbf{Q}_p)$ lie in the same $W_{ip}$ if, after a suitable linear transformation of variables, their reductions mod$p$ are both nonsingular points on the reduction of $V$ mod $p$. This does not quite constitute a decision procedure for finding the closure of $V(\mathbf{Q})$, but it seems to be good enough for practical purposes

Summing up, we can now put the Heath-Brown heuristic formula into its final form, still under the restriction that $r = 1$. This asserts that the constant $C$ in (1) and (2) is given by

$$C = \frac{1}{2}\tau(V,h)\lim(L(V,s)/(s-2)); \tag{19}$$

here $\tau(V,h)$ is a kind of Tamagawa number, defined by

$$\tau(V,h) = \iiint \left(\frac{dxdydz}{\partial f/\partial t},(\prod_i(1-\lambda_{ip}p^{-2}))\right)$$

where the integral is taken over that part of the closure of the affinisation of $V(\mathbf{Q})$ which satisfies $h_v(x,y,z,t) \le 1$ for all $v$. The products on the right are the necessary convergence factors.

There are two expressions which one would not have been surprised to see in a formula for $C$, though they do not appear in this particular one. One is the determinant of the intersection-number matrix of a base for $\operatorname{Pic} V$; under the condition that $V$ is a nonsingular cubic surface with $r = 1$, the value of this is necessarily 3, but it becomes more interesting once the set-up is further generalized. The other is the order of [Br $V$/Br $\mathbf{Q}$], which is also necessarily 3 for the particular surfaces (3) but can take the values $1, 2, 3, 4$ or $9$ for suitable cubic surfaces $V$ — probably even when $r = 1$. I do not myself expect either of these to appear in the formula for $C$, but the numerical evidence currently available is not sufficient to rule out the appearance of a power of their quotient.

If we are interested in the behaviour of the counting function $H(s, V)$, it is natural to write (19) in the alternative form

$$H(s, V) \sim \frac{1}{2}\tau(V, h)L(V, s+1) \quad \text{for } r = 1 \tag{20}$$

as $s \to 1$. The Tate conjecture, which is known to be true for the 2-cohomology of any Fano variety, says that $L(V, s+1)$ has a pole of order $r$ at $s = 1$. In the light of (2), it is natural to wish to drop the condition $r = 1$ in (20) — or more precisely, to replace it by the fundamental Manin condition that $V$ contains no exceptional curves of the first kind defined over $\mathbf{Q}$. In doing this, we must leave open the possibility that we have to insert some simple function of $r$ as an additional factor in (20). The second example in the next section suggests that when $r = 2$ we need to replace the factor $\frac{1}{2}$ in (20) by $\frac{1}{4}$; but on this evidence it would be rash to do more than propose a relation of the form

$$H(s, V) \sim \mu_r \tau(V, h)L(V, s+1) \tag{21}$$

as $s \to 1$, where $\mu_1 = \frac{1}{2}$ and $\mu_2 = \frac{1}{4}$. The simplest functions that fit this pattern are of course $\mu_r = 2^{-r}$ or $\mu_r = 1/2r$; but a comparison of (1) and (2) suggests that the presence of a further factor $(r-1)!$ should not be ruled out without some empirical evidence.

It is straightforward to generalize these conjectures to other Del Pezzo surfaces, and even to Fano varieties more generally — though the broader the generalization the more need there is to include undetermined 'fudge factors'. But it is worth remarking that the questions posed in the opening paragraph of this paper are not always the most natural ones even for surfaces. For example, if $V$ is a surface with a fibration $V \to \mathbf{P}^1$ by conics, the most natural thing to count is not the rational points on $V$ but the points on $\mathbf{P}^1$ whose associated fibre contains rational points.

4. There are some Fano surfaces for which $H(s)$ can be written down explicitly. In this section we do this for two particular examples, partly as confirmatory evidence for the natural generalization of the conjecture above, but primarily because it limits what we can hope for about the analytic continuability of $H(s)$. In each case we start with a rational surface $V'/\mathbf{Q}$ which is not Fano in the classical sense, because it is not embedded in projective space by means of the anti-canonical divisor class; and we take $h'$ to be a suitable height function on $V'$. There is then a Fano surface $V/\mathbf{Q}$ biregularly equivalent to $V'$, and a height function $h$ on $V$ in the sense of (10) and (11) which is simply a power of $h'$; thus if we know the function $H'$ derived from $h'$, we can write down $H$.

For our first example we take $V' = \mathbf{P}^2$ with coordinates $x, y, z$; and we take $h'$ to be the height derived from the linear forms $x, y, z$. There are 13 points on

$\mathbf{P}^2$ with $h' = 1$, and

$$12N^2 \prod (1 - p^{-2}) \quad \text{points with } h' = N > 1$$

where the product is taken over the distinct primes $p$ dividing $N$; and it follows easily that

$$H'(s) = 1 + 12\zeta(s-2)/\zeta(s).$$

For $V$ we take the image of the Veronese embedding $\mathbf{P}^2 \to \mathbf{P}^9$, where the coordinate functions are the monomials of degree 3 in $x, y, z$; the obvious height on $V$ is $h = (h')^3$ and for this we have

$$H(s) = 1 + 12\zeta(3s-2)/\zeta(3s).$$

Thus $H(s)$ can be analytically continued to the entire $s$-plane, but only as a meromorphic function with poles in somewhat unexpected places; and though it does satisfy a functional equation of a kind, it is not one which a respectable number theorist would wish to have anything to do with.

In our previous notation there are no primes of bad reduction and $N_p = p^2 + p + 1$; hence the $L$-series associated with the 2-cohomology is

$$L(V, s) = \zeta(s-1).$$

The Tamagawa number in this case is

$$\iiint (dx dy dz, (1 - p^{-1})) = 8 \prod (1 - p^{-3}) = 8/\zeta(3)$$

so that the analogues of (19) and (20) do indeed hold.

For our second example we take $V'$ to be the quadric surface in $\mathbf{P}^3$ given by $xy = zt$; and we take $h'$ to be the height derived from the linear forms $x, y, z, t$. If the standard map $\mathbf{P}^1 \times \mathbf{P}^1 \to V'$ is written $P_1 \times P_2 \mapsto P'$ then we have $h'(P') = h''(P_1)h''(P_2)$ for the obvious height function $h''$ on $\mathbf{P}^1$; and it follows easily that

$$H'(s) = 16\zeta^2(s-1)/\zeta^2(s).$$

For $V$ we take the image of $V'$ under the Veronese embedding $\mathbf{P}^3 \to \mathbf{P}^9$, where the coordinate functions are the monomials of degree 2 in $x, y, z, t$; the obvious height on $V$ is $h = (h')^2$ and for this we have

$$H(s) = 16\zeta^2(2s-1)/\zeta^2(2s).$$

In our previous notation there are no primes of bad reduction and $N_p = (p+1)^2$; hence the $L$-series associated with the 2-cohomology is

$$L(V, s) = \zeta^2(s-1)$$

and the Tamagawa number is

$$\iiint (z^{-1} dx dy dz, (1 - p^{-2})) = \sigma_\infty \prod (1 - p^{-2})^{-2} = \sigma_\infty/\zeta^2(2)$$

in our previous notation. An easy calculation gives $\sigma_\infty = 16$, so that in this case (for which $r = 2$), we have

$$H(s,V) \sim \frac{1}{4}\tau(V,h)L(V,s+1).$$

5. I am indebted to Dr P. M. H. Wilson for a number of valuable discussions.

# References

[1] Batyrev, V. V., and Manin, Yu. I., Sur le nombre des points rationnels de hauteur borné des variétés algébriques, *Math. Ann.* 286 (1990), 27-43.

[2] Colliot-Thélène, J-L, Kanevsky, D. and Sansuc, J. J., Arithmétique des surfaces cubiques diagonales, in Diophantine Approximation and Transcendence Theory (*Springer Lecture Notes*, 1290) (1987), 1 - 108.

[3] Franke, J., Manin, Yu. I. and Tschinkel, Yu., Rational points of bounded height on Fano varieties, *Invent. Math.* 95 (1989), 421-435.

[4] Heath-Brown, D. R., The density of zeros of forms for which weak approximation fails, (in course of publication).

[5] Hulsbergen, W. W. J., Conjectures in arithmetic algebraic geometry, (Aspect of Mathematics, vol E18) (Vieweg, 1992).

[6] Serre, J.-P., Facteurs locaux des functions zêta des variétés algébriques (définitions et conjectures), Sém. Delange-Pisot-Poitou 1969/70, exp. 19.

DPMMS
16 Mill Lane
Cambridge

Contemporary Mathematics
Volume **162**, 1994

# Miyaoka-Yau Inequality for Normal Surfaces and Local Analogues[1]

Jonathan Wahl

ABSTRACT: We prove a Miyaoka-Yau inequality for a log- terminal surface (X,D), combining Miyaoka's approach with some purely local results on complex normal surface singularities. Applied to a pair arising form a smoothing of negative weight of a quasi-homogenous Gorenstein surface singularity, this result yields an inequality of certain "local chern numbers" of that singularity. We formulate and motivate a general conjecture about smoothable Gorenstein surface singularities, saying that a local $c_1^2$ is $\leq 3 \cdot$ local $c_2$.

## Introduction

Let X be a smooth projective complex surface. If the Kodaira dimension $\kappa(X) \geq 0$ (i.e., X not birationally ruled), then the (Bogomolov)-Miyaoka-Yau inequality of Chern numbers states

$$c_1^2(X) \leq 3c_2(X)$$

([Mi1], [Y]). Various log-versions of this result

[1]1991 Mathematics Subject Classification. Primary 14J10, 32J15.
Partially supported by NSF Grant DMS-9103604

were proved by F. Sakai [S1] and Miyaoka [Mi2] (cf. also [CY], [KNS]); for instance, one has the

**Theorem** ([**Mi2**]): Let X be a projective surface with only quotient singularities $\{p_i\}$, $f:\widetilde{X}\to X$ the minimal resolution, with $E_i=f^{-1}(p_i)_{red}$. Suppose $K_X$ is nef. Then

$$c_2(\widetilde{X})-(\frac{1}{3})c_1^2(\widetilde{X}) \geq \Sigma\nu(p_i),$$

where $\nu(p_i)>0$ is given for each $p_i$ as in [Mi2].

In this paper we generalize Miyaoka's result, in particular extending the last result (under appropriate hypotheses) to normal surfaces; at the same time we clarify (and repair) some of his arguments. It seems natural to make purely local arguments about singularities in lieu of Miyaoka's ad hoc construction in [Mi2], §4; this has been developed in two papers [W1], [W2], whose results are used. But, the main skeleton of the argument here is exactly Miyaoka's.

To state the main theorem, we say (X,D) is a **log-terminal pair** in the sense of Shokurov (weakly log-terminal, in Kawamata's work) if one has the following: X is a projective surface, with only quotient singularities, and S=Sing(X). D is a reduced normal crossings divisor on X, for which $S''=S\cap D$ consists of smooth points of D; write $S'=S-S''$. Let $\pi:\widetilde{X}\to X$ be the minimal resolution, $\widetilde{D}$ the proper transform of D, $E=E'\cup E''$ the reduced exceptional divisor (E' and E'' lie over S' and S'', respectively). Then each $p\in S''$ is a cyclic quotient singularity, and $\widetilde{D}$ meets the chain $\pi^{-1}(p)$ transversally at one point of a component at the end of the chain.

Denote the Euler characteristic of a space Z by e(Z), and the local fundamental group of a singular point p of a surface by $G_p$. For $p\in S'$, denote $\nu(p)=e(E_p)-\frac{1}{|G_p|}$.

**Main Theorem**: Let (X,D) be a log-terminal pair as above. Assume

(*) $K_{\tilde{X}}+\tilde{D}+E$ has a Zariski decomposition with positive part P and negative part N+N'+N", where Supp(N')⊂E', Supp(N")⊂E", and Supp(N∩(E'∪E"))=∅.

Then

$$\text{(M-Y)} \qquad \sum_{p\in S'} \nu(p) \leq c_2(\tilde{X}) - e(\tilde{D}\cup E'') - (\tfrac{1}{3})P\cdot P - (\tfrac{1}{4})N\cdot N.$$

(Miyaoka's Theorem [Mi2] is identical except that S"=∅ there.) The above formulation was suggested to us by David Morrison. There is an analytic version of this theorem ([KNS]) which treats a more general orbifold situation (with D fractional); that proof is metric in nature, and the Zariski-decomposition hypothesis is stronger.

The Main Theorem provides an inequality of invariants of certain normal surfaces. One associates to a germ (X,p) of a normal surface singularity two rational numbers which are "characteristic" in that they multiply by degree under (Y,q)→(X,p) finite and unramified off {p}: these are $(-P\cdot P)_p$ (which is 0 exactly for log- canonical p [W2]) and $\frac{1}{|G_p|}$ (which is 0 for a non-quotient singularity). Next define a rational number $c_1^2(X)$ for a compact normal surface by making sense of $f^*K_X\cdot f^*K_X$, where $f:\tilde{X}\to X$ is any resolution. Blowing-down the negative part of the Zariski decomposition of $K_{\tilde{X}}$+(exceptional divisor of f) gives the log-canonical resolution (Y,D)→X. Then (M-Y) for (Y,D) translates, under appropriate hypotheses on a projective normal X, to

$$e(X) - \sum\left(1 - \frac{1}{|G_q|}\right) \geq \frac{1}{3}\left\{ c_1^2(X) - \sum -(P\cdot P)_q \right\},$$

the sum over all q∈Sing(X). We write this inequality as

$$\text{(M-Y)'} \qquad \tilde{e}(X) \geq \frac{1}{3}\tilde{c}_1^2(X),$$

noting that the new invariants $\tilde{e}$ and $\tilde{c}_1^2$ thus defined are also "characteristic" in that they multiply by degree under f:Y→X finite and unramified off $f^{-1}$(Sing X). Note $\tilde{e}$ is a generalized orbifold Euler characteristic for a normal surface.

As in Miyaoka's work ([Mi2], [Iv]), these inequalities imply bounds on the configuration of singular points of hypersurfaces in $\mathbb{P}^3$. Consistently with Miyaoka's definition for quotients, define for a smoothable Gorenstein surface singular point p of Milnor number $\mu$:

$$\nu(p)=1+\mu+(\frac{1}{3})P\cdot P-\frac{1}{|G_p|}.$$

**Corollary 1**: Let $X\subset\mathbb{P}^3$ be a hypersurface of degree d with isolated singularities for which (*) holds on the log-canonical resolution. Then

$$\sum_{p\in X}\nu(p)\leq\frac{2}{3}d(d-1)^2.$$

Miyaoka's Theorem is that the inequality is true for $d\geq 4$ when X has only RDP's (in which case (*) is certainly true). While both (*) and the inequality fail if X is a cone, it seems possible that the inequality holds for ali non-conical X once $d\geq 5$.

To find useful hypotheses for (*) or (M-Y)' to hold, assume each germ (X,p) is $\mathbb{Q}$-Gorenstein; thus, there is a nowhere-0 m-fold 2-form $\omega_m$ on X-{p}. One may define "log-adjoint" ideals $J_{nm}\subset\mathcal{O}_p$, so that $f\cdot(\omega_m)^{\otimes n}$ is a section of $nm(K_{\tilde{X}}+E)$ on the resolution iff $f\in J_{nm}$. Thus, $J_{km}=\mathcal{O}$ for all k iff (X,p) is log-canonical. We have the

**Corollary 2**: Let X be a projective surface with only $\mathbb{Q}$- Gorenstein singularities. Suppose that for some m>0, the linear system of m-forms with log-adjoint conditions is base-point-free off Sing X. Then the inequality (M-Y)' is valid.

**Corollary 3**: Let X be a projective surface with only log- canonical singularities, and with $K_X$ nef. Then the inequality (M-Y)' holds.

The last Corollary is proved in [Mi2] when the singularities are quotients, cusps, or simple-elliptic.

Let $\Gamma\subset SU(2,1)$ be a discrete subgroup, which acts properly discontinuously and with finite covolume and isolated fixed points on the open 2-ball $B^2$ in $\mathbb{P}^2$. Then $B^2/\Gamma$ admits a compactification $X=(B^2/\Gamma)^*$ by adding finitely many cusps. X is a normal projective surface with log-canonical singularities: quotient singularities (from fixed points of $\Gamma$) and simple elliptic singularities plus their cyclic quotients (from the cusps). It follows from Hirzebruch-proportionality in the non-compact case [Mu] that the inequality (M-Y)' is in fact an equality (and, $K_X$ is ample). Further, [KNS] have generalized Yau's methods to prove that if X has only log-canonical singularities and $K_X$ is ample, then equality in (M-Y)' implies X is of the form $(B^2/\Gamma)^*$. We expect that any X (with obvious exceptions) with equality in (M-Y)' must have only log-canonical singularities. Note that blowing down the dual cusps of a parabolic Inoue surface gives a (non-algebraic) compact normal surface with equality in (M- Y)' (each side equals 0).

In the last two sections, we formulate a purely local conjecture about deformations of smoothable Gorenstein surface singularities:

**Conjecture**: The singularity invariant $\nu(p)$ is semicontinuous.

We were looking for a local result which would imply Corollary 1, much as Varchenko's work on semi-continuity of the spectrum limits singular points on projective hypersurfaces. This Conjecture implies a slightly weaker inequality than Corollary 1 (see Proposition 4.4), but without hypotheses such as (*). An attractive consequence of the conjecture would be the inequality for smoothable Gorenstein singularities:

$$-P\cdot P\leq 3(1+\mu).$$

By studying negative weight smoothings of quasi-homogeneous singularities in §3, we show (Theorem 3.5) that this last inequality is equivalent to Miyaoka-Yau for an appropriate log-terminal pair.

The main technical contribution of the paper is Lemma 1.6; we could not follow Miyaoka's proof of it, even with $S''=\emptyset$. This point of Miyaoka's work has also been clarified, by making sense of Chern classes on varieties with quotient singularities, by Kawamata [Kw] and Megyesi [Me], via a different approach. We point out that Megyesi's method gives a version of the Main Theorem (including the case D fractional) with a more satisfactory hypothesis: $K_{\tilde{X}}+\tilde{D}+E$ has a Zariski decomposition.

We are grateful to Yoichi Miyaoka for many useful discussions on this subject, some of which took place at the 1992 conference Classification of Algebraic Varieties at L'Aquila. We are especially thankful to the organizers of that Conference, for tending so thoughtfully to the needs of the participants. This research was partially supported by grants from the National Science Foundation.

## 1. The main lemma

(1.1) To prove the main theorem, we first reduce to the case that the negative part of $K_{\tilde{X}}+\tilde{D}+E$ is supported on E.

**Lemma 1.2** ([Mi2], 4.1): There exists a generically finite surjective proper morphism $f:W\to X$, étale over $S=\text{Sing } X$, with

(1.2.1) $\text{Sing } W = f^{-1}(S)\equiv T$

(1.2.2) $f^{-1}(D)$ has normal crossings

(1.2.3) $(\frac{1}{2})f^*N$ is numerically equivalent to an integral effective divisor $\bar{N}$, disjoint from T.

Proof (Bloch-Gieseker): Let r be a positive integer for which $(\frac{r}{2})N=N_1$ is effective and integral. Choose an ample H so that $N_1+rH\equiv L$ is very ample, giving $X\subset\mathbb{P}^n$; via a generic projection, one gets $\Phi:X\to\mathbb{P}^2$, with $\Phi^*\mathcal{O}(1)=L$. Consider $\pi:\mathbb{P}^2\to\mathbb{P}^2$ given by $[x,y,z]\to[x^r,y^r,z^r]$, with $\pi^*\mathcal{O}(1)=\mathcal{O}(r)$; choose coordinates so that $\pi$ is unramified over $\Phi(S)$. Taking the fibre product and normalizing gives

$$\begin{array}{ccc} W_1 & \xrightarrow{f_1} & X \\ \varphi\downarrow & & \Phi\downarrow \\ \mathbb{P}^2 & \xrightarrow{\pi} & \mathbb{P}^2 \end{array}$$

Then $f_1^*L=\varphi^*(\pi^*\mathcal{O}(1))=\varphi^*\mathcal{O}(r)$. Thus, $f_1^*L=f_1^*N_1+rf_1^*H$ is r times a line bundle, hence so is $f_1^*N_1$. Let $f_2:W_2\to W_1\to X$ resolve the singularities of $f_1^{-1}$ (D) and of points of $W_1$ not lying over S. Then $f_2^*N_1$ is effective and integral, with support disjoint from $\text{Sing}(W_2)$, and is $\equiv rL'$, some invertible L' on $W_2$. Let s be the gcd of r and the coefficients of $f_2^* N_1$, r=st. Use the line bundle L' to extract a $t^{th}$ root of the integral divisor $(\frac{1}{s})f_2^* N_1$; resolving the (cyclic quotient) singularities of this connected cyclic cover of $W_2$ gives finally f:W→X, which is easily checked to have the desired properties.

**Lemma 1.3**: Let f:W→X be as in (1.2), $\pi:\tilde{X} \to X$ the minimal resolution, $\tilde{f}:\tilde{W}\to\tilde{X}$ the induced map of some resolution $\pi':\tilde{W}\to W$ into $\tilde{X}$. Let $\mathfrak{F} = \tilde{f}^*\Omega^1_{\tilde{X}}(\log(\tilde{D} + E))(-\pi'^*\bar{N})$; det $(\mathfrak{F})$ has Zariski decomposition $\tilde{f}^*(P) + \tilde{f}^*(N'+N'') = (\det \mathfrak{F})_+ + (\det \mathfrak{F})_-$. Then (M-Y) is equivalent to

$$(1.3.1) \qquad \sum_{p\in T'} \left(\frac{1}{|G_p|}\right) + c_2(\mathfrak{F}) \geq \left(\frac{1}{3}\right)(\det \mathfrak{F})_+^2.$$

Proof: Let d=deg f=deg $\tilde{f}$. A calculation shows

$$c_2(\mathfrak{F}) = \qquad d\{c_2(\Omega^1_{\tilde{X}}(\log(\tilde{D}+E))) - \frac{N\cdot N}{4}\}$$

$$(\det \mathfrak{F})_+^2 = \qquad d(P\cdot P)$$

$$c_2(\Omega^1_{\tilde{X}}(\log(\tilde{D}+E)))=c_2(\tilde{X})-\chi_{top}(\tilde{D}+E'')-\chi_{top}(E').$$

Since T→S is exactly d-to-1, direct substitution gives the result.

(1.4) We change notation slightly, basically replacing $\tilde{W}$ by $\tilde{X}$.

Let $\mathfrak{F}\subset\Omega^1_{\tilde{X}}(\log(\tilde{D}+E))$ be a rank 2 subbundle, equal to the bigger bundle in a neighborhood of E'∪E''. det($\mathfrak{F}$) has Zariski decomposition P+N, where N=N'+N'', and N'⊂Supp E', N''⊂Supp E''. Define

$$c = c_2(\mathfrak{F})+ \sum_{p\in S'} \left(\frac{1}{|G_p|}\right).$$

With this notation, the desired inequality (1.3.1) becomes

$$(1.4.1) \qquad c \geq \left(\frac{1}{3}\right)P^2.$$

The rest of the section is devoted to proving the following key result:

**Theorem 1.5**: Let L be a line bundle on $\tilde{X}$ for which $H^0(\tilde{X}, S^n\mathfrak{F} \otimes L^{-1}) \neq 0$. Then

$$L \cdot P \leq \max\{0, nc\}.$$

Proof: Following Bogomolov-Miyaoka [Mi1], a section gives a divisor $Z \subset \mathbb{P}(\mathfrak{F})$, of degree n over $\tilde{X}$. Taking an appropriate Galois extension F of the function field $K(\tilde{X})$, we form the diagram

$$\begin{array}{ccc} \mathbb{P}(\tilde{f}^*\mathfrak{F}) & \xrightarrow{\bar{f}} & \mathbb{P}(\mathfrak{F}) \\ \bar{p}\downarrow & & \downarrow p \\ \tilde{Y} & \xrightarrow{\tilde{f}} & \tilde{X} \\ \bar{\pi}\downarrow & & \downarrow\pi \\ Y & \xrightarrow{f} & X, \end{array}$$

where Y is the normalization of X in F; $\bar{\pi}: \tilde{Y} \to Y$ is a resolution making the bottom diagram commutative; and the top diagram is Cartesian. Further,

$$\bar{f}^* Z = \bar{Z}_1 + \ldots + \bar{Z}_n + \bar{p}^*(G),$$

where the $\bar{Z}_i \subset \mathbb{P}(\tilde{f}^*\mathfrak{F})$ are irreducible, of degree 1 over $\tilde{Y}$; and $G \subset \tilde{Y}$ is an effective divisor. As a divisor on $\mathbb{P}(\tilde{f}^*\mathfrak{F})$, $\bar{Z}_i \equiv \bar{H} - \bar{p}^* L_i$, where $\bar{H}$ is the tautological line bundle, and $L_i \subset \tilde{f}^*\mathfrak{F}$ is a line bundle; blowing up $\tilde{Y}$ further, we may suppose $L_i \subset \tilde{f}^*\mathfrak{F}$ has invertible quotient. Note that $\tilde{f}^*L = (\sum L_i) - G$ on $\tilde{Y}$. Letting $d = \deg \tilde{f} = \deg f$, one has

(1.5.1) $\quad dc_2(\mathfrak{F}) = c_2(\tilde{f}^*\mathfrak{F})$

(1.5.2) $\quad d(L \cdot P) = \tilde{f}^*L \cdot \tilde{f}^*P = (\sum L_i - G) \cdot \tilde{f}^*P \leq \sum (L_i \cdot \tilde{f}^*P),$

since $\tilde{f}^*P$ is still nef. Since $L_i \subset \tilde{f}^*\mathfrak{F}$ is a subbundle,

(1.5.3) $\quad c_2(\tilde{f}^*\mathfrak{F}) = L_i \cdot (\det \tilde{f}^*\mathfrak{F} - L_i).$

For each $q\in f^{-1}(S)$, let $C_q=\bar{\pi}^{-1}(q)$, a reduced, not necessarily connected, negative-definite curve on $\tilde{Y}$. For every $i=1,\ldots,n$, let $D_{i,q}$ be the rational combination of curves in Supp $C_q$ so that

$$(1.5.4)\qquad L_i\cdot C=D_{i,q}\cdot C,\ \text{all curves } C\subset \operatorname{Supp} C_q.$$

Also, let $d_q$=local degree of f at $q=\deg(\hat{\mathcal{O}}_{X,f(q)}\to\hat{\mathcal{O}}_{Y,q})$; thus, for $p\in S$ fixed,

$$(1.5.5)\qquad \sum_{f(q)=p} d_q = d.$$

The absolutely key technical assertion is

**Lemma 1.6**: (1) $D_{i,q}\geq 0$, i.e., $D_{i,q}$ is a non-negative rational combination of curves in Supp Cq.

(2) $$D_{i,q}\cdot(\tilde{f}^*N - D_{i,q}) \geq \begin{cases} \dfrac{-d_q}{|G_{f(q)}|}, & f(q)\in S' \\ 0, & f(q)\in S'' \end{cases}.$$

(1.7) Let us for the moment assume Lemma 1.6, and deduce Theorem 1.5. For each i, write $D_i=\sum D_{i,q}$, the sum over all q; by the Lemma, $D_i$ is an effective $\mathbb{Q}$-divisor. Next, let $L_i'=L_i-D_i$ in $\operatorname{Pic}(\tilde{Y})\otimes\mathbb{Q}$; by (1.5.4), $L_i'$ has trivial intersection with every exceptional curve, whence $L_i'\cdot\tilde{f}^*N = L_i'\cdot D_i=0$. Since $\det \tilde{f}^*\mathfrak{F} = \tilde{f}^*P + \tilde{f}^*N$ and $D_i\cdot\tilde{f}^*P=0$ (as $D_i\subset \operatorname{Supp}\tilde{f}^*N$), (1.5.3) implies

$$(1.7.1)\qquad c_2(\tilde{f}^*\mathfrak{F})=L_i'\cdot(\tilde{f}^*P-L_i') + D_i\cdot(\tilde{f}^*N-D_i).$$

Now, $L_i\subset\tilde{f}^*\mathfrak{F}\subset\Omega^1_{\tilde{Y}}(\log(\tilde{f}^{-1}(E+D)))$; by Deligne's version of Castelnuovo-Bogomolov, it follows that $L_i$ has D-dimension $\leq 1$, or

$$(1.7.2)\qquad h^0(L_i^{\otimes n}) = O(n).$$

Since some multiple $rD_i$ is effective and integral, $(L_i')^{\otimes m}\subset L_i^{\otimes m}$, so the first line bundle also cannot have quadratic growth of the global sections. Therefore,

$$(1.7.3)\qquad (L_i')^2\leq 0 \text{ or } L_i'\cdot\tilde{f}^*P\leq 0.$$

Next, by Lemma (1.6.2),

$$D_i\cdot(\tilde{f}^*N-D_i)=\sum_{f(q)\in S} D_{i,q}\cdot(\tilde{f}^*N-D_{i,q})\geq -\sum_{f(q)\in S'}\left(\frac{d_q}{|G_{f(q)}|}\right).$$

Combining with (1.5.5) yields

$$D_i\cdot(\tilde{f}^*N-D_i)\geq -d\sum_{p\in S'}\left(\frac{1}{|G_p|}\right). \tag{1.7.4}$$

Using this and (1.5.1), (1.7.1) becomes

$$L_i'\cdot(\tilde{f}^*P-L_i')\leq d\cdot c; \tag{1.7.5}$$

in light of (1.7.3),

$$L_i'\cdot\tilde{f}^*P\leq \max\{0,d\cdot c\}. \tag{1.7.6}$$

But $L_i'\cdot\tilde{f}^*P=L_i\cdot\tilde{f}^*P$; summing (1.7.6) over i from 1 to n, and using (1.5.2), gives

$$L\cdot P\leq \max\{0,nc\},$$

as asserted in Theorem 1.5.

(1.8) The proof of Lemma 1.6 remains. The assertions are purely local over points of S. Thus, we consider (X,o) a germ of a quotient singularity $(\mathbb{C}^2/G,o)$ corresponding to $p\in S$, with minimal resolution $\pi:(\tilde{X},E)\to(X,o)$, plus a rank 2 vector bundle $\mathfrak{F}$ on $\tilde{X}$. If $p\in S'$ (which we call case (i)), then $\mathfrak{F}=\Omega^1_{\tilde{X}}(\log E)$. For case (ii), or $p\in S''$, X is a cyclic quotient, and $\mathfrak{F}=\Omega^1_{\tilde{X}}(\log(E+\tilde{D}))$, where $\tilde{D}\subset\tilde{X}$ is the germ of a smooth curve meeting the chain E transversally at one end. In either case, N=det $\mathfrak{F}$ is an effective $\mathbb{Q}$-divisor. We have a commutative diagram

$$\begin{array}{ccc}(\tilde{Y},F) & \xrightarrow{\tilde{f}} & (\tilde{X},E)\\ \bar{\pi}\downarrow & & \downarrow\pi\\ (Y,0) & \xrightarrow{f} & (X,0),\end{array}$$

where f is finite, $\bar{\pi}$ is a resolution, and $\tilde{f}$ is generically finite. To prove Lemma 1.6, we must show that if $L\subset\tilde{f}^*\mathfrak{F}$ is a subline bundle, then

(1.8.1) $L \equiv \sum a_i F_i,\ a_i \geq 0$

(1.8.2) $$L\cdot(\det \tilde{f}^*\mathfrak{F} - L) \geq \begin{cases} \frac{-\deg f}{|G|} & \text{case (i)} \\ 0 & \text{case (ii)} \end{cases}.$$

(1.9) To give a context to these statements, recall first Sakai's local Zariski decomposition L=P+N of a line bundle on $\tilde{X}$ into $\mathbb{Q}$- divisors [S2]: here, P is nef, N is effective, and P·N=0. (1.8.1) says that L has no P-part; by [W1], Theorem 1.6, this would be implied by

(1.9.1) $H^0(\tilde{Y}, L^{\otimes n}) = H^0(\tilde{Y} - F, L^{\otimes n})$, all $n \geq 0$.

Next, recall the definition from [W2] of $c_2$ of a rank 2 vector bundle $\mathfrak{F}$ on a resolution $\tilde{X} \to X$ of a normal surface singularity:

$$c_2(\mathfrak{F}) = \inf \left\{ \frac{L\cdot(\det \tilde{f}^*\mathfrak{F} - L)}{\deg f} \mid \tilde{f}: \tilde{Y} \to \tilde{X} \text{ generically finite}, \right.$$

$$\left. L \subset \tilde{f}^*\mathfrak{F} \text{ a sub-line bundle} \right\}.$$

Therefore, (1.8.2) will follow from the assertion

(1.9.2) $$c_2(\mathfrak{F}) = \begin{cases} \frac{-1}{|G|} & \text{case (i)} \\ 0 & \text{case (ii)} \end{cases}.$$

The calculation of $c_2(\mathfrak{F})$in case (i) is in Theorem 2.11 of [W2]; we handle case (ii) similarly in (1.12) below.

(1.10) From the G-quotient map $h:(\mathbb{C}^2, o) \to (X, o)$, let $Z \to \mathbb{C}^2 \times_X \tilde{X}$ be the minimal resolution of the normalization of the fibre product. One has $\bar{\pi}:(Z, E') \to (\mathbb{C}^2, o)$, and $\tilde{h}: Z \to \tilde{X}$ is unramified off E'; thus

$$\tilde{h}^*\Omega^1_{\tilde{X}}(\log E) \cong \Omega^1_Z(\log E')$$

(e.g., see [W2], 2.10). Also, note that

$$\Omega^1_Z(\log E') \supset \bar{\pi}^*\Omega^1_{\mathbb{C}^2} \cong \mathcal{O}_Z^2.$$

Further, in case (ii), $\widetilde{D}\subset\widetilde{X}$ maps to smooth $D\subset X$, and $h^{-1}(D)=\bar{D}\subset\mathbb{C}^2$ is a smooth curve, with proper transform $D'\subset Z$. Then as above,

$$\widetilde{h}^*\Omega^1_{\widetilde{X}}(\log(E+\widetilde{D}))\cong\Omega^1_Z(\log(E'+D')). \tag{1.10.1}$$

Again, one has

$$\Omega^1_Z(\log(E'+D'))\supset\bar{\pi}^*\Omega^1_{\mathbb{C}^2}(\log\bar{D})\cong\mathcal{O}_Z{}^2,$$

since $\Omega^1_{\mathbb{C}^2}(\log\bar{D})$ is free of rank 2. Therefore, in cases (i) and (ii), there is an inclusion

$$\mathcal{O}_Z{}^2\subset\widetilde{h}^*\mathfrak{F}, \tag{1.10.2}$$

with equality off E'.

(1.11) Now, to prove (1.9.1) (or (1.8.1)), it suffices to check after a further pull-back to some $(Y',o)\rightarrow(Y,o)$, on an appropriate resolution of Y'. We may therefore assume that $f:(Y,o)\rightarrow(X,o)$ factors via $(\mathbb{C}^2,o)$. Pulling back (1.10.2) gives

$$\mathcal{O}_Y{}^2\subset\widetilde{f}^*\mathfrak{F}, \tag{1.11.1}$$

with equality on $\widetilde{Y}$-F. Since $H^0(\widetilde{Y},\mathcal{O}_{\widetilde{Y}})=H^0(\widetilde{Y}-F,\mathcal{O}_{\widetilde{Y}})$, we conclude that for every $n\geq 0$, one has

$$H^0(\widetilde{Y},S^n\widetilde{f}^*\mathfrak{F})=H^0(\widetilde{Y}-F,S^n\widetilde{f}^*\mathfrak{F}). \tag{1.11.2}$$

But $L^{\otimes n}$ is a subbundle of $S^n\widetilde{f}^*\mathfrak{F}$, with locally free quotient $\mathcal{Q}^n$; from the diagram

$$\begin{array}{ccccccc}
0\rightarrow & H^0(\widetilde{Y},L^{\otimes n}) & \rightarrow & H^0(\widetilde{Y},S^n\widetilde{f}^*\mathfrak{F}) & \rightarrow & H^0(\widetilde{Y},\mathcal{Q}^n) \\
& \cap & & \| & & \cap \\
0\rightarrow & H^0(\widetilde{Y}-F,L^{\otimes n}) & \rightarrow & H^0(\widetilde{Y}-F,S^n\widetilde{f}^*\mathfrak{F}) & \rightarrow & H^0(\widetilde{Y}-F,\mathcal{Q}^n),
\end{array}$$

we readily conclude (1.9.1).

(1.12) As mentioned above, (1.9.2) has already been proved in case (i); so we return to the notation of (1.10). Since $c_2$ multiplies by degree under pull-back, it suffices by (1.10.1) to prove that $c_2(\Omega_Z^1(\log(E'+D')))=0$. As in [W2], 2.9, one may compute this $c_2$ from the corresponding bundle on any blow-up of $\mathbb{C}^2$. In particular, if $p:(Z,E')\to(\mathbb{C}^2,o)$ is a simple blow-up, then (2.9.1) of [Wa2] gives the exact sequence

$$0 \to p^*\Omega^1_{\mathbb{C}^2}(\log \bar{D}) \to \Omega_Z^1(\log(E'+D')) \to \mathcal{O}_{E'}(-1) \to 0.$$

The calculation that $c_2$ of the middle term is 0 now follows from (2.7) of [W2]. This completes the proof of (1.9.2), and hence of Theorem 1.5.

## 2. The end of the proof

(2.1) We maintain the notation of (1.4), to prove (1.4.1): $c\geq(\frac{1}{3})P^2$.

**Proposition 2.2**: Fix $a\in\mathbb{Q}$, and consider $n\in\mathbf{N}$ with $naP$ an integral divisor. Suppose

(i) $h^0(S^n\mathfrak{F}(-naP)) = O(n^2)$

(ii) $h^2(S^n\mathfrak{F}(-naP)) = O(n^2)$

(iii) for every effective Z supported on E,

$$h^2(S^n\mathfrak{F}(-naP-nZ))=O(n^2).$$

Then

$$(2.2.1) \qquad (1-3a+3a^2)P^2 \leq c.$$

Proof: Riemann-Roch gives

$$\chi(S^n\mathfrak{F}(-naP))=(\frac{n^3}{6})\{(1-3a+3a^2)P^2+N^2-c_2(\mathfrak{F})\} + O(n^2).$$

Combining with (i) and (ii) gives

$$(2.2.2) \qquad h^1(S^n\mathfrak{F}(-naP))=(\frac{n^3}{6})\{c_2(\mathfrak{F})-(1-3a+3a^2)P^2-N^2\} + O(n^2).$$

We claim there is an effective divisor Z supported on E so that

$$(2.2.3) \qquad h^1(S^n\mathfrak{F}(-naP)\otimes\mathcal{O}_{nZ})=(\frac{n^3}{6})\{-N^2-\sum(\frac{1}{|G_p|})\} + O(n^2),$$

the sum over $p \in S'$. Since $h^2(S^n\mathfrak{F}(-naP-nZ))=O(n^2)$ by (iii), a cohomology sequence plus (2.2.2) and (2.2.3) yields exactly the inequality (2.2.1).

Since N is supported on E, (2.2.3) may be verified separately for each $p \in S$; specifically, one proves it for the minimal resolution $(\tilde{X},E) \to (X,p)$ of a quotient singularity. Assume first $p \in S'$. Then P=0, so we consider $h^1(S^n\mathfrak{F} \otimes \mathcal{O}_{nZ})$. By Theorem 4.6 of [W2], $h^1(\tilde{X},S^n\mathfrak{F})$ has the desired growth rate $(\frac{n^3}{6})\{-N^2-\frac{1}{|G|}\}$; thus, one can pick a Z for which $h^1(S^n\mathfrak{F}(-nZ))=0$. Such Z's exist for any bundle $\mathfrak{F}$, e.g. by Lemma 3.12 of [W2]. Similarly, if $p \in S''$, a completely analogous argument works, but using now the formula (4.18.2) of [W2].

**Proposition 2.3**: Suppose $P^2>0$. Then $c \geq (\frac{1}{3})P^2$.

Proof: Let $\alpha = \max\{0, \frac{c}{P^2}\}$. We assume $\alpha < \frac{1}{3}$, and derive a contradiction. By Theorem 1.5,

(2.3.1) $\quad h^0(S^n\mathfrak{F}(-nL)) \neq 0$ implies $L \cdot P \leq \max\{0,c\} = \alpha P^2$.

Thus, if $a \in \mathbb{Q}$ is such that naP is integral, then

(2.3.2) $\quad h^0(S^n\mathfrak{F}(-naP))=0$ if $a > \alpha$.

Also, $h^2(S^n\mathfrak{F}(-naP))=h^0(S^n\mathfrak{F}(naP+K-n\cdot\det(\mathfrak{F})))$ by duality; thus,

(2.3.3) $\quad h^2(S^n\mathfrak{F}(-naP))=0$ if $P^2(1-a)-(1/n)K\cdot P > \alpha P^2$.

In particular, if $1-a>\alpha$ and n is large, one has vanishing in (2.3.3). Finally, since $P \cdot Z=0$ for exceptional Z, the same condition in (2.3.3) also guarantees the vanishing of $h^2(S^n\mathfrak{F}(-naP-nZ))$. Consequently, if $a \in (\alpha, 1-\alpha)$ (which is non-empty, since $\alpha < \frac{1}{3}$), Proposition 2.2 implies

$$(1-3a+3a^2)P^2 \leq c.$$

The inequality must remain true when $a=\alpha$, giving

(2.3.4) $$1-3\alpha+3\alpha^2 \leq \frac{c}{P^2}.$$

If $c \leq 0$, then $\alpha = \max\{0, \frac{c}{P^2}\}=0$, contradicting (2.3.4). Thus, $\frac{c}{P^2}=\alpha$, and (2.3.4) gives

$$3\alpha^2-4\alpha+1=(3\alpha-1)(\alpha-1)\leq 0.$$

As $\alpha$ was assumed $<\frac{1}{3}$, this is a contradiction.

(2.4) To prove $c\geq(\frac{1}{3})P^2$ in the remaining case $P^2=0$ (recall P is nef), we show $c\geq 0$, considering separately the cases $P\not\equiv 0$ and $P\equiv 0$.

**Lemma 2.5**: Let $\mathfrak{F}$ be a rank 2 vector bundle on a smooth curve C, and $D_1$, $D_2$ divisors on C. Then

$$h^0(S^n\mathfrak{F}(nD_1+D_2))=O(n^2).$$

Proof: Choose $D_3$ effective, with $D_3\geq D_2$; then $S^n\mathfrak{F}(nD_1+D_2)\subset S^n\mathfrak{F}(n(D_1+D_3))=S^n(\mathfrak{F}(D_1+D_3))$. So, assume $D_1=D_2=0$, and $\mathfrak{F}$ very positive. Now use Riemann-Roch.

**Proposition 2.6**: Suppose $P^2=0$, $P\not\equiv 0$. Then $c\geq 0$.

Proof: Assume $c<0$. Theorem 1.5 states that

(2.6.1) $\qquad L\cdot P>0$ implies $h^0(S^n\mathfrak{F}\otimes L^{-1})=0$.

Since $P\not\equiv 0$, we can choose a smooth curve C with $P\cdot C>0$ and $P\cdot C>-P\cdot K$; note also that $\det\mathfrak{F}\cdot P=P\cdot P=0$. Now, (2.6.1) implies that the 3 bundles $S^n\mathfrak{F}$, $S^n\mathfrak{F}(K+n\cdot\det\mathfrak{F})$, and $S^n\mathfrak{F}(K+n\cdot\det\mathfrak{F}+nZ)$, when tensored with $\mathcal{O}(-C)$, have no global sections; and by Lemma 2.5, their restrictions to C have $h^0$'s which are $O(n^2)$. By duality, $h^0(S^n\mathfrak{F})$, $h^2(S^n\mathfrak{F})$, and $h^2(S^n\mathfrak{F}(-nZ))$ are $O(n^2)$; so Proposition 2.2 with $a=0$ implies $c\geq 0$, a contradiction.

**Lemma 2.7**: Suppose $P\equiv 0$, $c<0$. Then $h^0(S^n\mathfrak{F}\otimes L^{-1})\neq 0$ implies $L\cdot D\leq 0$, for every nef D.

Proof: Follow the proof of Theorem 1.5. Line (1.7.5) implies

$$-(L_i')^2\leq d\cdot c;$$

since $c<0$, we conclude that $(L_i')^2>0$. As in (1.7.3), it follows that

$$L_i'\cdot D\leq 0, \text{ for all D nef.}$$

As in (1.5.2), we conclude that $L\cdot D\leq 0$.

**Lemma 2.8**: If $P \equiv 0$, then $c \geq 0$.

Proof: Assume $c<0$. Choose a smooth, very ample curve C, with $C \cdot C > -K \cdot C$ (i.e., of large genus). By Lemma 2.7, we conclude the same equalities as (2.6.2), from which a contradiction is deduced as in Lemma 2.6.

## 3. Some local analogues

(3.1) We recall some facts about quasi-homogeneous surface singularities (X,p) and deformations of negative weight [Pi],[LW]. Assume X is not log-canonical. The minimal good resolution of X consists of a smooth central curve C, of genus g and self-intersection -b; and n strings of rational curves emanating from C, determined by continued fraction expansions of some rational numbers $\{\frac{n_i}{q_i}\}$, $i=1,\ldots,n$. There is a natural singular $\mathbb{C}^*$-compactification of X. Resolving at $\infty$ gives a projective surface Y with "dual" curve configuration D; the central C now has self-intersection b-n, and the strings are given by $\{\frac{n_i}{p_i}\}$, where $p_i=n_i-q_i$. A deformation of negative weight of (X,o) gives a deformation of Y fixing the topology of the configuration D; the general fibre is a projective surface Z with curve configuration D at $\infty$, with Z-D containing the singularities adjacent to (X,o) in the deformation. If X is Gorenstein, then $K_Z$ is linearly equivalent to an integral divisor supported on D. We apply our version of Miyaoka-Yau to $K_Z+D$.

(3.2) Consider the dual graph of C+F, where F is a string:

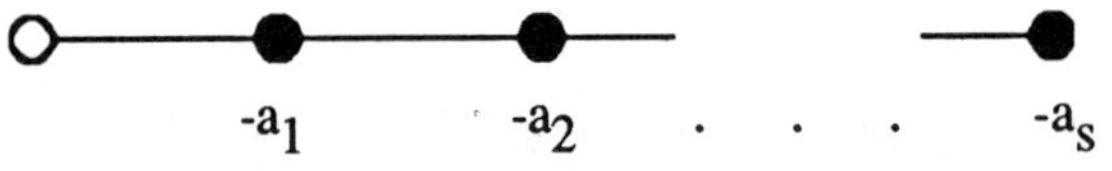


Let $\frac{n}{p}=[a_1,\ldots,a_s]$. In [W1], one defines the effective $\mathbb{Q}$-divisor N supported on F, by the property of dotting to 0 with all curves in |F| except the one on the right (for which it dots to -1). We define another $\mathbb{Q}$-divisor S, supported on F+C, containing C with multiplicity 1 and dotting to 0 with all curves in |F|; this has multiplicity 1/n for the right-hand curve of F, and p/n for the left-hand curve.

(3.3) Let Z be a projective surface, smooth near the reduced dual curve configuration $D=C+\sum F_i$. Consider two effective $\mathbb{Q}$-Cartier divisors supported on D: $\sum N_i$, where each Ni has been described in (3.2); and S, the divisor with multiplicity 1 at the central curve and the same multiplicities along the strings as described above. S is orthogonal to all the curves save the central C, hence

$$S\cdot S=S\cdot C=(b-n)+\sum\frac{p_i}{n_i}=b-\sum\frac{q_i}{n_i}\ .$$

This number is positive since X is not log-canonical; thus, S is effective and nef. As in [W1], 2.3, $K_Z+D-\sum N_i$ dots to 0 with all curves on the strings; thus, this divisor and some $\alpha S$ have the same intersection properties with all curves of D. Dotting with the central C gives (using [W1], (2.3.1))

$$2g-2+n-\sum\left(\frac{1}{n_i}\right)=\alpha\left(b-\sum\frac{q_i}{n_i}\right).$$

Computing using Theorem 3.2 of [W1], one finds

(3.3.1) $\qquad \alpha S\cdot\alpha S=-P_X\cdot P_X.$

(3.4) Suppose one has a deformation of negative weight of a Gorenstein X, so that the general fibre has only RDP singularities $\{p_j\}$. Let Z' be the projectivized general fibre, and $Z\to Z'$ the minimal resolution of the RDP's, with exceptional fibres $\{E_j\}$. Then $K_Z$ is linearly equivalent to a divisor supported on D (it is trivial near the $E_j$); hence one has equality of $\mathbb{Q}$-divisors on all of Z,

(3.4.1) $\qquad K_Z+D+\sum E_j=\alpha S+\left(\sum N_i+\sum E_j\right)$

which is a Zariski decomposition. Writing $D=C+\sum F_i$, where $F_i$ is the exceptional configuration of a cyclic quotient singularity, we may apply the Main Theorem with $\tilde{X}=Z$, $\tilde{D}=C$, $E'=\sum E_j$, $E''=N''=\sum F_i$, $P=\alpha S$, $N=0$, $N'=\sum N_i$. We conclude

(3.4.2) $\qquad \sum \nu(p_j)\le c_2(Z)-e(D)-\left(\frac{1}{3}\right)\alpha S\cdot\alpha S.$

By Mayer-Vietoris, $c_2(Z)-e(D)=e(Z-D)$; but since the resolution of an RDP is diffeomorphic to its Milnor fibre, we conclude Z-D is a Milnor fibre for a smoothing of X, so $e(Z-D)=1+\mu$, ($\mu$ is the Milnor number). Combining with (3.3.1), and checking the log-canonical cases, we have proved a special case of the "local Miyaoka-Yau inequality:"

**Theorem 3.5**: Suppose a quasi-homogeneous Gorenstein surface singularity X deforms with negative weight to some RDP's $\{p_j\}$. Then

$$\sum \nu(p_j) - (\frac{1}{3})P\cdot P \leq 1+\mu.$$

In particular, if X smooths negatively, then

$$-(\frac{1}{3})P\cdot P \leq 1+\mu.$$

**Conjecture 3.6 (Local Miyaoka-Yau):** If a Gorenstein X deforms to a configuration $\{p_j\}$ of RDP's, then

$$\sum \nu(p_j) - (\frac{1}{3})P\cdot P \leq 1+\mu.$$

In particular, if X is smoothable, then

$$-(\frac{1}{3})P\cdot P \leq 1+\mu.$$

Remarks (3.7.1) The conjectured inequality $-P\cdot P \leq 3(1+\mu)$ is unknown even for hypersurface singularities; however, our student Frieda Ganter has verified it for singularities of the form $z^n=f(x,y)$, where $f(x,y)$ is analytically irreducible [Ga]. The inequality would provide a new necessary condition for smoothability of X (cf. [LW]).

(3.7.2) One should view the "local chern numbers" of a smoothable Gorenstein singularity X as "$c_1^2$"$=-P\cdot P$, and "$c_2$"$=1+\mu-\frac{1}{|G|}$, where as usual G is the local fundamental group (see [W2] to interpret $-\frac{1}{|G|}$ as a chern number). The Conjecture then says that $\nu(p)=$"$c_2-(\frac{1}{3})c_1^2$" is semicontinuous under deformation to RDP's. We formulate therefore the very general

**Conjecture 3.8:** The invariant $\nu$ is semi-continuous under deformation of smoothable Gorenstein singularities: if (X,p) deforms to smoothable singularities $\{p_j\}$, then

$$\sum \nu(p_j) \leq \nu(p).$$

(3.9) If Z is a smoothing of negative weight of Gorenstein X as above, one can deduce a stronger inequality than (3.5) by considering the Zariski decomposition which omits C:

(3.9.1) $$K_Z+\sum F_i = \beta S + \sum N'_i.$$

Here, $N'_i$ is the effective $\mathbb{Q}$-divisor supported on $F_i$, equivalent to $K+F_i$ ($F_i$ supports the resolution of a cyclic quotient); a calculation shows that the coefficient of $N_i'$ at the curve adjacent to C is $\frac{(p_i+1)}{n_i}$. As above, one needs to solve (with $f\geq 0$)

$$2g-2-(b-n)+n-\sum\frac{(p_i+1)}{n_i} = f(b-\sum\frac{q_i}{n_i}).$$

Rewrite this equation as

$$(3.9.2) \qquad 2g-2+n-\sum\frac{1}{n_i} = (f+1)(b-\sum\frac{q_i}{n_i})$$

Recall the integer invariant t [Wt], which is the weight of a nowhere-0 holomorphic 2-form $\Sigma$ on X-{o}; equivalently, t-1 is the order of the pole of $\Sigma$ along the central curve in the MGR of X. $t>0$ unless X is log-canonical. In terms of the Zariski decomposition $K_{\tilde{X}}+E=P+N$ on the MGR of X itself, -t is the multiplicity of C in P; thus

$$-P\cdot P=t(P\cdot C).$$

Combining with [W1], (2.3.1) and (3.2.1), one finds that

$$(3.9.3) \qquad t = \frac{(2g-2+n-\sum\frac{1}{n_i})}{(b-\sum\frac{q_i}{n_i})}$$

This gives

$$\beta = t-1,$$

which is indeed $\geq 0$ by assumption. One may therefore apply Miyaoka-Yau in case (3.9.1), using in fact the version from Miyaoka's original paper [Mi2]; proceeding as in (3.4) yields the

**Theorem 3.10**: If a quasi-homogeneous Gorenstein X, not log- canonical, smooths with negative weight, then

$$-(\frac{1}{3})(P\cdot P)\{1+(\frac{1}{t})+(\frac{1}{t})^2\} \leq 1+\mu,$$

where $t\geq 1$ is the weight of a generator of the canonical module of X.

(3.11) The inequality of (3.10) is a good deal stronger than (3.5) for singularities with t=1 (for which the global smoothing Z would be a K-3 surface

[Pi]), while for large t the results are similar. One can strengthen this version further as in Theorem 3.5 to take into account RDP adjacencies. But a serious problem is that t is naturally defined only for quasi-homogeneous singularities; and, even in this case, the left-hand side of (3.10) is not a characteristic number for the singularity.

## 4. Local theorems give bounds for global hypersurfaces

(4.1) If $X \subset \mathbb{P}^n$ is a hypersurface with isolated singularites ($n \geq 2$), one may view X as the general fibre in a deformation of the cone Y over the (smooth) general hyperplane section $X \cap H$. One simply replaces a generic coordinate $z_0$ in the defining equation $G(z_0, z_1, \ldots, z_n)=0$ by $tz_0$, and views t as a deformation parameter. In fact, the deformation of Y is really just a deformation of negative weight of the singular germ of Y at its vertex. Therefore, restrictions on the singularities that can arise in (local) deformations of a cone carry over to projective hypersurfaces.

(4.2) This principle was used by A. Varchenko (e.g., [AGV], p. 417) to give the best general bounds for number of singular points (say, ordinary double points) on hypersurfaces of degree d in $\mathbb{P}^n$. The key is the semi-continuity of an invariant of a hypersurface singularity called the spectrum; it is a set of $\mu$ numbers which reflects the mixed Hodge structure on the cohomology of the Milnor fibre. Restricting to surfaces in $\mathbb{P}^3$, Varchenko's bound is not as good as Miyaoka's on the number of $A_1$'s (though it improves the classical bounds), but works better for $A_k$'s with k large, and can handle all singular points.

(4.3) One seeks a purely local theorem about deformations of surface singularities, which applied to deformations of a cone give Miyoaka's bound on the configurations of rational double points on $X^d \subset \mathbb{P}^3$. Computing $\nu$ for the vertex of a cone yields

**Proposition 4.4**: Conjecture 3.8 implies that for any hypersurface X of degree $d \geq 4$ in $\mathbb{P}^3$ with isolated singularities,

$$\sum_{p \in X} \nu(p) \leq \frac{2}{3} d(2d-3).$$

Remark (4.5) This inequality is slightly weaker than Miyaoka's (for the case X has only RDP's); his right hand term of $\frac{2}{3}d(d-1)^2$ is exceeded by ours by the quadratic term $\frac{d}{3}(d-2)$. Thus, the asymptotic bound for the number of ODP's is the same, $\frac{4}{9}d^3$. Corollary 1 in the Introduction suggests a stronger inequality might be true, once a few exceptions are made.

## BIBLIOGRAPHY


[AGV] V.I. Arnold, S.M. Gusein-Zade, and A.N. Varchenko, Singularities of Differentiable Maps, Vol. II, Birkhäuser, Boston, 1985

[CY] S.Y. Cheng and S.-T. Yau, On the existence of a complete Kähler metric on non-compact complex manifolds and the regularity of Fefferman's equation, Comm. Pure Appl. Math. **33** (1980), 507-544.

[Ga] F. Ganter, -P·P for surfaces $z^n=f(x,y)$, and inequalities involving -P·P, Ph. D. thesis, UNC Chapel Hill, 1993.

[Iv] K. Ivinskis, Normale Flächen und die Miyaoka-Kobayashi-Ungleichung, Diplomarbeit, Bonn (1985).

[Kw] Y. Kawamata, Abundance theorem for minimal threefolds, Invent. Math. **108** (1991), 229-246.

[KNS] R. Kobayashi, S. Nakamura, and F. Sakai, A numerical characterization of ball quotients for normal surfaces with branch loci, Proc. Japan Acad. Ser. A Math. Sci. **65** (1989), no. 7, 238-241.

[LW] E. Looijenga and J. Wahl, Quadratic functions and smoothing surface singularities, Topology **25** (1986), 261-291.

[Me] G. Megyesi, Inequalities between Chern numbers, Ph. D. thesis, University of Cambridge, 1993

[Mi1] Y. Miyaoka, On the Chern numbers of surfaces of general type, Invent. Math. **42** (1977), 225-237.

[Mi2] Y. Miyaoka, The maximal number of quotient singularities on surfaces with given numerical invariants, Math. Ann. **268** (1984), 159-171.

[Mu] D. Mumford, Hirzebruch's proportionality theorem in the non-compact case, Invent. Math. **42** (1977), 239-272.

[Pi] H. Pinkham, Deformations of normal singularities with $\mathbb{C}^*$ action, Math. Ann. **232** (1978), 65-84.

[S1] F. Sakai, Semi-stable curves on algebraic surfaces and logarithmic pluricanonical maps, Math. Ann. **254** (1980), 89-120.

[S2] F. Sakai, Anticanonical models of rational surfaces, Math. Ann. **269** (1984), 389-410.

[W1] J. Wahl, A characteristic number for links of surface singularities, Journal of the A.M.S. 3 (1990), 625-637.

[W2] J. Wahl, Second Chern class and Riemann-Roch for vector bundles on resolutions of surface singularities, Math. Ann. **295** (1992), 81-110.

[Wt] K. Watanabe, Some remarks concerning Demazure's construction of normal graded rings, Nagoya Math. J. **83** (1981), 203-211.

[Y] S.-T. Yau, Calabi's conjecture and some new results in algebraic geometry, Proc. Natl. Acad. Sci. USA **74** (1977), no. 5, 1798-1799.

Department of Mathematics
University of North Carolina
Chapel Hill, North Carolina USA 27514
E-mail address: jw@math.unc.edu

Contemporary Mathematics
Volume **162**, 1994

# MINIMAL MODELS OF CALABI-YAU THREEFOLDS

P.M.H. WILSON

## Introduction

In this paper I shall consider Calabi–Yau models (i.e. normal projective images of Calabi–Yau threefolds under birational morphisms - cf [13]) which admit no further birational contractions. Such a threefold $X$ will be called a *Calabi–Yau minimal model.* By definition $X$ has a resolution $\tilde{X}$ which is a smooth Calabi–Yau threefold, and the second Chern class of $\tilde{X}$ induces an integral linear form $c_2$ on $\mathrm{Pic}(X)$. The aim is to use the form $c_2$ for classification purposes in an analogous way to the use of the canonical divisor in the classification of minimal surfaces.

As with previous papers [13, 14, 16], the methods employed involve studying the nef cone $\overline{\mathcal{K}} \subset \mathrm{Pic}_{\mathbb{R}}(X)$ and its boundary $\partial\overline{\mathcal{K}}$, together with the cubic cone $W^* = \{D \,:\, D^3 = 0\} \subset \mathrm{Pic}_{\mathbb{R}}(X)$ and the hyperplane $\{D \,:\, D \cdot c_2 = 0\}$, where $X$ is a Calabi–Yau threefold (or more generally a Calabi–Yau model). The form $c_2$ has the property of being non-negative on $\overline{\mathcal{K}}$ (cf. Remarks 1.1 from [14]). For a given Calabi–Yau threefold $X$, we see in Section 1 that its nef cone $\overline{\mathcal{K}}$ is locally rational polyhedral away from the cubic cone $W^*$, and that the minimal models for $X$ correspond to minimal faces of $\overline{\mathcal{K}}$ not contained in $W^*$ (i.e. faces whose boundary rays are in $W^*$ but interior rays not).

With $X$ now a Calabi–Yau minimal model, the first possibility to consider is that of the form $c_2$ being trivial. In the case when $X$ is smooth, this is well-known to imply that $X$ is an étale quotient of a three-dimensional torus. In fact, it can be shown that in general $X$ will still be a finite quotient of a torus, allowing however fixed locus for the action – this is proved in a forthcoming preprint by the author and N.I. Shepherd-Barron. There should moreover be a classification of such quotients (cf. [11] for the case of cyclic quotients). The next possibility is that the form $c_2$ is non-trivial but not strictly positive on $\overline{\mathcal{K}} \setminus \{0\}$; we see however in Section 2 that this can only happen under rather restrictive

1991 *Mathematics Subject Classification.* Primary 14J10, 14J15; Secondary 14J30, 14J50.
This paper is in final form and no version of it will be submitted for publication elsewhere.

circumstances,and we conjecture that in this case a fibre space structure on $X$ will alwas exist.

The general case will be when $c_2$ is strictly positive on $\overline{\mathcal{K}}\setminus\{0\}$, and in this case minimal models turn out to have pleasant properties: in particular $c_2$ determines a canonical polarization in a natural way. Normally we should expect a minimal model to have $\rho = 1$ (for which the above fact is obvious), but $\rho \geq 2$ can also occur. The polarization then yields an embedding canonically associated with $c_2$. Finally, one conjectures that a Calabi–Yau threefold $X$ should only have finitely many Calabi–Yau minimal models *up to isomorphism.* This leads us to considering the automorphism group of $X$, which turns out to be finite unless the cubic and Kähler cones and the hyperplane $c_2 = 0$ satisfy certain exceptional conditions.

Much of this work was presented in a talk by the author at the conference in whose proceedings the paper appears. He would like to thank the organisers for what was a stimulating and very enjoyable conference.

## 1. Some basic facts on structure

Let $X$ denote a Calabi–Yau threefold (or more generally a Calabi–Yau model) with nef cone $\overline{\mathcal{K}}$ and cubic cone $W^*$.

**Fact A.** Away from $W^*$, the nef cone $\overline{\mathcal{K}}$ is locally rational polyhedral. The codimension one faces of $\overline{\mathcal{K}}$ not contained in $W^*$ correspond to primitive (see Sect. 1 of [14] for definition) birational contractions $\phi : X \to \overline{X}$, the exceptional loci of which are covered by rational curves (see Sect. 2 of [14] for a classification of primitive contractions when $X$ is smooth). The primitive contraction is just $\phi_{nD}$, where $D$ is any divisor representing an integral interior point of the face and $n$ is a suitably large integer.

**Fact B.** Non-singular rational points $D$ on $W^*$ which lie on $\partial\overline{\mathcal{K}}$ but not on the hyperplane $c_2 = 0$ give rise to elliptic fibre space structures $\phi : X \to S$ with $S$ a normal rational surface.

**Fact C.** Singular rational points $D$ on $W^*$ which lie on $\partial\overline{\mathcal{K}}$ but not on the hyperplane $c_2 = 0$ give rise to fibre space structures $\phi : X \to \mathbb{P}^1$, with general fibre a K3 surface.

The proofs of these Facts are essentially already in the literature - here we shall just refer to the appropriate sources and add any further necessary explanation.

Fact A (called Fact 1 in [14]) is essentially proved in [14] (or indeed in the earlier paper [13]). The proof given on page 565 of [14] proves the required rationality of $\partial\overline{\mathcal{K}}$ away from $W^*$, but needs a small modification to yield locally polyhedral as well (i.e. to prove a discreteness property on the codimension one faces). This however follows by the argument in [14] if we can show for every $D \in \partial\overline{\mathcal{K}} \setminus W^*$ that we can find a rational divisor class $D_0$ not in $\overline{\mathcal{K}}$ but representing an *effective* divisor on $X$, and a class $L \in \mathcal{K}$ with the line segment $D_0L$ meeting $\partial\overline{\mathcal{K}}$ at $D$. The argument of [14] only needs modifying in the case when for $H$ a hyperplane section, $D|_H$ is not ample. Since $D^2 \cdot H > 0$, we can apply an argument from page 144 of [13] to show that $D$ must be rational (Riemann–Roch applied to the surface $H$ enables us to find rational $D'$ near to $D$ but not in $\overline{\mathcal{K}}$ for which $h^0(\mathcal{O}_H(nD')) > 0$ for some integer n). Taking $D$ therefore to be integral, we consider the divisor class $nD - H$, clearly non-ample since $D \cdot C = 0$ for some curve $C$ on $H$. An easy approximation argument however shows that $h^2(nD - H)$ is of order $n$ for large $n$, and so Riemann–Roch on $X$ yields that $D_0 = nD - H$ is effective for $n$ sufficiently large. Since $nD = D_0 + H$, the required polyhedral structure of $\overline{\mathcal{K}}$ in a neighbourhood of $D$ follows as in the proof of Claim 2 on page 145 of [13], i.e. using the Theorem of the Cone for log terminal varieties.

The fact that the minimal models of a Calabi–Yau threefold $X$ correspond to the minimal faces of $\overline{\mathcal{K}}$ not contained in $W^*$ follows easily from Fact A (cf. the argument on page 564 of [14]).

For Fact B (called Fact 2 in [14]) we merely refer to page 566 of [14], or alternatively (3.2)′ of [13]. The reason why the surface $S$ has to be rational is explained in the proof of Proposition 3 from [16].

*Proof of Fact C.* Although not explicitly stated in the literature, Fact C follows as a consequence of existing results. The fact that $D$ represents a singularity of $W^*$ says that its square $D^2$ is a numerically trivial 1-cycle (see Remarks 1.1 of [14]). By pulling back to a smooth model, we can moreover assume that $X$ itself is smooth. If we can show that some integral multiple of $D$ on $X$ is effective, the result follows from the work of Oguiso [10], which in turn is based on Kawamata's proof of the Abundance Conjecture.

To prove the required effectivity, we assume that $D$ is integral and apply arguments from [6], in particular the Compensation Theorem (2.1). Let $\imath = \imath(D)$ denote the greatest integer $\leq \frac{1}{2}\,\mathrm{PI}(D : N(X)_{\mathbb{Z}})$, where PI denotes the primitive index as defined in [6]. By (3.2) of the cited paper, if $|H|$ is a very ample linear system on $X$ with $H$ a sufficiently general member, then

$$\mathrm{PI}(D|_H : N(H)_{\mathbb{Z}}) = \mathrm{PI}(D : N(X)_{\mathbb{Z}})$$

and so in particular the left-hand side is independent of $|H|$.

If $X$ does not have finite fundamental group, then from [1] we know that some finite étale cover of $X$ is of the form $E \times S$, with $E$ a elliptic curve and $S$ a K3 surface (the cover is not a torus since $c_2(X) \neq 0$). This being the case, we let $\tilde{D}$ denote the pullback of $D$ to $E \times S$ ; $\tilde{D}$ is then a nef divisor on $E \times S$ with $\tilde{D}^2 \equiv 0$ and $\tilde{D} \cdot c_2(E \times S) > 0$. It is then easily checked that $h^0(\mathcal{O}_{E\times S}(\tilde{D})) > 0$, and hence that $h^0(\mathcal{O}_X(D)) > 0$.

From now on therefore we can assume that $\pi_1(X)$ is finite, and set $g = |\pi_1(X)|$. By (3.1) of [6], we know that this is the same as $|\pi_1^{\mathrm{alg}}(H)|$ when $H$ is chosen general in its linear system $|H|$. We now chose a very ample divisor $H$ (general in its linear system) such that $H - 2\imath! gD$ is also very ample. If $h^2(D) = 0$, then

$$h^0(D) - h^1(D) = \chi(\mathcal{O}_X(D)) = \frac{1}{12} D \cdot c_2 > 0$$

and so $D$ is effective as required. From the exact sequence of sheaves

$$0 \to \mathcal{O}_X(-D-H) \to \mathcal{O}_X(-D) \to \mathcal{O}_H(-D) \to 0$$

we observe using Kodaira Vanishing that

$$h^2(\mathcal{O}_X(D)) = h^1(\mathcal{O}_X(-D)) = h^1(\mathcal{O}_H(-D)).$$

If however $h^1(\mathcal{O}_H(-D)) > 0$, then the Compensation Theorem ([6], Theorem 2.1) implies that $h^0(\mathcal{O}_H(2\imath! gD)) > 0$. From the exact sequence of sheaves

$$0 \to \mathcal{O}_X(2\imath! gD - H) \to \mathcal{O}_X(2\imath! gD) \to \mathcal{O}_H(2\imath! gD) \to 0$$

we deduce that $h^0(\mathcal{O}_X(2\imath! gD)) > 0$, thus completing the proof of Fact C.

*Remark.* We would also like statements concerning rational points $D$ lying on $\partial\overline{\mathcal{K}}$ and the hyperplane $c_2 = 0$. If for instance $D^2 \equiv 0$ (i.e. $D$ is a singular point of $W^*$), then we might expect $D$ to correspond to an abelian surface fibre space structure on $X$ ; the difficulty here lies in showing that some multiple of $D$ is effective. It is remarked in [10] that the existence of a rational divisor $D_1$ with $D_1^2 \equiv 0$ and $D_1 \cdot c_2 = 0$ implies the existence of a rational divisor $D_2$ giving rise either to a birational contraction morphism as in Fact A or an elliptic fibre space structure as in Fact B. Indeed, for $H$ any very ample divisor, simply consider the intersection of the line segment $D_1H$ with $\partial\overline{\mathcal{K}}$ ; if $D_2 \not\equiv D_1$ is a point of this intersection, then it satisfies the conditions for either (A) or (B).

## 2. The case when $c_2$ is not strictly positive

Let $X$ be a minimal model of a Calabi–Yau threefold, and consider $c_2$ as an integral linear form on $\mathrm{Pic}(X)$. In the case when $c_2 = 0$, we consider a general element $H_m$ in $|2^m H|$, where $H$ is a very ample divisor on $X$ and $m$ a sufficiently large integer. The surface $H_m$ has at worst rational double point singularities. We can consider now the sheaf $\tilde{\Omega}^1_X|_{H_m} = \mathcal{F}$ on $H_m$ (where $\tilde{\Omega}^1_X$ denotes double dual of the sheaf of Kähler differentials), a reflexive sheaf with $c_1(\mathcal{F}) = 0$, $c_2(\mathcal{F}) = 0$ (Chern classes here being $\mathbb{Q}$-Chern classes in the sense of Fulton [3, 8]). Moreover, from a result of Miyaoka [5,12], the sheaf $\tilde{\Omega}^1_X$ is semi-stable with respect to any nef divisors $H_1$, $H_2$. When $X$ is smooth, one can show (using the methods of [7]) that for any polarization $H'$ on X, $\mathcal{F}$ is $H'$-stable provided that $m$ is sufficiently large, from which we deduce as in [7] that $\mathcal{F}$ is a flat bundle, coming from a representation of $\pi_1(H_m)$. From Lefschetz Theory we have $\pi_1(X) \cong \pi_1(H)$, and so $\Omega^1_X$ comes from a representation of

$\pi_1(X)$ . The general case poses technical problems, but the above arguments can be generalized to obtain the result that a Calabi–Yau minimal model $X$ with $c_2(X) = 0$ is a finite quotient of a three-dimensional torus (cf. [1] for the smooth case, when the covering map is étale, and a forthcoming preprint by the author and N.I. Shepherd-Barron for the general case). Examples of such quotients have been studied in a number of papers, including the recent preprint [11]; moreover, there should be a classification of all such quotients along the lines described there for cyclic quotients.

The next possibility is that of $c_2 \neq 0$ but not strictly positive on $\overline{\mathcal{K}} \setminus \{0\}$. Since $X$ is assumed to be a minimal Calabi–Yau model, we recall from Fact A that $\partial\overline{\mathcal{K}} \subset W^*$. The intersection $\overline{\mathcal{K}} \cap \{c_2 = 0\}$ is a non-empty closed cone in the hyperplane $\{c_2 = 0\}$, and in particular there are rational rays in this intersection; the intersection might well just be a single ray along which $\{c_2 = 0\}$ is tangent to $W^*$, but this ray would then itself have to be rational. Any rational ray in $W^*$ corresponds to a rational point in the cubic hypersurface $W \subset \mathbb{P}^{\rho-1}(\mathbb{R})$.

We now recall the arguments from [13] (expanded further in the proof of Proposition 3 from [16]) which show if $\rho(X) > 3$ and $\partial\overline{\mathcal{K}} \subset W^*$, then the rational points cannot be dense in $W$. Furthermore, we noted in Section 1 of [13] that the hypersurface $W$ is not a cone, and hence when $\rho > 3$ contains at most one hyperplane.

**Proposition 2.1.** *If $X$ is a Calabi–Yau minimal model on which $c_2$ is non-trivial but not strictly positive on $\overline{\mathcal{K}} \setminus \{0\}$, then $\rho(X) \leq 5$. If $\rho(X) = 4$ or 5, then $W = \Lambda \cup Q$ with $\Lambda$ a hyperplane and $Q$ a quadric of rank $r$ ($\rho - 1 \leq r \leq 4$); in the case when $Q$ is a quadric cone, it has vertex $P \notin \Lambda$. If $\Lambda \neq \{c_2 = 0\}$, $X$ will have the structure of an elliptic fibre space as in Fact B. If $\Lambda$ is the hyperplane $\{c_2 = 0\}$, and $Q$ is a quadric cone , then $X$ has the structure of a K3 fibre space $\phi : X \to \mathbb{P}^1$ as in Fact C.*

*Remark.* This Proposition could be made far more precise if we had a Fact D in Section 1 concerning nef divisors $D$ on $X$ with $D \cdot c_2 = 0$ and $D^2$ not numerically trivial. What we need is a criterion for $h^0(nD) > 0$ for sufficiently large $n$, since the results of [10] would then ensure that $|nD|$ is free for some $n$, with $\phi_{nD} : X \to S$ defining an elliptic fibre space of rather special type.

*Proof of (2.1).* We suppose that $c_2$ is not strictly positive on $\overline{\mathcal{K}} \setminus \{0\}$ and that $\rho(X) \geq 4$, and consider separately the cases when $W$ is reducible and irreducible.

If $W$ is reducible and $\rho(X) \geq 4$, then $W$ splits up into a hyperplane $\Lambda$ and a hyperquadric $Q$. Letting $\overline{\mathbb{K}}$ denote the convex body in $\mathbb{P}^{\rho-1}(\mathbb{R})$ corresponding to $\overline{\mathcal{K}}$, we have that $\partial\overline{\mathbb{K}} \subset \Lambda \cup Q$. If $\Lambda$ is not the hyperplane $\{c_2 = 0\}$, then the argument sketched above yields a rational point $P$ on the hyperquadric $Q$. Since the rational points on $Q$ are not allowed to be dense, we deduce that $Q$ is a quadric cone with single vertex $P$ not on $\Lambda$ (since $W$ is not a cone). The intersection of $Q$ with $\Lambda$ is a non-singular quadric with real points but no rational points. By a standard result from the theory of quadric forms, we must therefore have $\rho(X) \leq 5$. Moreover, a general rational point on $\Lambda \cap \overline{\mathbb{K}}$ gives rise (using Fact B) to an elliptic fibre space structure $\phi : X \to S$ on $X$ with $\rho(S) \leq 4$ (in fact, it follows from the arguments of the Corrigendum to [13], expanded as in the proof of Proposition 3 from [16], that $\rho(S) = \rho(X) - 1$).

If $W$ does contain the hyperplane $\Lambda = \{c_2 = 0\}$, then $W = \Lambda \cup Q$ where $Q$ is either non-singular or a quadric cone with single vertex $P$ not on $\Lambda$. In the case when $Q$ is non-singular, it cannot contain any rational points and so we have $\rho(X) = 4$. If $Q$ is a quadric cone, the vertex then gives rise to a divisor satisfying the conditions of Fact C, and so a K3 fibre space structure is obtained.

We are therefore reduced to showing that under the assumptions that $c_2$ is non-trivial but not strictly positive on $\overline{\mathcal{K}}$ and that W is irreducible, we must have $\rho(X) \leq 3$. We show that for $\rho(X) \geq 4$, our assumptions imply that the rational points on $W$ are dense, contradicting the conclusions of the Corrigendum to [13] and Proposition 3 from [16].

We saw above that the fact of $c_2$ not being strictly positive implied the existence of a rational point $P$ on $W \cap \partial\overline{\mathbb{K}}$, which we may assume is a non-singular point of $W$ (since otherwise the density follows immediately). Since $\partial\overline{\mathbb{K}} \subset W$, we are in the case when the smooth locus $W_0$ of $W$ has two components, and that $\partial\overline{\mathbb{K}}\backslash$Sing(W) is contained in the bounded component of $W_0$. We claim now that the rational points in $W$ will have to be dense. We can clearly reduce down to the case when $W$ is a surface in $\mathbb{P}^3$ with isolated singularities (and not a cone). It is then a standard folklore result that, by means of the rational point $P$ on the bounded component of $W_0$, we can generate a dense subset of rational points in $W$ (cf. pages 51-53 of [4] for the general idea of the argument). In the case when $\rho = 3$, this argument fails ; if, for instance, W is an elliptic curve with two real components, and $P$ is a rational point on the bounded component, the tangent to the curve at $P$ might intersect the curve again at an inflexion point, and there is then no guarantee of any further rational points on the curve.

## 3. The case when $c_2$ is strictly positive

The cases described in the previous section of Calabi–Yau minimal models for which $c_2$ is not strictly positive on $\overline{\mathcal{K}}\backslash\{0\}$ should be regarded as being of 'special type', and eventually there should be structure theorems for all such threefolds. The Calabi–Yau minimal models with $c_2$ strictly positive on $\overline{\mathcal{K}} \setminus \{0\}$ should be regarded as constituting the general case, analogous in some ways to surfaces of general type.

Using the methods of [16], it follows that such minimal models can only exist for $\rho \leq 13$, and one suspects that a rather sharper bound will hold here. There are simple examples of Calabi–Yau minimal models with $\rho = 3$, but the author knows of none with $\rho = 4$. In Example 4.6 of [14], we have a smooth Calabi–Yau threefold $X_1$ with $\rho = 4$; this threefold has precisely four minimal models (corresponding to the faces of the tetrahedron), all of which have $\rho = 3$. The degeneration $\overline{X}$ of $X_1$ described in that example is a Calabi–Yau model with precisely four minimal models, only one of which however has $\rho = 3$, the other three having $\rho = 2$.

If now we are in the case of $c_2 > 0$ on $\overline{\mathcal{K}}\backslash\{0\}$, then a canonically defined class in $\mathrm{Pic}_{\mathbb{Z}}$(X) $=$ Pic$(X) \otimes \mathbb{Z}$ exists, since we can merely look for non-zero elements of $\mathrm{Pic}_{\mathbb{Z}}$(X)$\cap\mathcal{K}$ on which $c_2$ takes its minimum value. Our assumption on $c_2$ implies that there are only finitely many such elements , and by adding these classes together, we can form an element which has been intrinsically determined.

This yields an intrinsically determined ample divisor class on $X$ (if $\text{Pic}(X)$ has torsion, we sum over the finitely many elements of $\text{Pic}(X)$ with the given class in $\text{Pic}_{\mathbb{Z}}(X)$). From the results from [9], this gives an intrinsically determined projective embedding of the threefold.

**Lemma 3.1.** *Let $X$ be any Calabi–Yau model for which $c_2$ is strictly positive on the non-zero elements of $\overline{\mathcal{K}}$; then $X$ has finite automorphism group.*

*Proof.* The above argument gives an intrinsically defined embedding of $X$ (without any assumption of minimality) into some projective space $\mathbb{P}^N$. Thus any automorphism of $X$ comes from a linear automorphism of $\mathbb{P}^N$, and in particular it follows that $\text{Aut}(X)$ has only finitely many components. It is however easily checked for any Calabi–Yau model that $\text{Aut}(X)$ is a discrete group (since there are no global vector fields); hence $\text{Aut}(X)$ is finite for all $X$ as in the lemma.

*Remark.* There are other circumstances under which $\text{Aut}(X)$ is always finite. If for instance $X$ is a Calabi–Yau model for which $\partial\overline{\mathcal{K}} \cap W^* = \emptyset$, then $\overline{\mathcal{K}}$ is rational polyhedral (immediate from Fact A), and we can define an invariant polarization on $X$ by summing together the finitely many primitive generators of $\overline{\mathcal{K}}$ (the generators of course corresponding to the different minimal models for $X$). Thus, in this case, we shall also have a finite automorphism group. In general, we observe that an automorphism of $X$ gives rise to an automorphism of the pair of cubic $W$ and hyperplane $\{c_2 = 0\}$, and in many cases (e.g. $W$ non-singular) the automorphism group of this pair will be finite. This enables us to define a polarization on $X$ invariant under $\text{Aut}(X)$, and hence again we deduce that $\text{Aut}(X)$ is finite.

There certainly do however exist Calabi–Yau threefolds with infinite automorphism group. A very interesting such example (associated with a desingularization of the quintic threefold with 125 nodes obtained from the Horrocks-Mumford bundle in $\mathbb{P}^4$ – studied originally by C. Schoen) is described in [2]. In this example, the ray generated by $c_2$ in the dual cone to $\overline{\mathcal{K}}$ (i.e. the closed cone of effective 1-cycles $\overline{\text{NE}}(X)$) is an accumulation ray for the generators of $\overline{\text{NE}}(X)$ determined by the codimension one faces of $\overline{\mathcal{K}}$. The threefold $X$ has the structure of a fibre space of abelian surfaces over $\mathbb{P}^1$. One suspects that these properties might hold for any Calabi–Yau model with infinite automorphism group.

**Conjecture.** *A Calabi–Yau threefold $X$ has only finitely many Calabi–Yau minimal models up to automorphisms of $X$. In particular, if $c_2$ is strictly positive on $\overline{\mathcal{K}}$, then $X$ will have only finitely many Calabi–Yau minimal models.*

If $X$ is a Calabi–Yau model and $L$ is any ample divisor on $X$, it follows from Riemann–Roch and the definition of the $\Delta$-genus $\Delta(X, L)$ that $12(\Delta(X, L) - 3) = 10L^3 - L \cdot c_2$. Provided that $\Delta(X, L) > 2$ (those Calabi–Yau models with $\Delta(X, L) \leq 2$ can be classified - see [9]), it follows that $L \cdot c_2 \leq 10L^3$. An inequality of a similar type may be stated for $c_3$. If however $X$ is a Calabi–Yau minimal model for which $c_2$ is strictly positive on non-zero elements of the closed cone $\overline{\mathcal{K}}$, and $L$ is the canonically defined polarization on $X$, we would hope for further inequalities involving $L^3$, $L \cdot c_2$ and the topological invariants of the threefold; in particular these should give upper bounds for $L^3$. The author hopes to return in a later paper to this and the other questions which have been raised in this article.

## REFERENCES

1. Beauville, A.: Variétés Kähleriennes dont la première classe de Chern est nulle. J. Differ. Geom. **18** (1983), 755-782

2. Borcea, C.: On desingularized Horrocks-Mumford quintics. J. Reine Angew. Math. **421** (1991), 23-41

3. Fulton, W.: Rational equivalence on singular varieties. Publ. IHES **36** (1975), 147-167

4. Manin, Yu. I. : *Cubic Forms (2nd edition)*, North Holland, Amsterdam, 1986.

5. Miyaoka, Y.: The Chern classes and Kodaira dimension of a minimal variety. In : Oda, T. (ed.) *Algebraic Geometry, Sendai 1985*, Adv. Stud. Pure Math. Vol. 10, pp. 449-476. North Holland, Amsterdam, 1987

6. Miyaoka, Y.: On the Kodaira dimension of minimal threefolds. Math. Ann. **281** (1988), 325-332

7. Mehta, V.B., Ramanathan, A.: Restriction of stable sheaves and representations of the fundamental group. Invent. Math. **77** (1984), 163-172

8. Mumford, D.: Towards an enumerative geometry of the moduli space of curves. In : Artin, M., Tate, J. (eds) *Arithmetic and Geometry, papers dedicated to I.R. Shafarevich, Volume II Geometry* (Progr. Math., Vol. 36, pp. 271-326). Birkhäuser, Boston Basel Stuttgart, 1983.

9. Oguiso, K.: On polarized Calabi–Yau 3-folds. J. Fac. Sci. Univ. Tokyo Sect. 1A, Math. **38** (1991), 395-429

10. Oguiso, K.: On algebraic fiber space structures on a Calabi–Yau 3-fold. Univ. of Tokyo preprint, Tokyo, 1992

11. Roan, S.-S.: Complex 3-Tori with an order 7 special automorphism. Max-Planck-Institut preprint, Bonn 1992

12. Shepherd-Barron, N.I.: Miyaoka's Theorems on the generic semi-negativity of $T_X$. In : *Flips and Abundance for algebraic 3-folds, Ch. 9* . Astérisque (1992) to appear

13. Wilson, P.M.H.: Calabi–Yau manifolds with large Picard number. Invent. math. **98** (1989), 139-155

14. Wilson, P.M.H.: The Kähler cone on Calabi–Yau threefolds. Invent. math. **107** (1992), 561-583

15. Wilson, P.M.H.: Elliptic Ruled surfaces on Calabi–Yau threefolds. Math. Proc. Camb. Phil. Soc. **112** (1992), 45-52

16. Heath-Brown, D.R., Wilson, P.M.H.: Calabi–Yau threefolds with $\rho > 13$. Math. Ann. **294** (1992), 49-57

Department of Pure Mathematics, University of Cambridge,
16 Mill Lane, Cambridge CB2 1SB, ENGLAND
*E-mail address* : pmhw@pmms.cam.ac.uk